Texts in Applied Mathematics 32

For other titles published in this series, go to
http://www.springer.com/series/1214

Dale R. Durran

Numerical Methods for Fluid Dynamics

With Applications to Geophysics

Second Edition

Dale R. Durran
University of Washington
Department of Atmospheric Sciences
Box 341640
Seattle, WA 98195-1640
USA
drdee@uw.edu

Series Editors

J.E. Marsden
Control and Dynamical Systems, 107-81
California Institute of Technology
Pasadena, CA 91125
USA
marsden@cds.caltech.edu

L. Sirovich
Division of Applied Mathematics
Brown University
Providence, RI 02912
USA
lawrence.sirovich@mssm.edu

S.S. Antman
Department of Mathematics
and
Institute for Physical Science
and Technology
University of Maryland
College Park, MD 20742-4015
USA
ssa@math.umd.edu

ISSN 0939-2475
ISBN 978-1-4614-2685-1 ISBN 978-1-4419-6412-0 (eBook)
DOI 10.1007/978-1-4419-6412-0
Springer New York Dordrecht Heidelberg London

Mathematics Subject Classification (2010): 65-01, 65L04, 65L05, 65L06, 65L12, 65L20, 65M06, 65M08, 65M12, 65T50, 76-01, 76M10, 76M12, 76M22, 76M29, 86-01, 86-08

Cover illustration: Passive-tracer concentrations in a circular flow with deforming shear, plotted at three different times. This simulation was conducted using moderately coarse numerical resolution. A similar problem is considered in Section 5.9.5.

Printed on acid-free paper

Springer is part of Springer Science+Business Media (www.springer.com)

To every hand that's touched the Wall

Preface

This book is a major revision of *Numerical Methods for Wave Equations in Geophysical Fluid Dynamics*; the new title of the second edition conveys its broader scope. The second edition is designed to serve graduate students and researchers studying geophysical fluids, while also providing a non-discipline-specific introduction to numerical methods for the solution of time-dependent differential equations.

Changes from the first edition include a new Chapter 2 on the numerical solution of ordinary differential equations (ODEs), which covers classical ODE solvers as well as more recent advances in the design of Runge–Kutta methods and schemes for the solution of stiff equations. Chapter 2 also explores several characterizations of numerical stability to help the reader distinguish between those conditions sufficient to guarantee the convergence of numerical solutions to ODEs and the stronger stability conditions that must be satisfied to compute reasonable solutions to time-dependent partial differential equations with finite time steps. Chapter 3 (formerly Chapter 2) has been reorganized and now covers finite-difference schemes for the simulation of one-dimensional tracer transport due to advection, diffusion, or both. Chapter 4, which is devoted to finite-difference approximations to more general partial differential equations, now includes an improved discussion of skew-symmetric operators. Chapter 5, "Conservation Laws and Finite-Volume Methods," includes new sections on essentially nonoscillatory and weighted essentially nonoscillatory methods, the piecewise-parabolic method, and limiters that preserve smooth extrema. A section on the discontinuous Galerkin method now concludes Chapter 6, which continues to be rounded out with discussions of the spectral, pseudospectral, and finite-element methods. Chapter 7, "Semi-Lagrangian Methods," now includes discussions of "cascade interpolation" and finite-volume integrations with large time steps. More minor modifications and updates have been incorporated throughout the remaining chapters.

The majority of the schemes presented in this text were introduced in either the applied mathematics or the atmospheric science literature, but the focus is not on the details of particular atmospheric models but on fundamental numerical methods that have applications in a wide range of scientific and engineering disciplines.

The prototype problems considered include tracer transport, chemically reacting flow, shallow-water waves, and the evolution of internal waves in a continuously stratified fluid.

A significant fraction of the literature on numerical methods for these problems falls into one of two categories: those books and papers that emphasize theorems and proofs, and those that emphasize numerical experimentation. Given the uncertainty associated with the messy compromises actually required to construct numerical approximations to real-world fluid-dynamics problems, it is difficult to emphasize theorems and proofs without limiting the analysis to classical numerical schemes whose practical application may be rather limited. On the other hand, if one relies primarily on numerical experimentation, it is much harder to arrive at conclusions that extend beyond a specific set of test cases. In an attempt to establish a clear link between theory and practice, I have tried to follow a middle course between the theorem-and-proof formalism and the reliance on numerical experimentation. There are no formal proofs in this book, but the mathematical properties of each method are derived in a style familiar to physical scientists. At the same time, numerical examples are included that illustrate these theoretically derived properties and facilitate the comparison of various methods.

A general course on numerical methods for time-dependent problems might draw on portions of the material presented in Chapters 1–6, and I have used sections from these chapters in a graduate course entitled "Numerical Analysis of Time-Dependent Problems" that is jointly offered by the Department of Applied Mathematics and the Department of Atmospheric Sciences at the University of Washington. The material in Chapters 7 and 9 is not specific to geophysics, and appropriate portions of these chapters could also be used in courses in a wide range of disciplines. Both theoretical and applied problems are provided at the end of each chapter. Those problems requiring numerical computation are marked by an asterisk.

The portions of the book that are most explicitly related to atmospheric science are portions of Chapter 1, the treatment of spherical harmonics in Chapter 6, and Chapter 8. The beginning of Chapter 1 discusses the relation between the equations governing geophysical flows and other types of partial differential equations. Switching gears, Chapter 1 then concludes with a short overview of the strategies for numerical approximation that are considered in detail throughout the remainder of the book. Chapter 8 examines schemes for the approximation of slow-moving waves in fluids that support physically insignificant fast waves. The emphasis in Chapter 8 is on atmospheric applications in which the slow wave is an internal gravity wave and the fast waves are sound waves, or the slow wave is a Rossby wave and the fast waves are both gravity waves and sound waves.

Many numerical methods for the simulation of internally stratified flow require the repeated solution of elliptic equations for pressure or some closely related variable. Owing to the limitations of my own expertise and to the availability of other excellent references, I have not discussed the solution of elliptic partial differential equations in any detail. A thumbnail sketch of some solution strategies is provided in Section 8.1.3; the reader is referred to Chapter 5 of Ferziger and Perić (1997) for an excellent overview of methods for the solution of elliptic equations arising in computational fluid dynamics and to Chapters 3 and 4 of LeVeque (2007) for a very accessible and somewhat more detailed discussion.

I have attempted to provide sufficient references to allow the reader to further explore the theory and applications of many of the methods discussed in the text, but the reference list is far from encyclopedic and certainly does not include every worthy paper in the atmospheric science or applied mathematics literature. References to the relevant literature in other disciplines and in foreign language journals are rather less complete.[1]

The first edition of this book could not have been written without the generous assistance of several colleagues. Christopher Bretherton, in particular, provided many perceptive answers to my endless questions. J. Ray Bates, Byron Boville, Michael Cullen, Marcus Grote, Robert Higdon, Randall LeVeque, Christoph Schär, William Skamarock, Piotr Smolarkiewicz, and David Williamson all provided very useful comments on individual chapters. Many students used earlier versions of this manuscript in my courses in the Department of Atmospheric Sciences at the University of Washington, and their feedback helped improve the clarity of the manuscript. Two students to whom I am particularly indebted are Craig Epifanio and Donald Slinn. I am also grateful to Jim Holton for encouraging me to write the first edition.

Peter Blossey made many important contributions to the second edition, including performing the computations for Figures 5.24–5.28 and 7.3. Comments by Catherine Mavriplis and Ram Nair helped improve the new section on discontinuous Galerkin methods. Joel Thornton helped me better grasp the fundamentals of atmospheric ozone chemistry. Additional invaluable input was provided from many readers of the first edition who were kind enough to send their comments and help identify typographical errors.

It is my pleasure to acknowledge the many years of support for my numerical modeling efforts provided by the Physical and Dynamic Meteorology Program of the National Science Foundation. Additional significant support for my research on numerical methods for atmospheric models has been provided by the Office of Naval Research. Part of the first edition was completed while I was on sabbatical at the Laboratoire d'Aérologie of the Université Paul Sabatier in Toulouse, France, and I thank Daniel Guedalia and Evelyne Richard for helping make that year productive and scientifically stimulating.

As errors in the text are identified, they will be posted on the Web at http:// www. atmos.washington.edu/numerical.methods, which can be accessed directly or via Springer's home page at http://www.springer-ny.com. I would be most grateful to be advised of any typographical or other errors by electronic mail at drdee@uw.edu.

Seattle, Washington, USA *Dale R. Durran*

---

[1] Those not familiar with the atmospheric science literature may be surprised by the number of references to *Monthly Weather Review*, which, despite its title, has become the primary American journal for the publication of papers on numerical methods in atmospheric science.

Contents

Chapter 1
Introduction

The possibility of deterministic weather prediction was suggested by Vilhelm Bjerknes as early as 1904. Around the time of the First World War, Lewis Richardson actually attempted to produce such a forecast by manually integrating a finite-difference approximation to the equations governing atmospheric motion. Unfortunately, his calculations did not yield a reasonable forecast. Moreover, the human labor required to obtain this disappointing result was so great that subsequent attempts at deterministic weather prediction had to await the introduction of a high-speed computational aid. In 1950, a team of researchers, under the direction of Jule Charney and John von Neumann at the Institute for Advanced Study, at Princeton, journeyed to the Aberdeen Proving Ground, where they worked for approximately 24 h to coax a one-day weather forecast from the first general-purpose electronic computer, the ENIAC.[1] The first computer-generated weather forecast was surprisingly good, and its success led to the rapid growth of a new meteorological subdiscipline, "numerical weather prediction." These early efforts in numerical weather prediction also began a long and fruitful collaboration between numerical analysts and atmospheric scientists.[2] The use of numerical models in atmospheric and oceanic science has subsequently expanded into almost all areas of active research. Numerical models are currently employed to study phenomena as diverse as global climate change, the interaction of ocean currents with bottom topography, the evolution of atmospheric pollutants within an urban airshed, and the development of rotation in tornadic thunderstorms.

Weather forecasting is an initial-value problem, and the focus of this book is the study of numerical methods for the solution of time-dependent differential equations. The simplest time-dependent differential equations are ordinary differential equations. These arise, for example, when simulating chemical reactions in an isolated "box." Yet in most geochemical problems it is also necessary to consider the influence of fluid transport, and the inclusion of transport processes substantially complicates the governing equations. Such transport can be produced by molecular

[1] ENIAC is an acronym for Electronic Numerical Integrator and Calculator.

[2] Further details about these early weather prediction efforts may be found in Bjerknes (1904), Richardson (1922), Charney et al. (1950), Burks and Burks (1981), and Thompson (1983).

D.R. Durran, *Numerical Methods for Fluid Dynamics: With Applications to Geophysics*, Texts in Applied Mathematics 32, DOI 10.1007/978-1-4419-6412-0_1,

diffusion or by the flow of macroscopic fluid elements (advection). Except in boundary layers near interfaces between the atmosphere, the oceans, and the earth, advection plays a much larger role than diffusion in atmospheric and oceanic transport.

The advection of a passive tracer is governed by a simple "wave equation." Perhaps surprisingly, the choice of the best method to solve even the one-dimensional advection equation is not clear-cut. Hundreds of papers have been published examining various techniques for solving the advection equation and for generalizing those techniques to more complex problems with wavelike solutions. Here the adjective "wavelike" is used in the general sense suggested by Whitham (1974), who defined a wave as "any recognizable signal that is transferred from one part of a medium to another with a recognizable velocity of propagation."

This book presents the fundamental mathematical aspects of a wide variety of numerical methods for the simulation of wavelike flow. The methods considered are typically those that have seen some use in real-world atmospheric or ocean models, but the focus is on the essential properties of each method and not on the details of any specific model. The fundamental character of each scheme will be examined in standard fluid-dynamical problems such as tracer transport, chemically reacting flow, shallow-water waves, and waves in an internally stratified fluid. These are the same prototypical problems familiar to many applied mathematicians, fluid dynamicists, and practitioners in the larger discipline of computational fluid dynamics.

Most of the problems under investigation in the atmospheric and oceanic sciences involve fluid systems with low viscosity and weak dissipation. The equations governing these flows are often nonlinear, but their solutions almost never develop energetic shocks or discontinuities. Nevertheless, regions of scale collapse do frequently occur as the velocity field stretches and deforms an initially compact fluid parcel. The numerical methods that will be examined in this book may therefore be distinguished from the larger family of algorithms in computational fluid mechanics in that they are particularly appropriate for low-viscosity flows, but are not primarily concerned with the treatment of shocks.

It is assumed that the reader has already been exposed to the derivation of the equations describing fluid flow and tracer transport. These derivations are given in a general fluid-dynamical context in Batchelor (1967), Yih (1977), and Bird et al. (1960), and in the context of atmospheric and oceanic science in Gill (1982), Pedlosky (1987), Holton (2004), and Vallis (2006). The mathematical properties of these equations and commonly used simplifications, such as the Boussinesq approximation, will be briefly reviewed in this chapter. The chapter concludes with a brief overview of the numerical methods that will be considered in more detail throughout the remainder of the book.

1.1 Partial Differential Equations: Some Basics

Different types of partial differential equations require different solution strategies. It is therefore helpful to begin by reviewing some of the terminology used to describe various types of partial differential equations. The *order* of a partial differential equation is the order of the highest-order partial derivative that appears in the equation. Numerical methods for the solution of time-dependent problems are often designed to solve systems of partial differential equations in which the time derivatives are of first order. These numerical methods can be used to solve partial differential equations containing higher-order time derivatives by defining new unknown functions equal to the lower-order time derivatives of the original unknown function and expressing the result as a system of partial differential equations in which all time derivatives are of order 1. For example, the second-order partial differential equation

$$\frac{\partial^2 \psi}{\partial t^2} + \psi \frac{\partial \psi}{\partial x} = 0$$

can be expressed as the first-order system

$$\begin{aligned} \frac{\partial \upsilon}{\partial t} + \psi \frac{\partial \psi}{\partial x} &= 0, \\ \frac{\partial \psi}{\partial t} - \upsilon &= 0. \end{aligned}$$

In geophysical applications it is seldom necessary to actually formulate first-order-in-time equations using this procedure, because suitable first-order-in-time systems can usually be derived from fundamental physical principles.

The accurate numerical solution of equations describing wavelike flow becomes more difficult if the solution develops significant perturbations on spatial scales close to the shortest scale that can be resolved by the numerical model. The possibility of waves developing small-scale perturbations from smooth initial data increases as the governing partial differential equation becomes more nonlinear. A partial differential equation is *linear* if it is linear in the unknown functions and their derivatives, in which case the coefficients multiplying each function or derivative depend only on the independent variables. As an example,

$$\frac{\partial u}{\partial t} + x^3 \frac{\partial u}{\partial x} = 0$$

is a linear first-order partial differential equation, whereas

$$\left(\frac{\partial u}{\partial t}\right)^2 + \sin\left(u \frac{\partial u}{\partial x}\right) = 0$$

is a nonlinear first-order partial differential equation.

Analysis techniques and solution procedures developed for linear partial differential equations can be generalized most easily to the subset of nonlinear partial differential equations that are quasi-linear. A partial differential equation of order p is *quasi-linear* if it is linear in the derivatives of order p; the coefficient multiplying each pth derivative can depend on the independent variables and all derivatives of the unknown function through order $p-1$. Two examples of quasi-linear partial differential equations are

$$\frac{\partial u}{\partial t} + u^3 \frac{\partial u}{\partial x} = 0$$

and the vorticity equation for two-dimensional nondivergent flow

$$\frac{\partial \nabla^2 \psi}{\partial t} + \frac{\partial \psi}{\partial x}\frac{\partial \nabla^2 \psi}{\partial y} - \frac{\partial \psi}{\partial y}\frac{\partial \nabla^2 \psi}{\partial x} = 0,$$

where $\psi(x, y, t)$ is the stream function for the nondivergent velocity field and

$$\nabla^2 = \frac{\partial^2}{\partial x^2} + \frac{\partial^2}{\partial y^2}.$$

1.1.1 First-Order Hyperbolic Equations

Many waves can be mathematically described as solutions to hyperbolic partial differential equations. One simple example of a hyperbolic partial differential equation is the general first-order quasi-linear equation

$$A(x,t,u)\frac{\partial u}{\partial t} + B(x,t,u)\frac{\partial u}{\partial x} = C(x,t,u), \tag{1.1}$$

where A, B, and C are real-valued functions with continuous first derivatives. This equation is hyperbolic because there exists a family of real-valued curves in the x–t plane along which the solution can be locally determined by integrating ordinary differential equations. These curves, called *characteristics*, may be defined with respect to the parameter s by the relations

$$\frac{dt}{ds} = A \quad \text{and} \quad \frac{dx}{ds} = B. \tag{1.2}$$

The identity

$$\frac{du}{ds} = \frac{\partial u}{\partial t}\frac{dt}{ds} + \frac{\partial u}{\partial x}\frac{dx}{ds}$$

can then be used to express (1.1) as the ordinary differential equation

$$\frac{du}{ds} = C. \tag{1.3}$$

Given the value of u at some arbitrary point (x_0, t_0), the coordinates of the characteristic curve passing through (x_0, t_0) can be determined by integrating the ordinary differential equations (1.2). The solution along this characteristic can be obtained by integrating the ordinary differential equation (1.3). A unique solution to (1.1) can be determined throughout some local region of the x–t plane by specifying data for u along any noncharacteristic line.

In physical applications where the independent variable t represents time, the particular solution of (1.1) is generally determined by specifying initial data for u along the line $t = 0$. In such applications A is nonzero, and any perturbation in the distribution of u at the point (x_0, t_0) translates through a neighborhood of x_0 at the speed

$$\frac{dx}{dt} = \frac{B}{A}.$$

The solutions to (1.1) are wavelike in the general sense that the perturbations in u travel at well-defined velocities even though they may distort as they propagate. The evolution of the solution is particularly simple when $C = 0$ and B/A is some constant value c, in which case (1.1) reduces to

$$\frac{\partial u}{\partial t} + c\frac{\partial u}{\partial x} = 0. \tag{1.4}$$

If $u(x, 0) = f(x)$, the solution to the preceding equation is $f(x - ct)$, implying that the initial perturbations in u translate without distortion at a uniform velocity c. Equation (1.4), which is often referred to as the *one-way wave equation* or the *constant-wind-speed advection equation*, is the simplest mathematical model for wave propagation. Although it is quite simple, (1.4) is a very useful prototype problem for testing numerical methods because solutions to more complex linear hyperbolic systems can often be expressed as the superposition of individual waves governed by one-way wave equations.

A system of partial differential equations in two independent variables is hyperbolic if it has a complete set of characteristic curves that can in principle be used to locally determine the solution from appropriately prescribed initial data. As a first example, consider a constant-coefficient linear system of the form

$$\frac{\partial u_r}{\partial t} + \sum_{s=1}^{n} a_{rs}\frac{\partial u_s}{\partial x} = 0, \qquad r = 1, 2, \ldots, n. \tag{1.5}$$

This system may be alternatively written as

$$\frac{\partial \mathbf{u}}{\partial t} + \mathbf{A}\frac{\partial \mathbf{u}}{\partial x} = \mathbf{0},$$

where uppercase boldface letters represent matrices and lowercase boldface letters denote vectors. The system is hyperbolic if there exist bounded matrices $\mathbf{T}$ and $\mathbf{T}^{-1}$ such that $\mathbf{T}^{-1}\mathbf{A}\mathbf{T} = \mathbf{D}$, where $\mathbf{D}$ is a diagonal matrix with real eigenvalues d_{jj}. When the system is hyperbolic, it can be transformed to

$$\frac{\partial \mathbf{v}}{\partial t} + \mathbf{D}\frac{\partial \mathbf{v}}{\partial x} = \mathbf{0} \tag{1.6}$$

by defining $\mathbf{v} = \mathbf{T}^{-1}\mathbf{u}$. Since $\mathbf{D}$ is a diagonal matrix, each element v_j of the vector of unknown functions may be determined by solving a simpler scalar equation of the form (1.4). Each diagonal element of $\mathbf{D}$ is associated with a family of characteristic curves along which the perturbations in v_i propagate at speed $\mathrm{d}x/\mathrm{d}t = d_{jj}$. The wavelike character of the solution can be demonstrated by Fourier transforming (1.6) with respect to x to obtain

$$\frac{\partial \hat{\mathbf{v}}}{\partial t} + \mathrm{i}k\mathbf{D}\hat{\mathbf{v}} = \mathbf{0}, \tag{1.7}$$

where $\hat{\mathbf{v}}$ is the Fourier transform of $\mathbf{v}$ and k is the wave number, or dual variable. To satisfy (1.7), the jth component of $\mathbf{v}$ must be a wave of the form $\exp \mathrm{i}k(x - d_{jj}t)$. Every solution to the original system (1.5) is a linear superposition of these waves.

Now consider the general first-order linear system

$$\frac{\partial \mathbf{u}}{\partial t} + \mathbf{A}\frac{\partial \mathbf{u}}{\partial x} + \mathbf{B}\mathbf{u} + \mathbf{c} = \mathbf{0},$$

where the coefficient matrices are smooth functions of x and t. This system is hyperbolic throughout some region R of the x–t plane if for all x and t in R there exist bounded matrices $\mathbf{T}^{-1}$ and $\mathbf{T}$ such that $\mathbf{D}(x,t) = \mathbf{T}^{-1}(x,t)\mathbf{A}(x,t)\mathbf{T}(x,t)$ is a diagonal matrix with real eigenvalues. Again, let $\mathbf{u} = \mathbf{T}\mathbf{v}$. Then,

$$\frac{\partial \mathbf{v}}{\partial t} + \mathbf{D}\frac{\partial \mathbf{v}}{\partial x} + \tilde{\mathbf{B}}\mathbf{v} + \mathbf{T}^{-1}\mathbf{c} = \mathbf{0}, \tag{1.8}$$

where

$$\tilde{\mathbf{B}} = \mathbf{T}^{-1}\left(\frac{\partial \mathbf{T}}{\partial t} + \mathbf{A}\frac{\partial \mathbf{T}}{\partial x} + \mathbf{B}\mathbf{T}\right).$$

The solution to (1.8) may be obtained via the iteration

$$\frac{\partial \mathbf{v}^{n+1}}{\partial t} + \mathbf{D}\frac{\partial \mathbf{v}^{n+1}}{\partial x} + \tilde{\mathbf{B}}\mathbf{v}^{n} + \mathbf{T}^{-1}\mathbf{c} = \mathbf{0} \tag{1.9}$$

(Courant and Hilbert 1953, p. 476). Since $\mathbf{D}$ is diagonal, the preceding equation is a set of decoupled scalar relations for the components v_i^{n+1}, each of which is a simple first-order hyperbolic partial differential equation.

To generalize the preceding definition of a hyperbolic system to problems with three or more independent variables, consider the system of partial differential equations

$$\frac{\partial \mathbf{u}}{\partial t} + \left(\mathbf{A}_1\frac{\partial}{\partial x_1} + \mathbf{A}_2\frac{\partial}{\partial x_2} + \cdots + \mathbf{A}_m\frac{\partial}{\partial x_m}\right)\mathbf{u} = \mathbf{0} \tag{1.10}$$

and suppose that the coefficient matrices are constant. Unlike the two-independent-variable case, it is not usually possible to find a transformation that simultaneously diagonalizes all the coefficient matrices in (1.10) and thereby generates a set of decoupled scalar equations. Instead, take the Fourier transform of (1.10) with respect to each spatial coordinate to obtain

$$\frac{\partial \hat{\mathbf{u}}}{\partial t} + \mathrm{i}\mathbf{P}(\mathbf{k})\hat{\mathbf{u}} = \mathbf{0}, \qquad \text{where} \qquad \mathbf{P}(\mathbf{k}) = \sum_{q=1}^{m} \mathbf{A}_q k_q$$

and $\mathbf{k} = (k_1, k_2, \ldots, k_m)$ is a real-valued vector of the wave number (or dual variable) with respect to each spatial coordinate.

The system (1.10) will be hyperbolic if all its solutions are the linear superposition of waves of the form $\exp i(\mathbf{k} \cdot \mathbf{x} - \omega t)$, where $\omega(\mathbf{k})$ is a real-valued frequency. This will be the case if $\mathbf{P}(\mathbf{k})$ has a complete set of real eigenvalues for any nonzero wave number vector $\mathbf{k}$, or equivalently, if for every $\mathbf{k}$ such that $\|\mathbf{k}\| = 1$ there exist bounded matrices $\mathbf{T}^{-1}(\mathbf{k})$ and $\mathbf{T}(\mathbf{k})$ such that $\mathbf{D}(\mathbf{k}) = \mathbf{T}^{-1}(\mathbf{k})\mathbf{P}(\mathbf{k})\mathbf{T}(\mathbf{k})$ is a diagonal matrix with real eigenvalues.

The definition of a hyperbolic system in several space dimensions is extended to the case where the coefficient matrices in (1.10) are smooth functions of $\mathbf{x}$ and t by requiring that at every point $(\mathbf{x}, t)$ throughout some domain R there exist bounded matrices $\mathbf{T}^{-1}(\mathbf{k}, \mathbf{x}, t)$ and $\mathbf{T}(\mathbf{k}, \mathbf{x}, t)$ such that for all real vectors $\mathbf{k}$ of unit length, $\mathbf{T}^{-1}(\mathbf{k}, \mathbf{x}, t)\mathbf{P}(\mathbf{k}, \mathbf{x}, t)\mathbf{T}(\mathbf{k}, \mathbf{x}, t)$ is a diagonal matrix with real eigenvalues (Gustafsson et al. 1995, p. 221). Since all symmetric matrices may be transformed to real-valued diagonal matrices, the matrix $\mathbf{P}$ will be symmetric and the original system (1.10) will be hyperbolic if all the coefficient matrices $\mathbf{A}_q$ are symmetric. The easiest way to show that many multidimensional systems are hyperbolic is to transform them to equivalent systems in which all the coefficient matrices are symmetric.

1.1.2 Linear Second-Order Equations in Two Independent Variables

Not all waves are solutions to hyperbolic equations. Hyperbolic equations can be compared with two other fundamental types of partial differential equations, *parabolic* and *elliptic* equations, by considering the general family of linear second-order partial differential equations in two independent variables:

$$a(x, y)u_{xx} + 2b(x, y)u_{xy} + c(x, y)u_{yy} + L(x, y, u, u_x, u_y) = 0. \qquad (1.11)$$

In the preceding equation, the subscripts denote partial derivatives, and L is a linear function of u, u_x, and u_y whose coefficients may depend on x and y. New independent variables η and ξ can be defined that transform (1.11) into one of three canonical forms. The particular form that can be achieved depends on the number of families of characteristic curves associated with (1.11). In those regions of the x–y plane where $b^2 - ac > 0$ there are two independent families of characteristic curves; the equation is hyperbolic, and it can be transformed to the canonical form

$$u_{\xi\eta} + \tilde{L}(\xi, \eta, u, u_\xi, u_\eta) = 0. \qquad (1.12)$$

There is one family of characteristic curves and the equation is parabolic in those regions where $b^2 - ac = 0$, in which case (1.11) can be transformed to

$$u_{\eta\eta} + \tilde{L}(\xi, \eta, u, u_\xi, u_\eta) = 0. \tag{1.13}$$

In those regions where $b^2 - ac < 0$, there are no real-valued characteristic curves; the equation is elliptic, and it transforms to

$$u_{\xi\xi} + u_{\eta\eta} + \tilde{L}(\xi, \eta, u, u_\xi, u_\eta) = 0. \tag{1.14}$$

In each of the preceding equations, $\tilde{L}(\xi, \eta, u, u_\xi, u_\eta)$ is a linear function of u, u_ξ, and u_η with coefficients that may depend on ξ and η.

To carry out the transformation, the various partial derivatives of u with respect to x and y in (1.11) must be replaced by derivatives with respect to ξ and η. Differentiating $u[\xi(x, y), \eta(x, y)]$ yields

$$\begin{aligned}
u_x &= u_\xi \xi_x + u_\eta \eta_x, \\
u_{xx} &= u_{\xi\xi}\xi_x^2 + 2u_{\xi\eta}\xi_x\eta_x + u_{\eta\eta}\eta_x^2 + u_\xi \xi_{xx} + u_\eta \eta_{xx}, \\
u_{xy} &= u_{\xi\xi}\xi_x\xi_y + u_{\xi\eta}\left(\xi_x\eta_y + \xi_y\eta_x\right) + u_{\eta\eta}\eta_x\eta_y + u_\xi \xi_{xy} + u_\eta \eta_{xy},
\end{aligned}$$

along with similar expressions for u_y and u_{yy} that may be substituted into (1.11) to obtain

$$A(\xi, \eta)u_{\xi\xi} + 2B(\xi, \eta)u_{\xi\eta} + C(\xi, \eta)u_{\eta\eta} + \tilde{L}(\xi, \eta, u, u_\xi, u_\eta) = 0, \tag{1.15}$$

where

$$\begin{aligned}
A(\xi, \eta) &= a\xi_x^2 + 2b\xi_x\xi_y + c\xi_y^2, \\
B(\xi, \eta) &= a\xi_x\eta_x + b\left(\xi_x\eta_y + \xi_y\eta_x\right) + c\xi_y\eta_y, \\
C(\xi, \eta) &= a\eta_x^2 + 2b\eta_x\eta_y + c\eta_y^2.
\end{aligned}$$

The new coordinates must be chosen such that the Jacobian

$$\xi_x\eta_y - \xi_y\eta_x$$

is nonzero throughout the domain to guarantee that the transformation between (x, y) and (ξ, η) is unique and has a unique inverse. This coordinate transformation does not change the classification of the partial differential equation as hyperbolic, parabolic, or elliptic because, as can be shown by direct substitution,

$$B^2 - AC = (b^2 - ac)(\xi_x\eta_y - \xi_y\eta_x)^2, \tag{1.16}$$

implying that for nonsingular transforms the sign of $b^2 - ac$ is inherited by $B^2 - AC$.

Now consider the hyperbolic case, for which the canonical form (1.12) is obtained by choosing ξ and η to make $A(\xi, \eta) = C(\xi, \eta) = 0$. $A(\xi, \eta)$ will be zero when

$$a\xi_x^2 + 2b\xi_x\xi_y + c\xi_y^2 = 0,$$

or if $a \neq 0$,

$$\frac{\xi_x^2}{\xi_y^2} + 2\frac{b}{a}\frac{\xi_x}{\xi_y} + \frac{c}{a} = 0. \tag{1.17}$$

Assuming again that $a \neq 0$, the condition $C(\xi, \eta) = 0$ requires that η_x/η_y be a root of the same quadratic equation, i.e.,

$$\frac{\eta_x^2}{\eta_y^2} + 2\frac{b}{a}\frac{\eta_x}{\eta_y} + \frac{c}{a} = 0. \tag{1.18}$$

Since $b^2 - ac > 0$, these roots are real and distinct. Denoting the roots by $-v_1$ and $-v_2$, one may choose transformed coordinates such that $\mathrm{d}y/\mathrm{d}x = v_1$ along lines of constant ξ and $\mathrm{d}y/\mathrm{d}x = v_2$ along lines of constant η. This choice of coordinates satisfies (1.17) and (1.18) because

$$\left.\frac{dy}{dx}\right|_\xi = -\frac{\xi_x}{\xi_y} = v_1 \quad \text{and} \quad \left.\frac{dy}{dx}\right|_\eta = -\frac{\eta_x}{\eta_y} = v_2.$$

Those curves along which either ξ or η is constant are the characteristic curves for the hyperbolic equation (1.11).

After zeroing A and C, one obtains the canonical form (1.12) by dividing (1.15) by $B(\xi, \eta)$, which must be nonzero by (1.16) because

$$\xi_x\eta_y - \xi_y\eta_x = \xi_y\eta_y(v_2 - v_1) \neq 0.$$

In the case $a = 0$, a similar expression for the transformed coordinates can be obtained by dividing the relations $A(\xi, \eta) = 0$ and $C(\xi, \eta) = 0$ by c instead of a. If both a and c are zero, the partial differential equation is placed in canonical form simply by dividing by b (which is nonzero because $b^2 - ac > 0$).

If $\tilde{L}$ is zero, the canonical hyperbolic equation (1.12) has solutions of the form $g(\xi)$ and $h(\eta)$. One circumstance in which $\tilde{L}$ is zero occurs when a, b, and c are constant and $L = 0$ in (1.11). Then the characteristics are the straight lines

$$\xi = y - v_1 x \quad \text{and} \quad \eta = y - v_2 x,$$

and there exist solutions of the form $g(y - v_1 x)$ and $h(y - v_2 x)$. When (1.11) serves as a mathematical model for wave-propagation problems, it usually includes a second-order derivative with respect to time. Suppose, therefore, that $a \neq 0$ and that x represents the time coordinate. Then the speed of signal propagation along the characteristics is given by their slope in the y–x plane, which is v_1 for the constant-ξ characteristics and v_2 for the constant-η characteristics.

In the parabolic case with $a \neq 0$, the quadratic equation (1.17) has the double root $-b/a$, and there is a single characteristic defined such that

$$\left.\frac{dy}{dx}\right|_\xi = -\frac{\xi_x}{\xi_y} = \frac{b}{a}. \tag{1.19}$$

Let $\eta(x, y)$ be any simple function such that

$$\xi_x \eta_y - \xi_y \eta_x \neq 0.$$

These choices for ξ and η imply that $A = 0$ and $B^2 - AC = 0$, which in turn implies that $B = 0$. The canonical parabolic form (1.13) is obtained by dividing (1.15) by C, which must be nonzero, or else neither (1.11) nor (1.15) will be a second-order differential equation. If, on the other hand, $a = 0$ in (1.11), a similar transformation can be performed after dividing through by c, which must be nonzero if a second-order partial derivative is present in (1.11) because $b^2 = b^2 - ac = 0$.

When parabolic partial differential equations describe time-dependent physical systems, such as the diffusion of heat along a rod, the second-order partial derivative is usually computed with respect to a spatial coordinate. Letting x represent the spatial coordinate and y the time coordinate, the one-dimensional heat equation becomes

$$\frac{\partial^2 \psi}{\partial x^2} - \frac{\partial \psi}{\partial y} = 0,$$

which is in the general form (1.11) with $b = c = 0$. According to (1.19), the characteristic curves for the heat equation have slope $\mathrm{d}y/\mathrm{d}x = 0$, i.e., they are lines parallel to the spatial coordinate (which in contrast to the hyperbolic example is now x).

If the partial differential equation is elliptic, then $b^2 - ac < 0$, and there are no real-valued functions that satisfy (1.17) and (1.18). Provided that a, b, and c are analytic,[3] a transformation can always be found that zeros B and sets $A = C = 1$, thereby obtaining the canonical form (1.14). (See Carrier and Pearson 1988 or Kevorkian 1990 for further details.) If a, b, and c are constant, the transformation to canonical form may be accomplished by choosing

$$\xi = \frac{bx - ay}{(ac - b^2)^{1/2}}, \qquad \eta = x,$$

and dividing the resulting equation by a.

Since elliptic partial differential equations do not have real-valued characteristics, their solutions do not generally include wavelike perturbations that propagate through the domain at well-defined velocities. Nevertheless, elliptic equations describing the spatial distribution of a physical parameter such as pressure can be coupled with other time-dependent equations to yield a problem with wave-like solutions. As noted by Whitham (1974), linearized surface gravity waves in

[3] Let $z = x + \mathrm{i}y$ be a complex variable in which x and y are real-valued. The function $f(z)$ is *analytic* if its derivative

$$\frac{df}{dz} = \lim_{\Delta z \to 0} \frac{f(z + \Delta z) - f(z)}{\Delta z}$$

exists and is uniquely defined as Δz goes to zero along any arbitrary path in the complex plane. If $f = u + \mathrm{i}v$, where u and v are real-valued, a necessary condition for f to be analytic is that u and v satisfy the Cauchy–Riemann conditions

$$u_x = v_y, \qquad u_y = -v_x.$$

a flat-bottomed basin of infinite horizontal extent and depth H are governed by the elliptic partial differential equation

$$\frac{\partial^2 p}{\partial x^2} + \frac{\partial^2 p}{\partial y^2} + \frac{\partial^2 p}{\partial z^2} = 0, \tag{1.20}$$

subject to the upper and lower boundary conditions

$$\frac{\partial^2 p}{\partial t^2} + g\frac{\partial p}{\partial z} = 0 \quad \text{at} \quad z = 0,$$
$$\frac{\partial p}{\partial z} = 0 \quad \text{at} \quad z = -H.$$

The wave-like character of the solution is produced by the time-dependent upper boundary condition.

The elliptic nature of (1.20) does not follow from the preceding classification scheme, which requires the evaluation of $b^2 - ac$ and is directly applicable only to linear second-order partial differential equations in two independent variables. In order to generalize this classification scheme to equations with n independent variables, consider the family of linear second-order partial differential equations of the form

$$\sum_{i=1}^{n}\sum_{j=1}^{n} a_{ij}\frac{\partial^2 u}{\partial x_i \partial x_j} + \sum_{i=1}^{n} b_i \frac{\partial u}{\partial x_i} + cu + d = 0. \tag{1.21}$$

If a_{ij}, b_i, c, and d are constants, there exists a one-to-one transformation to a new set of independent variables ξ_i such that the second-order terms in the preceding equation become

$$\sum_{i=1}^{n} A_{ii}\frac{\partial^2 u}{\partial \xi_i{}^2}.$$

If all the A_{ii} are nonzero and have the same sign, (1.21) is elliptic. If all the A_{ii} are nonzero and all but one have the same sign, (1.21) is hyperbolic. If at least one of the A_{ii} is zero, (1.21) is parabolic.

1.2 Wave Equations in Geophysical Fluid Dynamics

The wave-like motions of primary interest in geophysical fluid dynamics are the physical transport of scalar variables by the motion of fluid parcels, oscillatory motions associated with buoyancy perturbations (gravity waves), and oscillatory motions associated with potential vorticity perturbations (Rossby waves). Acoustic waves (sound waves) also propagate through all geophysical fluids, but in many applications these are small-amplitude perturbations whose detailed structure is of no interest. Both inviscid tracer transport and the propagation of sound waves are mathematically described by hyperbolic partial differential equations. Gravity waves and Rossby waves are also solutions to hyperbolic systems of partial differential equations, but some of the fluid properties essential for the support of these waves are represented in the governing equations by terms involving the zero-order derivatives of the unknown variables. These zero-order terms play no role in the classification of the governing equations as hyperbolic, and simpler nonhyperbolic systems of partial differential equations, such as the Boussinesq equations, can be derived whose solutions closely approximate the gravity-wave and Rossby-wave solutions to the original hyperbolic system. These simpler systems will be referred to as *filtered* equations.

1.2.1 Hyperbolic Equations

The concentration of a nonreactive chemical constituent is approximately governed by the first-order linear hyperbolic equation

$$\frac{\partial \psi}{\partial t} + u\frac{\partial \psi}{\partial x} + v\frac{\partial \psi}{\partial y} + w\frac{\partial \psi}{\partial z} = S, \tag{1.22}$$

where $\psi(x, y, z, t)$ is the mixing ratio of the chemical (in nondimensional units such as grams per kilogram or parts per billion) and $S(x, y, z, t)$ is the sum of all sources and sinks. This equation is an approximation because the molecular diffusivity of air is assumed to be negligible, in which case the transport of ψ is produced entirely by the velocity field. The characteristic curves associated with (1.22) are identical to the fluid parcel trajectories determined by the ordinary differential equations

$$\frac{dx}{dt} = u, \qquad \frac{dy}{dt} = v, \qquad \frac{dz}{dt} = w. \tag{1.23}$$

In geophysics, the transport of a quantity by the velocity field is commonly referred to as *advection*;[4] both (1.22) and the one-way wave equation (1.4) are "advection equations."

Equations describing the inviscid transport and chemical reactions among a family of chemical constituents can be written as the system

$$\frac{\partial \mathbf{c}}{\partial t} + u\frac{\partial \mathbf{c}}{\partial x} + v\frac{\partial \mathbf{c}}{\partial y} + w\frac{\partial \mathbf{c}}{\partial z} = \mathbf{s},$$

where $\mathbf{c}$ is a vector whose components are the concentration of each individual chemical species and $\mathbf{s}$ is a vector whose components are the net sources and sinks of each species. In general, the sources and sinks depend on $\mathbf{c}$ but not on the derivatives

[4] In many disciplines the terms "convection" and "advection" are essentially interchangeable. In geophysics, however, the term "convection" is generally reserved for the description of thermally forced circulations.

of **c**, so the preceding equation is a first-order linear hyperbolic system whose solution could be obtained by integrating a coupled system of ordinary differential equations along the family of characteristic curves defined by (1.23).

When diffusion is included, the mathematical model for nonreactive chemical transport becomes

$$\frac{\partial \psi}{\partial t} + u\frac{\partial \psi}{\partial x} + v\frac{\partial \psi}{\partial y} + w\frac{\partial \psi}{\partial z} - S = \frac{\partial}{\partial x}\left(\kappa\frac{\partial \psi}{\partial x}\right) + \frac{\partial}{\partial y}\left(\kappa\frac{\partial \psi}{\partial y}\right) + \frac{\partial}{\partial z}\left(\kappa\frac{\partial \psi}{\partial z}\right), \tag{1.24}$$

which is a linear second-order parabolic partial differential equation. If this equation is derived strictly from first principles, κ represents a molecular diffusivity. The molecular diffusivities of air and water are so small that the contribution from the terms involving the second derivatives are important only when the fluctuations in ψ occur on much smaller scales than those of primary interest in most geophysical problems. Thus, in most geophysical applications the solution to (1.24) is essentially identical to that for the inviscid problem, and the numerical techniques suitable for the approximation of (1.24) are almost identical to those for the purely hyperbolic problem (1.22).

When computing numerical solutions to either (1.22) or (1.24), there will be limits on the spatial and temporal scales at which the velocity field can be represented in any finite data set. The influence of the unresolved velocity perturbations on the distribution of the tracer is not directly computable, but is often parameterized by replacing κ by an *eddy diffusivity*, κ_e. The eddy diffusivity is supposed to represent the tendency of random unresolved velocity fluctuations to spread the distribution of ψ away from the centerline of the smooth air-parcel trajectories computed from the resolved-scale velocity field. Eddy diffusivities are much larger than the molecular diffusivity, but even when κ is replaced by a typical eddy diffusivity, the terms on the right side of (1.22) remain relatively small, and the basic character of the solution is still wavelike. Nevertheless, some eddy-diffusivity parameterizations do generate large values for κ_e in limited regions of the flow. High values of κ_e might, for example, be found in the planetary boundary layer where strong subgrid-scale motions are driven by thermal and mechanical turbulence. Large κ_e might also be parameterized to develop in regions where vigorous subgrid-scale motions are generated through Kelvin–Helmholtz instability. In these limited areas of high eddy diffusivity, the solutions to the parameterized problem may no longer be wavelike.

Now consider the nonlinear shallow-water equations

$$\frac{\partial u}{\partial t} + u\frac{\partial u}{\partial x} + v\frac{\partial u}{\partial y} + g\frac{\partial h}{\partial x} - fv = 0, \tag{1.25}$$

$$\frac{\partial v}{\partial t} + u\frac{\partial v}{\partial x} + v\frac{\partial v}{\partial y} + g\frac{\partial h}{\partial y} + fu = 0, \tag{1.26}$$

$$\frac{\partial h}{\partial t} + u\frac{\partial h}{\partial x} + v\frac{\partial h}{\partial y} + h\left(\frac{\partial u}{\partial x} + \frac{\partial v}{\partial y}\right) = 0, \tag{1.27}$$

where u and v are the horizontal velocity components, h is the fluid depth, and f is the Coriolis parameter. This is a system of quasi-linear first-order differential equations. If one is concerned only with smooth solutions, the fundamental properties of the shallow-water system may be determined from the linearized versions of (1.25)–(1.27). Consider, therefore, a geostrophically balanced basic-state flow such that

$$fV = g\frac{\partial H}{\partial x} \quad \text{and} \quad fU = -g\frac{\partial H}{\partial y},$$

where U and V are constant and H is linear in x and y. The first-order perturbations satisfy

$$\frac{\partial \mathbf{u}}{\partial t} + \mathbf{A}_1\frac{\partial \mathbf{u}}{\partial x} + \mathbf{A}_2\frac{\partial \mathbf{u}}{\partial y} + \mathbf{B}\mathbf{u} = \mathbf{0}, \tag{1.28}$$

where

$$\mathbf{u} = \begin{pmatrix} u' \\ v' \\ h' \end{pmatrix}, \qquad \mathbf{B} = \begin{pmatrix} 0 & -f & 0 \\ f & 0 & 0 \\ fV/g & -fU/g & 0 \end{pmatrix},$$

$$\mathbf{A}_1 = \begin{pmatrix} U & 0 & g \\ 0 & U & 0 \\ H & 0 & U \end{pmatrix}, \qquad \mathbf{A}_2 = \begin{pmatrix} V & 0 & 0 \\ 0 & V & g \\ 0 & H & V \end{pmatrix}.$$

As discussed in connection with (1.10), the preceding system will be hyperbolic if any linear combination of the coefficient matrices, $k_1\mathbf{A}_1 + k_2\mathbf{A}_2$, can be transformed to a real diagonal matrix through multiplication by bounded transformation matrices. Such transformation matrices always exist when the coefficient matrices are symmetric. Thus, an easy way to demonstrate that the preceding system is hyperbolic is to perform a change of variables that renders $\mathbf{A}_1$ and $\mathbf{A}_2$ symmetric. A suitable transformation is obtained by letting $\mathbf{v} = \mathbf{S}^{-1}\mathbf{u}$, where

$$\mathbf{S}^{-1} = \begin{pmatrix} c & 0 & 0 \\ 0 & c & 0 \\ 0 & 0 & g \end{pmatrix},$$

and $c(x, y) = \sqrt{gH}$. Then (1.28) becomes

$$\frac{\partial \mathbf{v}}{\partial t} + \tilde{\mathbf{A}}_1\frac{\partial \mathbf{v}}{\partial x} + \tilde{\mathbf{A}}_2\frac{\partial \mathbf{v}}{\partial y} + \tilde{\mathbf{B}}\mathbf{v} = \mathbf{0},$$

where

$$\tilde{\mathbf{A}}_1 = \mathbf{S}^{-1}\mathbf{A}_1\mathbf{S} = \begin{pmatrix} U & 0 & c \\ 0 & U & 0 \\ c & 0 & U \end{pmatrix}, \qquad \tilde{\mathbf{A}}_2 = \mathbf{S}^{-1}\mathbf{A}_2\mathbf{S} = \begin{pmatrix} V & 0 & 0 \\ 0 & V & c \\ 0 & c & V \end{pmatrix},$$

$$\tilde{\mathbf{B}} = \mathbf{S}^{-1}\left[\mathbf{A}_1\frac{\partial \mathbf{S}}{\partial x} + \mathbf{A}_2\frac{\partial \mathbf{S}}{\partial y} + \mathbf{B}\mathbf{S}\right] = \begin{pmatrix} 0 & -f & 0 \\ f & 0 & 0 \\ \frac{1}{2}fV/c & -\frac{1}{2}fU/c & 0 \end{pmatrix}.$$

The symmetries of $\tilde{\mathbf{A}}_1$ and $\tilde{\mathbf{A}}_2$ imply that the linearized shallow-water equations are a hyperbolic system.

The wave solutions to this hyperbolic system do not, however, propagate exactly along the characteristic curves unless f is zero. The relationship between the paths followed by propagating waves and the characteristics is most easily investigated by considering plane waves propagating parallel to the x-axis in a basic state with no mean flow. Let the Coriolis parameter have the constant value f_0 and define a vector of new unknown functions

$$\mathbf{v} = \begin{pmatrix} u - gh/c \\ v \\ u + gh/c \end{pmatrix},$$

which transforms (1.28) to

$$\frac{\partial \mathbf{v}}{\partial t} + \begin{pmatrix} -c & 0 & 0 \\ 0 & 0 & 0 \\ 0 & 0 & c \end{pmatrix} \frac{\partial \mathbf{v}}{\partial x} + \begin{pmatrix} 0 & -f_0 & 0 \\ f_0/2 & 0 & f_0/2 \\ 0 & -f_0 & 0 \end{pmatrix} \mathbf{v} = \mathbf{0}.$$

The characteristics for this system are the curves satisfying $\mathrm{d}x/\mathrm{d}t = \pm c$ and $\mathrm{d}x/\mathrm{d}t = 0$.

Wave solutions to (1.28) have the form

$$(u', v', h') = \Re\left\{(u_0, v_0, h_0)\mathrm{e}^{\mathrm{i}(kx-\omega t)}\right\}, \tag{1.29}$$

provided that the frequency ω and wave number k satisfy the dispersion relation

$$\omega^2 = c^2k^2 + f_0^2, \tag{1.30}$$

as may be demonstrated by substituting (1.29) into (1.28). Lines of constant phase, such as the locations of the troughs and crests, propagate at the *phase speed* ω/k, which from (1.30) is

$$\frac{\omega}{k} = \pm c\left(1 + \frac{f_0^2}{c^2k^2}\right)^{1/2}.$$

A compact group of waves travels at the group velocity $\partial\omega/\partial k$, which can also be computed from (1.30):

$$\frac{\partial \omega}{\partial k} = \pm c\left(1 + \frac{f_0^2}{c^2k^2}\right)^{-1/2}.$$

In the limit $|k| \gg f_0/c$, the phase speed and group velocity both approach the slope of a characteristic along which $|\mathrm{d}x/\mathrm{d}t| = c$. Nevertheless, for any finite value of k,

$$\left|\frac{\partial \omega}{\partial k}\right| < c < \left|\frac{\omega}{k}\right|,$$

and neither the lines of constant phase nor the wave groups follow trajectories that coincide with the characteristic curves. Note that the magnitude of the group velocity, which is the rate at which energy propagates in a wave, is bounded by c. The maximum rate of energy propagation can therefore be determined without considering the zero-order coefficient matrix $\mathbf{B}$ in (1.28).

The loose connection between wave propagation and the characteristics in the preceding example can disappear altogether if the Coriolis parameter is a function of the spatial coordinate. Then a second type of wave, the Rossby wave, may appear as an additional solution. If f increases linearly in proportion to y, Rossby-wave solutions may exist with phase speeds in the negative-x direction (Holton 1992; Pedlosky 1987). Neither the phase speeds nor the group velocities of these waves have any relation to the characteristic curves. It is not surprising that Rossby waves do not propagate along the characteristics, because the terms involving the undifferentiated functions of u and v play no role in the determination of the characteristics of (1.28), yet those same terms are essential for the maintenance of the Rossby waves.

The Euler equations governing inviscid isentropic motion in a density stratified fluid provide another example of a hyperbolic system that supports a type of wave whose propagation is completely unrelated to the characteristics. The Euler equations for the inviscid isentropic motion of a perfect gas can be expressed in the form

$$\frac{d\mathbf{v}}{dt} + \frac{1}{\rho}\nabla p = -g\mathbf{k}, \tag{1.31}$$

$$\frac{\partial \rho}{\partial t} + \nabla \cdot (\rho \mathbf{v}) = 0, \tag{1.32}$$

$$\frac{d\theta}{dt} = 0, \tag{1.33}$$

where Coriolis forces have been neglected,

$$\frac{d(\;)}{dt} = \frac{\partial(\;)}{\partial t} + \mathbf{v} \cdot \nabla(\;),$$

$\mathbf{v}$ is the three-dimensional velocity vector, ρ is density, p is pressure, g is the gravitational acceleration, $\mathbf{k}$ is a unit vector directed opposite to the gravitational restoring force, and θ is the potential temperature, which is related to the entropy, S, such that

$$S = c_p \ln \theta + \text{constant}.$$

Conservation of momentum is required by (1.31), conservation of mass by (1.32), and conservation of entropy by (1.33).

As written above, the Euler equations constitute a system of five equations involving six unknowns. In atmospheric applications, the system may be closed using the equation of state for a perfect gas,

$$p = \rho R T, \tag{1.34}$$

the definition of the potential temperature,

$$\theta = T\,(p/p_0)^{-R/c_p},$$

and the identity $R = c_p - c_v$ to arrive at the diagnostic equation

$$p = p_0\left(\frac{R}{p_0}\rho\theta\right)^{c_p/c_v}. \tag{1.35}$$

In the preceding equation, T is the temperature, p_0 is a constant reference pressure, R is the gas constant for dry air, c_p is the specific heat at constant pressure, and c_v is the specific heat at constant volume.

The Euler equations are a quasi-linear system of first-order partial differential equations. The fundamental character of the smooth solutions to this system can be determined by linearizing these equations about a horizontally uniform isothermally stratified basic state. Simpler basic states can be obtained by neglecting gravitational forces and the density stratification (Gustafsson et al. 1995, p. 136; see also Problem 3), but the isothermal basic state is of more geophysical relevance. As a preliminary step, p and ρ can be eliminated from (1.31)–(1.33) by introducing the nondimensional *Exner function* pressure defined as

$$\pi = (p/p_0)^{R/c_p}. \tag{1.36}$$

It follows that

$$\frac{1}{\rho}\nabla p = c_p\theta\nabla\pi,$$

so the momentum equation may be written as

$$\frac{d\mathbf{v}}{dt} + c_p\theta\nabla\pi = -g\mathbf{k}. \tag{1.37}$$

It also follows from (1.35) and (1.36) that

$$\pi = \left(\frac{R}{p_0}\rho\theta\right)^{R/c_v};$$

thus,

$$\frac{d}{dt}\ln(\pi) = \frac{R}{c_v}\left[\frac{d}{dt}\ln(\rho) + \frac{d}{dt}\ln(\theta)\right],$$

or, using (1.32) and (1.33),

$$\frac{d\pi}{dt} + \frac{R\pi}{c_v}\nabla\cdot\mathbf{v} = 0. \tag{1.38}$$

Equations (1.33), (1.37), and (1.38) constitute a closed system of five equations in the five unknown variables, θ, π, and the three components of $\mathbf{v}$.

The essential properties of this system can be more simply examined in a two-dimensional context. Let x and z be the horizontal and vertical coordinates, and decompose the thermodynamic fields into a vertically varying basic state and a perturbation such that

$$\begin{aligned}\pi(x,z,t) &= \overline{\pi}(z) + \pi'(x,z,t),\\ \theta(x,z,t) &= \overline{\theta}(z) + \theta'(x,z,t),\\ c_p\overline{\theta}\frac{d\overline{\pi}}{dz} &= -g.\end{aligned} \tag{1.39}$$

The velocity components are decomposed as

$$u(x,z,t) = U + u'(x,z,t), \qquad w(x,z,t) = w'(x,z,t). \tag{1.40}$$

The basic-state vertical velocity is zero to ensure that the basic state is a steady solution to the nonlinear equations. Substituting these expressions for u, w, π, and θ into the two-dimensional versions of (1.33), (1.37), and (1.38), and neglecting second-order terms in the perturbation variables under the assumption that the perturbations are small-amplitude, one obtains the linear system

$$\left(\frac{\partial}{\partial t} + U\frac{\partial}{\partial x}\right)u' + c_p\overline{\theta}\frac{\partial \pi'}{\partial x} = 0, \tag{1.41}$$

$$\left(\frac{\partial}{\partial t} + U\frac{\partial}{\partial x}\right)w' + c_p\overline{\theta}\frac{\partial \pi'}{\partial z} = g\frac{\theta'}{\overline{\theta}}, \tag{1.42}$$

$$\left(\frac{\partial}{\partial t} + U\frac{\partial}{\partial x}\right)\theta' + \frac{\overline{\theta}}{g}N^2 w' = 0, \tag{1.43}$$

$$\left(\frac{\partial}{\partial t} + U\frac{\partial}{\partial x}\right)\pi' + w'\frac{\partial \overline{\pi}}{\partial z} + \frac{R\overline{\pi}}{c_v}\left(\frac{\partial u'}{\partial x} + \frac{\partial w'}{\partial z}\right) = 0, \tag{1.44}$$

where

$$N^2 = \frac{g}{\overline{\theta}}\frac{d\overline{\theta}}{dz}$$

is the square of the Brunt–Väisälä frequency.

Suppose that the reference state is isothermal. Then N^2 and the speed of sound $c_s = (c_p RT/c_v)^{1/2}$ are constant, and the preceding system can be simplified by removing the influence of the decrease in the mean density with height via the transformation

$$\tilde{u} = \left(\frac{\overline{\rho}}{\rho_0}\right)^{1/2} u', \qquad \tilde{\pi} = \left(\frac{\overline{\rho}}{\rho_0}\right)^{1/2} \frac{c_p\overline{\theta}}{c_s}\pi', \tag{1.45}$$

$$\tilde{w} = \left(\frac{\overline{\rho}}{\rho_0}\right)^{1/2} w', \qquad \tilde{\theta} = \left(\frac{\overline{\rho}}{\rho_0}\right)^{1/2} \frac{g}{N\overline{\theta}}\theta'. \tag{1.46}$$

Note that $\tilde{\theta}$ represents a scaled buoyancy and $\tilde{\pi}$ a scaled pressure. Let

$$\mathbf{v} = \left(\tilde{u}\ \tilde{w}\ \tilde{\theta}\ \tilde{\pi}\right)^{\mathrm{T}};$$

then the transformed equations have the form

$$\frac{\partial \mathbf{v}}{\partial t} + \mathbf{A}_1 \frac{\partial \mathbf{v}}{\partial x} + \mathbf{A}_2 \frac{\partial \mathbf{v}}{\partial z} + \mathbf{B}\mathbf{v} = \mathbf{0}, \tag{1.47}$$

in which

$$\mathbf{A}_1 = \begin{pmatrix} U & 0 & 0 & c_s \\ 0 & U & 0 & 0 \\ 0 & 0 & U & 0 \\ c_s & 0 & 0 & U \end{pmatrix}, \quad \mathbf{A}_2 = \begin{pmatrix} 0 & 0 & 0 & 0 \\ 0 & 0 & 0 & c_s \\ 0 & 0 & 0 & 0 \\ 0 & c_s & 0 & 0 \end{pmatrix},$$

$$\mathbf{B} = \begin{pmatrix} 0 & 0 & 0 & 0 \\ 0 & 0 & -N & -S \\ 0 & N & 0 & 0 \\ 0 & S & 0 & 0 \end{pmatrix}, \quad S = c_s \left[\frac{1}{2\overline{\rho}} \frac{d\overline{\rho}}{dz} + \frac{1}{\overline{\theta}} \frac{d\overline{\theta}}{dz} \right].$$

Since the coefficient matrices for the first-order derivatives in (1.47) are symmetric, the linearized Euler equations are a hyperbolic system. The eigenvalues of $\mathbf{A}_1$ are U, U, $U + c_s$, and $U - c_s$; those of $\mathbf{A}_2$ are 0, 0, c_s, and $-c_s$. The eigenvalues involving c_s give the speed at which sound waves in an unstratified fluid propagate parallel to the x and z coordinate axes. As will be demonstrated below, sound waves in an isothermally stratified atmosphere actually propagate at slightly different speeds owing to the influence of the zero-order term in (1.47). The remaining eigenvalues relate to the speed at which fluid parcels are advected horizontally and vertically by the mean flow. These eigenvalues have no relation to the propagation of gravity (or buoyancy) waves, which are the second type of fundamental wave motion supported by (1.47).

When the basic state is isothermal, S is constant, and wave solutions to (1.47) exist in the form

$$(\tilde{u}, \tilde{w}, \tilde{\theta}, \tilde{\pi}) = \Re\left\{(u_0, w_0, \theta_0, \pi_0)e^{i(kx+\ell z-\omega t)}\right\}, \tag{1.48}$$

provided that ω, k, and ℓ satisfy the dispersion relation

$$(\omega - Uk)^2 = \frac{c_s^2}{2}\left(k^2 + \ell^2 + \frac{N^2 + S^2}{c_s^2}\right) \pm \frac{c_s^2}{2}\left[\left(k^2 + \ell^2 + \frac{N^2 + S^2}{c_s^2}\right)^2 - \frac{4N^2k^2}{c_s^2}\right]^{1/2}, \tag{1.49}$$

which is obtained by substituting (1.48) into (1.47). As will be discussed in Sect. 8.2.4, the second term inside the square root is much smaller than the first term in most applications, so (1.49) can be separated into a pair of approximate dispersion relations for the sound waves and the gravity waves. The dispersion relation for the sound waves,

$$(\omega - Uk)^2 = c_s^2\left(k^2 + \ell^2\right) + S^2 + N^2,$$

is obtained by taking the positive root in (1.49). In a manner analogous to the effect of the Coriolis force on gravity waves in the shallow-water system, the terms involving the product of N or S with the zero-order derivatives of the unknown variables introduce a slight discrepancy between the phase speeds and group velocities of the actual sound waves and those that might be suggested by the eigenvalues of $\mathbf{A}_1$ and $\mathbf{A}_2$.

The dispersion relation for the gravity waves is obtained by taking the negative root in (1.49), which to a good approximation yields

$$(\omega - Uk)^2 = \frac{N^2k^2}{k^2 + \ell^2 + (S^2 + N^2)/c_s^2}. \tag{1.50}$$

Neither the phase speeds nor the group velocities of these waves have any relation to the eigenvalues of $\mathbf{A}_1$ and $\mathbf{A}_2$. Unlike sound waves, gravity waves do not even approximately propagate along the characteristics. There is no relation between the characteristics and the paths of the gravity waves because some of the physical processes essential for gravity-wave propagation are mathematically represented by undifferentiated functions of the unknown variables, and as such exert no influence on the shape of the characteristics.

1.2.2 Filtered Equations

The Euler equations support sound waves, but sound waves have no direct influence on many types of atmospheric and oceanic motion. Analytic simplicity can often be achieved by approximating the Euler equations with alternative sets of *filtered* governing equations that do not support sound waves. As will be discussed in Chap. 8, eliminating the sound waves may also allow the resulting system of equations to be numerically integrated using a much larger time step than that which would be required for a similar numerical integration of the original Euler equations. These sets of filtered equations are not hyperbolic systems, but they support gravity waves that closely approximate the gravity-wave solutions to the full Euler equations. If the latitudinal variation of the Coriolis parameter is included, the filtered equations also support Rossby waves. The Coriolis parameter will, however, be neglected in the following discussion in order to present the essential ideas in the simplest context.

According to (1.35), the pressure perturbations in a perfect gas arise from variations in density and entropy. Variations in entropy play no fundamental role in the physics of sound wave propagation. Indeed, for the general class of fluids described by an equation of state of the form

$$\rho \equiv \rho(p, S),$$

the speed of sound is given by the square root of $(\partial p/\partial \rho)_S$ (Batchelor 1967, p. 166). To filter sound waves from the governing equations, it is therefore necessary to sever the link between density perturbations and pressure perturbations. This can be accomplished through any one of a family of related approximations that neglect terms involving the time variation of the density in the mass continuity equation (1.32).

One approximation that will filter sound waves is obtained by assuming that the flow is incompressible, in which case

$$\nabla \cdot \mathbf{v} = 0, \tag{1.51}$$

and mass conservation is replaced by volume conservation. The approximation of (1.32) by (1.51) is widely referred to as the *Boussinesq approximation*. Unfortunately, the term "Boussinesq approximation" has been used in two different senses. In some disciplines, the Boussinesq approximation refers only to the approximation of mass conservation by volume conservation. In the atmospheric and oceanic sciences, the Boussinesq approximation is generally understood to include both the preceding and additional approximations in the momentum equations that will be discussed in connection with (1.60). The latter definition, encompassing approximations to both the mass continuity and the momentum equations, appears to be consistent with the actual approximations employed by Boussinesq (1903, pp. 157, 174), and will be the form of the Boussinesq approximation referred to throughout this book.

A second approximation to the full compressible continuity equation is *anelastic* compressibility (Ogura and Phillips 1962; Lipps and Hemler 1982),

$$\nabla \cdot (\overline{\rho}\mathbf{v}) = 0, \tag{1.52}$$

in which the density involved in the mass budget is a steady reference-state density $\overline{\rho}(z)$ that varies only along the coordinate axis parallel to the gravitational restoring force. A third approximation is *pseudo-incompressibility* (Durran 1989),

$$\frac{\partial \hat{\rho}}{\partial t} + \nabla \cdot (\hat{\rho}\mathbf{v}) = 0, \tag{1.53}$$

in which $\hat{\rho}$ is determined by the time-varying potential temperature and the pressure in a steady reference state $\tilde{p}(x, y, z)$ via the equation of state

$$\tilde{p} = p_0 \left(\frac{R}{p_0} \hat{\rho} \theta \right)^{c_p/c_v}.$$

The pseudo-incompressible approximation neglects the influence of perturbation pressure on perturbation density in the mass budget. According to the preceding definition of $\hat{\rho}$, the term $\partial \hat{\rho}/\partial t$ in (1.53) is entirely determined by $\partial \theta/\partial t$. The pseudo-incompressible continuity equation may be written in the obviously diagnostic form

$$\nabla \cdot (\tilde{\rho}\tilde{\theta}\mathbf{v}) = 0 \tag{1.54}$$

by using the thermodynamic equation (1.33) to eliminate $\partial\theta/\partial t$ from (1.53) and defining steady reference fields of density $\tilde{\rho}(x, y, z)$ and potential temperature $\tilde{\theta}(x, y, z)$ such that the reference fields satisfy the equation of state,

$$\tilde{p} = p_0 \left(\frac{R}{p_0} \tilde{\rho}\tilde{\theta} \right)^{c_p/c_v}.$$

Note that if F_θ represents any thermal forcing or viscous terms that might appear on the right side of the thermodynamic equation in more general applications, (1.53) is unchanged but (1.54) becomes

$$\nabla \cdot (\tilde{\rho}\tilde{\theta}\mathbf{v}) = \hat{\rho} F_\theta.$$

The pseudo-incompressible system can be rigorously derived through scale analysis by assuming that the Mach number (U/c_s) and the perturbation of the total pressure about the reference pressure, $\tilde{p}$, are both small (Durran 2008).

For an approximate set of governing equations to provide a physically acceptable approximation to the dynamics of the unapproximated system, the approximate equations should conserve energy in the sense that the domain integral of the total energy should be equal to the divergence of an energy flux through the boundaries of the domain. The energy equation for the full compressible system is

$$\frac{\partial E}{\partial t} + \nabla \cdot [(E + p)\mathbf{v}] = 0, \tag{1.55}$$

where

$$E = \rho \left(\frac{\mathbf{v} \cdot \mathbf{v}}{2} + gz + c_v T \right)$$

is the total energy (kinetic plus potential plus internal) per unit volume in a compressible fluid. Similar energy equations can be obtained using the incompressible or pseudo-incompressible continuity equations without introducing additional approximations in the momentum equations.

If the flow is incompressible, the mass continuity equation breaks into the two separate relations

$$\frac{d\rho}{dt} = 0 \tag{1.56}$$

and (1.51); the thermodynamic equation is no longer required to close the system, and the governing equations are simply (1.31), (1.51), and (1.56). The energy equation for this system has the same form as that for the compressible system (1.55) except that the energy,

$$E_\mathrm{i} = \rho \left(\frac{\mathbf{v} \cdot \mathbf{v}}{2} + gz \right),$$

does not include the term representing internal energy. The pseudo-incompressible system, which consists of (1.33), (1.37), and (1.54), conserves

$$E_\mathrm{pi} = \hat{\rho} \left(\frac{\mathbf{v} \cdot \mathbf{v}}{2} + gz \right) + c_v \tilde{\rho} \tilde{T},$$

according to the energy equation

$$\frac{\partial E_{\mathrm{pi}}}{\partial t} + \nabla \cdot \left[\frac{\hat{\rho}}{\rho} (E + p) \, \mathbf{v} \right] = 0.$$

Both the mechanical energy $\hat{\rho}(\mathbf{v} \cdot \mathbf{v}/2 + gz)$ and the energy flux densities in the pseudo-incompressible energy equation differ from those in the exact system only by a factor of $\hat{\rho}/\rho$. The internal energy density in the pseudo-incompressible system $\tilde{\rho} c_v \tilde{T}$ is just the internal energy of the reference state.

In contrast to the situation for the incompressible and pseudo-incompressible approximations, the pressure gradient terms in the momentum equations must be linearized and modified to obtain an energy-conservative system of anelastic equations. As a first step toward developing such a system, the thermodynamic variables are decomposed into a vertically varying reference state and a perturbation. This decomposition is also quite useful outside the context of the anelastic equations because in many geophysical fluids the gravitational acceleration and the vertical pressure gradient are nearly in balance. Both numerical accuracy and physical insight can therefore be enhanced by splitting the pressure and density fields into steady hydrostatically balanced vertical profiles and finite-amplitude perturbations about those reference profiles such that

$$\begin{aligned} p(x, y, z, t) &= \overline{p}(z) + p'(x, y, z, t), \\ \rho(x, y, z, t) &= \overline{\rho}(z) + \rho'(x, y, z, t), \\ \frac{d\,\overline{p}}{dz} &= -\overline{\rho} g. \end{aligned}$$

After the hydrostatically balanced component of the pressure has been removed, the momentum equation (1.31) may be written without approximation as

$$\frac{d\mathbf{v}}{dt} + \frac{1}{\rho} \nabla p' = -g \frac{\rho'}{\rho} \mathbf{k}. \tag{1.57}$$

If the pressure gradients in the momentum equation are expressed in terms of π and θ, the hydrostatic reference state is removed by defining

$$\begin{aligned} \pi(x, y, z, t) &= \overline{\pi}(z) + \pi'(x, y, z, t), \\ \theta(x, y, z, t) &= \overline{\theta}(z) + \theta'(x, y, z, t), \\ c_p \overline{\theta} \frac{d\,\overline{\pi}}{dz} &= -g, \end{aligned} \tag{1.58}$$

in which case (1.37) becomes

$$\frac{d\mathbf{v}}{dt} + c_p \theta \nabla \pi' = g \frac{\theta'}{\overline{\theta}} \mathbf{k}. \tag{1.59}$$

The term on the right side of either (1.57) or (1.59) represents a buoyancy force. Note that since no approximations have been introduced in these equations, the pressure gradient terms in (1.57) and (1.59) remain nonlinear.

In addition to the previously discussed modifications to the mass continuity equation, the Boussinesq and anelastic approximations include additional simplifications to the momentum equations that linearize the pressure gradient terms in (1.57) and (1.59). The form of the *Boussinesq approximation* that is most common in geophysical fluid dynamics *neglects the effects of density variations on the mass balance in the continuity equation and on inertia in the momentum equations, but includes the effect of density variations on buoyancy forces* (Gill 1982, p. 130). Letting ρ_0 be a constant reference density, one may write the Boussinesq form of the momentum equations as

$$\frac{d\mathbf{v}}{dt} + \frac{1}{\rho_0}\nabla p' = -g\frac{\rho'}{\rho_0}\mathbf{k}, \tag{1.60}$$

where the perturbation density continues to be defined as $\rho - \overline{\rho}(z)$ (rather than $\rho - \rho_0$). The resulting Boussinesq system, consisting of (1.51), (1.56), and (1.60), can be concisely expressed in terms of the Boussinesq pressure, buoyancy, and Brunt–Väisälä frequency,

$$P = \frac{p}{\rho_0}, \qquad b = -g\frac{\rho - \overline{\rho}}{\rho_0}, \qquad \text{and} \qquad N_b^2 = -\frac{g}{\rho_0}\frac{d\overline{\rho}}{dz},$$

respectively, as

$$\frac{d\mathbf{v}}{dt} + \nabla P = b\mathbf{k}, \tag{1.61}$$

$$\frac{db}{dt} + N_b^2 w = 0, \tag{1.62}$$

$$\nabla \cdot \mathbf{v} = 0, \tag{1.63}$$

where, as before, w is the vertical velocity component. The Boussinesq system is governed by an energy equation of the form (1.55), except that the total "Boussinesq" energy is

$$E_b = \rho_0\frac{\mathbf{v}\cdot\mathbf{v}}{2} + \rho g z.$$

Although the Boussinesq approximation provides a qualitatively correct mathematical model for the study of buoyancy effects in fluids, it is not quantitatively accurate in situations where there is a significant change in mean density over the depth of the fluid, as would be the case in any atmospheric layer that is more than a couple of kilometers deep. Somewhat better quantitative agreement between the Boussinesq equations and atmospheric flows can be obtained using the same Boussinesq system (1.61)–(1.63) with the pressure, buoyancy, and Brunt–Väisälä frequency defined as

$$P = c_p\theta_0\pi', \qquad b = g\frac{\theta - \overline{\theta}}{\theta_0}, \qquad \text{and} \qquad N_b^2 = \frac{g}{\theta_0}\frac{d\overline{\theta}}{dz},$$

respectively, where θ_0 is a constant reference temperature. Using these definitions for P and b, the full momentum equation (1.59) will be well approximated by

(1.61) whenever the full and basic-state potential temperatures are close to θ_0. In atmospheric applications, it is often easier to satisfy this constraint than to demand that ρ_0 be a good approximation to ρ in (1.57). Even if the reference state is nearly isentropic, some quantitative error in the Boussinesq solution will still be introduced by the incompressible continuity equation. The quantitative errors associated with Boussinesq approximations to deep atmospheric flows can be greatly diminished using either the anelastic or the pseudo-incompressible approximations.

An energy-conservative form of the anelastic equations was derived by Lipps and Hemler (1982) by writing the momentum equations in the form

$$\frac{d\mathbf{v}}{dt} + c_p \nabla(\overline{\theta}\pi') = g\frac{\theta'}{\overline{\theta}}\mathbf{k}, \tag{1.64}$$

where the hydrostatically balanced components of the Exner function pressure and the potential temperature have been removed using (1.58). The anelastic system can be derived from the pseudo-incompressible system by choosing a horizontally uniform hydrostatically balanced reference state, approximating the total pressure gradient as $c_p\overline{\theta}\nabla\pi'$, and neglecting the vertical derivative of the reference potential temperature in both (1.54) and the momentum equations. The same approximation can be obtained by a rigorous, if somewhat delicate, scaling argument (Lipps 1990). The anelastic system consisting of (1.33), (1.52), and (1.64) provides a good approximation to the full compressible equations. The Lipps–Hemler anelastic system satisfies the energy equation

$$\frac{\partial E_\mathrm{a}}{\partial t} + \nabla \cdot [(E_\mathrm{a} + \check{p})\mathbf{v}] = 0,$$

where

$$E_\mathrm{a} = \overline{\rho}\left(\frac{\mathbf{v}\cdot\mathbf{v}}{2} + gz + c_p\overline{\pi}\theta'\right) + c_v\overline{\rho}\overline{T}$$

and $\check{p} = \overline{p} + c_p\overline{\rho}\overline{\theta}\pi' \approx \overline{p} + p' = p$.

Simple wave solutions to the preceding filtered systems can be obtained by linearizing the two-dimensional form of each system about an appropriate basic-state flow with a constant horizontal wind speed U. Solutions to the two-dimensional Boussinesq system exist in the form

$$(u, w, P, b) = \Re\left\{(u_0, w_0, P_0, b_0)\mathrm{e}^{\mathrm{i}(kx+\ell z-\omega t)}\right\},$$

provided that N_b^2 is constant and

$$(\omega - Uk)^2 = \frac{N_\mathrm{b}^2 k^2}{k^2 + \ell^2}.$$

These solutions are gravity waves, as may be seen by comparing the preceding dispersion relation with (1.50) in the limit $c_\mathrm{s} \to \infty$. There are no sound-wave solutions to the Boussinesq equations.

If the basic state is isothermally stratified, the prognostic variables in the two-dimensional anelastic and pseudo-incompressible systems can be transformed as per (1.45) and (1.46) to yield constant-coefficient linear systems of partial differential equations with wave solutions of the form (1.48). In the case of the anelastic equations, these waves satisfy the dispersion relation (1.50), which is an excellent approximation to the dispersion relation for gravity waves in the full compressible system. In the case of the pseudo-incompressible equations, the waves satisfy the dispersion relation

$$(\omega - Uk)^2 = \frac{N^2k^2}{k^2 + \ell^2 + S^2/c_s^2}.$$

Since in most applications $k^2 + \ell^2$ is much larger than the remaining terms in the denominator of (1.50), the preceding equation is also a very good approximation to the gravity-wave dispersion relation for the full compressible equations. The relative accuracy of the anelastic and pseudo-incompressible approximations cannot be judged solely on the basis of their dispersion relations. Nance and Durran (1994) and Nance (1997) compared the accuracy of several different systems of filtered equations and found that the pseudo-incompressible system and the anelastic system suggested by Lipps and Hemler are the most accurate, and that the anelastic system performs slightly better in the hydrostatic limit, whereas the pseudo-incompressible system gives slightly better accuracy when the flow is farther from hydrostatic balance.

1.3 Strategies for Numerical Approximation

A wide variety of different methods have been employed to obtain numerical solutions to the systems of partial differential equations discussed in the preceding sections of this chapter. Before delving into the details of these methods, we conclude this introductory chapter by comparing some of the most general properties of the various methods, including the manner in which each method approximates the value of the unknown function and estimates its derivatives. We will also consider some of the fundamental differences between the numerical algorithms used to solve elliptic and hyperbolic partial differential equations.

1.3.1 Approximating Calculus with Algebra

Digital computers are not designed to solve differential equations directly. Although the digital computer can perform algebraic operations such as addition and multiplication, it does not have any intrinsic ability to differentiate and integrate functions. As a consequence, every numerical method is designed to convert the original differential equation into a set of solvable algebraic equations. As part of this task the

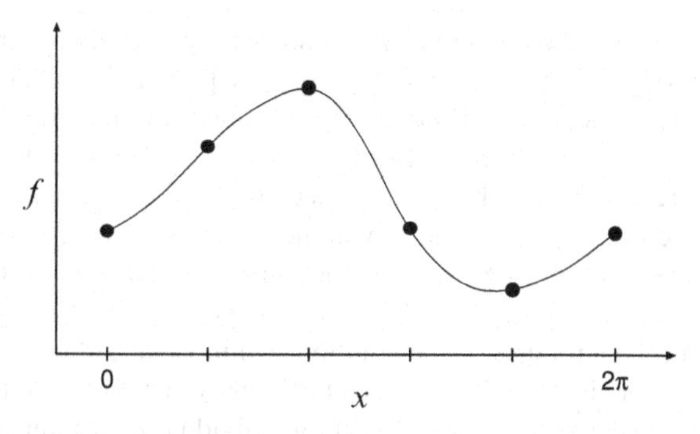

Fig. 1.1 Grid-point approximation of a periodic function on the interval $[0, 2\pi]$. Individual points show the function values at intervals of $2\pi/5$

continuous functions associated with the original problem must be represented by a finite set of numbers that can be stored in a computer's memory or on a disk. There are therefore two basic problems that must be addressed by every numerical scheme: how to represent the solution by a finite data set and how to compute derivatives. There are also two basic solution strategies: grid-point methods and series-expansion methods.

In grid-point methods, each function is described by its value at a set of discrete grid points. Figure 1.1 shows how $f(x)$, a periodic function on the interval $[0, 2\pi]$, might be represented by its exact value at five different points along the x-axis. The spacing of the grid points can be chosen arbitrarily, although any variations in the grid spacing will affect the accuracy of the approximation. If a priori knowledge of the function's periodicity is available, a natural choice for the five pieces of information would be $(f(2\pi/5), f(4\pi/5), \ldots, f(2\pi))$. No assumption is made about the value of the approximate solution between the points on the numerical mesh. These methods are usually called *finite-difference* methods because derivatives are approximated using formulae such as

$$\frac{df}{dx}(x_0) \approx \frac{f(x_0 + \Delta x) - f(x_0 - \Delta x)}{2\Delta x},$$

which is a centered finite difference computable from data on a uniform mesh with grid interval Δx. Finite-difference methods will be discussed in Chaps. 2–4.

Finite-volume methods are an important variation of the basic grid-point approach in which some assumption is made about the structure of the approximate solution between the grid points. In a finite-volume method the grid-point value f_j represents the average of the function $f(x)$ over the interval (or grid cell) $[(j - 1/2)\Delta x, (j + 1/2)\Delta x]$. Finite-volume methods are very useful for approximating solutions that contain discontinuities. If the solution being approximated is smooth, finite-difference and finite-volume methods yield essentially the same numerical schemes. It is sometimes mistakenly supposed that all grid-point methods necessarily generate approximations to the grid-cell average; however, only finite-volume methods have this property.

To completely define the numerical algorithm arising from a conventional finite-difference approximation, it is necessary to specify particular formulae for the finite differences (e.g., centered differencing, one-sided differencing, or one of the other options described in Chap. 2). In finite-volume methods, on the other hand, the derivatives are determined by the assumed structure of the approximate solution within each cell. In practice, finite-volume methods often require the computation of the fluxes through the edges of each grid cell rather than the evaluation of derivatives, but to compute these fluxes it is once again necessary to make an assumption about the structure of the solution within each grid cell. The approximate solution cannot simply be the piecewise-linear function that interpolates the grid-point values, because then the value at an individual grid point will not equal the average of the piecewise-linear approximation over the surrounding grid cell. Two possible finite-volume approximations to $f(x)$ are shown in Fig. 1.2. Accounting for periodicity, we again use five pieces of information to construct these approximations. Piecewise-constant functions are used in the approximation shown in Fig. 1.2a; Fig. 1.2b shows the approximation obtained using piecewise-linear functions defined such that

$$f(x) \approx f_j + \sigma_j(x - j\Delta x) \quad \text{for all} \quad x \in \left((j-\tfrac{1}{2})\Delta x,\, (j+\tfrac{1}{2})\Delta x\right),$$

where f_j is the average of the approximate solution over the grid cell centered at $j\Delta x$, and $\sigma_j = (f_{j+1} - f_j)/\Delta x$. The accuracy of the numerical approximations

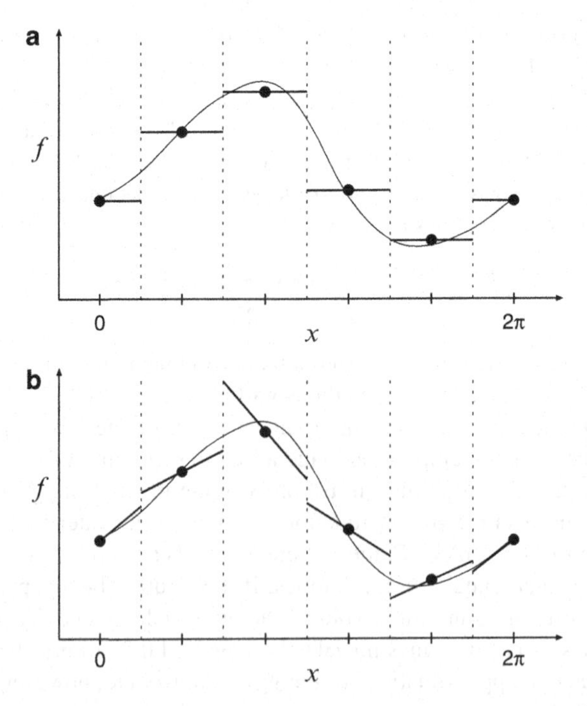

Fig. 1.2 Finite-volume approximation of a periodic function on the interval $[0, 2\pi]$ using **a** piecewise-constant functions, and **b** piecewise-linear functions

shown in Figs. 1.1 and 1.2 is poor because only five data points are used to resolve $f(x)$. Between 12 and 20 data points would be required to obtain a minimally acceptable approximation in most practical applications. Finite-volume methods will be discussed in Chap. 5.

In series-expansion methods, the unknown function is approximated by a linear combination of a finite set of continuous expansion functions, and the data set describing the approximated function is the finite set of expansion coefficients. Derivatives are computed analytically by differentiating the expansion functions. When the expansion functions form an orthogonal set, the series-expansion approach is a *spectral method.* If the preceding periodic function were to be approximated by a spectral method using five pieces of data, a natural choice would be the truncated Fourier series

$$a_1 + a_2 \cos x + a_3 \sin x + a_4 \cos 2x + a_5 \sin 2x. \tag{1.65}$$

The five Fourier coefficients $(a_1, a_2, \ldots, a_5)$ need not be chosen such that the value of the Fourier series exactly matches the value of $f(x)$ at any specific point in the interval $0 \le x \le 2\pi$. Nevertheless, one possible way to choose the coefficients would be to require that (1.65) be identical to $f(x)$ at each of the five points used by the grid-point methods discussed previously. Another useful strategy is to choose the coefficients to minimize the x-integral of the square of the difference between the approximation expansion (1.65) and $f(x)$.

If the expansion functions are nonzero in only a small part of the total domain, the series-expansion technique is a *finite-element method.* In the finite-element approach, the function $f(x)$ is again approximated by a finite series of functions of the form $b_0 s_0(x) + b_1 s_1(x) + \cdots + b_5 s_5(x)$, but the functions s_n differ from the trigonometric functions in the spectral method because each individual function is zero throughout most of the domain. The simplest finite-element expansion functions are piecewise-linear functions defined with respect to some grid. Each function is unity at one grid point, or node, and zero at all the other nodes. The values of the expansion function between the nodes are determined by linear interpolation using the values at the two nearest nodes. Six linear finite-element expansion functions suitable for approximating $f(x)$ might appear as shown in Fig. 1.3. Accounting for periodicity, the five pieces of information describing $f(x)$ would be the coefficients $(b_1, b_2, \ldots, b_5)$. When finite elements are constructed with piecewise-linear functions, the resulting numerical expressions are often similar to those obtained using grid-point methods. If finite elements are constructed from piecewise-quadratic or piecewise-cubic functions, however, the resulting formulae are quite different from those that arise naturally through finite differencing. Series-expansion methods will be studied in Chap. 6.

The numerical solution is defined throughout the entire spatial domain at every time step, but in time-dependent problems the approximate solution is typically available at only a few time levels at any given step of the numerical simulation. As a consequence, the use of series expansions is generally restricted to the representation of functional variations along spatial coordinates. Time derivatives are almost always approximated by finite differences.

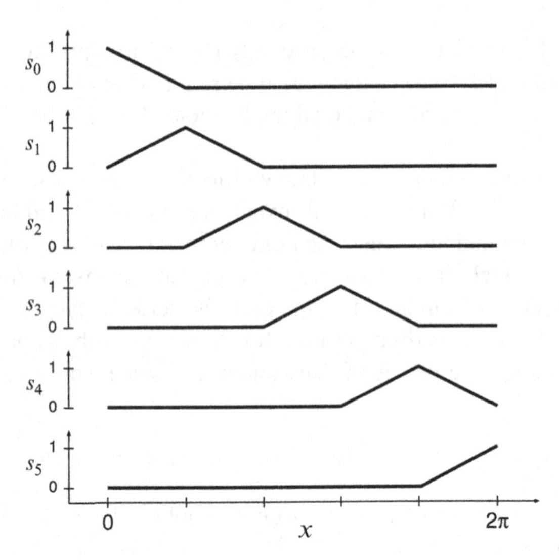

Fig. 1.3 Six finite-element expansion functions, $s_0(x), s_1(x), \ldots, s_5(x)$

1.3.2 Marching Schemes

Suppose that numerical solutions are sought to the first-order linear system

$$\frac{\partial u}{\partial y} + \frac{\partial v}{\partial x} = 0, \qquad \frac{\partial v}{\partial y} + \gamma \frac{\partial u}{\partial x} = 0 \tag{1.66}$$

throughout the domain $0 \le x \le 2\pi$, $0 \le y \le Y$. Let the domain be periodic in x and suppose that boundary conditions are specified for $u(x, 0)$ and $v(x, 0)$. One possible finite-difference approximation to the preceding system is

$$\frac{u_j^{n+1} - u_j^n}{\Delta y} + \frac{v_{j+1}^n - v_{j-1}^n}{2\Delta x} = 0, \tag{1.67}$$

$$\frac{v_j^{n+1} - v_j^n}{\Delta y} + \gamma \frac{u_{j+1}^{n+1} - u_{j-1}^{n+1}}{2\Delta x} = 0, \tag{1.68}$$

where u_j^n and v_j^n denote the numerical approximations to $u(j\Delta x, n\Delta y)$ and $v(j\,\Delta x, n\Delta y)$. The boundary conditions on u and v at $y = 0$ can be used to specify u_j^0 and v_j^0. The numerical solution along the line $y = \Delta y$ can then be calculated by solving (1.67) for u_j^1 at every j, and using these values of u_j^1 to compute v_j^1 from (1.68). In principle, this procedure can be repeated to compute approximations to the solution at $y = 2\Delta y, 3\Delta y, \ldots$ and thereby sequentially evaluate the numerical solution throughout the entire domain.

Under what circumstances will this procedure yield an accurate approximation to the true solution? This question can be answered without any detailed knowledge of numerical analysis when $\gamma < 0$. The linear second-order partial differential equation

$$\frac{\partial^2 u}{\partial y^2} - \gamma \frac{\partial^2 u}{\partial x^2} = 0 \tag{1.69}$$

can be obtained by eliminating v from (1.66). As per the discussion of (1.11), this equation is hyperbolic if $\gamma > 0$, and it is elliptic if $\gamma < 0$. Suppose the boundary conditions on u and v are

$$u(x,0) = \frac{1}{N^2}\sin(Nx), \qquad v(x,0) = 0, \tag{1.70}$$

where N is a positive integer. In the limit $N \to \infty$, the preceding boundary conditions become

$$u(x,0) = 0, \qquad v(x,0) = 0, \tag{1.71}$$

for which the exact solution to (1.66) is simply $u(x,y) = v(x,y) = 0$.

When (1.66) forms an elliptic system (i.e., when $\gamma < 0$), the exact solution subject to the boundary conditions (1.70) is

$$u(x,y) = \frac{1}{N^2}\sin(Nx)\cosh(\beta N y), \qquad v(x,y) = \frac{\beta}{N^2}\cos(Nx)\sinh(\beta N y),$$

where $\beta = \sqrt{-\gamma}$ is a real constant. As $N \to \infty$, the difference between the boundary conditions (1.70) and (1.71) disappears, but the difference between the solutions generated by each boundary condition increases without bound along any line $y = y_0 > 0$. Arbitrarily small changes in the amplitude of the imposed boundary values can produce arbitrarily large changes in the amplitude of the interior solution. Under such circumstances there is no hope of accurately approximating the true solution by the finite-difference method (1.67) and (1.68) because the round-off errors incurred as (1.70) is evaluated to obtain numerical values for the grid points along $y = 0$ may generate arbitrarily large perturbations in the interior solution.

The mathematical problem of solving (1.66) subject to boundary conditions specified for $u(x,0)$ and $v(x,0)$ is not well posed whenever $\gamma < 0$. A *well-posed problem* is one in which a unique solution to a given partial differential equation exists and depends continuously on the initial- and boundary-value data. When $\gamma < 0$, the preceding problem is not well posed because the solution does not depend continuously on the boundary data. On the other hand, when $\gamma > 0$, the problem is hyperbolic, and the solution subject to (1.70) is

$$u(x,y) = \frac{1}{N^2}\sin(Nx)\cos\left(\sqrt{\gamma}Ny\right), \quad v(x,y) = -\frac{\sqrt{\gamma}}{N^2}\cos(Nx)\sin\left(\sqrt{\gamma}Ny\right).$$

In this case $u(x,y) \to 0$ and $v(x,y) \to 0$ as $N \to \infty$. The interior solutions associated with the boundary conditions (1.70) and (1.71) approach each other as the difference between the two boundary conditions goes to zero, and small changes in the amplitude of the boundary data produce only small changes in the amplitude of the interior solution. As demonstrated in Gustafsson et al. (1995), the hyperbolic problem is well posed. When $\gamma > 0$, it is possible to obtain good approximations to the correct solution using (1.67) and (1.68), although, as will be discussed in Sect. 4.1.2, the quality of the result depends on the parameter $\sqrt{\gamma}\Delta y/\Delta x$.

Physicists seldom worry about well-posedness, since properly formulated mathematical models of the physical world are almost always well posed. The preceding example may be recognized as an initial-value problem in which y represents time and x is the spatial coordinate. In contrast to their hyperbolic cousins, elliptic partial differential equations describe steady-state physical systems and do not naturally arise as initial-value problems. When a real-world system is governed by an elliptic equation, physical considerations usually provide data for the dependent variables or their normal derivatives along each boundary, and the additional boundary-value data lead to a well-posed problem. The fact that elliptic partial differential equations are not well posed as initial-value problems may therefore be irrelevant to the physicist – but it is not irrelevant to the numerical analyst. Given a well-posed elliptic problem, such as (1.69) with $\gamma < 0$ and u specified at $y = 0$ and $y = Y$, could one expect to compute an accurate approximate solution on some numerical grid by starting with the known values along one boundary and stepping across the grid, one point at a time? The answer is no; an approach of this type is numerically unstable – indeed it mimics the not-well-posed formulation of an elliptic partial differential equation as an initial-value problem. Practical methods for the numerical solution of elliptic partial differential equations are therefore not "marching" schemes. Instead of computing the solution at one point and then proceeding to the next, one must simultaneously adjust all the grid-point values (perhaps through some iterative process) in order to adequately satisfy the governing differential equation and the boundary conditions. In contrast, hyperbolic partial differential equations do lend themselves to numerical solution via marching techniques.[5]

Another major difference in the numerical treatment of elliptic and hyperbolic equations arises in the specification of boundary conditions. As suggested by the preceding example, boundary conditions are usually imposed at every boundary as part of the natural formulation of an elliptic problem. Moreover, the incorporation of these boundary data into a numerical algorithm is generally straightforward. On the other hand, if one attempts to compute the solution to a hyperbolic problem in a limited spatial domain, the numerical algorithm may require boundary conditions in regions where none should actually be specified (i.e., at a boundary where all the characteristic curves are directed out of the domain). Improper boundary conditions may lead to instabilities or to nonuniqueness in the numerical solution of a hyperbolic system. Further discussion of boundary conditions will be presented in Chap. 9.

[5] L.F. Richardson, who explored the numerical solution of a variety of partial differential equations prior to his celebrated attempt at numerical weather prediction, coined the terms "jury" and "marching" methods to describe the basic difference between the numerical techniques suitable for the solution of elliptic equations and hyperbolic equations. The adjective "jury" alluded to the idea that one needed to adjust all the values in the numerical solution until the whole was "judged" to constitute a satisfactory approximation.

Problems

1. Suppose that

$$\frac{\partial^2 u}{\partial t^2} + a\frac{\partial^2 u}{\partial x \partial t} + b\frac{\partial^2 u}{\partial x^2} = 0$$

is a hyperbolic partial differential equation and that a and b are constants. Show that this equation can be transformed to a decoupled pair of first-order wave equations. What are the propagation speeds of the solutions to these first-order wave equations?

2. Show that when (1.11) is hyperbolic, it can be transformed to the alternative canonical form

$$u_{\xi\xi} - u_{\eta\eta} + \tilde{L}(\xi, \eta, u, u_\xi, u_\eta) = 0,$$

where $\tilde{L}(\xi, \eta, u, u_\xi, u_\eta)$ is once again a linear function of u, u_ξ, and u_η with coefficients that may depend on ξ and η. (*Hint:* Start with (1.12) and define new independent variables equal to $\xi + \eta$ and $\xi - \eta$.)

3. If gravity and density stratification are neglected, the two-dimensional Euler equations for inviscid isentropic flow reduce to a system of four equations in the unknowns (u, w, ρ, p). Linearize this system about a basic state with constant (u_0, w_0, ρ_0, p_0) and show that the linearized system is hyperbolic. (*Hint:* Transform the perturbation thermodynamic variables to $p'/(c\rho_0)$ and $\rho' - p'/c^2$, where $c^2 = \partial p/\partial \rho$ is the square of the speed of sound in the basic state.)

4. The pressure and density changes in compressible isentropic flow satisfy the relation

$$\frac{d\rho}{dt} = \frac{1}{c_s^2}\frac{dp}{dt}.$$

(a) Derive the preceding relationship.

(b) Show that the preceding relationship is approximated as

$$\frac{d\rho}{dt} = 0$$

in the incompressible system, as

$$\frac{d\rho}{dt} = \frac{\rho}{\overline{\rho}}\frac{d\overline{\rho}}{dt}$$

in the anelastic system, and as

$$\frac{d\rho}{dt} = \frac{\rho}{\tilde{\rho}}\frac{1}{\tilde{c}_s^2}\frac{d\tilde{p}}{dt}$$

in the pseudo-incompressible system (where the tilde denotes the steady reference field).

5. Show that the backward heat equation,

$$\frac{\partial \psi}{\partial t} = -\frac{\partial^2 \psi}{\partial x^2},$$

and the initial condition $\psi(x, 0) = f(x)$ do not constitute a well-posed problem on the domain $-\infty < x < \infty$, $t > 0$.

Chapter 2
Ordinary Differential Equations

Although the fundamental equations governing the evolution of geophysical fluids are partial differential equations, ordinary differential equations arise in several contexts. The trajectories of individual fluid parcels in an inviscid flow are governed by simple ordinary differential equations, and systems of ordinary differential equations may describe chemical reactions or highly idealized dynamical systems. Since basic methods for the numerical integration of ordinary differential equations are simpler than those for partial differential equations, and since the time-differencing formulae used in the numerical solution of partial differential equations are closely related to those used for ordinary differential equations, this chapter is devoted to the analysis of methods for the approximate solution of ordinary differential equations (ODE solvers). Nevertheless some approaches to the solution of partial differential equations, such as Lax–Wendroff and finite-volume methods, arise from fully discretized approximations in both space and time that cannot be correctly analyzed by considering the spatial and temporal differencing in isolation. These fully discretized approaches will be discussed in subsequent chapters.

Most of this chapter will focus on ODE solvers potentially suitable for the numerical integration of time-dependent partial differential equations. In comparison with typical ODE solvers, the methods used to integrate partial differential equations are relatively low order. Low-order schemes are used for two basic reasons. First, the approximation of the time derivative is not the only source of error in the solution of partial differential equations; other errors arise through the approximation of the spatial derivatives. In many circumstances the largest errors in the solution are introduced through the numerical evaluation of the spatial derivatives, so it is pointless to devote additional computational resources to higher-order time differencing. The second reason for using low-order methods is that practical limitations on computational resources often leave no other choice.

Consider the initial-value problem

$$\frac{d\psi}{dt} = F(\psi, t), \quad \psi(0) = \psi_0, \tag{2.1}$$

D.R. Durran, *Numerical Methods for Fluid Dynamics: With Applications to Geophysics*, Texts in Applied Mathematics 32, DOI 10.1007/978-1-4419-6412-0_2,

which will have a unique solution provided the function F is sufficiently smooth (in particular, F must satisfy a Lipschitz condition[1]). Let ϕ_n denote the numerical approximation to the true solution at some set of discrete time levels $t_n = n\Delta t$, $n = 0, 1, 2, \ldots$. Virtually all numerical methods for the solution of (2.1) replace the differential equation for ψ with algebraic equations for the ϕ_n that are solved repeatedly to step the solution forward from the initial condition $\phi_0 = \psi_0$.

It is often helpful to consider the algebraic equations used to generate the approximate solution as arising from one of two approaches. In the first approach, the time derivative in (2.1) is replaced with a finite difference. In the second approach, (2.1) is integrated over a time interval Δt

$$\psi(t_{n+1}) = \psi(t_n) + \int_{t_n}^{t_{n+1}} F\big(\psi(t), t\big)\, dt, \tag{2.2}$$

and the algebraic equations that constitute the numerical method can be most easily interpreted as providing an approximation to the integral of F. Backward difference methods for stiff equations are examples of the first approach in which the time derivative in (2.1) is replaced by finite differences. Runge–Kutta and linear multistep methods are more naturally understood as arising from approximations to the integral in (2.2). Some of the simplest schemes can be easily interpreted using either approach.

2.1 Stability, Consistency, and Convergence

The basic goal when computing a numerical approximation to the solution of a differential equation is to obtain a result that indeed approximates the true solution. In this section we will examine the relationship between three fundamental concepts characterizing the quality of the numerical solution in the limit where the separation between adjacent nodes on a numerical mesh approaches zero: consistency, stability, and convergence.

2.1.1 Truncation Error

The derivative of a function $\psi(t)$ at time t_n could be defined either as

$$\frac{d\psi}{dt}(t_n) = \lim_{\Delta t \to 0} \frac{\psi(t_n + \Delta t) - \psi(t_n)}{\Delta t}, \tag{2.3}$$

[1] The *Lipschitz condition* is that $|F(x,t) - F(y,t)| \leq L|x - y|$ for all x and y, and all $t \geq 0$, where $L > 0$ is a real constant. One way to satisfy this condition is if $|\partial F/\partial x|$ is bounded.

or as

$$\frac{d\psi}{dt}(t_n) = \lim_{\Delta t \to 0} \frac{\psi(t_n + \Delta t) - \psi(t_n - \Delta t)}{2\Delta t}. \tag{2.4}$$

If the derivative of $\psi(t)$ is continuous at t_n, both expressions produce the same unique answer. In practical applications, however, it is impossible to evaluate these expressions with infinitesimally small Δt. The approximations to the true derivative obtained by evaluating the algebraic expressions on the right side of (2.3) and (2.4) using finite Δt are known as *finite differences*. When Δt is finite, the preceding finite-difference approximations are not equivalent; they differ in their accuracy, and when they are substituted for derivatives in differential equations, they generate different algebraic equations. The differences in the structure of these algebraic equations can have a great influence on the stability of the numerical solution.

Which of the preceding finite-difference formulae is likely to be more accurate when Δt is small but finite? If $\psi(t)$ is a sufficiently smooth function of t, this question can be answered by expanding the terms $\psi(t_n \pm \Delta t)$ in Taylor series about t_n and substituting these expansions into the finite-difference formula. For example, substituting

$$\psi(t_n + \Delta t) = \psi(t_n) + \Delta t \frac{d\psi}{dt}(t_n) + \frac{(\Delta t)^2}{2}\frac{d^2\psi}{dt^2}(t_n) + \frac{(\Delta t)^3}{6}\frac{d^3\psi}{dt^3}(t_n) + \ldots$$

into (2.3), one finds that

$$\frac{\psi(t_n + \Delta t) - \psi(t_n)}{\Delta t} - \frac{d\psi}{dt}(t_n) = \frac{\Delta t}{2}\frac{d^2\psi}{dt^2}(t_n) + \frac{(\Delta t)^2}{6}\frac{d^3\psi}{dt^3}(t_n) + \ldots. \tag{2.5}$$

The right side of the preceding equation is the *truncation error* of the finite difference. The lowest power of Δt in the truncation error determines the *order of accuracy* of the finite difference. Inspection of (2.5) shows that the one-sided difference is first-order accurate. In contrast, the truncation error associated with the centered difference (2.4) is

$$\frac{(\Delta t)^2}{6}\frac{d^3\psi}{dt^3}(t_n) + \frac{(\Delta t)^4}{120}\frac{d^5\psi}{dt^5}(t_n) + \ldots,$$

and the centered difference is therefore second-order accurate. If the higher-order derivatives of ψ are bounded in some interval about t_n, (i.e., if ψ is "smooth") and Δt is repeatedly reduced, the error in the second-order difference (2.4) will approach zero more rapidly than the error in the first-order difference (2.3). The fact that the truncation error of the centered difference is higher order does not, however, guarantee that it will always generate a more accurate estimate of the derivative. If the function is sufficiently rough and Δt is sufficiently coarse, neither formula is likely to produce a good approximation, and the superiority of one over the other will be largely a matter of chance.

Euler's method (often called the forward-Euler method) approximates the derivative in (2.1) with the forward difference (2.3) to give the formula

$$\frac{\phi_{n+1} - \phi_n}{\Delta t} = F(\phi_n, t_n). \tag{2.6}$$

Clearly this formula can be used to obtain ϕ_1 from the initial condition $\phi_0 = \psi_0$, and then be applied recursively to obtain ϕ_{n+1} from ϕ_n. How well does this simple method perform?

One way to characterize the accuracy of this method is through its *truncation error, defined as the residual by which smooth solutions to the continuous problem fail to satisfy the discrete approximation* (2.6). Let τ_n denote the truncation error at time t_n; then from (2.5),

$$\frac{\psi(t_{n+1}) - \psi(t_n)}{\Delta t} - F\big(\psi(t_n), t_n\big) = \frac{d\psi}{dt}(t_n) + \tau_n - F\big(\psi(t_n), t_n\big) = \tau_n, \quad (2.7)$$

where the second equality holds because ψ is a solution to the continuous problem and

$$\tau_n = \frac{\Delta t}{2}\frac{d^2\psi}{dt^2}(t_n) + O\left[(\Delta t)^2\right].$$

It is not necessary to explicitly consider the higher-order terms in the truncation error to bound $|\tau_n|$; if ψ has continuous second derivatives, the mean-value theorem may be used to show

$$|\tau_n| \leq \frac{\Delta t}{2} \max_{t_n \leq s \leq t_{n+1}} \left|\frac{d^2\psi}{dt^2}(s)\right|. \quad (2.8)$$

Euler's method is *consistent of order 1* because the lowest power of Δt appearing in t_n is unity. If the centered difference (2.4) were used to approximate the time derivative in (2.6), the resulting method would be consistent of order 2.

2.1.2 Convergence

A *consistent* method is one for which the truncation error approaches zero as $\Delta t \to 0$. The order of the consistency determines the rate at which the solution of a *stable* finite-difference method *converges* to the true solution as $\Delta t \to 0$. To examine the relation between consistency and convergence, we define the *global error* at time t_n as $E_n = \phi_n - \psi(t_n)$. From (2.7),

$$\psi(t_{n+1}) = \psi(t_n) + \Delta t F\big(\psi(t_n), t_n\big) + \Delta t\, \tau_n, \quad (2.9)$$

which implies that if we start with the true solution at t_n, the *local* or *one-step* error generated by Euler's method in approximating the solution at t_{n+1} is $\Delta t\, \tau_n$, which is one power of Δt higher than the truncation error itself. One might suppose that the global error in the solution at time T is bounded by the maximum local error times the number of time steps $(\max_n |\Delta t\, \tau_n|)(T/\Delta t)$, which, like τ_n itself, is $O(\Delta t)$. This would be a welcome result because it would imply the error becomes arbitrarily small as the time step approaches zero, but such reasoning is incorrect because it does not account for the difference between ϕ_n and $\psi(t_n)$ arising from the accumulation of local errors over the preceding time steps. The increase in the global error generated over a single step satisfies

$$E_{n+1} = E_n + \Delta t \left[F(\phi_n, t_n) - F\big(\psi(t_n), t_n\big)\right] - \Delta t\, \tau_n, \tag{2.10}$$

which may be obtained by solving (2.6) for ϕ_{n+1} and subtracting (2.9). As apparent from (2.10), the global error will remain $O(\Delta t)$, and the numerical solution will converge to the true solution, provided $F(\phi_n, t_n) - F\big(\psi(t_n), t_n\big)$ remains finite in the limit $\Delta t \to 0$.

It is easy to show that Euler's method converges for the special case where

$$F(\psi, t) = \lambda\psi + g(t), \tag{2.11}$$

where λ is a constant.[2] We will examine this special case because it reveals the relatively weak stability condition required to ensure convergence to the true solution in the limit $\Delta t \to 0$. Substituting (2.11) into (2.10) gives

$$E_{n+1} = (1 + \lambda\Delta t)E_n - \Delta t\, \tau_n. \tag{2.12}$$

Note that $g(t)$, the part of $F(\psi, t)$ that is independent of ψ, drops out and has no impact on the growth of the global error. Suppose that $N = T/\Delta t$ is the number of time steps required to integrate from the initial condition at $t = 0$ to some fixed time T.

From (2.12)

$$\begin{aligned} E_N &= (1 + \lambda\Delta t)E_{N-1} - \Delta t\, \tau_{N-1} \\ &= (1 + \lambda\Delta t)\,[1 + \lambda\Delta t)E_{N-2} - \Delta t\, \tau_{N-2}] - \Delta t\, \tau_{N-1} \end{aligned}$$

and by induction,

$$E_N = (1 + \lambda\Delta t)^N E_0 - \Delta t \sum_{m=1}^{N} (1 + \lambda\Delta t)^{N-m} \tau_{m-1}.$$

Let

$$\tau_{\max} = \frac{\Delta t}{2} \max_{0 \le s \le t_N} \left| \frac{d^2\psi}{dt^2}(s) \right|,$$

which from (2.8) is an upper bound on $|\tau_n|$ for all n independent of the choice of time step used to divide up the interval $[0, T]$. Assuming the initial error E_0 is zero (although an $O(\Delta t)$ error would not prevent convergence), and noting that for $\Delta t > 0$, $1 + |\lambda|\Delta t \le \mathrm{e}^{|\lambda|\Delta t}$, one obtains

$$|E_N| \le N\Delta t\, (1 + |\lambda|\Delta t)^N \tau_{\max} = T\mathrm{e}^{|\lambda|T} \tau_{\max}. \tag{2.13}$$

[2] More general conditions sufficient to guarantee the convergence of Euler's method are that F is an analytic function (Iserles 1996, p. 7) or that the first two derivatives of ψ are continuous (Hundsdorfer and Verwer 2003, p. 24).

Since $Te^{|\lambda|T}$ has some finite value independent of the numerical discretization and τ_{max} is $O(\Delta t)$, the global error at time T must approach zero in proportion to the first power of Δt.

2.1.3 Stability

The foundation for the theory of numerical methods for differential equations is built on the theorem that *consistency of order p and stability imply convergence of order p* (Dalhquist 1956; Lax and Richtmyer 1956). Evidently Euler's method satisfies some type of stability condition since it is consistent and is convergent of order unity. The relation (2.12), which states that previous global errors amplify by a factor of $(1 + \lambda\Delta t)$ over each individual time step, provides the key for bounding the growth of the global error over a finite time interval. Define the *amplification factor* A as the ratio of the approximate solution at two adjacent time steps,

$$A = \phi_{n+1}/\phi_n. \tag{2.14}$$

A two-time-level method is stable in the sense that, if it is also consistent, it will converge in the limit $\Delta t \to 0$ provided that for some constant η (independent of the properties of the numerical discretization)

$$|A| \leq 1 + \eta\Delta t. \tag{2.15}$$

In the previous simple example, $\eta = \lambda$ is just the coefficient of ψ in the forcing $F(\psi, t)$. When Euler's method is applied to more general problems, η is a constant associated with the Lipschitz condition on $F(\psi, t)$. Essentially all consistent *two-time-level* ODE solvers satisfy this stability condition, but as discussed in Sect. 2.2.4, bounds similar to (2.15) are not satisfied by many potentially reasonable approximations to time-dependent partial differential equations.

2.2 Additional Measures of Stability and Accuracy

Although Euler's method is sufficiently stable to yield solutions that converge in the limit $\Delta t \to 0$, it may nevertheless generate a sequence $\phi_0, \phi_1, \ldots$ that blows up in a completely nonphysical manner when the computations are performed with finite values of Δt. Again suppose $F(\psi, t) = \lambda\psi$; if $\lambda < 0$, the true solution $\psi_0 e^{\lambda t}$ is bounded by ψ_0 for all time and approaches zero as $t \to \infty$. Yet if $\lambda\Delta t < -2$, then $A = 1 + \lambda\Delta t < -1$, and the numerical solution changes sign and amplifies geometrically every time step, diverging wildly from the true solution.

2.2.1 A-Stability

How can we characterize the stability of a consistent numerical method to give an indication whether a solution computed using finite Δt is likely to blow up in such an "unstable" manner? Clearly there are many physical problems where the true solution does amplify rapidly with time, and of course any convergent numerical method must be able to capture such amplification. On the other hand, there are also many problems in which the norm of the solution is bounded or decays with time. It is not practical to consider every possible case individually, but it is very useful to evaluate the behavior of schemes on the simple test problem

$$\frac{d\psi}{dt} = \gamma\psi, \qquad \psi(0) = \psi_0, \tag{2.16}$$

where, in contrast to our previous examples, ψ and γ are complex-valued. Breaking γ into its real and imaginary parts, such that $\gamma = \lambda + i\omega$ with λ and ω real, the solution to (2.16) is

$$\psi(t) = \psi_0 e^{\lambda t} e^{i\omega t},$$

showing that $\Re\{\gamma\} = \lambda$ determines rate of change of the magnitude (or modulus) of ψ, whereas $\Im\{\gamma\} = \omega$ governs the rate of change of its phase (or argument).

Despite its simplicity, (2.16) is prototypical of the time variations found in many important fluid-dynamical problems. For example the concentration χ of a passive tracer in a flow moving at speed c and diffusing with a diffusivity M along one spatial dimension is given by the partial differential equation

$$\frac{\partial \chi}{\partial t} + c\frac{\partial \chi}{\partial x} = M\frac{\partial^2 \chi}{\partial x^2}. \tag{2.17}$$

Suppose the spatial domain is $|x| \le 1$ and periodic, then the solution may be determined as the superposition of Fourier modes, each of which may be expressed in the form $b_k(t)e^{ikx}$, where b_k is a complex number determining the amplitude and phase of each mode and $k = n\pi,\ n = 0, \pm 1, \pm 2, \dots$ is the *wave number*. The wave number is inversely proportional to the *wavelength*, $L = 2\pi/k$, which is the distance over which a wave's shape repeats. Substituting an arbitrary Fourier mode into (2.17) yields the following ordinary differential equation of the form (2.16):

$$\frac{db_k}{dt} = -\left(Mk^2 + ick\right) b_k. \tag{2.18}$$

Note that in the context of the advection–diffusion problem, $\Re\{\gamma\}$ determines the changes in amplitude produced by diffusion and $\Im\{\gamma\}$ governs changes in phase produced by advection.

Numerical solutions to (2.16) computed with some specific value of Δt are *absolutely stable* if $|\phi_n| \le |\phi_0|$ for all n, or equivalently, if $|A| \le 1$. The amplification

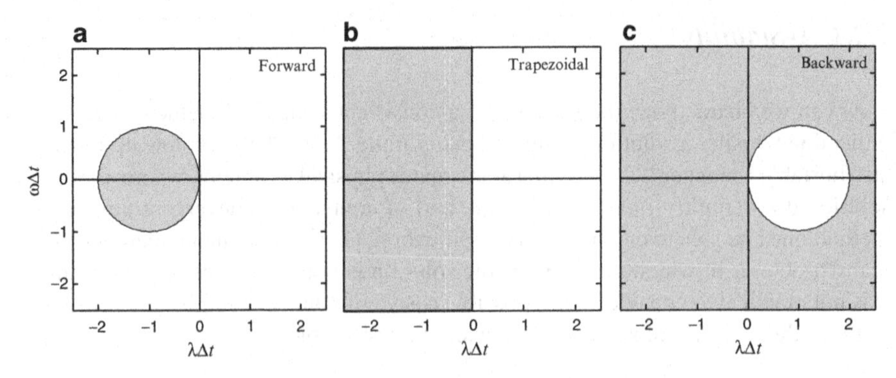

Fig. 2.1 Absolute stability regions (shaded) for **a** forward-Euler differencing, **b** trapezoidal differencing, and **c** backward-Euler differencing

factor for Euler's-method solutions to (2.16) is $1+\gamma\Delta t$, so the values of $(\lambda\Delta t, \omega\Delta t)$ for which $|A| \leq 1$ satisfy the inequality

$$(1 + \lambda\Delta t)^2 + (\omega\Delta t)^2 \leq 1.$$

This region of absolute stability, which is the interior of a unit circle centered at (−1,0) in the $\lambda\Delta t$–$\omega\Delta t$ plane, is plotted in Fig. 2.1a.

The true solution to (2.16) is nonamplifying for all $\lambda \leq 0$. Regardless of the time step, this behavior will be captured by numerical methods that are A-stable. *A numerical method is A-stable if it is absolutely stable for all* $\lambda\Delta t \leq 0$. Forward differencing is not A-stable but, as will be discussed in Sect. 2.2.3, the other methods whose absolute stability regions are shown in Fig. 2.1 are A-stable.

2.2.2 Phase-Speed Errors

When $M = 0$, the prototype equation (2.18) reduces to the *oscillation equation*

$$\frac{d\psi}{dt} = \mathrm{i}\omega\psi, \tag{2.19}$$

which serves as an important model for many nondissipative dynamical systems. The oscillation equation may also be derived from a two-component real-valued system of ordinary differential equations, such as those representing Coriolis accelerations:

$$\begin{aligned}\frac{du}{dt} - fv &= 0\\ \frac{dv}{dt} + fu &= 0.\end{aligned}$$

The preceding equations may be expressed in the form (2.19) by setting $\psi = u + iv$ and $\omega = -f$.

Integrating the oscillation equation over a time Δt yields

$$\psi(t_0 + \Delta t) = \mathrm{e}^{\mathrm{i}\omega\Delta t}\psi(t_0) \equiv A_\mathrm{e}\,\psi(t_0). \tag{2.20}$$

Here the last relation defines an "exact amplification factor" A_e, which in the case of the oscillation equation, is a complex number of modulus 1. Over a time interval Δt, ψ moves $\omega\Delta t$ radians around a circle of radius $|\psi(t_0)|$ centered at the origin in the complex plane.

Hundreds of papers have been written investigating techniques for solving (2.17) with $M = 0$ (see, for example, the extensive review in Rood 1987). The vastness of this body of literature is a testament to the subtle trade-offs involved in the selection of the "best" numerical method for even very simple equations. It might be supposed that the relative accuracy of different methods for the $M = 0$ problem could be easily determined by comparing their respective truncation errors. The analysis of truncation error is, however, most effective at predicting the behavior of well-resolved oscillations whose periods are orders of magnitude larger than a single time step. Yet the most serious errors are typically found in the poorly resolved components of the solution, which oscillate over periods between $2\Delta t$ and $10\Delta t$. These errors typically appear in both the amplitude and the phase of the solution.

It can therefore be helpful to examine the amplitude and phase errors as a function of numerical resolution, both of which may be evaluated from the amplification factor. If A is expressed in modulus-argument form $|A|\mathrm{e}^{\mathrm{i}\theta}$, where

$$|A| = (\Re\{A\}^2 + \Im\{A\}^2)^{1/2} \quad \text{and} \quad \theta = \arctan\left(\Im\{A\}/\Re\{A\}\right),$$

phase errors may be characterized by the relative phase change, $R = \theta/\omega\Delta t$, which is the ratio of the phase advance produced by one time step of the numerical scheme divided by the change in phase experienced by the true solution over the same time interval. If $R > 1$, the method is *accelerating*; if $R < 1$, the scheme is *decelerating*. Phase errors accumulate over the period of integration and can become quite large over long time periods

In a nondissipative system, amplitude errors represent spurious sinks or sources of energy. Amplitude errors arise from the difference between the magnitude of the approximate amplification factor $|A|$ and the correct value of unity. When $|A| = 1$, the scheme is *neutral*, if $|A| < 1$, the scheme is *damping*, and if $|A| > 1$, it is *amplifying*. The range of values of Δt for which a given approximation to the oscillation equation is not amplifying are given by the intersection of the absolute stability region for the scheme and the imaginary $(\omega\Delta t)$ axis, which in the case of Euler's method (Fig. 2.1a) is the origin.

The relationship between a scheme's order of accuracy and the orders of the amplitude and phase error is not entirely intuitive. According to (2.20), the exact amplification factor for solutions to the oscillation equation is

$$A_e = \mathrm{e}^{\mathrm{i}\omega\Delta t} = 1 + \mathrm{i}\omega\Delta t - \frac{(\omega\Delta t)^2}{2} - i\frac{(\omega\Delta t)^3}{6} + \frac{(\omega\Delta t)^4}{24} + \ldots$$

The amplification factor, A, of an nth-order time-differencing scheme will match all terms in the preceding expression through order $(\omega\Delta t)^n$. The amplitude error and the phase error characterize the errors in the modulus and the argument of A, respectively; and as such their order of accuracy may differ from the general order of accuracy of the scheme. In particular, since amplitude and phase errors are special aspects of the total error, it is possible for either of these quantities to be smaller than the total error. The general relationship between the truncation error and the amplitude and phase errors may be stated as follows (Durran 1991):

> If the oscillation equation (2.19) is integrated using a linear finite-difference scheme and if the truncation error of the resulting finite-difference approximation to the oscillation equation is order r, then as $\omega\Delta t \rightarrow 0$ the amplitude change in each step of the numerical solution is no worse than
>
> $$1 + O[(\omega\Delta t)^n], \quad \text{where} \quad \begin{cases} n = r+1, & \text{if } r \text{ is odd;} \\ n \geq r+2, & \text{if } r \text{ is even;} \end{cases}$$
>
> and the relative phase change is no worse than
>
> $$1 + O[(\omega\Delta t)^m], \quad \text{where} \quad \begin{cases} m \geq r+1, & \text{if } r \text{ is odd;} \\ m = r, & \text{if } r \text{ is even.} \end{cases}$$

Switching from an even- to an odd-order scheme increases the order of accuracy of the relative phase change without improving the order of accuracy of the amplitude error. Switching from odd to even order reduces the asymptotic amplitude error without altering the order of the error in the relative phase change.

2.2.3 Single-Stage, Single-Step Schemes

The simplest techniques for the solution of the ordinary differential equation (2.1) are members of the general family of single-stage, single-step schemes, which may be written in the form

$$\frac{\phi_{n+1} - \phi_n}{\Delta t} = (1-\alpha)F(\phi_n, t_n) + \alpha F(\phi_{n+1}, t_{n+1}). \tag{2.21}$$

Euler's method is obtained by setting $\alpha = 0$; the backward-Euler method corresponds to the case $\alpha = 1$, and the trapezoidal method is obtained when $\alpha = 1/2$. Substituting the true solution ψ into (2.21), expanding all terms in Taylor series about t_n, and using

$$F(\psi(t_{n+1}), t_{n+1}) = \frac{d\psi}{dt}(t_{n+1}) = \frac{d\psi}{dt}(t_n) + \Delta t \frac{d^2\psi}{dt^2}(t_n) + \frac{(\Delta t)^2}{2}\frac{d^3\psi}{dt^3}(t_n) + \ldots,$$

one may show the truncation error for all members of this family of schemes is $O(\Delta t)$, except for the trapezoidal method, for which it is $O\left[(\Delta t)^2\right]$.

Application of (2.21) to the test equation for absolute stability (2.16) yields

$$A = \frac{\phi_{n+1}}{\phi_n} = \frac{1 + (1-\alpha)\gamma\Delta t}{1 - \alpha\gamma\Delta t}. \tag{2.22}$$

For the backward-Euler method, $A = (1 - \gamma\Delta t)^{-1} = (1 - \lambda\Delta t - \mathrm{i}\omega\Delta t)^{-1}$. Multiplying A by its complex conjugate gives

$$|A|^2 = \frac{1}{(1-\lambda\Delta t)^2 + (\omega\Delta t)^2},$$

implying that backward-Euler differencing will produce absolutely stable solutions for all $(\lambda\Delta t, \omega\Delta t)$ *outside* the circle:

$$(1-\lambda\Delta t)^2 + (\omega\Delta t)^2 \leq 1. \tag{2.23}$$

This region is shown in Fig. 2.1c, and since it includes the half-plane $\lambda\Delta t \leq 0$, backward-Euler differencing is A-stable. Although it generates physically appropriate solutions for $\lambda < 0$, the backward-Euler method can produce large errors if $\lambda > 0$. If $\lambda > 0$ and $(\lambda\Delta t, \omega\Delta t)$ is not inside the circle (2.23), the numerical solution will decay but the true solution should grow exponentially with time.

The amplification factor for the trapezoidal method is

$$A = \frac{1 + \gamma\Delta t/2}{1 - \gamma\Delta t/2}, \tag{2.24}$$

from which

$$|A|^2 = \frac{(1+\lambda\Delta t/2)^2 + (\omega\Delta t)^2}{(1-\lambda\Delta t/2)^2 + (\omega\Delta t)^2}.$$

Thus, the absolute stability region for the trapezoidal method (shown in Fig. 2.1b) is the precisely the half-plane $\lambda\Delta t \leq 0$, and the trapezoidal method is A-stable.

Now consider the behavior of these schemes in the purely oscillatory case; then $\gamma = \mathrm{i}\omega$, and from (2.22)

$$|A|^2 = \frac{1 + (1-\alpha)^2(\omega\Delta t)^2}{1 + \alpha^2(\omega\Delta t)^2} = 1 + (1-2\alpha)\frac{(\omega\Delta t)^2}{1 + \alpha^2(\omega\Delta t)^2}. \tag{2.25}$$

Inspection of the preceding equation shows that the scheme is amplifying when $\alpha < 1/2$, neutral when $\alpha = 1/2$, and damping when $\alpha > 1/2$. These results are consistent with the locations of the boundaries of the absolute stability regions in Fig. 2.1.

The amplitude and phase errors in the approximate solution are functions of the *numerical resolution*. The solution to the governing differential equation (2.19) oscillates with a period $T = 2\pi/\omega$. An appropriate measure of numerical resolution is the number of time steps per oscillation period, $T/\Delta t$. The numerical resolution is improved by decreasing the step size. In the limit of very good numerical resolution,

$T/\Delta t \to \infty$ and $\omega\Delta t \to 0$. Assuming good numerical resolution, the Taylor series expansion

$$(1+x)^{1/2} = 1 + \frac{x}{2} - \frac{x^2}{8} + \dots, \quad \text{for} \quad |x| < 1,$$

may be used to reduce (2.25) to

$$|A| \approx 1 + \tfrac{1}{2}(1-2\alpha)(\omega\Delta t)^2.$$

It follows that

$$|A|_{\text{forward}} \approx 1 + \tfrac{1}{2}(\omega\Delta t)^2 \qquad \text{and} \qquad |A|_{\text{backward}} \approx 1 - \tfrac{1}{2}(\omega\Delta t)^2, \tag{2.26}$$

indicating that, as expected for first-order schemes, the spurious amplitude changes introduced by both the forward-Euler method and the backward-Euler method are $O[(\omega\Delta t)^2]$.

The relative phase change in the family of single-stage, two-level schemes is

$$R = \frac{1}{\omega\Delta t}\arctan\left(\frac{\omega\Delta t}{1-\alpha(1-\alpha)(\omega\Delta t)^2}\right).$$

Thus,

$$R_{\text{forward}} = R_{\text{backward}} = \frac{\arctan\omega\Delta t}{\omega\Delta t}, \tag{2.27}$$

which ranges between 0 and 1, implying that both the forward-Euler scheme and the backward-Euler scheme are decelerating. Assuming, once again, that the numerical solution is well resolved, we may approximate the preceding expression for the phase-speed error using the Taylor series expansion

$$\arctan x = x - \frac{x^3}{3} + \frac{x^5}{5} - \dots \quad \text{for} \quad |x| < 1$$

to obtain

$$R_{\text{forward}} = R_{\text{backward}} \approx 1 - \frac{(\omega\Delta t)^2}{3}.$$

The phase-speed error, like the amplitude error, is $O[(\Delta t)^2]$. The relative phase change for the trapezoidal scheme is

$$R_{\text{trapezoidal}} = \frac{1}{\omega\Delta t}\arctan\left(\frac{\omega\Delta t}{1-\omega^2\Delta t^2/4}\right),$$

which for small values of $\omega\Delta t$ is

$$R_{\text{trapezoidal}} \approx \frac{1}{\omega\Delta t}\arctan\left(\omega\Delta t\left(1+\frac{\omega^2\Delta t^2}{4}\right)\right) \approx 1 - \frac{\omega^2\Delta t^2}{12}.$$

As with the forward- and backward-Euler methods, the trapezoidal scheme retards the phase change of well-resolved oscillations. However, the deceleration is only one quarter as great as that produced by the other schemes.

Although the trapezoidal scheme is second-order accurate and A-stable, it has the disadvantage in that it requires the evaluation of $F(\phi_{n+1})$ during the computation of ϕ_{n+1}. A scheme such as the trapezoidal method, in which the calculation of ϕ_{n+1} depends on $F(\phi_{n+1})$, is an *implicit* method. If the calculation of ϕ_{n+1} does not depend on $F(\phi_{n+1})$, the scheme is *explicit*. In the case of the test problem (2.16), implicitness is a trivial complication. However, if F is a nonlinear function, any implicit finite-difference scheme will convert the differential equation into a nonlinear algebraic equation for ϕ_{n+1}. In the general case, the solution to this nonlinear equation must be obtained by some iterative technique. Thus, implicit finite-difference schemes generally require much more computation per individual time step than do similar explicit methods. Nevertheless, in problems where accuracy considerations do not demand a short time step, the extra computation per implicit time step can be more than compensated by using a much larger time step than that required to maintain the stability of comparable explicit schemes.

2.2.4 Looking Ahead to Partial Differential Equations

Consider once again the advection–diffusion equation (2.17) that motivated the selection of (2.16) as a prototype ordinary differential equation. According to (2.18), each individual Fourier coefficient b_k oscillates at the frequency ck. The highest frequency resolved by any completely discrete approximation to (2.17) will be associated with the highest-wave-number or, equivalently, the shortest-horizontal-wavelength captured by the spatial discretization. As a concrete example, suppose the spatial derivatives are replaced by finite differences, then the maximum resolved k scales like $(\Delta x)^{-1}$. Let us temporarily suppose that the physical viscosity M is zero, and that the finite-difference approximation to $\partial\psi/\partial x$ does not introduce "numerical diffusion" (such diffusion can be avoided by using centered spatial differences; see Sect. 3.3.1). Then if Euler's method is used to approximate the time derivative, the frequency of the most rapidly varying Fourier component $\omega_{\max}$ will be $O(c/\Delta x)$, and over each time step its Fourier coefficient $b_{k_{\max}}$ will change by a factor $A_{k_{\max}} = 1 + iO(c\Delta t/\Delta x)$.

When attempting to obtain converged solutions to partial differential equations, one typically reduces the spatial and the temporal resolution simultaneously. But if Δt and Δx are both repeatedly halved, $\Delta t/\Delta x$ remains a constant finite value, and if Euler's method is used to integrate the numerical solution over a fixed physical time $T = N\Delta t$, the inequality

$$|A_{k_{\max}}| \leq 1 + |\omega_{\max}\Delta t| = 1 + O(|c\Delta t/\Delta x|)$$

cannot be used to bound $|A_{k_{\max}}|^N$ as $\Delta t \to 0$. Thus, the approach used to prove the convergence of Euler's method for ordinary differential equations in Sect. 2.1.2 fails, and, as will be considered more rigorously in Sect. 3.4, forward-in-time, centered-in-space approximations to the pure advection problem are unstable. Those time stepping schemes suitable for use with centered-in-space approximations to the advection equation are ones for which the point $(0, \omega_{\max}\Delta t)$ lies in the scheme's region of absolute stability whenever $|c\Delta t/\Delta x|$ is less than some constant.

Now consider finite-difference approximations to the pure diffusion problem, for which (2.18) reduces to

$$\frac{db_k}{dt} = -Mk^2 b_k. \tag{2.28}$$

If the time derivative is approximated by Euler's method, the amplification factor for the Fourier coefficient of the shortest-wavelength, most rapidly decaying component of the solution becomes $1 - O[M\Delta t/(\Delta x)^2]$, which approaches negative infinity if Δt and Δx are both repeatedly halved in an effort to obtain a convergent approximation. In most practical applications involving diffusion-dominated problems, $\Delta t/(\Delta x)^2$ becomes unbounded as the numerical resolution is refined, and it is therefore advantageous to approximate their temporal evolution using A-stable schemes, all of which are implicit.

Explicit time differences may, nevertheless, yield good results in the special case where M represents an "eddy diffusivity" M_e rather than a true physical diffusivity. Eddy diffusivities are designed to parameterize the effects of mixing by fluid motions whose scale is too small to be captured on the numerical mesh, and M_e is typically proportional to the spatial grid interval. Thus, $M_e\Delta t/(\Delta x)^2$ remains constant as $\Delta t,\ \Delta x \to 0$ with $\Delta t/\Delta x$ fixed, and it becomes practical to satisfy conditions such as $0 \le M_e\Delta t/(\Delta x)^2 \le 1$, which would allow Euler's method to be used to stably integrate those terms representing parameterized diffusion.

2.2.5 L-Stability

A-stability is not always sufficient to guarantee good behavior in practical applications involving systems of equations in which the individual components decay at very different rates. When A-stable trapezoidal time differencing is used in conjunction with finite-difference approximations to the spatial derivative in the diffusion equation, the amplification factor for the Fourier coefficient of the shortest-wavelength mode may be obtained by replacing $\lambda/2$ in (2.24) with $-\sigma M/(\Delta x)^2$, to give

$$A_{k_{\max}} = \frac{1 - \sigma M\Delta t/(\Delta x)^2}{1 + \sigma M\Delta t/(\Delta x)^2},$$

where σ is a positive constant determined by the exact finite-difference formulation.

In some applications it may not be necessary to follow the precise behavior of the most rapidly decaying, shortest-wavelength modes, and a time step appropriate for the accurate and efficient simulation of other aspects of the problem (for example, the slower diminution of the longer-wavelength components) can make

$M\Delta t/(\Delta x)^2 \gg 1$. Yet in the limit $M\Delta t/(\Delta x)^2 \to \infty$, $A_{k_{\max}} \to -1$, in which case the short-wavelength components of the trapezoidal integration will flip sign every time step without significant loss of amplitude. Although large time steps will not produce an unstable amplification of the shortest-wavelength modes, sufficiently large steps do prevent those modes from decaying properly.

The correct behavior in the limit $M\Delta t/(\Delta x)^2 \to \infty$ is recovered if backward-Euler differencing is used to approximate the time derivative. Then the amplification factor for the Fourier coefficient of the shortest-wavelength mode becomes

$$A_{k_{\max}} = \frac{1}{1 + 2\sigma M\Delta t/(\Delta x)^2},$$

and the amplification factor approaches zero as $\Delta t/(\Delta x)^2$ becomes arbitrarily large. Backward-Euler differencing is an example of an *L-stable* method. L-stable methods are defined in the context of the prototype problem (2.16) as those schemes that are A-stable and satisfy the additional property that $A \to 0$ as $\Re\{\gamma\}\Delta t \to -\infty$. As will be discussed in connection with stiff problems in Sect. 2.5, L-stable methods are also of great use in simulation of systems in which chemical reactions occur over a broad range of timescales.

2.3 Runge–Kutta (Multistage) Methods

Definite integrals are often evaluated numerically through quadrature formulae

$$\int_a^b f(t)\,\mathrm{d}t \approx \sum_{j=1}^{s} b_j f(c_j), \tag{2.29}$$

where the *weights* b_j and the *nodes* c_j are independent of the function f (Iserles 1996, p. 33). A similar strategy may be used to step the solution of an ordinary differential equation forward over a time interval Δt by approximating the integral in (2.2) such that

$$\psi(t_{n+1}) \approx \psi(t_n) + \Delta t \sum_{j=1}^{s} b_j F(\psi(t_n + c_j\Delta t), t_n + c_j\Delta t). \tag{2.30}$$

In contrast to the situation with the simple quadrature formula (2.29), however, the values of $\psi(t_n + c_j\Delta t)$ required for the evaluation of (2.30) are not known at time t_n, and must therefore be estimated numerically through a series of preliminary calculations, or *stages*. An *explicit* s-stage Runge–Kutta scheme iteratively builds an approximation to (2.30) as follows

$$\xi_1 = \phi_n \tag{2.31}$$

$$\xi_2 = \phi_n + \Delta t\, a_{2,1} F(\xi_1, t_n) \tag{2.32}$$

$$\xi_3 = \phi_n + \Delta t\,[a_{3,1}F(\xi_1, t_n) + a_{3,2}F(\xi_2, t_n + c_2\Delta t)] \tag{2.33}$$

$$\vdots$$

$$\xi_s = \phi_n + \Delta t \sum_{j=1}^{s-1} a_{s,j} F(\xi_j, t_n + c_j\Delta t) \tag{2.34}$$

$$\phi_{n+1} = \phi_n + \Delta t \sum_{j=1}^{s} b_j F(\xi_j, t_n + c_j\Delta t) \tag{2.35}$$

By convention, we ensure that ξ_j is at least a first-order approximation to $\psi(t_n + c_j\Delta t)$ by setting $c_1 = 0$ and

$$c_j = \sum_{k=1}^{j-1} a_{j,k} \qquad j = 2, 3, \ldots, s. \tag{2.36}$$

Runge–Kutta schemes are explicit if $a_{j,k}$ is zero for $k \geq j$. Implicit s-stage schemes are obtained by replacing (2.31)–(2.34) with

$$\xi_j = \phi_n + \Delta t \sum_{k=1}^{s} a_{j,k} F(\xi_k, t_n + c_k\Delta t), \tag{2.37}$$

where, in general, all the $a_{j,k}$ may be nonzero. The order conditions given above, and in the next two sections, apply both to implicit and to explicit Runge–Kutta methods.

2.3.1 Explicit Two-Stage Schemes

Taylor series expansions may be used to arrive at the additional conditions Runge–Kutta methods must satisfy to achieve a given level of accuracy. First-order accuracy requires

$$\sum_{j=1}^{s} b_j = 1. \tag{2.38}$$

For a single-stage method, the unique solution to (2.38) is $b_1 = 1$ and (2.31)–(2.35) reduce to the forward-Euler method. Second-order accuracy requires (2.36), (2.38), and

$$\sum_{j=1}^{s} b_j c_j = \frac{1}{2}. \tag{2.39}$$

For an explicit two-stage scheme, these accuracy requirements reduce to

$$c_2 = a_{2,1}, \quad b_1 + b_2 = 1, \quad b_2 c_2 = 1/2,$$

which is a system of three equations in four unknowns whose solution is not unique, but may be expressed in terms of the free parameter $a_{2,1}$. One well-known second-order two-stage scheme is the *Heun* method, for which $a_{2,1} = 1$ (and therefore

$b_1 = b_2 = 1/2$, $c_2 = 1$). The Heun method creates a trapezoidal-like approximation to the integral of F, but differs from the true trapezoidal method because $F(\phi_{n+1}, t_{n+1})$ is replaced by the estimate $F(\xi_2, t_{n+1})$. Another second-order two-stage scheme is the *midpoint* method, in which $a_{2,1} = 1/2$. Also of note is the *first-order* two-stage forward–backward scheme (Matsuno 1966b) in which $a_{2,1} = b_2 = c_2 = 1$ and $b_1 = 0$.

One important difference among the basic explicit Runge–Kutta schemes is whether they generate nonamplifying solutions in purely oscillatory problems. If the oscillation equation (2.19) is approximated using a two-stage scheme of at least first order, the result may be written as

$$\phi_{n+1} = \phi_n + b_2 \mathrm{i}\omega\Delta t\,(\phi_n + a_{2,1}\mathrm{i}\omega\Delta t\,\phi_n) + (1 - b_2)\mathrm{i}\omega\Delta t\,\phi_n. \tag{2.40}$$

The amplification factor is

$$A = 1 + \mathrm{i}\omega\Delta t - a_{2,1}b_2(\omega\Delta t)^2,$$

and

$$|A|^2 = 1 + (1 - 2a_{2,1}b_2)(\omega\Delta t)^2 + (a_{2,1}b_2)^2(\omega\Delta t)^4, \tag{2.41}$$

which shows that the set of second-order schemes (i.e., those schemes for which $a_{2,1}b_2 = 1/2$) have $O[(\Delta t)^4]$ amplitude error, whereas the amplitude error in first-order two-stage schemes is $O[(\Delta t)^2]$. Unfortunately, all the second-order two-stage explicit Runge–Kutta schemes are amplifying, since in the limit of good numerical resolution

$$|A|_{\mathrm{RKe2}} \approx 1 + \tfrac{1}{8}(\omega\Delta t)^4.$$

Although these schemes are amplifying, the growth is $O[(\Delta t)^4]$. At a given step size, the erroneous amplification produced by a second-order two-stage scheme will be much weaker than the $O[(\Delta t)^2]$ growth produced by forward time differencing (or alternatively, the first-order single-stage Runge–Kutta method, see (2.26)).

Many physical systems contain several different modes, each oscillating at a different frequency. When these systems are simulated, the highest-frequency components of the numerical solution are likely to be most seriously in error because of their poor numerical resolution. It is precisely these poorly resolved features that are amplified most rapidly by the second-order two-stage methods. In contrast, nonamplifying solutions in which the high-frequency components are strongly damped can be obtained using Matsuno's forward–backward scheme, for which

$$|A|^2_{\mathrm{Matsuno}} = 1 - (\omega\Delta t)^2 + (\omega\Delta t)^4. \tag{2.42}$$

The Matsuno scheme damps the solution whenever $0 < \omega\Delta t < 1$. Differentiation of (2.42), with respect to $\omega\Delta t$, shows that the maximum damping occurs when $\omega\Delta t = 1/\sqrt{2}$. Thus, if the time step is chosen such that $0 \le \omega\Delta t \le 1/\sqrt{2}$ for all frequencies ω in the physical system, Matsuno time differencing will preferentially damp the highest-frequency waves. The damping properties of the Matsuno scheme

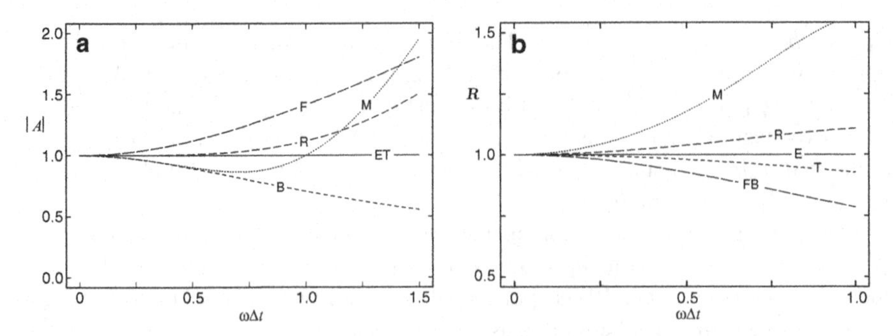

Fig. 2.2 The modulus of the amplification factor (**a**) and the relative phase change (**b**) as a function of temporal resolution $\omega\Delta t$ for the true solution and five two-level schemes: exact solution (E) and trapezoidal method (T), forward-Euler scheme (F), backward-Euler scheme (B), two-stage second-order Runge–Kutta scheme (R), and Matsuno scheme (M)

have been exploited to eliminate high-frequency gravity waves generated during the initialization of weather prediction models. The standard Matsuno scheme produces too much damping, however, for most nonspecialized applications. The fourth-order Runge–Kutta scheme (see Sect. 2.3.2) may also be used to preferentially damp high-frequency modes, and in most instances it would be a better choice than the Matsuno scheme because it is more efficient and far more accurate.

The amplitude errors generated by the preceding Runge–Kutta schemes are compared with those produced by backward-Euler and trapezoidal differencing in Fig. 2.2. The strong damping associated with the backward-Euler and Matsuno schemes is evident, along with the rapid amplification produced by forward-Euler differencing. These relatively large errors may be contrasted with the significantly weaker amplification produced by the second-order Runge–Kutta methods, and the neutral amplification of the trapezoidal method.

The relative phase change associated with the general two-stage explicit Runge–Kutta method (2.40) is

$$R = \frac{1}{\omega\Delta t}\arctan\left(\frac{\omega\Delta t}{1 - a_{2,1}b_2(\omega\Delta t)^2}\right).$$

In the limit of good numerical resolution, the relative phase changes produced by second-order schemes and the Matsuno scheme are

$$R_{\mathrm{RKe2}} \approx 1 + \tfrac{1}{6}(\omega\Delta t)^2, \qquad R_{\mathrm{Matsuno}} \approx 1 + \tfrac{2}{3}(\omega\Delta t)^2.$$

The relative phase change for these schemes is plotted as a function of temporal resolution in Fig. 2.2, along with that for forward-Euler, backward-Euler, and trapezoidal differencing. The Matsuno and second-order Runge–Kutta schemes are accelerating, whereas the forward-Euler, backward-Euler, and trapezoidal schemes are decelerating.

2.3.2 Explicit Three- and Four-Stage Schemes

Runge–Kutta schemes satisfying

$$\sum_{j=1}^{s} b_j c_j^2 = \frac{1}{3} \quad \text{and} \quad \sum_{j=1}^{s}\sum_{k=1}^{s} b_j a_{j,k} c_k = \frac{1}{6}, \tag{2.43}$$

as well as (2.36), (2.38), and (2.39) are third-order accurate. For explicit three-stage Runge–Kutta schemes, (2.43) reduces to

$$b_2 c_2 + b_3 c_3 = \frac{1}{3} \quad \text{and} \quad b_3 a_{3,2} c_2 = \frac{1}{6}.$$

As with the second-order methods there is no unique choice for the coefficients of a three-stage third-order scheme. On example is Heun's third-order method,

$$\xi_1 = \phi_n, \quad \xi_2 = \phi_n + \frac{\Delta t}{3} F(\xi_1, t_n), \quad \xi_3 = \phi_n + \frac{2\Delta t}{3} F(\xi_2, t_n + \tfrac{\Delta t}{3}),$$

$$\phi_{n+1} = \phi_n + \frac{\Delta t}{4}\Big[F(\xi_1, t_n) + 3F(\xi_3, t_n + \tfrac{2\Delta t}{3})\Big].$$

Another possibility is the low storage variant recommended by Williamson (1980) which may be written as

$$\begin{aligned} q_1 &= \Delta t F(\phi_n, t_n), & \phi_{(1)} &= \phi_n + q_1/3, \\ q_2 &= \Delta t F(\phi_{(1)}, t_n + \tfrac{\Delta t}{3}) - 5q_1/9, & \phi_{(2)} &= \phi_{(1)} + 15q_2/16, \\ q_3 &= \Delta t F(\phi_{(2)}, t_n + \tfrac{5\Delta t}{12}) - 153q_2/128, & \phi_{n+1} &= \phi_{(2)} + 8q_3/15. \end{aligned}$$

In practical applications involving time-dependent partial differential equations, ϕ_n may be an extremely long vector of unknown variables (e.g., the velocity, temperature, pressure, and mixing ratio of chemical species at every node on a large three-dimensional mesh). It may, therefore, be difficult to store several copies of ϕ and $F(\phi)$ in the in-core memory of a digital computer. If m is the number of unknowns in ϕ, the Williamson–Runge–Kutta scheme economizes on storage by allowing the integration to proceed using only $2m$ storage locations, divided between arrays q and ϕ, which are overwritten three times during each integration step.

In addition to (2.36)–(2.39) and (2.43), fourth-order Runge–Kutta methods must satisfy four additional equations (Hundsdorfer and Verwer 2003, p. 141). Once again, the solutions for the coefficients of a four-stage explicit method are not unique. The most well-known four-stage fourth-order method is the classical Runge–Kutta formulation,

$$\begin{aligned} \xi_1 &= \phi_n, & \xi_2 &= \phi_n + \frac{\Delta t}{2} F(\xi_1, t_n), \\ \xi_3 &= \phi_n + \frac{\Delta t}{2} F(\xi_2, t_n + \tfrac{\Delta t}{2}), & \xi_4 &= \phi_n + \Delta t\, F(\xi_3, t_n + \tfrac{\Delta t}{2}), \end{aligned} \tag{2.44}$$

$$\phi_{n+1} = \phi_n + \frac{\Delta t}{6}\Big[F(\xi_1, t_n) + 2F(\xi_2, t_n + \tfrac{\Delta t}{2}) + 2F(\xi_3, t_n + \tfrac{\Delta t}{2}) + F(\xi_4, t_{n+1})\Big].$$

Low-storage variants also exist for fourth-order schemes (Blum 1962), but in contrast to the third-order methods, they require $3m$ storage locations to advance an m-dimensional vector of unknowns forward in time.

Fifth- or higher-order explicit Runge–Kutta schemes are relatively unattractive because the number of stages required to achieve order s exceeds s for all $s > 4$. Nevertheless, the simple s-stage scheme

$$\xi_0 = \phi_n; \quad \xi_j = \phi_n + \frac{\Delta t}{s - j + 1} F(\xi_{j-1}), \quad 1 \leq j \leq s; \quad \phi_{n+1} = \xi_s, \tag{2.45}$$

is accurate to order s when $F(\psi)$ is linear in ψ (as would be the case in many applications involving time-dependent partial differential equations). When F is nonlinear, (2.45) is no better than second-order accurate.

Figure 2.3 shows the amplification factor for third- and fourth-order Runge–Kutta solutions to the oscillation equation (2.19) plotted as a function of temporal resolution. As shown in Fig. 2.3, once the time step exceeds the maximum stable time step for the third-order scheme, the fourth-order method becomes highly damping. In some circumstances it may be desirable to selectively damp the highest-frequency modes, and in such cases the fourth-order Runge–Kutta method would be clearly preferable to the first-order Matsuno method. On the other hand, if one wishes to avoid excessive damping of the high-frequency components, it will not be possible to use the full stable time step of the fourth-order Runge–Kutta scheme.

As was the case for the two-stage first-order Matsuno method, the stability of explicit Runge–Kutta solutions to the oscillation equation may be enhanced by adding extra stages if one is willing to settle for first- or second-order accuracy. In particular, the stability condition $\max|\omega\Delta t| = s - 1$ may be obtained for an s-stage scheme that will be second-order accurate if s is odd, and first-order accurate when s is even (Hundsdorfer and Verwer 2003, p. 150). Note that despite their high-order accuracy, explicit fourth-order four-stage Runge–Kutta methods are stable for $\max|\omega\Delta t| < 2.82$, which is very close to the optimal limit of $\max|\omega\Delta t| = 3$ obtainable using a *first-order* four-stage explicit method.

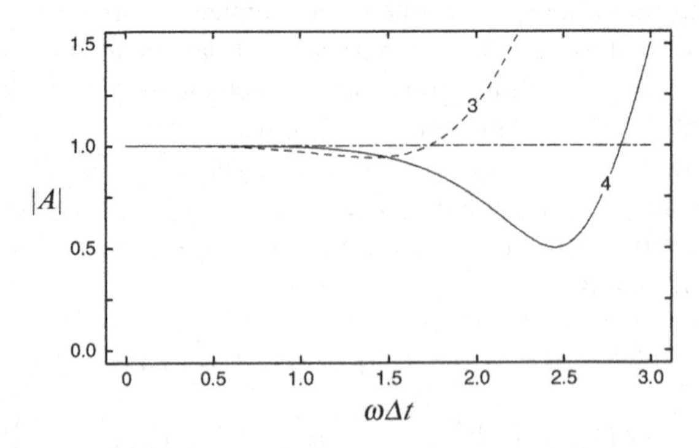

Fig. 2.3 Modulus of the amplification factor as a function of temporal resolution $\omega\Delta t$ for third-order three-stage (*dashed line*) and fourth-order four-stage (*solid line*) explicit Runge–Kutta solutions to the oscillation equation

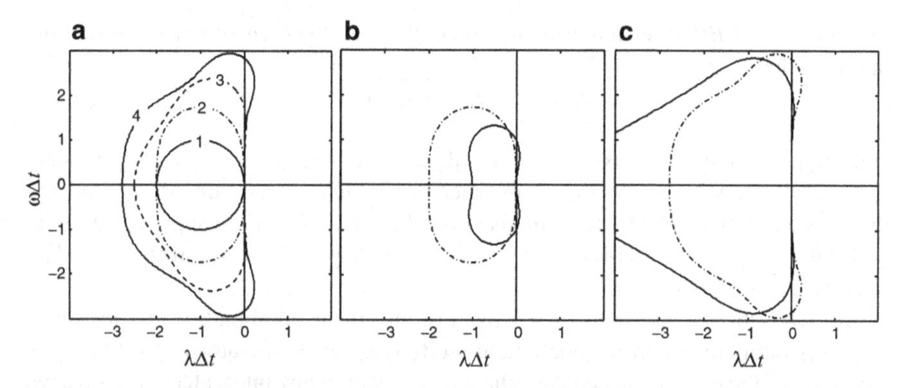

Fig. 2.4 Absolute stability regions for explicit Runge–Kutta schemes: **a** of equal orders and stages, 1–4; **b** two-stage methods – Matsuno (*solid line*) and second-order (*dashed line*); **c** four-stage methods – Spiteri and Ruuth's third-order strong-stability-preserving Runge–Kutta scheme (*solid line*) and fourth-order (*dashed line*). In each case, the region of absolute stability lies inside the curve. When $\omega = 0$, the absolute stability region for the Spiteri–Ruuth scheme extends to roughly $\lambda\Delta t = -5.15$

Absolute stability regions for explicit Runge–Kutta schemes of orders 1–4 are plotted in Fig. 2.4a. Consistent with the behaviors of the amplification factors for the oscillation equation shown in Figs. 2.2 and 2.3, the third- and fourth-order methods are the only ones for which the absolute stability regions include a finite segment of the imaginary axis. None of these methods and indeed no explicit Runge–Kutta scheme is A-stable. Nevertheless, in contrast to the behavior of linear multistep methods that will be discussed in Sect. 2.4.3, the area of absolute stability increases as the order of accuracy of the scheme increases.

Figure 2.4b compares the absolute stability region for a pair of explicit two-stage methods, the first-order Matsuno method and any second-order scheme. The increase in absolute stability along the real axis in the Matsuno scheme is achieved not only by sacrificing accuracy, but also by considerably reducing the overall region of absolute stability relative to the second-order schemes.

2.3.3 Strong-Stability-Preserving Methods

As will be discussed in Chap. 5, many methods for the numerical integration of conservation laws avoid the generation of spurious maxima and minima through the use of some type of flux limiter. The time differencing associated with such methods is often forward-Euler differencing. Strong-stability-perserving Runge–Kutta (SSPRK) schemes can be used to obtain higher-order accuracy in time while preserving the beneficial results of the flux limiter. To be more precise, suppose that U is a vector of unknowns at every point on the spatial mesh, and that $\|U\|$ represents a measure such as the maximum of $|U|$ or the total variation of U over all spatial grid points. Let $B(\phi)$ be an approximation to the flux divergences in a conservation law such that

$$U_{n+1} = (I + \Delta t B)\, U_n, \tag{2.46}$$

and suppose that the fluxes are limited so that $\|U_{n+1}\| \leq \|U_n\|$ provided $|c\Delta t/\Delta x| \leq 1$, where c is the phase speed at which signals are propagated by the conservation law. SSPRK methods allow the forward-in-time approximation in (2.46) to be replaced by a higher-order scheme while preserving the *strong-stability condition* that $\|U_{n+1}\| \leq \|U_n\|$.

SSPRK schemes are constructed by forming linear combinations of forward-Euler operators in which the coefficient multiplying each operator is positive. The positivity of the coefficients ensures that a conservation law integrated with the new scheme retains the strong-stability properties of the original forward-Euler approximation (2.46). The precise value of the positive coefficients is chosen to obtain some combination of high-order accuracy and a favorable maximum stable time step. A two-stage second-order SSPRK method is

$$\begin{aligned}
\phi_{(1)} &= \phi_n + \Delta t B(\phi_n), \\
\phi_{(2)} &= \phi_{(1)} + \Delta t B(\phi_{(1)}), \\
\phi_{n+1} &= \tfrac{1}{2}\left(\phi_n + \phi_{(2)}\right),
\end{aligned} \tag{2.47}$$

and a three-stage third-order scheme is

$$\begin{aligned}
\phi_{(1)} &= \phi_n + \Delta t B(\phi_n), \\
\phi_{(2)} &= \tfrac{3}{4}\phi_n + \tfrac{1}{4}\left[\phi_{(1)} + \Delta t B(\phi_{(1)})\right], \\
\phi_{n+1} &= \tfrac{1}{3}\phi_n + \tfrac{2}{3}\left[\phi_{(2)} + \Delta t B(\phi_{(2)})\right].
\end{aligned} \tag{2.48}$$

Both of these schemes, which were proposed by Shu and Osher (1988), are strong-stability preserving for $|c\Delta t/\Delta x| \leq 1$.

Schemes (2.47) and (2.48) are optimal in the sense that no two-stage second-order or three-stage third-order SSPRK scheme exists that allows a larger maximum time step (Gottlieb and Shu 1998). Nevertheless, in some applications the four-stage third-order SSPRK scheme

$$\begin{aligned}
\phi_{(1)} &= \phi_n + \tfrac{1}{2}\Delta t B(\phi_n), \\
\phi_{(2)} &= \phi_{(1)} + \tfrac{1}{2}\Delta t B(\phi_{(1)}), \\
\phi_{(3)} &= \tfrac{2}{3}\phi_n + \tfrac{1}{3}\left[\phi_{(2)} + \tfrac{1}{2}\Delta t B(\phi_{(2)})\right], \\
\phi_{n+1} &= \phi_{(3)} + \tfrac{1}{2}\Delta t B(\phi_{(3)})
\end{aligned} \tag{2.49}$$

proposed by Spiteri and Ruuth (2002) may be more efficient, because it is strong-stability preserving for $|c\Delta t/\Delta x| \leq 2$, allowing one to double the time step while increasing the computational burden associated with the evaluation of B by only 33% relative to that required by (2.48).

It should be emphasized that these methods are only strong-stability preserving when flux-limiting ensures that the forward step (2.46) yields a strongly stable result. Amplifying solutions are produced if (2.47) is applied directly to the oscillation equation (2.19). Since (2.47) is an explicit two-stage second-order method and since (2.48) is an explicit three-stage third-order scheme, their absolute stability regions are exactly those shown for the second- and third-order methods in Fig. 2.4a. On the other hand, as shown in Fig. 2.4c, the four-stage third-order scheme (2.49) has a different, and generally larger, region of absolute stability than the family of four-stage fourth-order Runge–Kutta methods. More information about strong-stability-preserving time-differencing schemes may be found in the reviews by Gottleib et al. (2001) and Gottleib (2005).

2.3.4 Diagonally Implicit Runge–Kutta Methods

Diagonally implicit Runge–Kutta schemes are obtained when the implicit coupling in (2.37) is limited by requiring $a_{j,k} = 0$ whenever $k > j$. In comparison with methods with more extensive implicit coupling, the relative efficiency of diagonally implicit schemes makes them more attractive for applications involving partial differential equations or large systems of ordinary differential equations. Backward-Euler differencing is a first-order accurate, single-stage diagonally implicit Runge–Kutta scheme. The implicit midpoint method,

$$\begin{aligned}\xi_1 &= \phi_n + \tfrac{1}{2}\Delta t\; F(\xi_1, t_n + \tfrac{1}{2}\Delta t),\\ \phi_{n+1} &= \phi_n + \Delta t\; F(\xi_1, t_n + \tfrac{1}{2}\Delta t),\end{aligned} \tag{2.50}$$

is a second-order accurate, single-stage scheme. The implicit midpoint method is A-stable; its amplification factor is identical to that for the trapezoidal method (2.24).

A family of two-stage diagonally implicit Runge–Kutta schemes of at least second-order accuracy may be written in terms of a single free parameter α as

$$\begin{aligned}\xi_1 &= \phi_n + \alpha\Delta t F(\xi_1, t_n + \alpha\Delta t),\\ \xi_2 &= \phi_n + (1-2\alpha)\Delta t F(\xi_1, t_n + \alpha\Delta t) + \alpha\Delta t F(\xi_2, t_n + (1-\alpha)\Delta t),\\ \phi_{n+1} &= \phi_n + \tfrac{1}{2}\Delta t\left[F(\xi_1, t_n + \alpha\Delta t) + F(\xi_2, t_n + (1-\alpha)\Delta t)\right].\end{aligned} \tag{2.51}$$

Third-order accuracy is obtained if $\alpha = 1/2 \pm \sqrt{3}/6$. According to Hundsdorfer and Verwer (2003), the preceding method was developed independently by Nørsett and by Crouzeix.

If one of the schemes defined by (2.51) is applied to the test problem (2.16), the resulting amplification factor is

$$A = \frac{1 + (1-2\alpha)\gamma\Delta t + (\frac{1}{2} - 2\alpha + \alpha^2)(\gamma\Delta t)^2}{(1-\alpha\gamma\Delta t)^2}. \tag{2.52}$$

These schemes are A-stable if and only if $\alpha \geq 1/4$, as may be easily appreciated in the particular case for which $\gamma\Delta t \to (-\infty, 0)$; then the leading-order behavior of $|A|$ is $(1/2-2\alpha+\alpha^2)/\alpha^2$, which is bounded by unity for $\alpha \geq 1/4$. The $(\gamma\Delta t)^2$ term in the numerator of (2.51) is zero, and the scheme is L-stable if $\alpha = 1 \pm 1/2\sqrt{2}$.

2.4 Multistep Methods

Multistep methods are an alternative to multistage methods in which information from several earlier time levels is incorporated into the integration formula. For example, the general form for an explicit two-step method is

$$\phi_{n+1} = \alpha_1\phi_n + \alpha_2\phi_{n-1} + \beta_1\Delta t F(\phi_n, t_n) + \beta_2\Delta t F(\phi_{n-1}, t_{n-1}). \tag{2.53}$$

In contrast to multistage methods, the forcing $F(\psi, t)$ is only evaluated at integer time steps and all the required values except $F(\phi_n, t_n)$ have already been calculated at previous time steps. Since the evaluation of $F(\psi, t)$ is often computationally intensive, storing and reusing these values has the potential to increase efficiency, although obviously it may also require more storage. Multistep methods also require special start-up procedures, because an n-step method requires data from the previous n time levels, but initial conditions for well-posed physical problems give information about the solution at only one time. Multistage or lower-order multistep methods must therefore be used for the first $n - 1$ steps of the integration.

2.4.1 Explicit Two-Step Schemes

In many geophysical applications, the memory required to store data from each time level is enormous, so let us begin by considering the family of two-step schemes (2.53), some of which have very modest storage requirements. When formulating a two-step scheme, one seeks to improve upon the single-step methods, so it is reasonable to require that the global truncation error be at least second order. Scheme (2.53) will be at least second order if

$$\alpha_1 = 1 - \alpha_2, \qquad \beta_1 = \tfrac{1}{2}(\alpha_2 + 3), \qquad \beta_2 = \tfrac{1}{2}(\alpha_2 - 1), \tag{2.54}$$

where the coefficient α_2 remains a free parameter.

If $\alpha_2 = 5$, the resulting scheme,

$$\phi_{n+1} = -4\phi_n + 5\phi_{n-1} + \Delta t\,(4F(\phi_n, t_n) + 2F(\phi_{n-1}, t_{n-1})), \tag{2.55}$$

is third-order accurate. One might suppose that (2.55) is superior to all other explicit two-step schemes because its truncation error is higher order, but in fact this scheme is useless. The problem is easily revealed in the trivial case where $F = 0$; then (2.38) reduces to the linear homogeneous difference equation

$$\phi_{n+1} = -4\phi_n + 5\phi_{n-1}. \tag{2.56}$$

As before, let the amplification factor be defined as $A = \phi_{n+1}/\phi_n$; since the coefficients of ϕ are constant in (2.56), A is independent of the time step and $\phi_{n+1} = A^2\phi_{n-1}$. Expressing each term in (2.56) in terms of ϕ_{n-1}, it follows that (except for the special case where $\phi_0 = 0$)

$$A^2 + 4A - 5 = 0, \tag{2.57}$$

whose roots are $A = 1$ or $A = -5$. Of course the true solution to the differential equation $\mathrm{d}\psi/\mathrm{d}t = 0$ is constant with time and this behavior is correctly captured by the physical root $A = 1$. The second root represents a spurious *computational mode*. Such computational modes arise in all multistep methods, and in all useful multistep methods the time step can be chosen to keep the amplitude of these modes from growing. In this case, however, the amplitude of the computational mode grows by a factor of 5 every time step. Even if the initial amplitude in the computational mode is only produced by round-off error, its wildly unstable growth soon generates numbers too large to represent in standard floating point arithmetic.

Since it is not practical to choose the coefficients in (2.54) to minimize the truncation error, the most important explicit two-step schemes are obtained by choosing α_2 to minimize the amount of data that must be stored and carried over from the $n-1$ time level, i.e., by setting $\alpha_2 = 1$, in which case $\beta_2 = 0$, or by setting $\alpha_2 = 0$. If α_2 is set to 1, (2.53) becomes the *leapfrog* scheme. The choice $\alpha_2 = 0$ gives the two-step *Adams–Bashforth* method. The remainder of this section will be devoted to an examination of the performance of these two schemes in problems with purely oscillatory solutions. More general applications will be considered in Sect. 2.4.3.

If the leapfrog scheme,

$$\phi_{n+1} = \phi_{n-1} + 2\Delta t F(\phi_n, t_n), \tag{2.58}$$

is used to integrate the oscillation equation (2.19), its amplification factor satisfies

$$A^2 - 2\mathrm{i}\omega\Delta t A - 1 = 0,$$

whose two roots are

$$A_\pm = \mathrm{i}\omega\Delta t \pm \left(1 - \omega^2\Delta t^2\right)^{1/2}. \tag{2.59}$$

In the limit of good numerical resolution, $\omega\Delta t \to 0$ and $A_+ \to 1$, which captures the correct behavior of the physical mode. The second root, for which $A_- \to -1$, represents the computational mode. If $|\omega\Delta t| \le 1$, the second term in (2.59) is real and $|A_+| = |A_-| = 1$, i.e., both the physical and the computational modes are stable and neutral. In the case $\omega\Delta t > 1$,

$$|A_+| = \left|\mathrm{i}\omega\Delta t + \mathrm{i}\left(\omega^2\Delta t^2 - 1\right)^{1/2}\right| > |\mathrm{i}\omega\Delta t| > 1,$$

and the scheme is unstable. When $\omega\Delta t < -1$, a similar argument shows that $|A_-| > 1$. Note that when $\omega\Delta t > 1$, A_+ lies on the positive imaginary axis in the complex plane, and thus each integration step produces a 90° shift in the phase of the oscillation. As a consequence, unstable leapfrog solutions grow with a period of $4\Delta t$.

The complete leapfrog solution can typically be written as a linear combination of the physical and computational modes. An exception occurs if $\omega\Delta t = \pm 1$, in which case $A_+ = A_- = \mathrm{i}\omega\Delta t$, and the physical and computational modes are not linearly independent. In such circumstances, the general solution to the leapfrog approximation to the oscillation equation has the form

$$\phi_n = C_1(\mathrm{i}\omega\Delta t)^n + C_2 n(\mathrm{i}\omega\Delta t)^n.$$

Since the magnitude of the preceding solution grows as a function of the time step, the leapfrog scheme is *not* stable when $|\omega\Delta t| = 1$. Nevertheless, the $O(n)$ growth of the solution that occurs when $\omega\Delta t = \pm 1$ is far slower than the $O(A^n)$ amplification that is produced when $|\omega\Delta t| > 1$.

The source of the computational mode is particularly easy to analyze in the trivial case of $\omega = 0$; then the analytic solution to the oscillation equation (2.19) is $\psi(t) = C$, where C is a constant determined by the initial condition at $t = t_0$. Under these circumstances, the leapfrog scheme reduces to

$$\phi_{n+1} = \phi_{n-1}, \tag{2.60}$$

and the amplification factor has the roots $A_+ = 1$ and $A_- = -1$. The initial condition requires $\phi_0 = C$, which, according to the difference scheme (2.60), also guarantees that $\phi_2 = \phi_4 = \phi_6 = \ldots = C$. The odd time levels are determined by a second, computational initial condition imposed on ϕ_1. In practice ϕ_1 is often obtained from ϕ_0 by taking a single time step with a single-step method, and the resulting approximation to $\psi(t_0 + \Delta t)$ will contain some error E. It is obvious that in our present example the correct choice for ϕ_1 is C, but to mimic the situation in a more general problem, suppose that $\phi_1 = C + E$. Then the numerical solution at any subsequent time will be the sum of two modes:

$$\phi_n = (A_+)^n\phi_+ + (A_-)^n\phi_- = (C + E/2) - (-1)^n(E/2).$$

Here, the first term represents the physical mode and the second term represents the computational mode. The computational mode oscillates with a period of $2\Delta t$, and does not decay with time.

In the previous example, the amplitude of the computational mode is completely determined by the error in the specification of the computational initial condition ϕ_1. Since there is no coupling between the physical and computational modes in solutions to linear problems, the errors in the initial conditions also govern the amplitude of the computational mode in leapfrog solutions to most linear equations. If the governing equations are nonlinear, however, the nonlinear terms introduce a coupling between ϕ_+ and ϕ_- that often amplifies the computational mode until it eventually dominates the solution. This spurious growth of the computational mode can be avoided by periodically discarding the solution at ϕ_{n-1} and taking a single time step with a two-level scheme, or by filtering the high-frequency components of the numerical solution. Various techniques for controlling the leapfrog scheme's computational mode will be discussed in Sect. 2.4.2.

The relative phase changes in the two leapfrog modes are

$$R_{\pm_{\text{leapfrog}}} = \frac{1}{\omega\Delta t}\arctan\left(\frac{\pm\omega\Delta t}{(1-\omega^2\Delta t^2)^{1/2}}\right).$$

The computational mode and the physical mode oscillate in opposite directions. In the limit of good time resolution,

$$R_{+_{\text{leapfrog}}} \approx 1 + \frac{(\omega\Delta t)^2}{6},$$

showing that leapfrog time differencing is accelerating.

Now consider the other fundamental explicit two-step method, the two-step Adams–Bashforth scheme,

$$\phi_{n+1} = \phi_n + \Delta t\left(\tfrac{3}{2}F(\phi_n, t_n) - \tfrac{1}{2}F(\phi_{n-1}, t_{n-1})\right). \tag{2.61}$$

Applying the preceding formula to the oscillation equation yields

$$\phi_{n+1} = \phi_n + \mathrm{i}\omega\Delta t\left(\tfrac{3}{2}\phi_n - \tfrac{1}{2}\phi_{n-1}\right).$$

The amplification factor associated with this scheme is given by the quadratic

$$A^2 - \left(1 + \frac{3\mathrm{i}\omega\Delta t}{2}\right)A + \frac{\mathrm{i}\omega\Delta t}{2} = 0,$$

in which case

$$A_{\pm} = \frac{1}{2}\left(1 + \frac{3\mathrm{i}\omega\Delta t}{2} \pm \left(1 - \frac{9(\omega\Delta t)^2}{4} + \mathrm{i}\omega\Delta t\right)^{1/2}\right). \tag{2.62}$$

As the numerical resolution increases, $A_+ \to 1$ and $A_- \to 0$. Thus, the Adams–Bashforth method damps the computational mode, which of course is highly desirable. Unfortunately the physical mode is weakly amplifying, as revealed if (2.62) is approximated under the assumption that $\omega\Delta t$ is small; then

$$A_+ = \left(1 - \frac{(\omega\Delta t)^2}{2} - \frac{(\omega\Delta t)^4}{8} - \dots\right) + \mathrm{i}\left(\omega\Delta t + \frac{(\omega\Delta t)^3}{4} + \dots\right),$$
$$A_- = \left(\frac{(\omega\Delta t)^2}{2} + \frac{(\omega\Delta t)^4}{8} + \dots\right) + \mathrm{i}\left(\frac{\omega\Delta t}{2} - \frac{(\omega\Delta t)^3}{4} - \dots\right),$$

and

$$|A_+|_{\text{A–B2}} \approx 1 + \tfrac{1}{4}(\omega\Delta t)^4, \qquad |A_-|_{\text{A–B2}} \approx \tfrac{1}{2}\omega\Delta t.$$

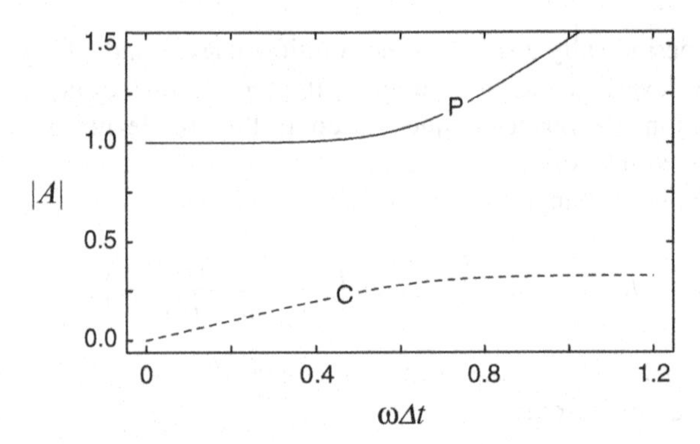

Fig. 2.5 Modulus of the amplification factors for the second-order Adams–Bashforth scheme as a function of temporal resolution $\omega\Delta t$. The *solid lines* and the *dashed lines* represent the physical and the computational modes, respectively

The modulus of the amplification factor of the physical mode exceeds unity by an $O[(\omega\Delta t)^4]$ term, as was the case for the two-stage second-order Runge–Kutta methods. The dependence of $|A_+|$ and $|A_-|$ upon temporal resolution is plotted in Fig. 2.5. The relative phase change in the physical mode in the Adams–Bashforth method is

$$R_{\text{A-B2}} \approx 1 + \frac{5}{12}(\omega\Delta t)^2,$$

where, as before, it is assumed that $\omega\Delta t \ll 1$. Comparisons between the absolute stability regions for the leapfrog and two-step Adams–Bashforth methods will be presented in Sect. 2.4.3.

The leapfrog and two-step Adams–Bashforth methods must be initialized using a single-step scheme to compute ϕ_1 from ϕ_0. In most instances, a simple forward step is adequate. Although forward differencing is amplifying, the amplification produced by a single step will generally not be large. Moreover, even though the truncation error of a forward difference is $O(\Delta t)$, the execution of a single forward time step does not reduce the $O\left[(\Delta t)^2\right]$ global accuracy of leapfrog and Adams–Bashforth integrations. The basic reason that $O\left[(\Delta t)^2\right]$ accuracy is preserved is that forward differencing is only used over a Δt-long portion of the total integration. The contribution to the total error produced by the accumulation of $O\left[(\Delta t)^2\right]$ errors in a stable scheme over a finite time interval is of the same order as the error arising from the accumulation of $O(\Delta t)$ errors over a time Δt.

2.4.2 Controlling the Leapfrog Computational Mode

The best explicit second-order scheme for the integration of oscillatory systems considered so far might appear to be the leapfrog scheme. The leapfrog scheme is nonamplifying (unlike the Adams–Bashforth and two-stage Runge–Kutta methods), and it requires only one function evaluation per time step (unlike two-or-higher-stage Runge–Kutta schemes). The weakness of the leapfrog scheme is its undamped computational mode, which slowly amplifies to produce "time-splitting" instability in simulations of nonlinear systems. Here we review several methods that have been used in practice to control such time splitting.

One way to control the growth of the computational mode is to periodically discard the data from the $n-1$ time level (or alternatively to average the n and $n-1$ time-level solutions) and to restart the integration using a two-time-level method. Forward differencing is often used to reinitialize leapfrog integrations because it is easy to implement, but being a first-order scheme, its application over a fixed percentage of the total integration time degrades the second-order accuracy of the unadulterated leapfrog method. In addition, forward differencing significantly amplifies the high-frequency components of the solution. Moreover, it is difficult to quantify these adverse effects since they vary according to the number of leapfrog steps between each forward step. Restarting with a second-order Runge–Kutta scheme is a far better choice since this preserves second-order accuracy and produces less unphysical amplification. The midpoint method is one second-order Runge–Kutta formulation that can be used to restart leapfrog integrations in complex numerical models without greatly complicating the coding. A midpoint method restart may be implemented by taking a forward step of length $\Delta t/2$ followed by a single leapfrog step of length $\Delta t/2$.

In atmospheric science, it is common, though questionable, practice to control the computational mode through the use of a second-order time filter. This filter is closely related to the centered second-derivative time filter,

$$\overline{\phi_n} = \phi_n + \gamma\,(\phi_{n+1} - 2\phi_n + \phi_{n-1}), \tag{2.63}$$

where ϕ_n denotes the solution at time $n\Delta t$ prior to time filtering, $\overline{\phi_n}$ is the solution after filtering, and γ is a positive real constant that determines the strength of the filter. The last term in (2.63) is the usual finite-difference approximation to the second derivative and preferentially damps the highest frequencies. Suppose that the unfiltered values are sampled from the exact solution to the oscillation equation, then

$$\overline{\phi_n} = \left(1 + \gamma\left(e^{i\omega\Delta t} - 2 + e^{-i\omega\Delta t}\right)\right)\phi_n.$$

Defining a *filter factor* $X = \overline{\phi_n}/\phi_n$, one obtains

$$X_{\text{centered}} = 1 - 2\gamma(1 - \cos\omega\Delta t). \tag{2.64}$$

Since X_{centered} is real, it does not produce any change in the phase of the solution. In the limit $\omega\Delta t \to 0$,

$$X_{\text{centered}} \approx 1 - \gamma(\omega\Delta t)^2,$$

showing that well-resolved oscillations undergo an $O[(\Delta t)^2]$ damping. The centered filter has the greatest impact on the most poorly resolved component of the solution, the $2\Delta t$ oscillation. According to (2.64), each filter application reduces the amplitude of the $2\Delta t$ wave by a factor of $1-4\gamma$. If γ is specified to be $1/4$, each filtering operation will completely eliminate the $2\Delta t$ oscillation.

Robert (1966) and Asselin (1972) suggested a scheme to control the leapfrog computational mode by incorporating an approximate second-derivative time filter into the time integration cycle such that each leapfrog step

$$\phi_{n+1} = \overline{\phi_{n-1}} + 2\Delta t F(\phi_n)$$

is followed by the filtering operation:

$$\overline{\phi_n} = \phi_n + \gamma\left(\overline{\phi_{n-1}} - 2\phi_n + \phi_{n+1}\right). \tag{2.65}$$

A filter parameter of $\gamma = 0.06$ is typically used in global atmospheric models. Values of $\gamma = 0.2$ are common in convective cloud models; indeed Schlesinger et al. (1983) recommend choosing γ in the range 0.25–0.3 for certain advection–diffusion problems.

If the Asselin-filtered leapfrog scheme is applied to the oscillation equation, the amplification factor is determined by the simultaneous equations

$$A^2\phi_{n-1} = \overline{\phi_{n-1}} + 2i\omega\Delta t A\phi_{n-1}, \tag{2.66}$$

$$\overline{A\phi_{n-1}} = A\phi_{n-1} + \gamma\left(\overline{\phi_{n-1}} - 2A\phi_{n-1} + A^2\phi_{n-1}\right). \tag{2.67}$$

Under the assumption that $\overline{(A\phi_n)} = A(\overline{\phi_n})$, whose validity will be discussed shortly, (2.67) may be written

$$(A-\gamma)\overline{\phi_{n-1}} = A\left((1-2\gamma) + A\gamma\right)\phi_{n-1}. \tag{2.68}$$

Eliminating $\overline{\phi_{n-1}}$ between (2.66) and (2.68) yields

$$A_{\pm} = \gamma + i\omega\Delta t \pm \left((1-\gamma)^2 - \omega^2\Delta t^2\right)^{1/2}, \tag{2.69}$$

which reduces to the result for the standard leapfrog scheme when $\gamma = 0$. In the limit of small $\omega\Delta t$, the amplification factor for the Asselin-filtered physical mode becomes

$$A_{\text{Asselin--LF}} = 1 + i\omega\Delta t - \frac{(\omega\Delta t)^2}{2(1-\gamma)} + O[(\omega\Delta t)^4].$$

A comparison of this expression with the asymptotic behavior of the exact amplification factor,

$$A_e = e^{i\omega\Delta t} = 1 + i\omega\Delta t - \frac{(\omega\Delta t)^2}{2} - i\frac{(\omega\Delta t)^3}{6} + O[(\omega\Delta t)^4],$$

shows that the *local* truncation error of the Asselin-filtered leapfrog scheme is $O[(\omega\Delta t)^2]$. In contrast, the local truncation error of the unfiltered leapfrog scheme ($\gamma = 0$) is $O[(\omega\Delta t)^3]$. Thus, Asselin filtering degrades the *global* truncation error of the leapfrog scheme from second order to first order.

The preceding derivation was based on the assumption that $\overline{(A\phi_n)} = A(\overline{\phi_n})$; is this justified? In practice, the initial condition is not time-filtered; one simply defines $\overline{\phi_0} \equiv \phi_0$. Thus,

$$\overline{(A\phi_0)} - A(\overline{\phi_0}) = \overline{\phi_1} - \phi_1 = \gamma\,(\phi_0 - 2\phi_1 + \phi_2) \neq 0.$$

Nevertheless, an application of the Asselin time filter to ϕ_{n+1} gives

$$\begin{aligned}\overline{(A\phi_n)} &= A\phi_n + \gamma\left(\overline{(A\phi_{n-1})} - 2A\phi_n + A^2\phi_n\right)\\ &= A\left(\phi_n + \gamma\left(\overline{\phi_{n-1}} - 2\phi_n + A\phi_n\right)\right) + \gamma\left(\overline{(A\phi_{n-1})} - A\overline{\phi_{n-1}}\right)\\ &= A(\overline{\phi_n}) + \gamma\left(\overline{(A\phi_{n-1})} - A\overline{\phi_{n-1}}\right),\end{aligned}$$

from which it follows that

$$\overline{(A\phi_n)} - A(\overline{\phi_n}) = \gamma^n[\overline{\phi_1} - \phi_1]. \tag{2.70}$$

In all cases of practical interest, $n \gg 1$ and $\gamma \ll 1$; therefore, (2.70) implies that A may be factored out of the filtering operation with negligible error, and that (2.68) is indeed equivalent to (2.67).

In the limit of $\omega\Delta t \ll 1$, the modulus of the amplification factor for the Asselin-filtered leapfrog scheme may be approximated as

$$|A_+|_{\text{Asselin–LF}} \approx 1 - \frac{\gamma}{2(1-\gamma)}(\omega\Delta t)^2,$$

$$|A_-|_{\text{Asselin–LF}} \approx (1-2\gamma) + \frac{\gamma}{2-6\gamma+4\gamma^2}(\omega\Delta t)^2.$$

Like other first-order schemes, such as forward differencing and the Matsuno method, the physical mode in the Asselin-filtered leapfrog scheme has an $O[(\Delta t)^2]$ amplitude error. The behavior of the computational mode is also notable in that $|A_-|$ does not approach zero as $|\omega\Delta t| \to 0$. The asymptotic behavior of the relative phase change in the physical mode is

$$R_{+\text{Asselin–LF}} \approx 1 + \frac{1+2\gamma}{6(1-\gamma)}(\omega\Delta t)^2.$$

Asselin–Robert filtering increases the phase error, doubling it as γ increases from 0 to 1/4.

The main problem with the Asselin-filtered leapfrog scheme is its first-order accuracy. Williams (2009) has recently proposed a modification of the Asselin–Robert filter for leapfrog integrations that almost conserves the three-time-level mean of the predicted field and greatly reduces the magnitude of the first-order truncation error. Two alternative techniques which control the leapfrog computational mode without sacrificing second-order accuracy are the leapfrog–trapezoidal method and the Magazenkov method. The leapfrog–trapezoidal method (Kurihara 1965; Zalesak

1979) is an iterative scheme in which a leapfrog predictor is followed by a trapezoidal correction step, i.e.,

$$\phi_* = \phi_{n-1} + 2\Delta t F(\phi_n), \tag{2.71}$$

$$\phi_{n+1} = \phi_n + \frac{\Delta t}{2}\left(F(\phi_n) + F(\phi_*)\right). \tag{2.72}$$

If this scheme is applied to the oscillation equation, the amplitude and relative phase changes in the physical mode are

$$|A|_{\mathrm{LF-trap}} \approx 1 - \frac{(\omega\Delta t)^4}{4}, \qquad R_{\mathrm{LF-trap}} \approx 1 - \frac{(\omega\Delta t)^2}{12},$$

where, as usual, these approximations hold for small $\omega\Delta t$. Leapfrog–trapezoidal integrations of the oscillation equation will be stable provided $\omega\Delta t \leq \sqrt{2}$. The amplitude error associated with the leapfrog–trapezoidal scheme is plotted as a function of temporal resolution in Fig. 2.6a.

Magazenkov (1980) suggested that the computational mode could be controlled by alternating each leapfrog step with a second-order Adams–Bashforth step. Since the Magazenkov method uses different schemes on the odd and even time steps, the amplification factor differs between the odd and even steps. To analyze the behavior of the Magazenkov method, it is therefore best to consider the averaged effect of a combined leapfrog–Adams–Bashforth cycle.

Thus, for analysis purposes, the scheme will be written as a system of equations that maps (ϕ_{n-2}, ϕ_{n-1}) into (ϕ_n, ϕ_{n+1}),

$$\phi_n = \phi_{n-2} + 2\Delta t F(\phi_{n-1}), \tag{2.73}$$

$$\phi_{n+1} = (\phi_{n-2} + 2\Delta t F(\phi_{n-1})) + \frac{\Delta t}{2}\left[3F\left(\phi_{n-2} + 2\Delta t F(\phi_{n-1})\right) - F(\phi_{n-1})\right]. \tag{2.74}$$

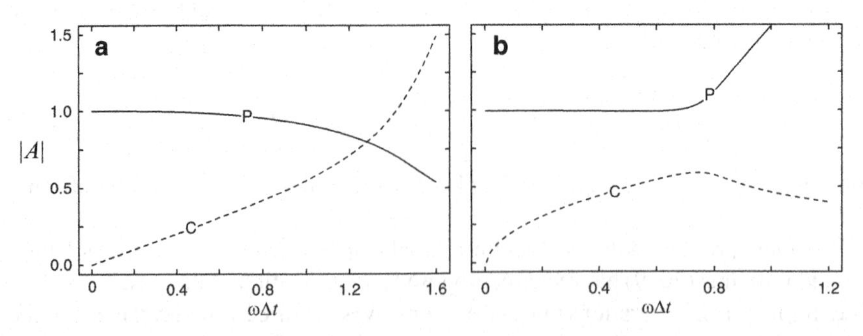

Fig. 2.6 Modulus of the amplification factor as a function of temporal resolution $\omega\Delta t$ for **a** the leapfrog–trapezoidal method and **b** the Magazenkov method. The *solid lines* and *dashed lines* represent the physical and the computational modes, respectively

When actually implementing the Magazenkov method, however, one would replace (2.74) by the equivalent expression (2.61). Application of (2.73) and (2.74) to the oscillation equation yields a system of two equations in two unknowns,

$$\begin{pmatrix} 1 & 2\mathrm{i}\omega\Delta t \\ 1+\frac{3}{2}\mathrm{i}\omega\Delta t & \frac{3}{2}\mathrm{i}\omega\Delta t - 3(\omega\Delta t)^2 \end{pmatrix}\begin{pmatrix} \phi_{n-2} \\ \phi_{n-1} \end{pmatrix} = \begin{pmatrix} \phi_n \\ \phi_{n+1} \end{pmatrix}.$$

The coefficient matrix in the preceding equation determines the combined amplification and phase shift generated by each pair of leapfrog and Adams–Bashforth time steps. The eigenvalues of the coefficient matrix are determined by the characteristic equation

$$\lambda^2 - \left(\frac{3\mathrm{i}\omega\Delta t}{2} - 3(\omega\Delta t)^2 + 1\right)\lambda - \frac{\mathrm{i}\omega\Delta t}{2} = 0.$$

The eigenvalues are distinct and have magnitudes less than 1 when $|\omega\Delta t| < 2/3$, implying that the method is conditionally stable. For well-resolved physical-mode oscillations, the average amplitude and relative phase change per single time step are

$$|A|_{\mathrm{Mag}} = (|\lambda|)^{1/2} \approx 1 - \frac{(\omega\Delta t)^4}{4}, \qquad R_{\mathrm{Mag}} = 1 + \frac{(\omega\Delta t)^2}{6}.$$

The average amplitude error per single time step is plotted as a function of temporal resolution in Fig. 2.6b.

2.4.3 Classical Multistep Methods

The leapfrog and two-step Adams–Bashforth methods discussed in Sect. 2.4.1 are just two examples from the large family of linear multistep methods. A general linear s-step method may be written in the form

$$\sum_{k=0}^{s} \alpha_k \phi_{n+k} = \Delta t \sum_{k=0}^{s} \beta_k F(\phi_{n+k}, t_{n+k}). \tag{2.75}$$

Here the notation is simplified by departing from the convention used in (2.53) and letting the index of the most advanced time level be $n+s$. The coefficients α_k and β_k are usually scaled so that $\alpha_s = 1$. If the method is explicit, β_s is zero.

As before, the truncation error is the residual by which smooth solutions ψ to the continuous problem fail to satisfy (2.75). If the method is $O\left[(\Delta t)^p\right]$, then

$$\sum_{k=0}^{s} \alpha_k \psi(t_{n+k}) - \Delta t \sum_{k=0}^{s} \beta_k \frac{d\psi}{dt}(t_{n+k}) = O\left[(\Delta t)^{p+1}\right]. \tag{2.76}$$

The right side of the preceding equation is the local error, which, as discussed for Euler's method in connection with (2.9), is one power of Δt higher than the truncation error. Expanding ψ and $d\psi/dt$ in Taylor series about time t_n and substituting into (2.76) yields the following conditions for pth-order accuracy:

$$\sum_{k=0}^{s} \alpha_k = 0, \qquad \sum_{k=0}^{s} \alpha_k k^m = m \sum_{k=0}^{s} \beta_k k^{m-1} \quad \text{for} \quad m = 1, 2, \ldots, p. \tag{2.77}$$

When ψ is a polynomial of degree p, the left side of (2.76) may be expressed as the sum of powers of Δt up to and including $(\Delta t)^p$, and this sum must be zero if (2.76) is satisfied. Thus, if a linear multistep method is of order p, it must be exact for all polynomials $\psi(t)$ of degree p or less; this can be checked relatively easily by trying each of the set of test functions

$$\psi_k(t) = (t - t_n)^k, \quad k = 0, 1, \ldots, p, \tag{2.78}$$

which span the $(p + 1)$-dimensional space of polynomials of degree p. The equivalence of these two criteria can be verified by substituting the functions (2.78) into (2.76) to obtain the order conditions (2.77).

Stability conditions for linear multistep methods can be concisely expressed in terms of the polynomials

$$\rho(z) = \sum_{k=0}^{s} \alpha_k z^k \quad \text{and} \quad \sigma(z) = \sum_{k=0}^{s} \beta_k z^k.$$

The linear multistep method (2.75) is stable enough to converge to the correct solution as $\Delta t \to 0$, provided that it is at least first-order accurate, that the starting values $\phi_1, \ldots, \phi_s$ converge as $\Delta t \to 0$, and that *the polynomial $\rho(z)$ satisfies the root condition* (Dalhquist 1956). A polynomial satisfies the *root condition* if the magnitude of each of its roots is bounded by unity and if all the roots of unit magnitude are distinct.

An example of the problems that arise when $\rho(z)$ does not satisfy the root condition was revealed in connection with the discussion of the third-order explicit two-step method (2.55). Recall that applying this method to the particularly simple case where $F(\psi, t) = 0$ reduced the differential equation to $d\psi/dt = 0$. The amplification factor for the $F = 0$ problem satisfies (2.57), or equivalently, $\rho(A) = 0$, whose roots are 1 and -5. The mode for which $A = -5$ undergoes rapid unstable amplification.

The method (2.55) is one in which the $2s - 1$ coefficients in an explicit s-step method are chosen to create a scheme of maximum order (namely, $2s - 1$), but the result is unstable. When the choice of coefficients is additionally constrained to require $\rho(z)$ to satisfy the root condition, the highest order than can be achieved by an *explicit* s-step method is s; for an *implicit* s-step method it is $s + 1$ when s is odd and $s + 2$ when s is even (Dalhquist 1959). These limitations on the maximum order of stable linear multistep methods constitute *Dahlquist's first barrier*.

As discussed in Sect. 2.2.1, merely ensuring a consistent scheme is sufficiently stable to converge in the limit $\Delta t \to 0$ does not guarantee against the spurious amplification of solutions computed using finite values of Δt. An indication of the behavior of solutions computed with finite Δt is provided by analyzing a scheme's region of absolute stability. If an s-step linear multistep method is applied to the test problem (2.16), the amplification factors for the physical mode and the $s-1$ computational modes satisfy the polynomial

$$\rho(A) - \gamma\Delta t\, \sigma(A) = 0. \tag{2.79}$$

The absolute stability region for the method is the set of $\gamma\Delta t$ for which the preceding polynomial satisfies the root condition.

Adams methods are one very important family of classical multistep methods. In an s-step Adams method $\alpha_s = 1$, $\alpha_{s-1} = -1$, and $\alpha_k = 0$ for $k < s-1$. Thus, $\rho(z) = z^{s-1}(z-1)$, and at least for sufficiently small Δt, all the computational modes (the roots of z^{s-1}) are strongly damped. The β_k are chosen to maximize the order of accuracy. Explicit s-step Adams schemes, known as Adams–Bashorth methods, achieve optimal s-order accuracy. For implicit s-step schemes, called Adams–Moulton methods, the $s+1$ available values of β_k allow (2.77) to be satisfied through order $s+1$.

The one-step Adams–Bashforth method is identical to the forward-Euler scheme; the two-step Adams–Bashforth method (2.61) was already introduced in Sect. 2.4.1. Returning to the convention that $n+1$ is the most advanced time level, the three-step Adams–Bashforth method is

$$\phi_{n+1} = \phi_n + \frac{\Delta t}{12}\left[23F(\phi_n) - 16F(\phi_{n-1}) + 5F(\phi_{n-2})\right]. \tag{2.80}$$

The amplitude error in the three-step Adams–Bashforth solution to the oscillation equation is plotted in Fig. 2.7. As with all Adams schemes, both computational modes are strongly damped in the limit $\omega\Delta t \to 0$. Unlike the two-step Adams–Bashforth method shown in Fig. 2.5, instability is not associated with unstable growth of the physical mode; instead, it is one of the computational modes which becomes unstable for $\omega\Delta t > 0.724$.

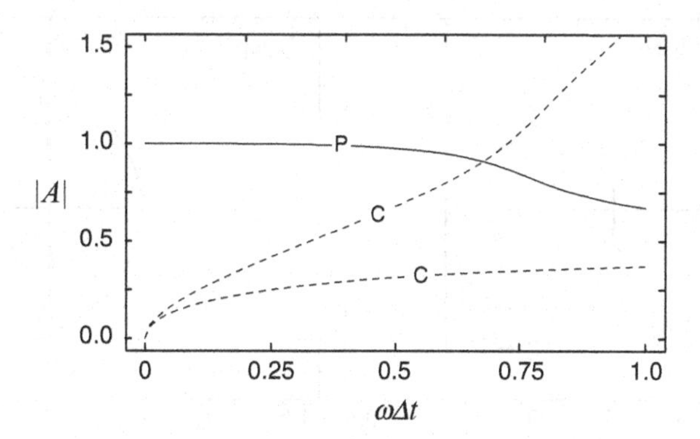

Fig. 2.7 Modulus of the amplification factor for the three-step Adams–Bashforth method plotted as a function of temporal resolution $\omega\Delta t$. The *solid line* represents the physical mode and the *dashed lines* represent the two computational modes

The one-step Adams–Moulton method is the trapezoidal scheme

$$\phi_{n+1} = \phi_n + \frac{h}{2}\left[F(\phi_{n+1}) + F(\phi_n)\right]. \tag{2.81}$$

The two-step and three-step Adams–Moulton methods are

$$\phi_{n+1} = \phi_n + \frac{h}{12}\left[5F(\phi_{n+1}) + 8F(\phi_n) - F(\phi_{n-1})\right], \tag{2.82}$$

$$\phi_{n+1} = \phi_n + \frac{h}{24}\left[9F(\phi_{n+1}) + 19F(\phi_n) - 5F(\phi_{n-1} + 1F(\phi_{n-1})\right]. \tag{2.83}$$

In practice, the implicit coupling in an Adams–Moulton method is often avoided using a predictor-corrector iteration. An s-step Adams–Bashforth scheme is used to estimate ϕ_{n+1} in the prediction step; then this estimate is used to evaluate $F(\phi^{n+1})$ as ϕ_{n+1} is recomputed using an s-step Adams–Moulton scheme in the corrector step. The resulting predictor-corrector iteration is accurate to order $s+1$. Note that, like the leapfrog–trapezoidal method (2.71) and (2.72), the Adams–Bashforth–Moulton predictor corrector cannot be expressed as a linear multistep method of the form (2.75).

The absolute stability regions for the two-step and three-step Adams–Bashforth methods are shown in Fig. 2.8. Consistent with the analysis in Sect. 2.4.1 for purely oscillatory problems, the imaginary axis (except for the origin) lies outside the region of absolute stability of the two-step scheme. On the other hand, the segment of the imaginary axis for which $|\omega\Delta t| < 0.72$ does lie within the absolute stability region of the three-step Adams–Bashforth method, and this scheme can be an attractive choice for the integration of problems with primarily oscillatory solutions.

The absolute stability regions for the two-step and three-step Adams–Moulton methods, shown in Fig. 2.9, are much larger than those for the two-step and three-step Adams–Bashforth methods (note the change in the scale of the coordinate axes), but are nevertheless too small to justify the extra work required to use an implicit

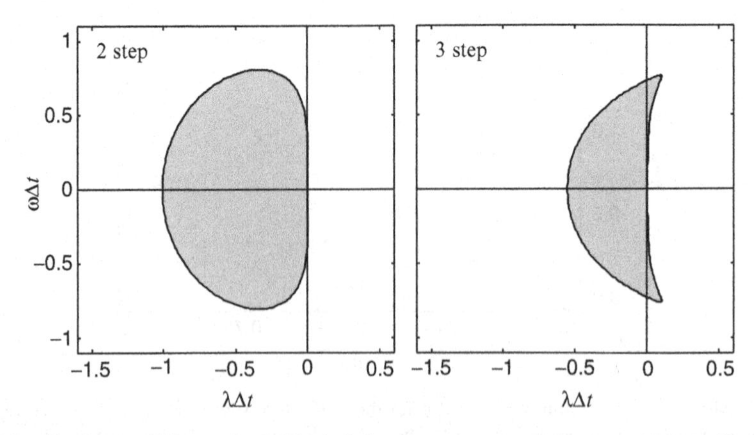

Fig. 2.8 Absolute stability regions (*shaded*) for the two-step and three-step Adams–Bashforth methods

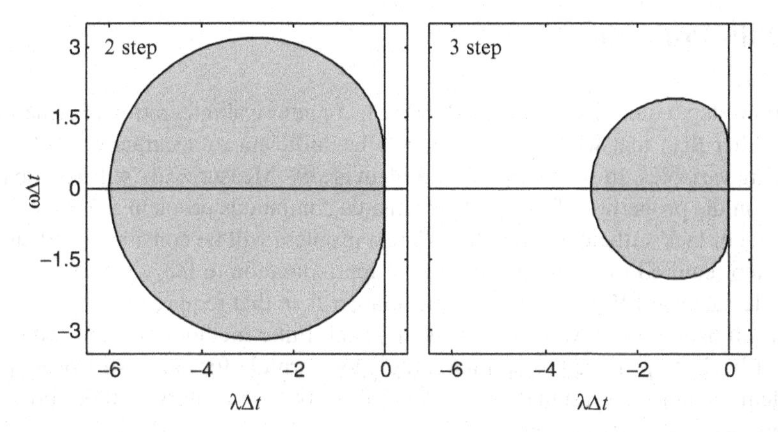

Fig. 2.9 Absolute stability regions (*shaded*) for the two-step and three-step Adams–Moulton methods

method in many practical applications. Neither method includes a finite segment of the imaginary axis or the point $(-\infty, 0)$ in the region of absolute stability. As a consequence, the two-step and three-step Adams–Moulton methods are not suitable for the solution of oscillatory problems or systems of equations where some components decay rapidly.

Another classical multistep method, the leapfrog scheme, was considered only in the context of the oscillation equation in Sect. 2.4.1. The absolute stability region for the leapfrog scheme can be determined by analyzing the two roots of (2.79), which are

$$A_{\mathrm{lf}} = \gamma\Delta t \pm \left[(\gamma\Delta t)^2 + 1\right]^{1/2}.$$

Define ζ, such that

$$\mathrm{i}\cos\zeta = \gamma\Delta t = \mathrm{i}\omega\Delta t + \lambda\Delta t.$$

Then

$$A_{\mathrm{lf}} = \mathrm{i}\cos\zeta \pm \sin\zeta = i\,\mathrm{e}^{\mp\mathrm{i}\zeta},$$

and, thus, the two amplification factors associated with the leapfrog scheme have magnitudes $|\mathrm{e}^{-\mathrm{i}\zeta}|$ and $|\mathrm{e}^{\mathrm{i}\zeta}|$. One of these will exceed unity unless ζ is real, and since

$$\zeta = \cos^{-1}(\omega\Delta t - \mathrm{i}\lambda\Delta t),$$

the condition for absolute stability reduces to

$$\lambda\Delta t = 0 \quad \text{and} \quad |\omega\Delta t| < 1.$$

The region of the $\lambda\Delta t$–$\omega\Delta t$ plane within which the leapfrog scheme is stable is just an open line segment along the $\omega\Delta t$-axis. The endpoints bounding this segment, $(\lambda\Delta t, \omega\Delta t) = (0, \pm 1)$, yield a pair of identical A_{lf} of unit magnitude and hence do not satisfy the root condition. Since its region of absolute stability is confined to the imaginary axis, the leapfrog scheme is only suitable for the simulation of nondissipative wavelike flows.

2.5 Stiff Problems

The time step required to maintain stability in a numerical integration is sometimes far smaller than that which might seem to be sufficient to accurately resolve the evolving variables. In such cases the problem is *stiff*. Measures of "stiffness" based solely on the properties of the unapproximated continuous problem are often cumbersome or lack sufficient generality. Here a problem will be considered stiff if the time step required to obtain a satisfactory approximation to the solution using the L-stable backward-Euler method is much larger than that required to obtain a similar result using the forward-Euler scheme. Both Euler methods are first order and are not likely to provide the optimal integration formula for either stiff or nonstiff problems, but the contrast in the behavior of these two elementary methods provides a handy way to assess stiffness.

In geophysical applications, stiffness often poses a problem when modeling chemical reactions. Stiffness can also create difficulties if the compressible equations of motion are used to simulate atmospheric flows in which acoustic waves have negligible amplitude. Later in this section we will consider a stiff problem that arises in very simple model of urban air pollution. Issues related to the efficient treatment of rapidly moving acoustic waves are discussed in Chap. 8.

As a simple starting point, consider the scalar equation

$$\frac{d\psi}{dt} = \lambda\,(\psi - g(t)) + \frac{dg}{dt}, \tag{2.84}$$

subject to the initial condition $\psi(0) = \psi_0$, whose solution is

$$\psi(t) = (\psi_0 - g(0))\,e^{\lambda t} + g(t). \tag{2.85}$$

Our interest is in the case $\lambda < 0$, for which the first term in (2.85) decays after an initial transient. If (2.84) is integrated using the forward-Euler method, the leading-order truncation error (see (2.8)) is proportional to

$$\frac{d^2\psi}{dt^2} = \lambda^2(\psi - g) + \frac{d^2 g}{dt^2}.$$

Thus, *after the initial transient decays*, $\lambda^2(\psi - g) \to 0$ and for a given Δt, the truncation error is determined solely by the forcing $g(t)$. Yet as discussed in Sect. 2.1.2, g has no influence on the step-to-step growth of the global error in the numerical solution. The behavior of the global error depends only on λ, and over each time step, it amplifies by a factor of $1 + \lambda\Delta t$. The restriction on Δt necessary to keep the global error bounded ($-2 \le \lambda\Delta t \le 0$) has therefore no relation to the choice of time step required to keep the truncation error below a given threshold, and when $|\lambda|$ is sufficiently large, the time step required to maintain stability can be arbitrarily smaller than that required solely by accuracy considerations.

Suppose $\lambda = -100$, $\psi_0 = 2$, $g(t) = \sin(t)$, and that (2.84) is integrated over the interval $[0, 2\pi]$ using the forward-Euler and backward-Euler methods. In this

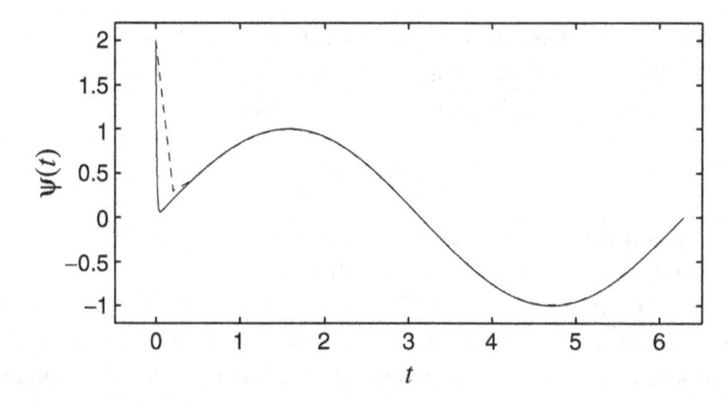

Fig. 2.10 Solutions to (2.84) with $\lambda = -100$, $\psi_0 = 2$, and $g(t) = \sin(t)$. The *solid line* shows the superposition of the exact solution and the solutions given by the forward and the backward methods using a variable step size. The *dashed line* is the backward solution with a constant $\Delta t = \pi/15$

example the initial and minimum time step is 0.002, and as a simple way to vary the step size in this particular example, whenever the estimated truncation error[3] drops below 0.01, the time step is doubled. The solutions from the two methods are shown in Fig. 2.10; both are indistinguishable from the correct solution at the resolution plotted. The backward scheme required 120 steps to complete the integration, whereas the forward scheme required 383 steps. The contrast between the number of steps required by each scheme becomes larger as $|\lambda|$ increases.

It should be emphasized that the preceding problem is *not stiff* in the immediate neighborhood of $t = 0$, where the solution is changing rapidly and a very small time step is required to correctly resolve the rapid variation. Rather, the problem is stiff for t greater than about 0.2, where the actual fluctuations in the true solution can easily be resolved using time steps far larger than the maximum step for which the forward-Euler method is stable. For example, the solution shown by the dashed line in Fig. 2.10 can be obtained with just 30 steps of uniform size using the L-stable backward-Euler method. In some computationally intensive applications it is not necessary to capture the behavior of rapid transients precisely, and the solution shown by the dashed line may be satisfactory.

2.5.1 Backward Differentiation Formulae

The key property allowing stable backward-Euler integrations to be performed with large time steps in the previous example is that the point $(\lambda\Delta t, 0)$ remains in the method's region of absolute stability as $\lambda\Delta t \to -\infty$. Any linear multistep method (2.75) having this property must be implicit, otherwise the order of the polynomial $\rho(A)$ will exceed that of $\sigma(A)$, and the magnitude of one root of (2.79) will approach

[3] The local contribution to the *global* truncation error was estimated as $0.5\Delta t|\psi''(t_n)|$ using the exact solution (2.84) to compute ψ''.

infinity as $\lambda \Delta t \to -\infty$. One simple strategy for ensuring that $(-\infty, 0)$ is included in the region of absolute stability of an s-step scheme is to set all the β_k in (2.75) to zero except β_s. Then, the amplification factor satisfies

$$\rho(A) - \gamma \Delta t \, \beta_s A^s = 0, \tag{2.86}$$

and since $\rho(A)$ is a polynomial of order s, as $\lambda \Delta t \to -\infty$ the s roots of (2.86) approach the roots of $\beta_s A^s = 0$, which are all zero.

Backward differentiation formulae (BDF) are linear multistep methods (2.75) for which all β_k are zero except β_s and the α_k are chosen to obtain the highest possible order of accuracy, which for an s-step method is s. The one-step BDF method is the backward-Euler method

$$\phi^{n+1} = \phi^n + \Delta t \, F(\phi^{n+1}, t_{n+1}). \tag{2.87}$$

The two-step, three-step, and four-step BDF methods are

$$\phi^{n+1} - \frac{4}{3}\phi^n + \frac{1}{3}\phi^{n-1} = \frac{2}{3}\Delta t \, F(\phi^{n+1}, t_{n+1}), \tag{2.88}$$

$$\phi^{n+1} - \frac{18}{11}\phi^n + \frac{9}{11}\phi^{n-1} - \frac{2}{11}\phi^{n-2} = \frac{6}{11}\Delta t \, F(\phi^{n+1}, t_{n+1}), \tag{2.89}$$

$$\phi^{n+1} - \frac{48}{25}\phi^n + \frac{36}{25}\phi^{n-1} - \frac{16}{25}\phi^{n-2} + \frac{3}{25}\phi^{n-3} = \frac{12}{25}\Delta t \, F(\phi^{n+1}, t_{n+1}). \tag{2.90}$$

As discussed in Sect. 2.2.5, L-stable methods are A-stable schemes which satisfy the additional criterion that $A \to 0$ as $\lambda \Delta t \to -\infty$; thus, by construction, BDF methods will be L-stable provided they are also A-stable. Yet, in contrast to implicit Runge–Kutta methods, which can be both high order and A-stable, e.g., (2.51) with $\alpha = 1/2 + \sqrt{3}/6$, no linear multistep method of order greater than 2 is A-stable (Dalhquist 1963). This limitation on the order of A-stable linear multistep methods is *Dahlquist's second barrier*. It follows that since the three-step BDF method is third order, it cannot be A-stable, nor strictly speaking L-stable.

The regions of absolute stability for the two-step, three-step and four-step BDF methods are shown in Fig. 2.11. The values of $\gamma \Delta t$ that spoil the A-stability of the three-step BDF method lie in a thin region just to the left of the imaginary axis. In practical applications that benefit from the use of an L-stable method, the relevant values of $\gamma \Delta t$ have negative real parts that are large in magnitude in comparison with $|\Im\{\gamma \Delta t\}|$ and therefore well within the absolute stability region for the three-step BDF method. For BDF methods of orders 3–6, there is a wedge expanding outward from the origin in the left half-plane throughout which the method is absolutely stable. Let the angle between the top or bottom of this wedge and the negative real axis be α. For the four-step BDF method, $\alpha = 73°$, and the wedge throughout which the scheme is absolutely stable is indicated by the dashed lines in Fig. 2.11. As the order of the BDF method increases, the wedge narrows; α is 86°for the three-step method, 51°for the five-step method, and just 17°for the six-step scheme. BDF methods using more than six steps are unstable.

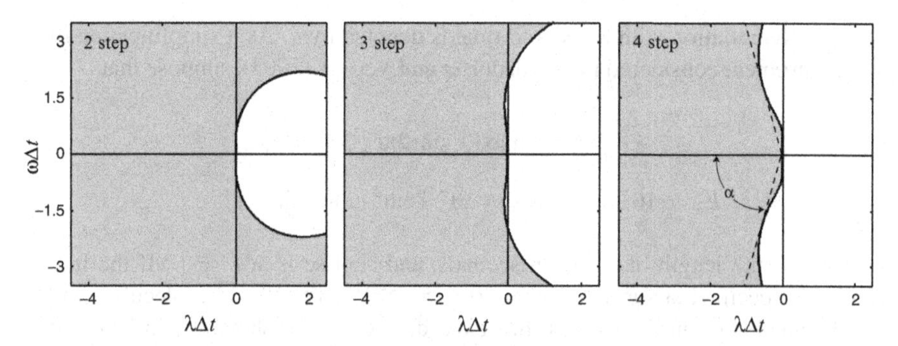

Fig. 2.11 Absolute stability regions (*shaded*) for the two-step, three-step, and four-step backward differentiation formula methods

Methods that are absolutely stable for all values of $\omega\Delta t$ and all $\lambda\Delta t \le 0$ lying in the wedge $|\arctan(\omega/\lambda)| \le \alpha$ are A(α)-stable. When choosing a high-order A(α)-stable BDF method for a particular problem with eigenvalues γ, one should take care that all γ with a negative real part satisfy $|\arctan(\Im\{\gamma\}/\Re\{\gamma\})| \le \alpha$.

2.5.2 Ozone Photochemistry

Now consider a classic example from atmospheric chemistry involving the following reactions between atomic oxygen (O), nitrogen oxides (NO and NO_2), and ozone (O_3):

$$
\begin{aligned}
NO_2 + h\upsilon &\xrightarrow{k_1} NO + O, \\
O + O_2 &\xrightarrow{k_2} O_3, \\
NO + O_3 &\xrightarrow{k_3} O_2 + NO_2.
\end{aligned}
\tag{2.91}
$$

Here $h\upsilon$ denotes a photon of solar radiation. Letting $\mathbf{c} = (c_1, \ldots, c_4)^{\mathrm{T}}$ represent the concentration in molecules per cubic centimeter of O, NO, NO_2, and O_3, respectively, and approximating the background concentration of O_2 as constant, the preceding set of reactions is governed by the system

$$\dot{c_1} = k_1c_3 - k_2c_1, \tag{2.92}$$
$$\dot{c_2} = k_1c_3 - k_3c_2c_4, \tag{2.93}$$
$$\dot{c_3} = k_3c_2c_4 - k_1c_3, \tag{2.94}$$
$$\dot{c_4} = k_2c_1 - k_3c_2c_4, \tag{2.95}$$

where differentiation with respect to time is denoted by $\dot{c}$. As a simplification of a similar problem considered in Hundsdorfer and Verwer (2003), suppose that

$$k_1 = 10^{-2} \max[0, \sin(2\pi t/t_d)] \text{ s}^{-1},$$

$$k_2 = 10^5 \text{ s}^{-1}, \quad k_3 = 10^{-16} \text{cm}^3 \text{ molecule}^{-1} \text{ s}^{-1},$$

where t_d is the length of 1 day in seconds, and sunrise is at $t = 0$. If the initial condition specified at sunrise is $\mathbf{c} = (0, 0, 5 \times 10^{11}, 8 \times 10^{11})^T$ molecules cm^{-3}, the chemical concentrations over the next 2 days evolve as shown by the thin solid curves in Fig. 2.12. The concentration of atomic oxygen spikes at sunrise, which triggers a sustained daytime increase in O_3 and NO and a decrease in NO_2. The solution shown by the solid lines in Fig. 2.12 was computed in MATLAB using a high-order BDF scheme with variable step sizes and a tight error tolerance. The solution shown by the dashed lines was computed using the second-order Rosenbrock Runge–Kutta method (2.113)–(2.115) with a constant time step of 15 min. Although it does not precisely capture the spike in atomic oxygen occurring just after sunrise, the Rosenbrock Runge–Kutta method efficiently models all other aspects of the system with reasonable fidelity.

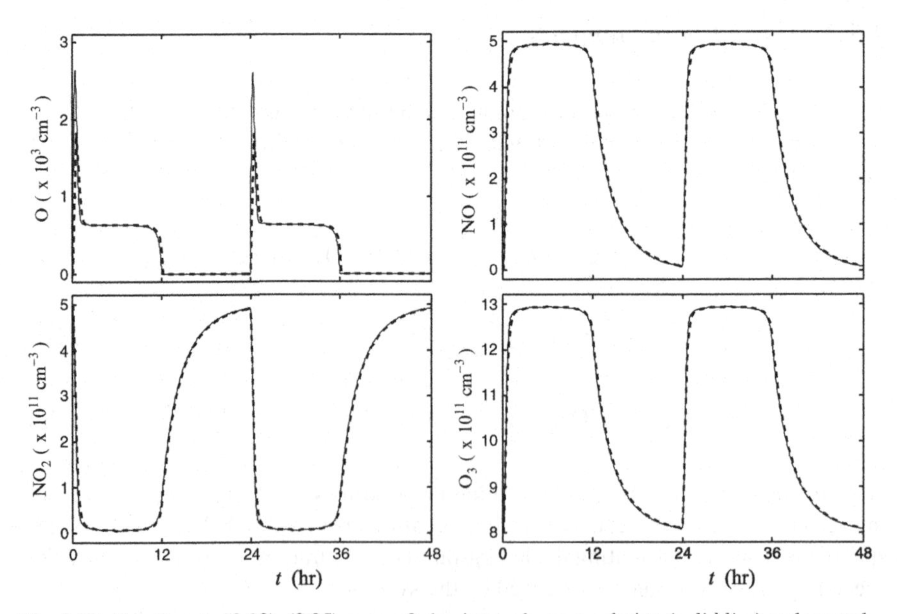

Fig. 2.12 Solutions to (2.92)–(2.95) over a 2-day interval: exact solution (*solid line*) and second-order Rosenbrock Runge–Kutta solution with a constant time step of 15 min (*dashed line*)

2.5.3 Computing Backward-Euler Solutions

The system (2.92)–(2.95) has the form

$$\dot{\mathbf{c}} = \mathbf{f}(\mathbf{c}, t), \tag{2.96}$$

where $\mathbf{c}$ and $\mathbf{f}$ are column vectors. Binary reactions, such as the reaction between nitric oxide and ozone in (2.91), make $\mathbf{f}$ a nonlinear function of the unknown variable $\mathbf{c}$. Suitable solvers for stiff problems are all implicit time-difference approximations, which when applied to nonlinear differential equations generate nonlinear algebraic equations. Accurate solutions to these nonlinear algebraic equations can often be efficiently obtained using Newton's method.

As an example, consider the autonomous system of differential equations[4]

$$\dot{\mathbf{c}} = \mathbf{f}(\mathbf{c}). \tag{2.97}$$

Letting $\boldsymbol{\phi}_n$ be the numerical approximation to $\mathbf{c}(n\Delta t)$, the backward-Euler approximation to (2.97) is

$$\boldsymbol{\phi}_{n+1} = \boldsymbol{\phi}_n + \Delta t\, f(\boldsymbol{\phi}_{n+1}). \tag{2.98}$$

To solve the preceding equation for ϕ_{n+1} using Newton's method, we seek a root of

$$\mathbf{g}(\boldsymbol{\phi}) = \boldsymbol{\phi} - \boldsymbol{\phi}_n - \Delta t\, f(\boldsymbol{\phi}). \tag{2.99}$$

Let $\mathbf{G}$ be the matrix whose ijth element is $\partial g_i/\partial \phi_j$. Differentiating (2.99) with respect to c_j yields

$$\mathbf{G}(\boldsymbol{\phi}) = \mathbf{I} - \Delta t\, \mathbf{J}(\boldsymbol{\phi}), \tag{2.100}$$

where $\mathbf{J}$ is the Jacobian matrix whose ijth element is $\partial f_i/\partial c_j$. Newton's method works best when a good first guess is available, and a very satisfactory first guess is generally provided by the solution at the previous time step. Beginning with $\boldsymbol{\phi}^1 = \boldsymbol{\phi}_n$, one may compute the kth Newton iterate by solving the linear system

$$\mathbf{G}(\boldsymbol{\phi}^k)\boldsymbol{\phi}^{k+1} = \mathbf{G}(\boldsymbol{\phi}^k)\boldsymbol{\phi}^k - \mathbf{g}\left(\boldsymbol{\phi}^k\right) \tag{2.101}$$

for $\boldsymbol{\phi}^{k+1}$.

One advantage of using Newton's method with the exact Jacobian to solve (2.98) is that, in contrast to many alternative iterative schemes, it preserves chemical conservation laws. For example, one such law is the conservation of the total number of nitrogen atoms by the preceding NO_x–O_3 reactions, which may be verified by adding (2.93) and (2.94) to obtain $\dot{c}_2 + \dot{c}_3 = 0$, thereby implying $c_2 + c_3$ is constant. A second relation involving the molecules that act as repositories for atomic

[4] The system is autonomous if $\mathbf{f}$ is not an explicit function of t. Any system can be made autonomous by adding to a vector of unknowns one additional variable $\tilde{c}$, satisfying $\dot{\tilde{c}} = 1$.

oxygen is the constancy of $c_1 + c_3 + c_4$, which is revealed by adding together (2.92), (2.94), and (2.95). General conservation laws for an N-dimensional chemical system take the form $\boldsymbol{\gamma} \cdot \mathbf{c} =$ constant, where $\boldsymbol{\gamma} = (\gamma_1, \ldots, \gamma_N)$ and the γ_i are nonnegative constants.

To demonstrate how chemical conservation laws are preserved when using Newton's method to evaluate solutions to backward-Euler differencing, first note that when $\boldsymbol{\gamma} \cdot \mathbf{c}$ is constant, $\boldsymbol{\gamma} \cdot \dot{\mathbf{c}} = 0$ and the governing equations imply $\boldsymbol{\gamma} \cdot \mathbf{f}(\mathbf{c}) = 0$. Differentiating $\boldsymbol{\gamma} \cdot \mathbf{f}(\mathbf{c}) = 0$ with respect to an arbitrary c_j, yields

$$0 = \frac{\partial}{\partial c_j} \left(\sum_{i=1}^{N} \gamma_i f_i(\mathbf{c}) \right) = \sum_{i=1}^{N} \gamma_i \frac{\partial f_i}{\partial c_j}(\mathbf{c}) \quad \text{for all } j,$$

or equivalently, $\boldsymbol{\gamma} \cdot \mathbf{J}(\mathbf{c}) = \mathbf{0}$.

Now suppose a chemical system in which $\boldsymbol{\gamma} \cdot \mathbf{c}$ is constant is integrated using the backward-Euler method, and that (2.98) is evaluated numerically using Newton's method. Dotting $\boldsymbol{\gamma}$ with (2.101), using (2.99), (2.100), and the relations $\boldsymbol{\gamma} \cdot \mathbf{f}(\mathbf{c}) = 0$ and $\boldsymbol{\gamma} \cdot \mathbf{J}(\mathbf{c}) = 0$ gives

$$\boldsymbol{\gamma} \cdot \boldsymbol{\phi}^{k+1} = \boldsymbol{\gamma} \cdot \boldsymbol{\phi}^{k} - \left(\boldsymbol{\gamma} \cdot \boldsymbol{\phi}^{k} - \boldsymbol{\gamma} \cdot \boldsymbol{\phi}_n \right) = \boldsymbol{\gamma} \cdot \boldsymbol{\phi}_n,$$

implying that regardless of the number of Newton iterations, $\boldsymbol{\gamma} \cdot \boldsymbol{\phi}_{n+1} = \boldsymbol{\gamma} \cdot \boldsymbol{\phi}_n$.

For comparison with Rosenbrock Runge–Kutta methods, which are described in the next section, note that if we implement the backward-Euler scheme with two iterations of Newton's method per time step and define $\boldsymbol{q}_1 = \boldsymbol{\phi}^1 - \boldsymbol{\phi}_n$, $\boldsymbol{q}_2 = \boldsymbol{\phi}_{n+1} - \boldsymbol{\phi}^1$, then

$$\boldsymbol{\phi}_{n+1} = \boldsymbol{\phi}_n + \boldsymbol{q}_1 + \boldsymbol{q}_2 \tag{2.102}$$

and the two iterations of (2.101) may be expressed as

$$(\mathbf{I} - \Delta t\, \mathbf{J}(\boldsymbol{\phi}_n))\, \mathbf{q}_1 = \Delta t\, \mathbf{f}(\boldsymbol{\phi}_n), \tag{2.103}$$

$$(\mathbf{I} - \Delta t\, \mathbf{J}(\boldsymbol{\phi}_n + \boldsymbol{q}_1))\, \mathbf{q}_2 = \Delta t\, \mathbf{f}(\boldsymbol{\phi}_n + \mathbf{q}_1) - \mathbf{q}_1. \tag{2.104}$$

2.5.4 Rosenbrock Runge–Kutta Methods

Rosenbrock (1963) suggested incorporating the Jacobian directly into the integration formula to obtain linearly implicit approximations to nonlinear differential equations. In applications where the solution needs only be computed with low or moderate precision, Rosenbrock Runge–Kutta schemes can be more efficient than BDF methods (Sandu et al. 1997). In addition, Rosenbrock Runge–Kutta schemes (unlike second- or higher-order BDF methods) are single-step schemes, which facilitates their use in time-split applications (see Sect. 4.3).

An s-stage Rosenbrock Runge–Kutta scheme for the solution of the autonomous system of ordinary differential equations (2.97) may be expressed in the form

$$\boldsymbol{\phi}_{n+1} = \boldsymbol{\phi}_n + \sum_{j=1}^{s} b_j \mathbf{q}_j, \tag{2.105}$$

where

$$\mathbf{q}_j = \Delta t\, \mathbf{f}\left(\boldsymbol{\phi}_n + \sum_{i=1}^{j-1} a_{ij}\mathbf{q}_i\right) + \Delta t\, \mathbf{J}_n \sum_{i=1}^{j} \alpha_{ij}\mathbf{q}_i, \tag{2.106}$$

where $\boldsymbol{\phi}_n$ is once again the numerical approximation to $\mathbf{c}(n\Delta t)$, and $\mathbf{J}_n = \mathbf{J}(\boldsymbol{\phi}_n)$. If the term involving the Jacobian were omitted, the preceding method would be equivalent to the explicit s-stage Runge–Kutta method (2.31)–(2.35), in which $\Delta t F(\xi_i)$ in the earlier notation is now written as $\mathbf{q}_i$.

We focus on those two-stage methods for which $\alpha_{11} = \alpha_{22} = \alpha$, in which case (2.106) may be expressed as

$$(\mathbf{I} - \alpha\Delta t\ \mathbf{J}_n)\,\mathbf{q}_1 = \Delta t\, \mathbf{f}(\boldsymbol{\phi}_n), \tag{2.107}$$

$$(\mathbf{I} - \alpha\Delta t\ \mathbf{J}_n)\,\mathbf{q}_2 = \Delta t\, \mathbf{f}(\boldsymbol{\phi}_n + a_{21}\mathbf{q}_1) + \alpha_{21}\Delta t\ \mathbf{J}_n\mathbf{q}_1. \tag{2.108}$$

Using the same value for α_{11} and α_{22} makes the matrix multiplying the unknown increment $\mathbf{q}$ identical in both stages, thereby reducing the work required to solve the pair (2.107) and (2.108) by LU decomposition or by iterative methods that use a preconditioner.

Noting that $\ddot{\mathbf{c}} = \mathbf{J}\dot{\mathbf{c}} = \mathbf{J}\mathbf{f}(\mathbf{c})$ and expanding $\mathbf{c}$ in a Taylor series about $\mathbf{c}_n = \mathbf{c}(n\Delta t)$, (2.107) and (2.108) imply that through $O\left[(\Delta t)^2\right]$

$$\mathbf{q}_1 = \Delta t\, \dot{\mathbf{c}}_n + (\Delta t)^2\alpha\ddot{\mathbf{c}}_n, \quad \mathbf{q}_2 = \Delta t\, \dot{\mathbf{c}}_n + (\Delta t)^2(a_{21} + \alpha_{21} + \alpha)\ddot{\mathbf{c}}_n.$$

As usual, the condition for second-order accuracy is that

$$\mathbf{c}_{n+1} = \mathbf{c}_n + \Delta t\, \dot{\mathbf{c}}_n + \frac{(\Delta t)^2}{2}\ddot{\mathbf{c}}_n + O\left[(\Delta t)^3\right],$$

from which it follows that the two-stage method defined by (2.105), (2.107), and (2.108) is second-order accurate provided

$$b_1 + b_2 = 1, \quad b_2 a_{21} + b_2\alpha_{21} + \alpha = \frac{1}{2}. \tag{2.109}$$

The basic stability properties of the preceding scheme can be assessed by applying the method to the simple scalar test problem (2.16). For any second-order two-stage method, i.e., for those methods with coefficients satisfying (2.109), the amplification factor is

$$A = \frac{1 + (1 - 2\alpha)\gamma\Delta t + (\frac{1}{2} - 2\alpha + \alpha^2)(\gamma\Delta t)^2}{(1 - \alpha\gamma\Delta t)^2}, \tag{2.110}$$

which is identical to (2.52). Thus, second-order two-stage Rosenbrock Runge–Kutta schemes have the same absolute stability characteristics as two-stage, second-order diagonally implicit Runge–Kutta methods, and the method will be L-stable when $\alpha = 1 \pm 1/2\sqrt{2}$. Verwer et al. (1999) recommend using $\alpha = 1 + 1/2\sqrt{2}$, noting it gives better stability properties in nonlinear problems and is more effective at preserving positive chemical concentrations than the choice $\alpha = 1 - 1/2\sqrt{2}$.

If the problem is not autonomous, (2.106) becomes

$$\mathbf{q}_j = \Delta t\, \mathbf{f}\left(\boldsymbol{\phi}_n + \sum_{i=1}^{j-1} a_{ij}\mathbf{q}_i, t_n + b\Delta t\right) + \Delta t\, \mathbf{J}_n \sum_{i=1}^{j} \alpha_{ij}\mathbf{q}_i + \beta(\Delta t)^2 \frac{\partial \mathbf{f}}{\partial t}(\boldsymbol{\phi}_n, t_n),$$

where

$$b = \sum_{i=1}^{j-1} a_{ij}, \quad \beta = \sum_{i=1}^{j} \alpha_{ij},$$

and $\mathbf{J}_n$ is the Jacobian evaluated at $(\boldsymbol{\phi}_n, t_n)$. Nevertheless, the evaluation of $\partial \mathbf{f}/\partial t$ can be avoided in two-stage second-order methods as follows. Suppose that $\mathbf{J}_n$ is replaced by an arbitrary matrix $\mathbf{W}$ having the same dimensions as the Jacobian, such that (2.107) and (2.108) become

$$(\mathbf{I} - \alpha\Delta t\ \mathbf{W})\,\mathbf{q}_1 = \Delta t\, \mathbf{f}(\boldsymbol{\phi}_n, t_n), \tag{2.111}$$

$$(\mathbf{I} - \alpha\Delta t\ \mathbf{W})\,\mathbf{q}_2 = \Delta t\, \mathbf{f}(\boldsymbol{\phi}_n + a_{21}\mathbf{q}_1, t_n + a_{21}\Delta t) + \alpha_{21}\Delta t\ \mathbf{W}\mathbf{q}_1. \tag{2.112}$$

Again expanding the true solution in a Taylor series about $\mathbf{c}_n$, one finds that through $O\left[(\Delta t)^2\right]$,

$$\mathbf{q}_1 = \Delta t\, \dot{\mathbf{c}}_n + (\Delta t)^2\alpha\mathbf{W}\dot{\mathbf{c}}_n, \quad \mathbf{q}_2 = \Delta t\, \dot{\mathbf{c}}_n + (\Delta t)^2\left(a_{21}\ddot{\mathbf{c}}_n + (\alpha_{21} + \alpha)\mathbf{W}\dot{\mathbf{c}}_n\right).$$

Despite neglecting $\partial \mathbf{f}/\partial t$ and replacing the Jacobian by the arbitrary matrix $\mathbf{W}$, one can still obtain a second-order method if

$$b_1 = 1 - b_2, \quad a_{21} = \frac{1}{2b_2}, \quad \alpha_{21} = -\frac{\alpha}{b_2}.$$

Although (2.111) and (2.112) are second-order accurate, their stability properties depend on choosing a $\mathbf{W}$ that provides a reasonable approximation to the true Jacobian. Schemes in which the exact Jacobian is replaced by an approximation are known as W-methods.

As noted by Verwer et al. (1999), the matrix multiplication in the last term in (2.112) can be avoided by choosing $b_2 = 1/2$ and replacing $\mathbf{q}_2$ with $\tilde{\mathbf{q}}_2 + 2\mathbf{q}_1$. If one also sets $\mathbf{W} = \mathbf{J}_n$ to obtain good stability characteristics, the result is

$$\boldsymbol{\phi}_{n+1} = \boldsymbol{\phi}_n + \tfrac{3}{2}\mathbf{q}_1 + \tfrac{1}{2}\tilde{\mathbf{q}}_2, \tag{2.113}$$

$$(\mathbf{I} - \alpha\Delta t\ \mathbf{J}_n)\,\mathbf{q}_1 = \Delta t\, \mathbf{f}(\boldsymbol{\phi}_n, t_n), \tag{2.114}$$

$$(\mathbf{I} - \alpha\Delta t\ \mathbf{J}_n)\,\tilde{\mathbf{q}}_2 = \Delta t\, \mathbf{f}(\boldsymbol{\phi}_n + \mathbf{q}_1, t_n + \Delta t) - 2\mathbf{q}_1. \tag{2.115}$$

Choosing $\alpha = 1 + 1/2\sqrt{2}$ yields a second-order L-stable method that has been shown to perform well in several atmospheric chemistry problems (Verwer et al. 1999).

The preceding Rosenbrock method has several clear advantages over the approximate backward-Euler formulae (2.102)–(2.104). In particular, it avoids any worries about the number of Newton iterations required to achieve acceptable accuracy, and in contrast to the first-order truncation error of the backward-Euler method, (2.113)–(2.115) are second order. In addition, this Rosenbrock method is more efficient since it only requires one evaluation of the Jacobian matrix per time step, whereas two evaluations of J are required to advance one time step using (2.102)–(2.104).

The method (2.113)–(2.115) was used to integrate (2.92)–(2.95) with a constant 15-min time step to produce the dashed curves shown in Fig. 2.12. As discussed previously, even with such a large time step this method performs very well except for errors in its representation of the spike in atomic oxygen that occurs near sunrise. The treatment of this spike can be greatly improved by using a shorter time step. (The BDF solution shown by the solid line in Fig. 2.12 was computed using a nonconstant time step that was as small as 5×10^{-10} s around sunrise.) Jannelli and Fazio (2006) discussed a simple strategy for controlling the step size in integrations using (2.113)–(2.115).

2.6 Summary

This chapter has provided an overview of time-differencing schemes, covering several basic topics concerning the numerical solution of ordinary differential equations. We have primarily focused on methods useful for the solution of time-dependent partial differential equations, and have therefore not discussed a few important issues such as automatic step-size control. We have also focused on relatively low order methods. In many applications involving partial differential equations, the errors in the numerical representation of the spatial derivatives dominate the time discretization error, and as a consequence, the accuracy of the solution cannot easily be improved by using very high order time differences, or by adjusting the time step without simultaneously refining the resolution of the spatial domain.

Indeed second-order time differences would be sufficiently accurate for the solution of many partial differential equations. The main reason we have also explored third- and fourth-order methods is that some of them have attractive stability characteristics in applications involving advection or wave propagation. Recall that the two-stage second-order Runge–Kutta and two-step Adams–Bashforth schemes produce amplifying solutions to the oscillation equation. On the other hand, the three-step Adams–Bashforth method is stable and, unlike the second-order leapfrog scheme, has a strongly damped computational mode. The classical three-stage third-order and the four-stage fourth-order explicit Runge–Kutta schemes, as well as the four-stage explicit third-order SSPRK scheme (2.49), are also potentially attractive methods.

The properties of several basic schemes are compared with respect to their suitability to solve the oscillation equation in Tables 2.1 and 2.2. Table 2.1 gives the order and finite-difference formulae for each scheme. Although it is not apparent from their most common names, most of the schemes shown in Table 2.1 are Adams–Bashforth, Adams–Moulton, or Runge–Kutta schemes. As previously mentioned, forward differencing is both the first-order Adams–Bashforth method and the one-stage first-order explicit Runge–Kutta method. Backward differencing is a one-stage first-order implicit Runge–Kutta scheme. The trapezoidal method is a one-step Adams–Moulton method. The particular two-stage second-order Runge–Kutta scheme appearing in Table 2.1 is also the second-order Adams–Bashforth–Moulton predictor corrector.

Several important properties of the schemes listed in Table 2.1 are given in Table 2.2. The column labeled "storage factor" indicates the number of full arrays that must be allocated for each unknown variable in order to implement each scheme. Storage factors are not provided for the implicit methods listed in Table 2.2 because the storage factor for implicit methods can vary from problem to problem, depending on the numerical algorithm used to solve the implicit system. Inspection of Table 2.2 clearly reveals the low-storage advantage of the three-stage third-order Runge–Kutta scheme. This advantage may, however, be slightly exaggerated since the storage factors listed in Table 2.2 are upper limits that allow each method to be programmed in a completely straightforward manner. In many instances, it is possible to utilize less memory than that suggested by the storage factor if newly computed quantities are initially placed in a small, temporary storage array. As an example, when integrating a partial differential equation with forward time differencing, one cannot generally write the newly computed ϕ_j^{n+1} directly into the storage occupied by ϕ_j^n, because ϕ_j^n may be required for the computation of ϕ_{j+1}^{n+1}. However, at some point in the integration cycle, ϕ_j^n will no longer be needed and at that stage it may be overwritten by ϕ_j^{n+1}. During the interim between the calculation of ϕ_j^{n+1} and the last use of ϕ_j^n, ϕ_j^{n+1} may be held in a temporary storage array. In many applications, the temporary storage array can be much smaller than the full array required to hold a complete set of ϕ_n, and use of such a temporary array will reduce the storage factor by almost one unit.

In applications where storage is not a problem, the third-order Adams–Bashforth scheme can be an attractive alternative. The primary advantage of the third-order Adams–Bashforth scheme is its relative efficiency. In most practical applications involving partial differential equations, the bulk of the computational effort is associated with the evaluation of F, the function that determines the time derivative. Thus, a rough measure of the comparative efficiency of each method may be obtained by defining an efficiency factor as the maximum stable time step with which the oscillation equation can be integrated divided by the number of evaluations of $F(\phi)$ that each scheme requires to perform a single integration step. Inspection of Table 2.2 shows that, with the exception of the leapfrog scheme and its time-filtered variant, the third-order Adams–Bashforth scheme has the highest efficiency factor.

Table 2.1 Summary of methods for the solution of ordinary differential equations. The second- and third-order Runge–Kutta methods are low-storage variants; $h = \Delta t$

Method	Order	Formulae
Forward	1	$\phi_{n+1} = \phi_n + hF(\phi_n)$
Backward	1	$\phi_{n+1} = \phi_n + hF(\phi_{n+1})$
Asselin leapfrog	1	$\phi_{n+1} = \overline{\phi_{n-1}} + 2hF(\phi_n)$, $\overline{\phi_n} = \phi_n + \gamma(\overline{\phi_{n-1}} - 2\phi_n + \phi_{n+1})$
Leapfrog	2	$\phi_{n+1} = \phi_{n-1} + 2hF(\phi_n)$
Adams–Bashforth	2	$\phi_{n+1} = \phi_n + \frac{h}{2}[3F(\phi_n) - F(\phi_{n-1})]$
Trapezoidal	2	$\phi_{n+1} = \phi_n + \frac{h}{2}[F(\phi_{n+1}) + F(\phi_n)]$
Runge–Kutta (2-step explicit)	2	$q_1 = hF(\phi_n)$, $\phi_1 = \phi_n + q_1$, $q_2 = hF(\phi_1) - q_1$, $\phi_{n+1} = \phi_1 + q_2/2$
Magazenkov	2	$\phi_n = \phi_{n-2} + 2hF(\phi_{n-1})$ $\phi_{n+1} = \phi_n + \frac{h}{2}[3F(\phi_n) - F(\phi_{n-1})]$
Leapfrog–trapezoidal	2	$\phi_1 = \phi_{n-1} + 2hF(\phi_n)$, $\phi_{n+1} = \phi_n + \frac{h}{2}[F(\phi_1) + F(\phi_n)]$
Adams–Bashforth	3	$\phi_{n+1} = \phi_n + \frac{h}{12}[23F(\phi_n) - 16F(\phi_{n-1}) + 5F(\phi_{n-2})]$
Adams–Moulton	3	$\phi_{n+1} = \phi_n + \frac{h}{12}[5F(\phi_{n+1}) + 8F(\phi_n) - F(\phi_{n-1})]$
Adams–Bashforth–Moulton predictor corrector	3	$\phi_1 = \phi_n + \frac{h}{2}[3F(\phi_n) - F(\phi_{n-1})]$, $\phi_{n+1} = \phi_n + \frac{h}{12}[5F(\phi_1) + 8F(\phi_n) - F(\phi_{n-1})]$
Runge–Kutta (3-step explicit)	3	$q_1 = hF(\phi_n)$, $\phi_1 = \phi_n + q_1/3$, $q_2 = hF(\phi_1) - 5q_1/9$, $\phi_2 = \phi_1 + 15q_2/16$, $q_3 = hF(\phi_2) - 153q_2/128$, $\phi_{n+1} = \phi_2 + 8q_3/15$
Runge–Kutta (4-step explicit)	4	$q_1 = hF(\phi_n)$, $q_2 = hF(\phi_n + q_1/2)$, $q_3 = hF(\phi_n + q_2/2)$, $q_4 = hF(\phi_n + q_3)$, $\phi_{n+1} = \phi_n + (q_1 + 2q_2 + 2q_3 + q_4)/6$

Table 2.2 Characteristics of the schemes listed in Table 2.1. The amplification factor and relative phase change are for well-resolved solutions to the oscillation equation, and $s = \omega\Delta t$. Max s is the maximum value of $\omega\Delta t$ for which the solution is nonamplifying. The storage and efficiency factors are defined in the text. No storage factor is given for implicit schemes

Method	Storage factor	Efficiency factor	Amplification factor	Phase error	Max s
Forward	2	0	$1+\frac{s^2}{2}$	$1-\frac{s^2}{3}$	0
Backward	–	∞	$1-\frac{s^2}{2}$	$1-\frac{s^2}{3}$	∞
Asselin leapfrog	3	<1	$1-\frac{\gamma s^2}{2(1-\gamma)}$	$1+\frac{(1+2\gamma)s^2}{6(1-\gamma)}$	<1
Leapfrog	2	1	1	$1+\frac{s^2}{6}$	1
Adams–Bashforth-2	3	0	$1+\frac{s^4}{4}$	$1+\frac{5}{12}s^2$	0
Trapezoidal	–	∞	1	$1-\frac{s^2}{12}$	∞
Runge–Kutta-2	2	0	$1+\frac{s^4}{8}$	$1+\frac{s^2}{6}$	0
Magazenkov	3	0.67	$1-\frac{s^4}{4}$	$1+\frac{s^2}{6}$	0.67
Leapfrog–trapezoidal	3	0.71	$1-\frac{s^4}{4}$	$1-\frac{s^2}{12}$	1.41
Adams–Bashforth-3	4	0.72	$1-\frac{3}{8}s^4$	$1+\frac{289}{720}s^4$	0.72
Adams–Moulton	–	0	$1+\frac{s^4}{24}$	$1-\frac{11}{720}s^4$	0
Adams–Bashforth–Moulton predictor corrector	4	0.60	$1-\frac{19}{144}s^4$	$1+\frac{1243}{8640}s^4$	1.20
Runge–Kutta-3	2	0.58	$1-\frac{s^4}{24}$	$1+\frac{s^4}{30}$	1.73
Runge–Kutta-4	4[a]	0.70	$1-\frac{s^6}{144}$	$1-\frac{s^4}{120}$	2.82

[a]A storage factor of 3 may be achieved following the algorithm of Blum (1962).

Now consider problems where the amplitude of the solution decays with time, due to diffusion with a diffusivity M. As suggested in Sect. 2.2.4, the maximum stable time step with which such problems can be integrated using consistent explicit schemes must generally satisfy a constraint similar to $M\Delta t/(\Delta x)^2 < O(1)$, where Δx is the spatial grid spacing, which requires Δt to decrease more rapidly than Δx as the temporal and spatial meshes are refined. In such cases it is often most efficient to compute solutions using A-stable methods, among which the trapezoidal scheme is frequently the most attractive alternative. In the special case where M is an "eddy diffusivity" that is proportional to Δx, explicit methods remain efficient alternatives, and a scheme whose region of absolute stability includes a large segment of the negative real axis, such as the four-stage explicit third-order SSPRK scheme (2.49), is particularly appropriate.

Finally, suppose the amplitude of some components of a solution vector decay very rapidly, but that these components are sufficiently small during most of the period of integration such that they have only a minor influence on the solution. In a stiff problem such as this, L-stable methods are generally most efficient. Backward differentiation formulae are one classical way to integrate stiff problems; however, if one is content with low-order accuracy, Rosenbrock Runge–Kutta schemes may be more efficient. By incorporating the Jacobian directly in the integration formula, Rosenbrock Runge–Kutta methods also allow one to avoid the additional complications and possible inaccuracies involved in numerically solving the nonlinear algebraic equations generated by most implicit time-differencing schemes.

Problems

1. Suppose the truncation error in a finite difference method is of order s and that the global error at some particular physical time $(n\Delta t)$ satisfies $|E| = \alpha(\Delta t)^s$, where α is a constant. A series of numerical simulations are conducted in which Δt is repeatedly halved, and the error $E(\Delta t)$ is plotted as a function of Δt on a log–log scale. Show that the result will be a straight line of slope s.

2. *Test the theoretical result from the preceding problem by plotting the error versus Δt on a log–log scale for solutions obtained using the one-, two-, and four-stage versions of the explicit Runge–Kutta scheme (2.45).

(a) First test solutions to

$$\frac{d\psi}{dt} = \psi, \quad \psi(0) = 1.$$

Show the error at $t = 1$ for all three schemes as three separate curves on the same log–log plot. Compute solutions for $\Delta t = 0.05/(2^n)$ for $n = 0, 1, \ldots, 4$. Compare the slopes of each curve with lines with slopes of 1, 2, and 4. Describe how your results compare with theory.

(b) Next examine solutions to

$$\frac{d\psi}{dt} = \frac{\psi^3}{3}, \quad \psi(0) = 1$$

at $t = 1$. Perform the same analysis of the error versus Δt as you did in part (a), again using values of Δt in the range 0.05–0.05/(2^5). Again describe how your results compare with theory.

3. Use the mean-value theorem to prove (2.8).

4. Explicit two-stage second-order Runge–Kutta schemes are determined to within one free parameter. Using the order conditions given in Sect. 2.3.2, determine the number of free parameters available when specifying the coefficients of a third-order three-stage explicit Runge–Kutta scheme.

5. Show that the implicit midpoint method (2.50) is algebraically equivalent to

$$\phi_{n+1} = \phi_n + \Delta t\, F\left[\tfrac{1}{2}(\phi_{n+1} + \phi_n), t_{n+\frac{1}{2}}\right].$$

Both the implicit midpoint method and the trapezoidal method are second-order accurate. Compare the coefficients of the leading order truncation error in these two methods. Which is smaller?

6. Compare and contrast the instabilities that arise when the oscillation equation is integrated using either forward (Euler) differencing or the third-order two-step scheme (2.55). If the integration is to be terminated at a fixed time t_{f}, can either scheme be used to obtain a numerical solution that converges to $\psi(t_{\mathrm{f}})$ as Δt is repeatedly decreased?

7. Show that the three-step SSPRK scheme (2.48) is mathematically equivalent to the simple three-stage scheme (2.45) *when applied to the elementary test problem* (2.16).

8. *Evaluate the performance of several numerical schemes for approximating the two-component system of ordinary differential equations

$$\frac{du}{dt} = fv, \qquad \frac{dv}{dt} = -fu,$$

subject to the initial conditions $u(0) = 1$, $v(0) = 0$. Set $f = \pi$.

(a) Compare the exact solution with the solutions obtained using the following schemes: (1) forward, (2) backward, (3) trapezoidal, (4) Matsuno, (5) Huen variant of second-order Runge–Kutta, (6) leapfrog, and (7) second-order Adams–Bashforth. Initialize the leapfrog and second-order Adams–Bashforth schemes by taking one forward time step and set $f\Delta t = \pi/6$. Submit plots of u as a function of t comparing the various methods with the exact solution over the interval $0 \le t \le 6$. Set the vertical scale to $-2 \le u \le 2$ and terminate the curve for wildly unstable schemes when u exceeds these limits.

(b) Compare the average damping or amplification and the average phase-speed error per time step in your solution with the theoretical value *for small* $f\Delta t$. Choose $f\Delta t = 0.2$ for this comparison and present your results in a table. The table should contain the amplification per time step as predicted by theory (in the limit of good numerical resolution) and as determined from the numerical simulation. The table should also contain the phase-speed error as predicted by theory and as determined by the numerical solution. In gathering data for the table, run the simulations for long enough to get reasonable estimates for each numerical scheme.

9. *Suppose that one hopes to compute an extremely accurate approximation to the derivative simply by making the grid spacing extremely small.

(a) Compare the error in the approximation of the derivative of $\psi(t) = \cos(\pi t)$ at $t_n = 3/4$ using the one-sided difference (2.3), the centered difference (2.4), and the centered fourth-order difference

$$\frac{4}{3}\left(\frac{\psi(t_n + \Delta t) - \psi(t_n - \Delta t)}{2\Delta t}\right) - \frac{1}{3}\left(\frac{\psi(t_n + 2\Delta t) - \psi(t_n - 2\Delta t)}{4\Delta t}\right)$$

as Δt varies over the range between 1 and 10^{-16}. Choose 50 evenly spaced values of $\log(\Delta t)$ and graph the errors for these values versus Δt on a log–log plot. (*Hint:* An easy way to initialize the values of Δt in MATLAB is with the command: `dt = logspace(-16,0);`.)

(b) Why do the errors increase as Δt becomes very small? Which method is capable of producing the smallest error for an optimally selected value of Δt?

10. *Compare solutions to the ozone photochemistry problem (2.92)–(2.95) generated by the second-order Rosenbrock method (2.113)–(2.115) with those produced by two Newton's method iterations of the backward-Euler scheme (2.102)–(2.104). Use a constant time step of 15 min. Discuss errors in both the phase and the amplitude of the ozone concentration.

9. Suppose that one wishes to compute an extremely accurate approximation to the derivative simply by using the grid spacing [illegible] small.

(a) Compare the error in the approximation as the [illegible]

[illegible]

Chapter 3
Finite-Difference Approximations for One-Dimensional Transport

As discussed in Chap. 1, one basic strategy for representing continuous functions on digital computers is through the set of values assumed by the function at a finite number of grid points. Such grid-point methods approximate derivatives of the original continuous function using finite differences. Finite differences were introduced in connection with the solution of ordinary differential equations in Sect. 2.1.1. In this chapter we examine the behavior of numerical schemes in which finite differences replace both time and space derivatives in time-dependent partial differential equations.

3.1 Accuracy and Consistency

Recall that the truncation error of a finite difference indicates how rapidly its value approaches that of the true derivative of a smooth function[1] as the interval over which the difference is computed approaches zero. For example, the truncation error in the two-point centered difference is second order, as may be verified by expanding the smooth function $f(x)$ in a Taylor series about x_0 to yield

$$\frac{f(x_0+\Delta x)-f(x_0-\Delta x)}{\Delta x}-\frac{df}{dx}(x_0)=O\left[(\Delta x)^2\right].$$

In comparison with the approximation of time derivatives, the formulation of higher-order finite-difference approximations to spatial derivatives is facilitated by the stepwise nature of numerical solutions to most time-dependent problems. After the nth step of the integration, the numerical solution ϕ will be known at every spatial grid point, and several grid-point values may be easily included in any finite-difference approximation to a spatial derivative. Time derivatives, on the other hand, are typically approximated using as few time levels as possible to minimize storage requirements, and the only time levels available are those from previous

[1] A smooth function is one that has many continuous derivatives.

D.R. Durran, *Numerical Methods for Fluid Dynamics: With Applications to Geophysics*, Texts in Applied Mathematics 32, DOI 10.1007/978-1-4419-6412-0_3,

time steps. Thus, higher-order finite-difference approximations to time derivatives are inherently one-sided, as evident, for example, in the backward differentiation formula methods (2.88)–(2.90).

A centered fourth-order approximation to $\mathrm{d}f/\mathrm{d}x$ can be obtained by determining the five coefficients $a, b, \ldots e$ satisfying

$$\frac{df}{dx}(x_0) = af(x_0 + 2\Delta x) + bf(x_0 + \Delta x) + cf(x_0) + df(x_0 - \Delta x) + ef(x_0 - 2\Delta x) + O\left[(\Delta x)^4\right]. \tag{3.1}$$

Expanding $f(x_0 \pm \Delta x)$ and $f(x_0 \pm 2\Delta x)$ in Taylor series, substituting those expansions into (3.1), and equating the coefficients of like powers of Δx yields five equations for the unknown coefficients:

$$\begin{aligned} a + b + c + d + e &= 0, \\ 2a + b - d - 2e &= 1/\Delta x, \\ 4a + b + d + 4e &= 0, \\ 8a + b - d - 8e &= 0, \\ 16a + b + d + 16e &= 0. \end{aligned}$$

The unique solution to this system requires $c = 0$ and yields an approximation to the derivative of the form

$$\frac{df}{dx}(x_0) = \frac{4}{3}\left(\frac{f(x_0 + \Delta x) - f(x_0 - \Delta x)}{2\Delta x}\right) - \frac{1}{3}\left(\frac{f(x_0 + 2\Delta x) - f(x_0 - 2\Delta x)}{4\Delta x}\right) + O\left[(\Delta x)^4\right]. \tag{3.2}$$

Similar procedures can be used to generate even-higher-order formulae, off-centered formulae, and formulae for irregular grid intervals.

As an alternative to the brute force manipulation of Taylor series, the derivation of higher-order finite-difference formulae can be facilitated by the systematic use of operator notation and simple lower-order formulae. A simpler derivation of (3.2) may be obtained by defining a finite-difference operator δ_{nx} such that

$$\delta_{nx} f(x) = \frac{f(x + n\Delta x/2) - f(x - n\Delta x/2)}{n\Delta x}. \tag{3.3}$$

Using this notation, the centered second-order difference satisfies

$$\delta_{2x} f = \frac{df}{dx} + \frac{(\Delta x)^2}{6}\frac{\mathrm{d}^3 f}{\mathrm{d}x^3} + O\left[(\Delta x)^4\right]. \tag{3.4}$$

From the definition of δ_{nx},

$$\delta_x^2 f = \delta_x(\delta_x f) = \frac{f(x + \Delta x) - 2f(x) + f(x - \Delta x)}{(\Delta x)^2},$$

and a conventional Taylor series analysis of the truncation error shows that

$$\delta_x^2 f = \frac{d^2 f}{dx^2} + O\left[(\Delta x)^2\right].$$

It follows that $\delta_{2x}\delta_x^2 f$ is a second-order approximation to the third derivative of f, since

$$\delta_{2x}\delta_x^2 f = \delta_{2x}\left(\frac{d^2 f}{dx^2} + O\left[(\Delta x)^2\right]\right) = \frac{d^3 f}{dx^3} + O\left[(\Delta x)^2\right].$$

Substitution of the preceding equation into (3.4) yields

$$\left(1 - \frac{(\Delta x)^2}{6}\delta_x^2\right)\delta_{2x} f = \frac{df}{dx} + O\left[(\Delta x)^4\right]. \tag{3.5}$$

Expansion of this formula via the operator definition (3.3) yields the centered fourth-order difference (3.2). Although it allows finite-difference equations to be expressed in a very compact form, operator notation will be reserved for complicated formulae that become unwieldy when written in expanded form. Most of the finite-difference schemes considered in the remainder of this chapter are sufficiently simple that they will be expressed without using operator notation.

When f represents the flux of some physical quantity, it is sometimes advantageous to approximate its spatial derivative as the difference between the fluxes at the right and left boundaries of a single grid cell. Let $F_{j+1/2}$ be the approximate flux at the boundary between the cells centered at x_j and x_{j+1}, and $F_{j-1/2}$ the flux at the boundary between cells at x_{j-1} and x_j; then the flux divergence may be estimated as

$$\left(\frac{df}{dx}\right)_j \approx \frac{F_{j+\frac{1}{2}} - F_{j-\frac{1}{2}}}{\Delta x}. \tag{3.6}$$

Setting $F_{j+1/2} = (f_{j+1} + f_j)/2$ yields a simple centered second-order approximation. Taylor series expansions can be employed to obtain expressions for fluxes that give higher-order approximations to the flux divergence. For example, if

$$F_{j+\frac{1}{2}} = \frac{1}{3}f_{j+1} + \frac{5}{6}f_j - \frac{1}{6}f_{j-1},$$

(3.6) is accurate to $O\left[(\Delta x)^3\right]$. Expressions for fluxes giving divergences of orders 2–5 can be found in Table 5.2.

We turn now from the consideration of individual finite differences to examine the accuracy of an entire finite-difference scheme. Suppose that an approximation to the advection equation

$$\frac{\partial \psi}{\partial t} + c\frac{\partial \psi}{\partial x} = 0 \tag{3.7}$$

is to be obtained at the grid points $(n\Delta t, j\Delta x)$, where n and j are integers. It is convenient to represent the numerical approximation to $\psi(n\Delta t, j\Delta x)$ in the shorthand notation ϕ_j^n. One possible finite-difference formula for the numerical approximation of (3.7) is

$$\frac{\phi_j^{n+1} - \phi_j^n}{\Delta t} + c\frac{\phi_j^n - \phi_{j-1}^n}{\Delta x} = 0; \tag{3.8}$$

when $c > 0$, this is known as the "upstream" or "donor-cell" scheme. The order of accuracy of a finite-difference scheme is characterized by the residual error with which the solution of the continuous equation fails to satisfy the finite-difference formulation. Under the assumption that ψ is sufficiently smooth, its value at adjacent grid points can be obtained from a Taylor series expansion about $(n\Delta t, j\Delta x)$ and substituted into (3.8). Using (3.7) to simplify the result gives

$$\frac{\psi_j^{n+1} - \psi_j^n}{\Delta t} + c\frac{\psi_j^n - \psi_{j-1}^n}{\Delta x} = \frac{\Delta t}{2}\left(\frac{\partial^2 \psi}{\partial t^2}\right)_j^n - c\frac{\Delta x}{2}\left(\frac{\partial^2 \psi}{\partial x^2}\right)_j^n + \ldots, \tag{3.9}$$

where $\psi_j^n = \psi(n\Delta t, j\Delta x)$. The right side of (3.9) is the truncation error of the finite-difference scheme. The order of accuracy of the scheme is determined by the lowest powers of Δt and Δx appearing in the truncation error. According to (3.9), the upstream scheme is first-order accurate in space and time. If the truncation error of the finite-difference scheme approaches zero as $\Delta t \to 0$ and $\Delta x \to 0$, the scheme is *consistent*. Inspection of (3.9) clearly shows that the upstream scheme is consistent. Although it is not difficult to design consistent difference schemes, this property should not be taken for granted. One sometimes encounters methods that require additional relations between Δt and Δx, such as $\Delta t/\Delta x \to 0$, to achieve consistency.

3.2 Stability and Convergence

The preceding measures of accuracy do not describe the difference between the numerical solution ϕ_j^n and the true solution ψ_j^n, which, of course, is the most direct measure of the quality of the numerical solution. The error $\psi_j^n - \phi_j^n$ is a grid-point function, and its size is most conveniently measured by either the maximum norm or the L_2 (or Euclidean) norm.

The maximum norm, defined as

$$\|\phi\|_\infty = \max_{1 \le j \le N} |\phi_j|, \tag{3.10}$$

is simply the extremum of the grid-point values. The Euclidean, or L_2, norm is defined as

$$\|\phi\|_2 \equiv \left(\sum_{j=1}^{N} |\phi_j|^2 \Delta x\right)^{1/2}. \tag{3.11}$$

If the constant scaling factor Δx is ignored, (3.11) is just the length of an N-dimensional vector (hence the name Euclidean norm). The inclusion of the

Δx factor makes (3.11) a numerical approximation to the square root of the spatial integral of the function times its complex conjugate, $\phi\phi^*$, whence the name L_2 norm.[2] It might appear that the maximum norm is the most natural one to compute when working with grid-point values; however, the L_2 norm is also useful because it is more closely related to conserved physical quantities, such as the total energy. When it is not necessary to specify a specific norm, the subscript identifying the norm is omitted.

A finite-difference scheme is said to be *convergent* of order (p, q) if at any fixed physical time $n\Delta t$

$$\|\psi^n - \phi^n\| = O\left[(\Delta t)^p\right] + O\left[(\Delta x)^q\right]$$

in the limit $\Delta x, \Delta t \to 0$. Here ψ^n, the exact solution at time $n\Delta t$, is evaluated at the same grid points $j\Delta x$ as the approximate solution. The relationship between convergence and consistency is described by the *Lax equivalence theorem*, which states that *if a finite-difference scheme is linear, stable, and accurate of order* (p, q), *then it is convergent of order* (p, q) (Lax and Richtmyer 1956). As was the case with numerical approximations to ordinary differential equations, mere consistency is not enough to ensure the convergence of a numerical method; the method must also be *stable*.

The fundamental definition of stability makes no reference to the properties of the true solution and only identifies the least-restrictive additional constraint that must be satisfied to ensure the convergence of solutions generated by a consistent finite-difference scheme. A consistent linear finite-difference scheme will be convergent, and the Lax equivalence theorem will be satisfied, provided that for any time T there exists a constant C_T such that

$$\|\phi^n\| \le C_T \|\phi^0\| \quad \text{for all} \quad n\Delta t \le T \tag{3.12}$$

and all sufficiently small values of Δt and Δx. In the preceding expression, C_T may depend on the time T, but not on Δt, Δx, or the number of time steps n. This definition leaves the numerical solution tremendous latitude for growth with time, but it rules out solutions that grow as a function of the number of time steps. If a difference scheme is unstable in the sense that it fails to satisfy (3.12), repeated reductions in Δt and Δx may generate an unbounded amplification in the numerical approximation to the true solution at time T. In such a situation, the numerical solution could hardly be expected to converge to the true solution in the limit Δx, $\Delta t \to 0$.

The practical shortcoming of the preceding definition of stability is that it says nothing about the quality of the solution that might be obtained when using finite

[2] To better appreciate the notation used to represent the maximum and L_2 norms, note that $\|\phi\|_\infty$ is essentially the integral

$$\left(\int |\phi|^\infty \, \mathrm{d}x\right)^{1/\infty}$$

and $\|\phi\|_2$ is

$$\left(\int |\phi|^2 \, \mathrm{d}x\right)^{1/2}.$$

values of Δt and Δx; it only ensures that an accurate solution will be obtained in the limit $\Delta x, \Delta t \rightarrow 0$. Schemes that are stable according to the criterion (3.12) may, nevertheless, generate solutions that "blow up" in practical applications (see Sect. 3.5.2 for an example). To ensure that the numerical solution is qualitatively similar to the true solution when Δx and Δt are finite, it is often useful to impose stability constraints that are more stringent than (3.12). This is reminiscent of the way that A-stability was introduced in the study of ordinary differential equations to provide more information about the qualitative character of "nonconverged" solutions computed at relatively coarse resolution.

In many wave propagation or diffusion problems, the norm of the true solution is nonincreasing with time, and in such instances it is appropriate to require that the numerical scheme satisfy

$$\|\phi^n\| \leq \|\phi^0\| \quad \text{for all} \quad n. \tag{3.13}$$

In contrast to (3.12), this condition is not necessary for convergence, and it cannot be sensibly imposed without specific knowledge about the boundedness of the solutions to the associated partial differential equation. Nevertheless, (3.13) will generally be taken as the practical stability condition required for problems for which the norm of the true solution does not grow with time.

It is relatively easy to formulate consistent difference schemes and to determine their truncation error and order of accuracy. The analysis of stability can, however, be far more difficult, particularly when the finite-difference scheme and the associated partial differential equation are nonlinear. Thus, our initial discussion of stability will be focused on the simplest case – linear finite-difference schemes for the approximation of linear partial differential equations with constant coefficients. Nonlinear equations and linear equations with variable coefficients will be considered in Chap. 4.

3.2.1 The Energy Method

In practice, the energy method is used much less frequently than the von Neumann method, which will be discussed in the next section. Nevertheless, the energy method is important because, unlike the von Neumann method, it can be applied to nonlinear equations and to problems without periodic boundaries. The basic idea behind the energy method is to find a positive-definite quantity like $\sum_j (\phi_j^n)^2$ and show that this quantity is bounded for all n. If $\sum_j (\phi_j^n)^2$ is bounded, the solution is stable with respect to the L_2 norm.

As an example, let us investigate the stability of the upstream finite-difference scheme (3.8). Defining $\mu = c\Delta t/\Delta x$, we may write the scheme as

$$\phi_j^{n+1} = (1 - \mu)\phi_j^n + \mu\phi_{j-1}^n. \tag{3.14}$$

Squaring both sides and summing over all j gives

$$\sum_j (\phi_j^{n+1})^2 = \sum_j \left[(1-\mu)^2(\phi_j^n)^2 + 2\mu(1-\mu)\phi_j^n\phi_{j-1}^n + \mu^2(\phi_{j-1}^n)^2\right]. \quad (3.15)$$

Assuming cyclic boundary conditions,[3]

$$\sum_j (\phi_{j-1}^n)^2 = \sum_j (\phi_j^n)^2, \quad (3.16)$$

and using the Schwarz inequality (which states that for two vectors $\mathbf{u}$ and $\mathbf{v}$, $|\mathbf{u}\cdot\mathbf{v}| \le \|\mathbf{u}\|\,\|\mathbf{v}\|$),

$$\sum_j \phi_j^n\phi_{j-1}^n \le \left[\sum_j (\phi_j^n)^2\right]^{1/2} \left[\sum_j (\phi_{j-1}^n)^2\right]^{1/2} = \sum_j (\phi_j^n)^2. \quad (3.17)$$

If $\mu(1-\mu) \ge 0$, all three coefficients in (3.15) are positive, and (3.16) and (3.17) may be used to construct the inequality

$$\sum_j (\phi_j^{n+1})^2 \le \left[(1-\mu)^2 + 2\mu(1-\mu) + \mu^2\right] \sum_j (\phi_j^n)^2 = \sum_j (\phi_j^n)^2, \quad (3.18)$$

which requires $\|\phi^{n+1}\|_2 \le \|\phi^0\|_2$ and implies that the scheme is stable. The condition used to obtain (3.18),

$$\mu(1-\mu) \ge 0, \quad (3.19)$$

is therefore a sufficient condition for stability. Under the assumption that $\mu > 0$, division of (3.19) by μ leads to the relation $\mu \le 1$, and the total constraint on μ is therefore $0 < \mu \le 1$. A similar treatment of the case $\mu < 0$ leads to the contradictory requirement that $\mu > 1$ and provides no additional solutions. Thus, recalling the definition of μ and noting that $\mu = 0$ satisfies (3.19), we may write the stability condition as

$$0 \le \frac{c\Delta t}{\Delta x} \le 1.$$

As is typical with most conditionally stable difference schemes, there is a maximum limit on the time step beyond which the scheme is unstable, and the stability limit becomes more severe as the spatial resolution is increased.

---

[3] If more general boundary conditions are imposed at the edges of the spatial domain, a rigorous stability analysis becomes much more difficult. The determination of stability in the presence of nonperiodic boundaries is discussed in Sect. 9.1.6.

3.2.2 Von Neumann's Method

One drawback of the energy method is that each new problem requires fresh insight to define an appropriate energy and to show that the finite-difference scheme preserves a bound on that energy. Von Neumann's method has the advantage that it can be applied by following a prescribed procedure; however, it is applicable only to linear finite-difference equations with constant coefficients.[4] The basic idea of the von Neumann method is to represent the discretized solution at some particular time step by a finite Fourier series of the form

$$\phi_j^n = \sum_{k=-N}^{N} a_k^n \mathrm{e}^{\mathrm{i}kj\Delta x},$$

and to examine the stability of the individual Fourier components. The total solution will be stable if and only if every Fourier component is stable. The use of finite Fourier series is strictly appropriate only if the spatial domain is periodic. When problems are posed with more general boundary conditions, a rigorous stability analysis is more difficult, but the von Neumann method still provides a useful way of weeding out obviously unsuitable schemes.

A key property of Fourier series is that individual Fourier modes are eigenfunctions of the derivative operator, i.e.,

$$\frac{d}{dx}\mathrm{e}^{\mathrm{i}kx} = ik\mathrm{e}^{\mathrm{i}kx}.$$

Finite Fourier series have an analogous property in that individual modes $\mathrm{e}^{\mathrm{i}kj\Delta x}$ are eigenfunctions of linear finite-difference operators. Thus, if the initial conditions for some linear, constant-coefficient finite-difference scheme are $\phi_j^n = \mathrm{e}^{\mathrm{i}kj\Delta x}$, after one iteration the solution will have the form

$$\phi_j^{n+1} = A_k \mathrm{e}^{\mathrm{i}kj\Delta x},$$

where the *amplification factor* A_k is a complex constant determined by the form of the finite-difference formula.[5] Since the analysis is restricted to linear constant-coefficient schemes, the amplification factor will not vary from time step to time step, and if a_k^n denotes the amplitude of the kth finite Fourier component at the nth time step, then

$$a_k^n = A_k a_k^{n-1} = (A_k)^n a_k^0.$$

It follows that the stability of each Fourier component is determined by the modulus of its amplification factor.

[4] To apply von Neumann's method to more general problems, the governing finite-difference equations must be linearized and any variable coefficients must be frozen at some constant value. The von Neumann stability of the family of linearized, frozen-coefficient systems may then be examined. See Sect. 4.4.

[5] Except in this section, the amplification factor will be written without the subscript k for conciseness. When several schemes are discussed together, subscripts in roman font will be used to distinguish between the amplification factors of the various schemes.

The *von Neumann stability condition*, which is necessary and sufficient for the stability of a two-time-level linear constant-coefficient finite-difference equation,[6] requires the amplification factor of every Fourier component resolvable on the grid to be bounded such that, for all sufficiently small Δt and Δx,

$$|A_k| \leq 1 + \eta \Delta t, \tag{3.20}$$

where η is a constant independent of k, Δt, and Δx. This condition ensures that a consistent finite-difference scheme satisfies the minimum stability criteria for convergence in the limit Δx, $\Delta t \to 0$, (3.12). In applications where the true solution is bounded by the norm of the initial data, it is usually advantageous to enforce the more stringent requirement that

$$|A_k| \leq 1, \tag{3.21}$$

which will guarantee satisfaction of the stability condition (3.13). When the von Neumann condition is satisfied, every finite Fourier component is stable, and the full solution, being a linear combination of the individual Fourier components, must also be stable.

As an illustration of the von Neumann method, consider once again the finite-difference equation (3.14). The solutions to the associated partial differential equation (3.7) do not grow with time, so we will require $|A_k| \leq 1$. Substitution of an arbitrary Fourier component, of the form $e^{ikj\Delta x}$, into (3.14) yields

$$A_k e^{ikj\Delta x} = (1 - \mu)e^{ikj\Delta x} + \mu e^{ik(j-1)\Delta x}.$$

Dividing out the common factor $e^{ikj\Delta x}$ gives

$$A_k = 1 - \mu + \mu e^{-ik\Delta x}. \tag{3.22}$$

The magnitude of A_k is obtained by multiplying by its complex conjugate and taking the square root. Thus,

$$\begin{aligned} |A_k|^2 &= (1 - \mu + \mu e^{-ik\Delta x})(1 - \mu + \mu e^{ik\Delta x}) \\ &= 1 - 2\mu(1 - \mu)(1 - \cos k\Delta x). \end{aligned} \tag{3.23}$$

The von Neumann condition (3.21) will therefore be satisfied if

$$1 - 2\mu(1 - \mu)(1 - \cos k\Delta x) \leq 1.$$

Since $1 - \cos k\Delta x > 0$ for all wave numbers except the trivial case $k = 0$, the preceding inequality reduces to

$$\mu(1 - \mu) \geq 0,$$

---

[6] The sufficiency of the von Neumann condition holds only for single equations in one unknown. The stability of systems of finite-difference equations in several unknown variables is discussed in Sect. 4.1.

which is identical to the condition (3.19) obtained using the energy method. As discussed previously in connection with (3.19), this stability condition may be expressed as

$$0 \leq \frac{c\Delta t}{\Delta x} \leq 1.$$

Inspection of (3.23) shows that the $2\Delta x$ wave grows most rapidly in any integration performed with an unstable value of μ. Thus, as it "blows up," an unstable solution becomes dominated by large-amplitude $2\Delta x$ waves. Most other finite-difference approximations to the advection equation exhibit the same tendency: When solutions become unstable, they usually become contaminated by large-amplitude short waves. The upstream scheme is, nevertheless, unusual in that all waves become unstable for the same critical value of μ. In many other schemes, such as the leapfrog-time centered-space formulation (3.59), there exist values of μ for which only a few of the shorter wavelengths are unstable. One might suppose that such nominally unstable values of μ could still be used in numerical integrations if the initial data were filtered to remove all amplitude from the unstable finite Fourier components; however, even if the initial data have zero amplitude in the unstable modes, round-off error in the numerical computations will excite the unstable modes and trigger the instability.

3.2.3 The Courant–Friedrichs–Lewy Condition

The basic idea of the Courant–Friedrichs–Lewy (CFL) condition is that the solution of a finite-difference equation must not be independent of the data that determine the solution to the associated partial differential equation. The CFL condition can be made more precise by defining the *domain of influence* of a point (x_0, t_0) as that region of the x–t plane where the solution to some particular partial differential equation is influenced by the solution at (x_0, t_0). A related concept, the *domain of dependence* of a point (x_0, t_0), is defined as the set of points containing (x_0, t_0) within their domains of influence. The domain of dependence of (x_0, t_0) will therefore consist of all points (x, t) at which the solution has some influence on the solution at (x_0, t_0). A similar concept applicable to the discretized problem is the *numerical domain of dependence* of a grid point $(n_0\Delta t, j_0\Delta x)$, which consists of the set of all nodes on the space–time grid $(n\Delta t, j\Delta x)$ at which the value of the numerical solution influences the numerical solution at $(n_0\Delta t, j_0\Delta x)$. *The CFL condition requires that the numerical domain of dependence of a finite-difference scheme include the domain of dependence of the associated partial differential equation.* Satisfaction of the CFL condition is a necessary condition for stability, but is not sufficient to guarantee stability.

The nature of the CFL condition can be illustrated by considering the advection equation (3.7), which has general solutions of the form $\psi(x - ct)$. Thus, the true

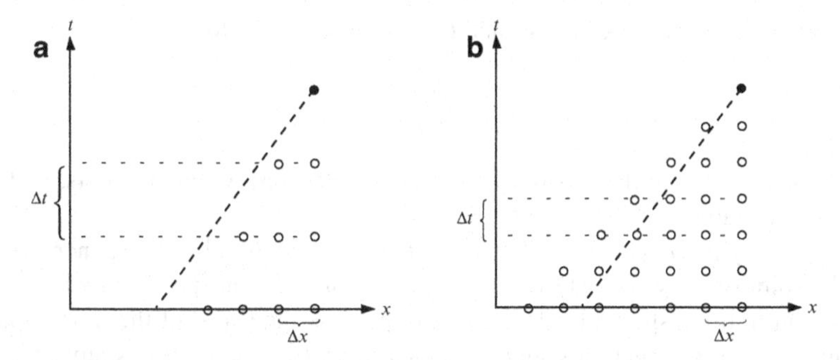

Fig. 3.1 The influence of the time step on the relationship between the numerical domain of dependence of the upstream scheme (*open circles*) and the true domain of dependence of the advection equation (*dashed line*): **a** unstable Δt, **b** stable Δt

domain of influence of a point (x_0, t_0) is the straight line

$$t = t_0 + \tfrac{1}{c}(x - x_0), \quad t \geq t_0.$$

The same "characteristic line" also defines the true domain of dependence of (x_0, t_0), except that one looks backward in time by requiring $t \leq t_0$. The true domain of dependence is plotted as a dashed line in Fig. 3.1, together with those grid points composing the numerical domain of dependence of the upstream finite-difference scheme (3.14). The two panels in this figure show the influence of two different time steps on the shape of the numerical domain of dependence. In Fig. 3.1a, the initial value of ψ along the x-axis, which determines the solution to the partial differential equation at $(n\Delta t, j\Delta x)$, plays no role in the determination of the finite-difference solution ϕ_j^n. The numerical solution can be in error by any arbitrary amount, and will not converge to the true solution as $\Delta t, \Delta x \to 0$ unless there is a change in the ratio $\Delta t/\Delta x$. Hence, the finite-difference method, which is consistent with the original partial differential equation, must be unstable (or else the Lax equivalence theorem would be violated).

The situation shown in Fig. 3.1b is obtained by halving the time step. Then the numerical domain of dependence contains the domain of dependence of the true solution, and it is possible for the numerical solution to be stable. In this example the CFL condition requires the slope of the characteristic curve to be greater than the slope of the left edge – and less than the slope of the right edge – of the domain of dependence. As evident from Fig. 3.1, the slope condition at the right edge of the domain is $1/c \leq \infty$, which is always satisfied. The slope condition at the left edge of the domain may be expressed as $\Delta t/\Delta x \leq 1/c$. If $c > 0$, this requires

$$c\Delta t/\Delta x \leq 1, \tag{3.24}$$

and the nonnegativity of Δt and Δx implies

$$c\Delta t/\Delta x \geq 0. \tag{3.25}$$

Simultaneous satisfaction of (3.24) and (3.25) is obtained when

$$0 \leq c\frac{\Delta t}{\Delta x} \leq 1. \tag{3.26}$$

In the case $c < 0$, similar reasoning leads to contradictory requirements, and the solution is unstable.

The quantity $|c\Delta t/\Delta x|$ is known as the *Courant number*. In more general problems the solution may consist of waves traveling at different speeds, in which case the Courant number should be defined such that c is the speed of the most rapidly moving wave. It is sometimes suggested that *the CFL condition* is simply the requirement $|c\Delta t/\Delta x| \leq 1$, but as demonstrated in the next example, the true CFL condition is typically not $|c\Delta t/\Delta x| \leq 1$. In fact, in some cases there is no point in trying to characterize the true CFL condition in terms of the Courant number because the Courant number has only minimal influence on the geometry of the numerical domain of dependence. For example, all stable semi-Lagrangian approximations to the advection equation satisfy the CFL condition, and do so without respecting any particular maximum value for $|c\Delta t/\Delta x|$ (see Chap. 7).

The condition (3.26) is identical to those stability conditions already obtained using the energy and von Neumann methods, but such agreement is actually rather unusual. The CFL condition is only a necessary condition for stability, and in many cases the sufficient conditions for stability are more restrictive than those required by the CFL condition. As an example, consider the following approximation to the advection equation,

$$\delta_{2t}\phi + c\left(\frac{4}{3}\delta_{2x}\phi - \frac{1}{3}\delta_{4x}\phi\right) = 0,$$

which uses the fourth-order-accurate approximation to the spatial derivative (3.2). Since the spatial difference utilizes a five-grid-point-wide stencil, the CFL condition is satisfied when

$$\left|c\frac{\Delta t}{\Delta x}\right| \leq 2.$$

Yet the actual sufficient condition for stability is the much more restrictive condition

$$\left|c\frac{\Delta t}{\Delta x}\right| \leq 0.728,$$

which may be derived via a von Neumann stability analysis.

3.3 Space Differencing for Simulating Advection

Having examined the errors associated with time differencing in Chap. 2, let us now consider the errors introduced when spatial derivatives are replaced with finite differences. To isolate the influence of the spatial differencing, the time dependence will not be discretized. Our investigation will focus on the constant-wind-speed advection equation,

$$\frac{\partial \psi}{\partial t} + c\frac{\partial \psi}{\partial x} = 0, \tag{3.27}$$

which both describes the important physical process of transport by macroscopic fluid motions and also provides the simplest prototype equation with wavelike solutions (as discussed in connection with (1.4)). If the x-domain is periodic (or unbounded), the spatial structure of the solution may be represented by a Fourier series (or a Fourier integral), and a solution for each individual mode may be sought in the form of a traveling wave:

$$\psi(x,t) = \mathrm{e}^{\mathrm{i}(kx-\omega t)}.$$

Here k, equal to 2π divided by the wavelength, is the *wave number*, and ω, equal to 2π divided by the period, is the *frequency*. Substitution of this assumed solution into (3.27) shows that the traveling wave will satisfy the governing equation only if its frequency satisfies the *dispersion relation*

$$\omega = ck.$$

The wave travels with constant amplitude at a *phase speed* $\omega/k = c$. These waves are *nondispersive*, meaning that their phase speed is independent of the wave number. The energy associated with an isolated "packet" of waves propagates at the group velocity $\partial\omega/\partial k = c$, which is also independent of wave number. Readers unfamiliar with the concept of group velocity may wish to consult Gill (1982) or Whitham (1974).

3.3.1 Differential–Difference Equations and Wave Dispersion

Suppose that the spatial derivative in the advection equation is replaced with a second-order centered difference. Then (3.27) becomes the *differential–difference* equation:[7]

$$\frac{d\phi_j}{dt} + c\left(\frac{\phi_{j+1} - \phi_{j-1}}{2\Delta x}\right) = 0. \tag{3.28}$$

Individual wavelike solutions to this equation may be obtained in the form

$$\phi_j(t) = \mathrm{e}^{\mathrm{i}(kj\Delta x - \omega_{2c} t)}, \tag{3.29}$$

[7] The set of differential–difference equations (3.28) for ϕ_j at every grid point constitute a large system of ordinary differential equations that could, in principle, be evaluated numerically using standard packages. This procedure, known as the *method of lines*, is usually not the most efficient approach.

where ω_{2c} denotes the frequency associated with centered second-order spatial differencing. Substitution of (3.29) into the differential–difference equation yields

$$-\mathrm{i}\omega_{2c}\phi_j = -c\left(\frac{\mathrm{e}^{\mathrm{i}k\Delta x} - \mathrm{e}^{-\mathrm{i}k\Delta x}}{2\Delta x}\right)\phi_j,$$

from which one obtains the dispersion relation

$$\omega_{2c} = c\frac{\sin k\Delta x}{\Delta x}. \tag{3.30}$$

Because ω_{2c} is real, there is no change in wave amplitude with time, and therefore no amplitude error. However, the phase speed,

$$c_{2c} \equiv \frac{\omega_{2c}}{k} = c\frac{\sin k\Delta x}{k\Delta x}, \tag{3.31}$$

is a function of k, so unlike the solutions to the original advection equation, these waves are dispersive. If the numerical resolution is good, $k\Delta x \ll 1$, and the Taylor series expansion $\sin x \approx x - x^3/6$ may be used to obtain

$$c_{2c} \approx c\left[1 - \tfrac{1}{6}(k\Delta x)^2\right],$$

showing that the phase-speed error is second order in $k\Delta x$. Although the error for a well-resolved wave is small, the phase-speed error does become significant as the spatial resolution decreases. The least well resolved wave on a numerical grid has wavelength $2\Delta x$ and wave number $k = \pi/\Delta x$. According to (3.31), *the phase speed of the* $2\Delta x$ *wave is zero*. Needless to say, this is a considerable error. The situation with the group velocity

$$\frac{\partial\omega_{2c}}{\partial k} = c\cos k\Delta x \tag{3.32}$$

is, however, even worse. The group velocity of well-resolved waves is approximately correct, but the group velocity of the poorly resolved waves is severely retarded. The group velocity of the $2\Delta x$ wave is $-c$; its energy propagates backward!

If the spatial derivative in the advection equation is replaced with a fourth-order centered difference, the resulting differential–difference equation

$$\frac{d\phi_j}{dt} + c\left[\frac{4}{3}\left(\frac{\phi_{j+1} - \phi_{j-1}}{2\Delta x}\right) - \frac{1}{3}\left(\frac{\phi_{j+2} - \phi_{j-2}}{4\Delta x}\right)\right] = 0 \tag{3.33}$$

has wave solutions of the form (3.29), provided that the frequency ω_{4c} satisfies the dispersion relation

$$\omega_{4c} = \frac{c}{\Delta x}\left(\frac{4}{3}\sin k\Delta x - \frac{1}{6}\sin 2k\Delta x\right).$$

As is the case for centered second-order differences, there is no amplitude error, only phase-speed error. Once again, the waves are dispersive, and the phase speed of the $2\Delta x$ wave is zero. The phase-speed error of a well-resolved wave is, however, reduced to $O\left[(k\Delta x)^4\right]$, since for $k\Delta x$ small,

$$c_{4c} = \frac{\omega_{4c}}{k} \approx c\left(1 - \frac{(k\Delta x)^4}{30}\right).$$

The group velocity

$$\frac{\partial \omega_{4c}}{\partial k} = c\left(\frac{4}{3}\cos k\Delta x - \frac{1}{3}\cos 2k\Delta x\right) \tag{3.34}$$

is also fourth-order accurate for well-resolved waves, but the group velocity of the $2\Delta x$ wave is $-5c/3$, an even greater error than that obtained using centered second-order differences.

The influence of spatial differencing on the frequency is illustrated in Fig. 3.2. As suggested by the preceding analysis, ω_{4c} approaches the true frequency more rapidly than ω_{2c} as $k\Delta x \to 0$, but both finite-difference schemes completely fail to capture the oscillation of $2\Delta x$ waves. The greatest advantages of the fourth-order difference over the second-order formulation are evident at "intermediate" wavelengths on the order of $3\Delta x$ to $8\Delta x$. The improvements in the frequencies of these intermediate waves also generates a considerable improvement in their phase speeds and group velocities. The variation in the phase speed of a Fourier mode as a function of wave number is shown in Fig. 3.3. The improvement in the phase speed associated

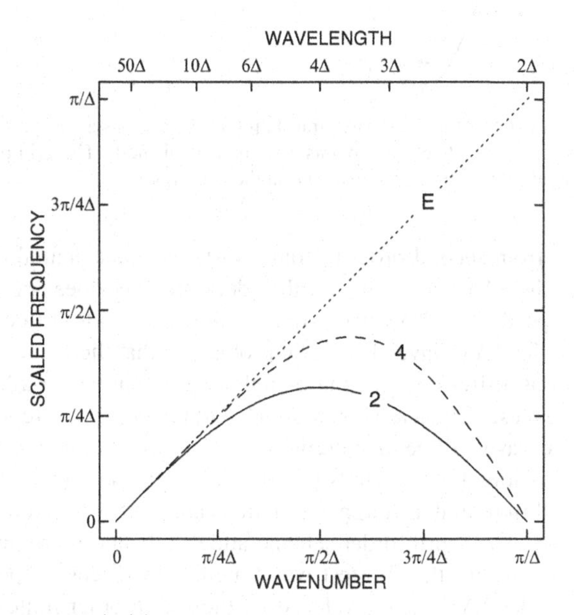

Fig. 3.2 Scaled frequency (ω/c) as a function of wave number for the analytic solution of the advection equation (*dotted line*) and for corresponding differential–difference approximations using second-order (*solid line*) and centered fourth-order (*dashed line*) differences

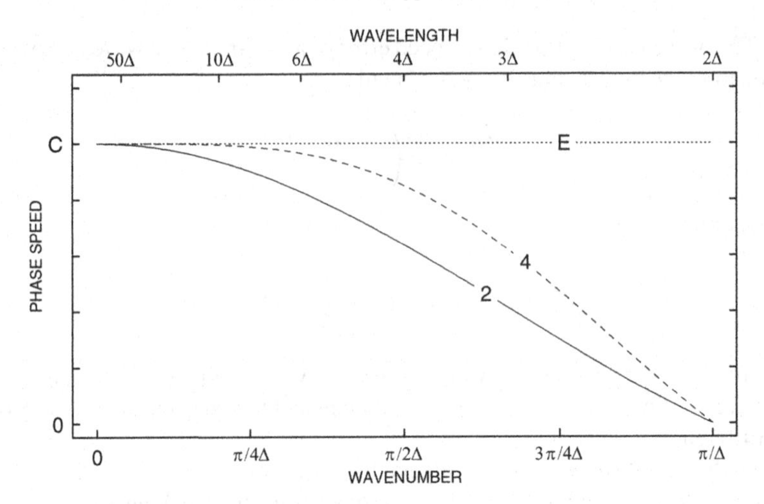

Fig. 3.3 Phase speed as a function of numerical resolution for the analytic solution of the advection equation (*dotted line*) and for corresponding differential–difference approximations using second-order (*solid line*) and centered fourth-order (*dashed line*) differences

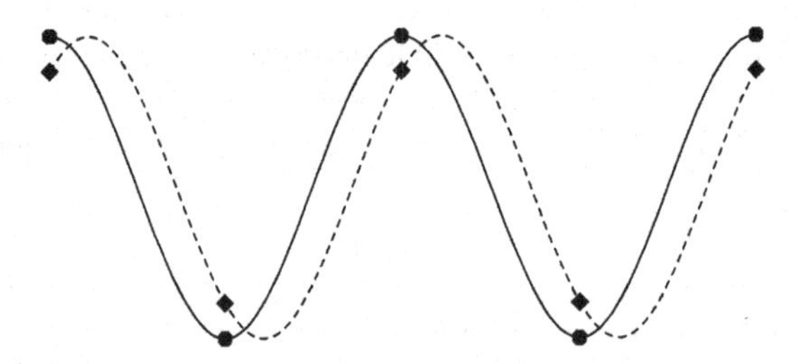

Fig. 3.4 Misrepresentation of a $2\Delta x$ wave translating to the right as a decaying standing wave when the wave is sampled at fixed grid points on a numerical mesh. The grid-point values are indicated by *dots* at the earlier time and *diamonds* at the later time

with an increase from second-order to fourth-order accurate spatial differences is apparent even in the $3\Delta x$ wave. The fourth-order difference does not, however, improve the phase speed of the $2\Delta x$ wave. In fact, almost all finite-difference schemes fail to propagate the $2\Delta x$ wave. The basic problem is that there are only two possible configurations, differing by a phase angle of 180°, in which $2\Delta x$ waves can appear on a finite mesh. Thus, as shown in Fig. 3.4, the grid-point representation of a translating $2\Delta x$ wave will be misinterpreted as a decaying standing wave.

The group velocities for the true solution and for the solutions of the second- and fourth-order differential–difference equations are plotted as a function of wave number in Fig. 3.5. The fourth-order scheme allows a better approximation of the group velocity for all but the shortest wavelengths. As discussed previously, the group velocity of the $2\Delta x$ wave produced by the fourth-order finite difference is actually worse than that obtained with the second-order method. The degradation of the $2\Delta x$ group velocity in the higher-order scheme – or equivalently, the increase

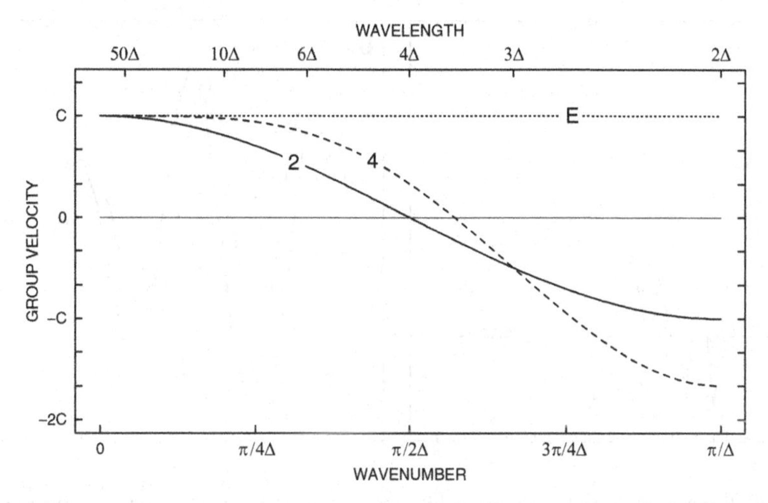

Fig. 3.5 Group velocity as a function of numerical resolution for the analytic solution of the advection equation (*dotted line*) and corresponding differential–difference approximations using second-order (*solid line*) and centered fourth-order (*dashed line*) differences

in $\partial\omega/\partial k$ at $k = \pi/\Delta x$ in Fig. 3.2 – is an unfortunate by-product of the inability of all finite-difference schemes to propagate the $2\Delta x$ wave and the otherwise desirable tendency of higher-order schemes to better approximate ω for wavelengths slightly longer than $2\Delta x$. In the absence of dissipation, the large negative group velocities associated with the $2\Delta x$ wave rapidly spread short-wavelength noise away from regions where $2\Delta x$ waves are forced.

One might attempt to improve the representation of extremely short waves by avoiding centered differences. If the spatial derivative in the advection equation is replaced with a first-order one-sided difference, (3.27) becomes

$$\frac{d\phi_j}{dt} + c\left(\frac{\phi_j - \phi_{j-1}}{\Delta x}\right) = 0. \tag{3.35}$$

Substitution of a wave solution of the form (3.29) into (3.35) yields the dispersion relation for the frequency associated with one-sided spatial differencing,

$$\omega_{1s} = \frac{c}{\mathrm{i}\Delta x}\left(1 - \mathrm{e}^{-ik\Delta x}\right) = \frac{c}{\Delta x}\big(\sin k\Delta x + \mathrm{i}(\cos k\Delta x - 1)\big). \tag{3.36}$$

The real part of ω_{1s} is identical to the real part of ω_{2c}, and hence one-sided spatial differencing introduces the same dispersive error as centered second-order spatial differencing. Unlike centered differencing, however, the one-sided difference also generates amplitude error through the imaginary part of ω_{1s}. The amplitude of the differential–difference solution will grow or decay at the rate

$$\exp\left(-\frac{c}{\Delta x}(1 - \cos k\Delta x)t\right).$$

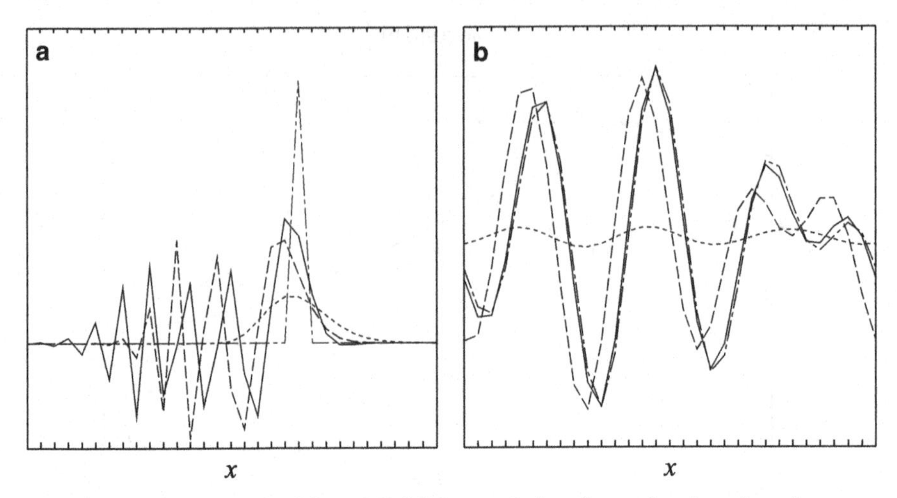

Fig. 3.6 Exact solution and differential–difference solutions for **a** advection of a spike over a distance of five grid points, and **b** advection of the sum of equal-amplitude $7.5\Delta x$ and $10\Delta x$ sine waves over a distance of 12 grid points. Exact solution (*dot-dashed line*), one-sided first-order solution (*short-dashed line*), centered second-order solution (*long-dashed line*), and centered fourth-order solution (*solid line*). The distribution is translating to the right. Grid-point locations are indicated by the *tick marks* at the top and bottom of the plot

Thus, poorly resolved waves change amplitude most rapidly. If $c > 0$, the solution damps; the solution amplifies when $c < 0$. Note that if $c < 0$, the numerical domain of dependence does not include the domain of dependence of the original partial differential equation, so instability could also be predicted from the CFL condition.

A comparison of the performance of first-order, second-order, and fourth-order spatial differencing is provided in Fig. 3.6, which shows analytic solutions to the advection equation and numerical solutions to the corresponding differential–difference problem. The differential–difference equations are solved numerically on a periodic spatial domain using a fourth-order Runge–Kutta scheme to integrate (3.28), (3.33), and (3.35) with a very small time step.

Figure 3.6a shows the distribution of ϕ that develops when the initial condition is a narrow spike, such that $\phi_j(t = 0)$ is zero everywhere except at the midpoint of the domain. Although the numerical domain is periodic, large-amplitude perturbations have not reached the lateral boundaries at the time shown in Fig. 3.6a. The narrow initial spike is formed by the superposition of many waves of different wavelengths; however, the Fourier components with largest amplitude are all of very short wavelength. The large diffusive error generated by one-sided differencing rapidly damps these short wavelengths and reduces the spike to a highly smoothed low-amplitude disturbance. The centered second- and fourth-order differences also produce a dramatic distortion in the amplitude of the solution. Although the centered schemes preserve the amplitude of each individual Fourier component, the various components propagate at different speeds, and thus the superposition of these components ceases to properly represent the true solution. Consistent with the values of the group velocity given by (3.32) and (3.34), the energy in the shortest waves propagates back upstream from the initial location of the spike. As predicted by theory, the upstream propagation of the $2\Delta x$ wave is most rapid for the fourth-order method. Switching to a higher-order scheme does not improve the performance of finite-difference methods when they are used to model poorly resolved features like the spike in Fig. 3.6a; in fact, in many respects the fourth-order solution is worse than the second-order result.

The spike test is an extreme example of a common problem for which many numerical schemes are poorly suited, namely, the task of properly representing solutions with near discontinuities. As such, the spike test provides a reference point that characterizes a scheme's ability to properly model poorly resolved waves. A second important reference point is provided by the test in Fig. 3.6b, which examines each scheme's ability to approximate features at an intermediate numerical resolution. The solution in Fig. 3.6b is the sum of equal-amplitude $7.5\Delta x$ and $10\Delta x$ waves; in all other respects the problem is identical to that in Fig. 3.6a. Unlike the situation with the spike test, the higher-order schemes are clearly superior in their treatment of the waves in Fig. 3.6b. Whereas the first-order difference generates substantial amplitude error and is distinctly inferior to the other two schemes, the second-order difference produces a reasonable approximation to the correct solution. Centered second-order differencing does, however, generate a noticeable lag in the phase speed of the disturbance (as in (3.31)). Moreover, since the phase lag of the $7.5\Delta x$ wave differs from that of the $10\Delta x$ wave in the second-order solution, the relative phase of the two waves changes during the simulation, and a significant error develops in the amplitude of the two rightmost wave crests. This example serves to emphasize that although centered differences do not produce amplitude errors in individual Fourier components, they still generate amplitude errors in the total solution. Finally, in contrast to the first- and second-order schemes, the errors introduced by fourth-order differencing are barely detectable at this time in the simulation.

The damping associated with the first-order upstream scheme (3.35) can be significantly reduced by using a higher-order one-sided difference. The differential–difference equation

$$\frac{d\phi_j}{dt} + \frac{c}{6}\left(\frac{2\phi_{j+1} + 3\phi_j - 6\phi_{j-1} + \phi_{j-2}}{\Delta x}\right) = 0 \tag{3.37}$$

may be obtained by replacing the spatial derivative in the advection equation with a third-order difference. The dispersion relation associated with this differential–difference equation is

$$\omega_{3s} = \frac{c}{\Delta x}\left[\left(\frac{4}{3}\sin k\Delta x - \frac{1}{6}\sin 2k\Delta x\right) - \frac{i}{3}(1 - \cos k\Delta x)^2\right]. \tag{3.38}$$

The real part of ω_{3s} is identical to that of ω_{4c}, and the phase-speed errors associated with the third- and fourth-order schemes are therefore identical. As was the case with first-order one-sided differencing, the sign of the imaginary part of ω_{3s} is determined by the sign of c such that solutions amplify for $c < 0$ and damp for $c > 0$.

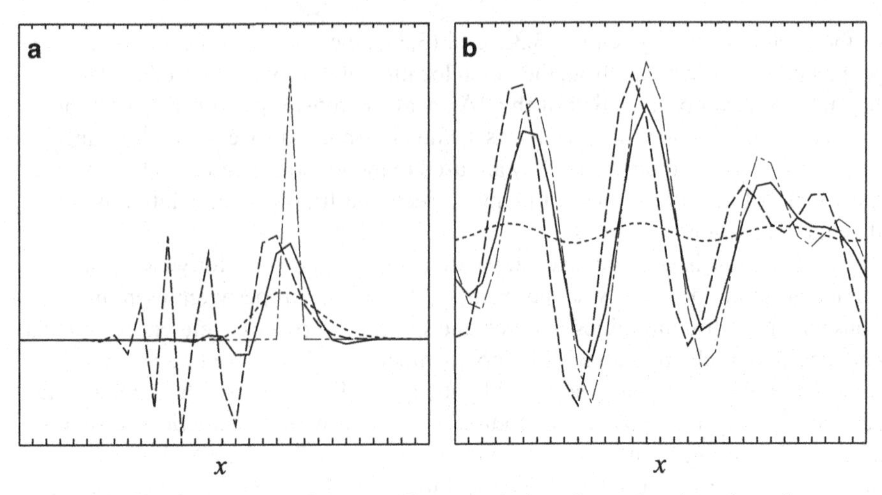

Fig. 3.7 Exact solution and differential-difference solutions for **a** advection of a spike over a distance of five grid points, and **b** advection of the sum of equal-amplitude $7.5\Delta x$ and $10\Delta x$ sine waves over a distance of 12 grid points. Exact solution (*dot-dashed line*), one-sided first-order solution (*short-dashed line*), centered second-order solution (*long-dashed line*), and one-sided third-order solution (*solid line*)

The damping associated with the third-order scheme is considerably less than that of the first-order scheme. According to (3.36) and (3.38),

$$\frac{\Im(\omega_{3s})}{\Im(\omega_{1s})} = \frac{1}{3}(1 - \cos k\Delta x).$$

As might be expected with a higher-order scheme, the well-resolved waves are damped much more slowly by the third-order approximation. Even the short waves show substantial improvement.

Some idea of the relative performance of the first-, second-, and third-order differences is provided in Fig. 3.7, which is identical to Fig. 3.6 except that the solid curve now represents the third-order solution. As indicated in Fig. 3.7, the damping produced by the third-order scheme is much weaker than that generated by first-order upstream differencing. Moreover, the third-order solution to the spike test is actually better than the second- and fourth-order results (compare Figs. 3.6a, 3.7a). In problems with extremely poor resolution, such as the spike test, the tendency of the third-order scheme to damp short wavelengths can be beneficial, because it largely eliminates the dispersive train of waves found in the centered difference solutions. On the other hand, the damping of intermediate wavelengths is sufficiently weak that the third-order solution retains almost the same amplitude in the region of the spike as the "nondamping" second- and fourth-order schemes. The situation in Fig. 3.7b is somewhat different, and it is not entirely obvious whether the third-order scheme should be preferred over the second-order scheme. The third-order scheme clearly exhibits less phase-speed error, but it also shows more amplitude error than the centered second-order method.

3.3.2 Dissipation, Dispersion, and the Modified Equation

One way to estimate phase-speed and amplitude error is to derive the differential–difference dispersion relation, as described in the preceding section. Another way to characterize the relative magnitude of the these errors is to examine the lowest-order terms in the truncation error of the finite-difference formula. The truncation errors for each of the finite-difference approximations considered in the preceding section are as follows:
One-sided, first-order:

$$\frac{\psi_j - \psi_{j-1}}{\Delta x} = \frac{\partial \psi}{\partial x} - \frac{\Delta x}{2}\frac{\partial^2 \psi}{\partial x^2} + \frac{(\Delta x)^2}{6}\frac{\partial^3 \psi}{\partial x^3} + O\left[(\Delta x)^3\right]. \tag{3.39}$$

Centered, second-order:

$$\frac{\psi_{j+1} - \psi_{j-1}}{2\Delta x} = \frac{\partial \psi}{\partial x} + \frac{(\Delta x)^2}{6}\frac{\partial^3 \psi}{\partial x^3} + O\left[(\Delta x)^4\right].$$

One-sided, third-order:

$$\frac{2\psi_{j+1} + 3\psi_j - 6\psi_{j-1} + \psi_{j-2}}{6\Delta x} = \frac{\partial \psi}{\partial x} + \frac{(\Delta x)^3}{12}\frac{\partial^4 \psi}{\partial x^4} - \frac{(\Delta x)^4}{30}\frac{\partial^5 \psi}{\partial x^5} + O\left[(\Delta x)^5\right].$$

Centered, fourth-order:

$$\frac{4}{3}\left(\frac{\psi_{j+1} - \psi_{j-1}}{2\Delta x}\right) - \frac{1}{3}\left(\frac{\psi_{j+2} - \psi_{j-2}}{4\Delta x}\right) = \frac{\partial \psi}{\partial x} - \frac{(\Delta x)^4}{30}\frac{\partial^5 \psi}{\partial x^5} + O\left[(\Delta x)^6\right].$$

If one of these formulae is used to determine the truncation error in a differential–difference approximation to the advection equation and the resulting scheme is $O\left[(\Delta x)^m\right]$ accurate, the same differential–difference scheme will approximate the *modified equation*

$$\frac{\partial \psi}{\partial t} + c\frac{\partial \psi}{\partial x} = a(\Delta x)^m \frac{\partial^{m+1} \psi}{\partial x^{m+1}} + b(\Delta x)^{m+1}\frac{\partial^{m+2} \psi}{\partial x^{m+2}} \tag{3.40}$$

to $O\left[(\Delta x)^{m+2}\right]$, where a and b are rational numbers determined by the particular finite-difference formula. Thus, as $\Delta x \to 0$, the numerical solution to the differential–difference equation will approach the solution to the modified equation more rapidly than it approaches the solution to the advection equation. A qualitative description of the effects of the leading-order errors in the differential–difference equation may therefore be obtained by examining the prototypical response generated by each of the forcing terms on the right side of the modified equation (3.40).

The term with the even-order derivative in (3.40) introduces a forcing identical to that in the prototypical equation

$$\frac{\partial \xi}{\partial t} = (-1)^{m+1}\frac{\partial^{2m} \xi}{\partial x^{2m}},$$

whose solutions

$$\xi(x,t) = C\mathrm{e}^{\mathrm{i}kx}\mathrm{e}^{-k^{2m}t}$$

become smoother with time because the shorter-wavelength modes decay more rapidly than the longer modes. Thus, the term with the lowest-order even derivative produces amplitude error, or *numerical dissipation*, in the approximate solution of the advection equation. The odd-order derivative on the right side of (3.40) introduces a forcing identical to that in the prototypical equation

$$\frac{\partial \xi}{\partial t} = -\frac{\partial^{2m+1}\xi}{\partial x^{2m+1}},$$

whose solutions are waves of the form

$$\xi(x,t) = C\mathrm{e}^{\mathrm{i}(kx-\omega t)}, \quad \text{where} \quad \omega = (-1)^m k^{2m+1}.$$

For $m > 0$, these waves are dispersive, because their phase speed ω/k depends on the wave number k. As a consequence, the lowest-order odd derivative on the right side of (3.40) produces a wave-number-dependent phase-speed error known as *numerical dispersion*.

Centered spatial differences do not produce numerical dissipation because there are no even derivatives in the truncation error of a centered difference scheme. Numerical dissipation is, however, produced by the leading-order term in the truncation error of the one-sided differences. There is a pronounced qualitative difference between the solutions generated by schemes with leading-order dissipative and leading-order dispersive errors. The modified equations associated with the preceding first- and second-order spatial differences both include identical terms in $\partial^3\psi/\partial x^3$. As a consequence, both schemes produce essentially the same dispersive error. The dispersive errors in the third- and fourth-order schemes are also very similar because the truncation error associated with each of these schemes includes identical terms in $\partial^5\psi/\partial x^5$. Yet, as was illustrated in Figs. 3.6 and 3.7, the impact of dispersion on even- and odd-order schemes is very different. Numerical dispersion is the only error in the centered even-order differences, so when short-wavelength modes are present, the dispersion is quite evident. In contrast, the numerical dispersion generated by the one-sided odd-order schemes is largely obscured by the lower-order dissipative errors that dominate the total error in these schemes.

3.3.3 Artificial Dissipation

As suggested by the test problems shown in Figs. 3.6 and 3.7, the lack of dissipation in centered spatial differences can sometimes be a disadvantage. In particular, the error produced by the dispersion of poorly resolved Fourier components is free to propagate throughout the solution without loss of amplitude. It is therefore often useful to add scale-selective dissipation to otherwise nondissipative schemes to damp the shortest resolvable wavelengths. Moreover, in nonlinear problems it is often necessary to remove energy from the shortest spatial scales to prevent the development of numerical instabilities that can arise through the nonlinear interaction of short-wavelength modes (see Sect. 4.5).

The centered finite-difference approximations to even spatial derivatives of order 2 or higher provide potential formulae for scale-selective smoothers. Consider the isolated effect of a second-derivative smoother in an equation of the form

$$\frac{d\phi_j}{dt} = \gamma_2 \left(\phi_{j+1} - 2\phi_j + \phi_{j-1}\right), \tag{3.41}$$

where γ_2 is a parameter that determines the strength of the smoother. Substitution of solutions of the form

$$\phi_j = b(t)e^{ikj\Delta x} \tag{3.42}$$

into (3.41) yields

$$\frac{db}{dt} = -2\gamma_2(1 - \cos k\Delta x)b,$$

implying that $2\Delta x$ waves are damped most rapidly, and that well-resolved waves undergo an $O[(k\Delta x)^2]$ dissipation. Indeed, if the second-derivative smoother is combined with the basic centered second-order difference,[8] the total truncation error in the smoothed difference becomes

$$\frac{\psi_{j+1} - \psi_{j-1}}{2\Delta x} - \gamma_2 \left(\psi_{j+1} - 2\psi_j + \psi_{j-1}\right)$$
$$= \frac{\partial\psi}{\partial x} - (\Delta x)^2 \left(\gamma_2 \frac{\partial^2\psi}{\partial x^2} - \frac{1}{6}\frac{\partial^3\psi}{\partial x^3}\right) + O\left[(\Delta x)^4\right].$$

Thus, the smoothed difference remains of second order, but the leading-order truncation error becomes both dissipative and dispersive. Note that as $\Delta x \to 0$, the preceding scheme will generate less dissipation than one-sided differencing, because as indicated by (3.39), one-sided differencing produces $O(\Delta x)$ dissipation. Furthermore, the addition of a separate smoother allows the dissipation rate to be explicitly controlled through the specification of γ_2.

Greater scale selectivity can be obtained using a fourth-derivative filter of the form

$$\frac{d\phi_j}{dt} = \gamma_4 \left(-\phi_{j+2} + 4\phi_{j+1} - 6\phi_j + 4\phi_{j-1} - \phi_{j-2}\right), \tag{3.43}$$

or the sixth-derivative filter

$$\frac{d\phi_j}{dt} = \gamma_6 \left(\phi_{j+3} - 6\phi_{j+2} + 15\phi_{j+1} - 20\phi_j + 15\phi_{j-1} - 6\phi_{j-2} + \phi_{j-3}\right). \tag{3.44}$$

[8] If a dissipative filter is used in conjunction with leapfrog time differencing, the terms involved in the filtering calculation must be evaluated at the $t - \Delta t$ time level to preserve stability. Time-differencing schemes appropriate for the simulation of diffusive processes are examined in Sect. 3.5.

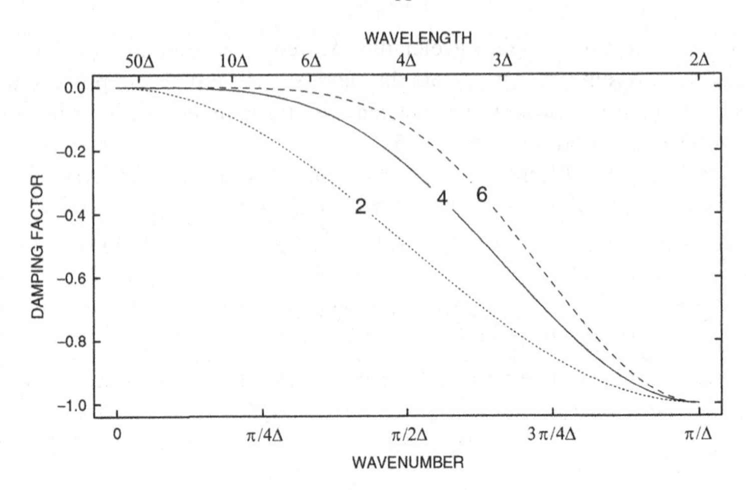

Fig. 3.8 Normalized damping rate as a function of horizontal wave number for second-order (*dotted line*), fourth-order (*solid line*), and sixth-order (*dashed line*) diffusive filters

Substituting a single wave of the form (3.42) into any of the preceding smoothers (3.41), (3.43), or (3.44) yields

$$\frac{db}{dt} = -\gamma_n \left[2(1 - \cos k\Delta x)\right]^{n/2} b, \tag{3.45}$$

where $n = 2, 4$, or 6 is the order of the derivative in each of the respective smoothers. In all cases, the $2\Delta x$ wave is damped most rapidly, and long waves are relatively unaffected. The actual scale selectivity of these filters is determined by the factor $(1 - \cos k\Delta x)^{n/2}$, which for well-resolved waves is $O\left[(k\Delta x)^n\right]$. This scale selectivity is illustrated in Fig. 3.8, in which the exponential decay rate associated with each smoother is plotted as a function of the wave number. To facilitate the comparison of these filters, the decay rate of the $2\Delta x$ wave has been normalized to unity by choosing $\gamma_n = 2^{-n}$.

The test problems shown in Figs. 3.6 and 3.7 were repeated using centered fourth-order differencing in combination with fourth-order and sixth-order spatial smoothers, and the results are plotted in Fig. 3.9. The filtering coefficients were set such that $\gamma_4 = 0.2$ and $\gamma_6 = \gamma_4/4$; this choice for γ_6 ensures that both filters will damp a $2\Delta x$ wave at the same rate. As evident in a comparison of Figs. 3.6a and 3.9a, both the fourth-order and the sixth-order filters remove much of the dispersive train of short waves that were previously present behind the isolated spike in the unfiltered solution. Those waves that remain behind the spike in the smoothed solutions have wavelengths near $4\Delta x$. Since γ_4 and γ_6 have been chosen to damp $2\Delta x$ waves at the same rate, the $4\Delta x$ waves in the dispersive train are not damped as rapidly by the sixth-order smoother, and as is evident in Fig. 3.9a, the sixth-order smoother leaves more amplitude in the wave train behind the spike. Although the scale selectivity of the sixth-order smoother interferes with the damping of the dispersive wave train behind the spike, it significantly improves the simulation of the moderately resolved waves shown in Fig. 3.9b. The solution obtained using the

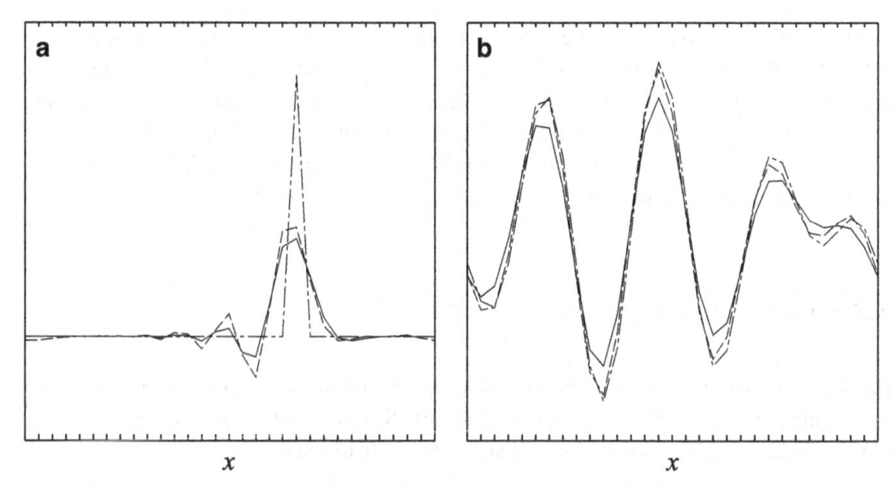

Fig. 3.9 Exact solution and differential–difference solutions for **a** advection of a spike over a distance of five grid points, and **b** advection of the sum of equal-amplitude $7.5\Delta x$ and $10\Delta x$ sine waves over a distance of 12 grid points. Exact solution (*dot-dashed line*), and fourth-order centered difference solutions in combination with a fourth-derivative filter (*solid line*) or a sixth-derivative filter (*dashed line*)

sixth-order filter is almost perfect, whereas the fourth-order filter generates significant damping. In fact, the general character of the solution obtained with the fourth-order filter is reminiscent of that obtained with the third-order one-sided finite-difference approximation. This similarity is not coincidental; the phase-speed errors produced by the third- and fourth-order finite differences are identical, and the leading-order numerical dissipation in the third-order difference, which is proportional to the fourth derivative, has the same scale selectivity as the fourth-order smoother.

Indeed, the proper choice of γ_4 will produce an exact equivalence between the solution obtained with the third-order scheme and the result produced by the combination of a centered fourth-order difference and a fourth-order smoother. The third-order differential–difference equation (3.37) can be expressed in a form that remains upstream independent of the sign of c as

$$\frac{d\phi_j}{dt} + \frac{c}{12\Delta x}\left(-\phi_{j+2} + 8(\phi_{j+1} - \phi_{j-1}) + \phi_{j-2}\right)$$
$$= -\frac{|c|}{12\Delta x}\left(\phi_{j+2} - 4\phi_{j+1} + 6\phi_j - 4\phi_{j-1} + \phi_{j-2}\right), \qquad (3.46)$$

which is the combination of a centered fourth-order spatial difference and a fourth-order filter with a filter coefficient $\gamma_4 = |c|/(12\Delta x)$. Note that the value of the fourth-derivative filter in the preceding equation is an inverse function of Δx. The implicit Δx-dependence of the filtering coefficient in (3.46) makes the scheme $O\left[(\Delta x)^3\right]$, whereas the dissipation introduced by the explicit fourth-order filter (3.43) is $O\left[(\Delta x)^4\right]$.

In practical applications, the time derivatives in (3.41) and (3.43) are replaced by finite differences, and the maximum values for γ_2 and γ_4 will be determined by stability considerations. If the differencing is forward in time, the maximum useful smoothing coefficients are determined by the relations $\gamma_2 \Delta t \leq 0.25$ and $\gamma_4 \Delta t \leq 0.0625$. When $\gamma_2 \Delta t = 0.25$ (or $\gamma_4 \Delta t = 0.0625$), any $2\Delta x$ wave will be completely removed by a single application of the second-order (or fourth-order) filter.

3.3.4 Compact Differencing

Further improvements in the filtered solutions shown in Fig. 3.9 can be obtained by using more accurate finite-difference schemes. Simply switching to a higher-order explicit scheme, such as the centered sixth-order difference

$$\frac{df}{dx} = \frac{3}{2}\delta_{2x} f - \frac{3}{5}\delta_{4x} f + \frac{1}{10}\delta_{6x} f + O\left[(\Delta x)^6\right] \tag{3.47}$$

(where the operator δ_{nx} is defined by (3.3)), provides only marginal improvement. More significant improvements can be obtained using *compact differencing*, in which the desired derivative is given implicitly by a matrix equation. Our attention will be restricted to compact schemes in which this implicit coupling leads to tridiagonal matrices, since tridiagonal systems can be evaluated with modest computational effort (see Appendix).

The simplest compact scheme is obtained by rewriting the expression for the truncation error in the centered second-order difference (3.4) in the form

$$\delta_{2x} f = \left(1 + \frac{(\Delta x)^2}{6}\delta_x^2\right)\frac{df}{dx} + O\left[(\Delta x)^4\right]. \tag{3.48}$$

Expanding the finite-difference operators in the preceding expression yields the following $O\left[(\Delta x)^4\right]$ accurate expression for the derivative:

$$\frac{f_{j+1} - f_{j-1}}{2\Delta x} = \frac{1}{6}\left[\left(\frac{df}{dx}\right)_{j+1} + 4\left(\frac{df}{dx}\right)_j + \left(\frac{df}{dx}\right)_{j-1}\right]. \tag{3.49}$$

This scheme allows fourth-order-accurate derivatives to be calculated on a three-point stencil. At intermediate numerical resolution, the fourth-order compact scheme is typically more accurate than the sixth-order explicit difference (3.47).

If one is going to the trouble to solve a tridiagonal matrix, it can be advantageous to do a little extra work and use the sixth-order tridiagonal scheme. The formula for the sixth-order tridiagonal compact scheme may be derived by first noting that the truncation error in the fourth-order explicit scheme (3.5) is

$$\left(1 - \frac{(\Delta x)^2}{6}\delta_x^2\right)\delta_{2x} f = \frac{df}{dx} - \frac{(\Delta x)^4}{30}\frac{\mathrm{d}^5 f}{\mathrm{d}x^5} + O\left[(\Delta x)^6\right],$$

and the truncation error in the fourth-order compact scheme (3.48) is

$$\delta_{2x} f = \left(1 + \frac{(\Delta x)^2}{6}\delta_x^2\right)\frac{df}{dx} - \frac{(\Delta x)^4}{180}\frac{d^5 f}{dx^5} + O\left[(\Delta x)^6\right].$$

Eliminating the $O\left[(\Delta x)^4\right]$ term between these two expressions, one obtains

$$\left(1 + \frac{(\Delta x)^2}{30}\delta_x^2\right)\delta_{2x} f = \left(1 + \frac{(\Delta x)^2}{5}\delta_x^2\right)\frac{df}{dx} + O\left[(\Delta x)^6\right].$$

Expanding the operators in the preceding yields the following $O\left[(\Delta x)^6\right]$ accurate tridiagonal system for df/dx:

$$\frac{1}{15}\left(14\delta_{2x} f_j + \delta_{4x} f_j\right) = \frac{1}{5}\left[\left(\frac{df}{dx}\right)_{j+1} + 3\left(\frac{df}{dx}\right)_j + \left(\frac{df}{dx}\right)_{j-1}\right]. \tag{3.50}$$

When compact schemes are used to approximate partial derivatives in complex equations in which one must compute several different spatial derivatives, such as the multidimensional advection equation, it is simplest to solve either (3.49) or (3.50) as a separate tridiagonal system for each derivative. However, in very simple problems, such as the one-dimensional advection equation (3.27), the spatial derivatives in the compact formulae may be replaced directly by $-(1/c)\partial\psi/\partial t$. Thus, to analyze the phase-speed error associated with compact spatial differencing, the fourth-order compact approximation to the advection equation may be written

$$\frac{\phi_{j+1} - \phi_{j-1}}{2\Delta x} = \frac{-1}{6c}\left[\left(\frac{d\phi}{dt}\right)_{j+1} + 4\left(\frac{d\phi}{dt}\right)_j + \left(\frac{d\phi}{dt}\right)_{j-1}\right]. \tag{3.51}$$

Substitution of a wave solution of the form (3.29) into the preceding equation yields the following expression for the phase speed of the differential–difference solution:

$$c_{4c} = \frac{\omega_{4p}}{k} = \frac{3c}{2 + \cos k\Delta x}\left(\frac{\sin k\Delta x}{k\Delta x}\right). \tag{3.52}$$

The phase speeds for the sixth-order compact scheme,

$$c_{6c} = \frac{c}{3(3 + 2\cos k\Delta x)}\left(14\frac{\sin k\Delta x}{k\Delta x} + \frac{\sin 2k\Delta x}{2k\Delta x}\right),$$

may be obtained through a similar derivation. These phase speeds are plotted as a function of $k\Delta x$, together with the curves for second-, fourth-, and sixth-order explicit centered differences, in Fig. 3.10. It is apparent that the compact schemes are superior to the explicit schemes. In particular, the phase speeds associated with the sixth-order compact differencing are almost perfect for wavelengths as short as $4\Delta x$. Note that although the order of accuracy of a scheme determines the rate at which the phase-speed curves in Fig. 3.10 asymptotically approach the correct value

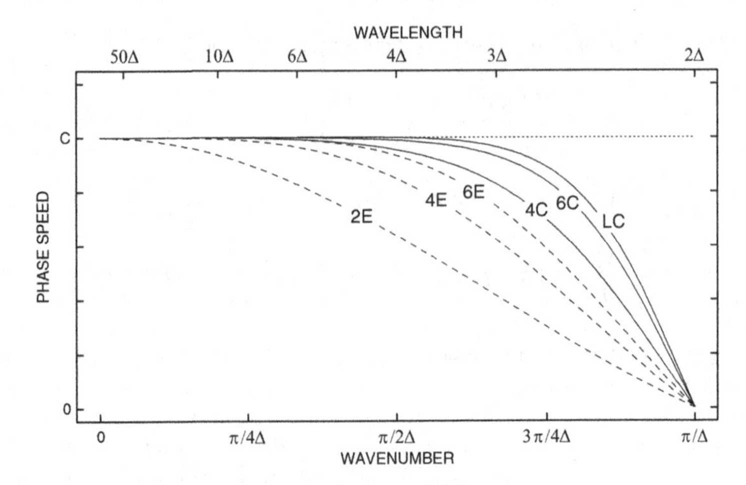

Fig. 3.10 Phase speed as a function of numerical resolution for the analytic solution of the advection equation (*dotted line*) and for corresponding differential–difference approximations using second-, fourth-, and sixth-order explicit differences (*dashed lines*), fourth- and sixth-order compact differences (*solid lines*), and the low-phase-speed-error fourth-order compact scheme of Lele (*solid line labeled LC*)

as $k\Delta x \to 0$, the order of accuracy does not reliably predict a scheme's ability to represent the poorly resolved waves. Lele (1992) observed that a better treatment of the shorter waves can be obtained by perturbing the coefficients in the sixth-order compact scheme to create the fourth-order method:

$$\frac{1}{12}\left(11\delta_{2x}f_j + \delta_{4x}f_j\right) = \frac{1}{24}\left[5\left(\frac{df}{dx}\right)_{j+1} + 14\left(\frac{df}{dx}\right)_j + 5\left(\frac{df}{dx}\right)_{j-1}\right]. \tag{3.53}$$

The phase speeds associated with this differencing scheme are plotted as the solid curve labeled "LC" in Fig. 3.10. Observe that Lele's compact scheme produces phase-speed errors in a $3\Delta x$ wave that are comparable to the errors introduced in a $6\Delta x$ wave by explicit fourth-order differences.

The performance of Lele's compact scheme on the test problems considered previously in connection with Figs. 3.6, 3.7, and 3.9 is illustrated in Fig. 3.11. Since they accurately capture the frequency of very short waves while still failing to detect any oscillations at $2\Delta x$, compact schemes propagate the energy in the $2\Delta x$ wave backward at very large group velocities (i.e., $-\partial\omega/\partial k$ is large near $k = 2\Delta x$). The preceding compact schemes are also nondamping because they are centered in space. It is therefore necessary to use a spatial filter in conjunction with these schemes when modeling problems with significant short-wavelength features. In these tests, a sixth-order filter (3.44) was used in combination with both the compact scheme (3.53) and the fourth-order explicit method. In all cases $\gamma_6 = 0.05$, which is the same value as that used in the computations shown in Fig. 3.9. In fact, the

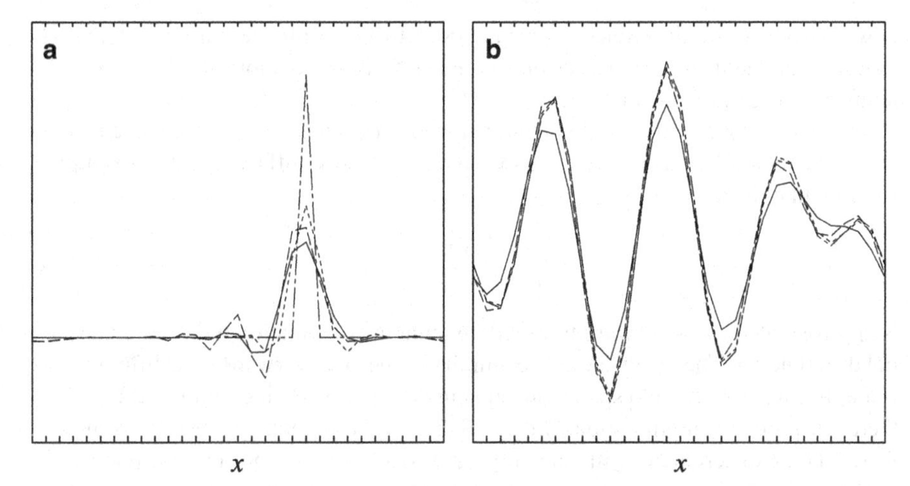

Fig. 3.11 Exact solution and differential-difference solutions for **a** advection of a spike over a distance of five grid points, and **b** advection of the sum of equal-amplitude $7.5\Delta x$ and $10\Delta x$ sine waves over a distance of 12 grid points. Exact solution (*dot-dashed line*), third-order one-sided solution (*solid line*), centered fourth-order explicit solution (*long dashed* line), and the solution obtained using Lele's low-phase-speed-error compact scheme (*short dashed line*). A sixth-order smoother, with $\gamma_6 = 0.05$, was used in combination with the fourth- and sixth-order differences

fourth-order solutions shown in these tests are identical to those shown previously in Fig. 3.9. Also plotted in Fig. 3.11 are the exact solution and the third-order one-sided solution (previously plotted in Fig. 3.7). As evident in Fig. 3.11a, the smoothed sixth-order compact scheme exhibits less of a $4\Delta x$ dispersive train than either the third-order or the fourth-order scheme. Since the dissipation applied to the compact solution is identical to that used with the fourth-order scheme (and less than that inherent in the third-order method), the relative absence of dispersive ripples in the compact solution indicates a relative lack of dispersive error at the $4\Delta x$ wavelength. This, of course, is completely consistent with the theoretical phase speed analysis shown in Fig. 3.10. The compact scheme also performs best on the two-wave test (Fig. 3.11b). Although the filtered compact scheme is the best-performing method considered in this section, it is also the most computationally burdensome. Other approaches to the problem of creating methods that can adequately represent short-wavelength features without sacrificing accuracy in smoother parts of the flow will be discussed in Chap. 5.

3.4 Fully Discrete Approximations to the Advection Equation

The error introduced by time differencing in ordinary differential equations was examined in Chap. 2. In Sect. 3.3, the error generated by spatial differencing was isolated and investigated through the use of differential–difference equations. We now consider finite-difference approximations to the complete partial differential equation and analyze the total error that arises from the combined effects of both temporal and spatial differencing.

In some instances, the fundamental behavior of a scheme can be deduced from the characteristics of its constituent spatial and temporal differences. For example, suppose that the advection equation

$$\frac{\partial \psi}{\partial t} + c\frac{\partial \psi}{\partial x} = 0 \tag{3.54}$$

is approximated using forward time differencing in combination with centered spatial differencing. The result should be amplifying because forward time differencing is amplifying and centered spatial differencing is neutral. Their combined effect will therefore produce amplification. On the other hand, it might be possible to combine forward time differencing with one-sided spatial differencing because the one-sided spatial difference is damping – provided that it is computed using upstream data. If this damping dominates the amplification generated by the forward time difference, it will stabilize the scheme. Further analysis would be required to determine the actual stability condition and the phase-speed error.

As another example, consider the use of leapfrog time differencing and centered spatial differencing to approximate the advection equation. Since both differences are neutral, it seems likely that such a scheme would be conditionally stable. Once again, further analysis is required to determine the exact stability condition and the phase-speed error. In the absence of such analysis, the sign of the phase-speed error is in doubt, since the leapfrog scheme is accelerating, whereas centered spatial differencing is decelerating. Finally, suppose that leapfrog differencing is combined with one-sided spatial differences. The result should be unstable because the leapfrog solution consists of two modes (the physical and computational modes) each propagating in the opposite direction. If the one-sided difference is upstream with respect to one mode, it will be downstream with respect to the second mode, and the second mode will amplify. Alternatively, one may note that the upstream difference approximation to the spatial derivative yields the differential–difference equation (3.35) whose solutions decay exponentially. Yet, as discussed in Sect. 2.4.3, the region of absolute stability for the leapfrog scheme is restricted to the imaginary axis, and therefore, if (3.35) is integrated using leapfrog differencing, the numerical solution will experience spurious growth.

Although as just noted, the forward-time and centered-space scheme

$$\frac{\phi_j^{n+1} - \phi_j^n}{\Delta t} + c\,\frac{\phi_{j+1}^n - \phi_{j-1}^n}{2\Delta x} = 0 \tag{3.55}$$

will produce a nonphysical amplification of the approximate solution to the advection problem, one might wonder whether this amplification is sufficiently weak that the scheme nevertheless satisfies the more general von Neumann stability condition

$$|A| \leq 1 + \eta \Delta t, \tag{3.56}$$

where η is a constant independent of k, Δt, and Δx. If so, then (3.55) will still generate convergent approximations to the correct solution in the limit $\Delta x \to 0$, $\Delta t \to 0$, because it is a consistent approximation to the advection equation. Recall that as discussed in Sect. 2.2.3, forward differencing produces amplifying solutions that nevertheless converge to the correct solution of the oscillation equation as $\Delta t \to 0$. The amplification factor arising from a von Neumann stability analysis of (3.55) satisfies

$$|A|^2 = 1 + \left(\frac{c \sin(k\Delta x)}{\Delta x}\right)^2 (\Delta t)^2.$$

Here, in contrast to the results obtained when ordinary differential equations are approximated with a forward difference, the coefficient of Δt includes a factor of $(\Delta x)^{-2}$ that cannot be bounded by a constant independent of k and Δx as $\Delta x \to 0$. As a consequence, the forward-time centered-space scheme does not satisfy the von Neumann condition (3.56) and is both unstable in the sense that it generates growing solutions to a problem where the true solution is bounded, and unstable in the more general sense that it does not produce convergent solutions as $\Delta x \to 0$ and $\Delta t \to 0$. Note in particular that after N time steps the amplitude of a $4\Delta x$ wave increases by a factor of $(1 + \mu^2)^{N/2}$ (where $\mu = c\Delta t/\Delta x$). Thus, if a series of integrations are performed in which the space–time grid is refined while holding μ constant, the cumulative amplification of the $4\Delta x$ wave occurring over a fixed interval of physical time increases as $\Delta x \to 0$ and $\Delta t \to 0$. As anticipated in the discussion in Sect. 2.2.4, the difference in the stability of forward difference approximations to the oscillation and advection equations arises because the frequency of solution to the oscillation equation is fixed, whereas that of the most rapidly varying component in the solution to the advection equation increases without bound as $\Delta x \to 0$.

3.4.1 The Discrete-Dispersion Relation

Although the preceding discussion suggests that useful deductions can be made by examining temporal and spatial differences independently, that discussion also reveals the need to rigorously analyze the combined effects of all finite differences in a specific formula to determine the complete behavior of the numerical solution. A useful tool in the analysis of errors in wave propagation problems is the *discrete-dispersion relation*, which is just the finite-difference analogue to the dispersion relation associated with the original continuous problem. The discrete-dispersion relation is obtained by substituting a traveling wave solution of the form

$$\phi_j^n = e^{i(kj\Delta x - \omega n\Delta t)} \tag{3.57}$$

into the finite-difference formula and solving for ω. If the frequency is separated into its real and imaginary parts $(\omega_r + i\omega_i)$, (3.57) becomes

$$\phi_j^n = e^{\omega_i n\Delta t} e^{i(kj\Delta x - \omega_r n\Delta t)} = |A|^n e^{i(kj\Delta x - \omega_r n\Delta t)}. \tag{3.58}$$

The determination of the imaginary part of ω is tantamount to a von Neumann stability analysis, since ω_i determines the amplification factor and governs the rate of numerical dissipation. Information about the phase-speed error can be obtained from ω_r.

Suppose that the advection equation is approximated with leapfrog-time and centered-second-order-space differencing such that

$$\frac{\phi_j^{n+1}-\phi_j^{n-1}}{2\Delta t}+c\frac{\phi_{j+1}^{n}-\phi_{j-1}^{n}}{2\Delta x}=0. \tag{3.59}$$

Substitution of (3.57) into this finite-difference scheme gives

$$\left(\frac{\mathrm{e}^{-\mathrm{i}\omega\Delta t}-\mathrm{e}^{\mathrm{i}\omega\Delta t}}{2\Delta t}\right)\phi_j^n=-c\left(\frac{\mathrm{e}^{\mathrm{i}k\Delta x}-\mathrm{e}^{-\mathrm{i}k\Delta x}}{2\Delta x}\right)\phi_j^n,$$

or equivalently,

$$\sin\omega\Delta t=\mu\sin k\Delta x, \tag{3.60}$$

where $\mu=c\Delta t/\Delta x$. Inspection of (3.60) demonstrates that if $|\mu|<1$, ω will be real and the scheme will be neutral. The scheme also appears to be neutral when $|\mu|=1$, but this is a special case. When $\mu=1$, (3.60) reduces to

$$\frac{\omega}{k}=\frac{\Delta x}{\Delta t}=c,$$

showing that the numerical solution propagates at the correct phase speed. Although there are no phase-speed errors when $|\mu|=1$, the two roots of (3.60) become identical if $k\Delta x=\pi/2$, and as a consequence of this double root, the scheme admits a weakly unstable $4\Delta x$ wave. When $\mu=1$, the weakly growing mode has the form

$$\phi_j^n=n\cos\left[\pi(j-n)/2\right]. \tag{3.61}$$

The distinction between the sufficient condition for stability $|\mu|<1$ and the more easily derived necessary condition $|\mu|\le 1$ is, however, of little practical significance because uncertainties about the magnitudes of the spatially and temporally varying velocities in real-world applications usually make it impossible to choose a time step such that $|\mu|=1$.

The frequencies resolvable in the discretized time domain lie in the interval $0\le\omega_r\le\pi/\Delta t$. Except for the special case just considered when $|\mu|=1$ and $k\Delta x=\pi/2$, there are two resolvable frequencies that satisfy (3.60). Dividing these frequencies by k gives the phase speed of the physical and computational modes

$$c_{\text{phys}}^{*}\equiv\frac{\omega_{\text{phys}}}{k}=\frac{1}{k\Delta t}\arcsin(\mu\sin k\Delta x)$$

and

$$c_{\text{comp}}^{*}\equiv\frac{\omega_{\text{comp}}}{k}=\frac{1}{k\Delta t}\left[\pi-\arcsin(\mu\sin k\Delta x)\right].$$

As in the differential–difference problem, the $2\Delta x$ physical mode does not propagate. The $2\Delta x$ computational mode flips sign each time step, or equivalently, it moves at the speed $\Delta x/\Delta t$. In the limit of good spatial resolution ($k\Delta x \to 0$), the Taylor series approximations

$$\sin x \approx x - \tfrac{1}{6}x^3 \quad \text{and} \quad \arcsin x \approx x + \tfrac{1}{6}x^3$$

can be used to obtain

$$c^*_{\text{phys}} \approx c\left(1 - \frac{k^2(\Delta x)^2}{6}(1-\mu^2)\right). \tag{3.62}$$

If the time step is chosen to ensure stability, then $\mu^2 < 1$, $|c^*| < |c|$, and the decelerating effect of centered spatial differencing dominates the accelerating effects of leapfrog time differencing. As suggested by (3.62), in practical computations the most accurate results are obtained using a time step such that the maximum value of $|\mu|$ is slightly less than 1.

Now consider the upstream scheme

$$\frac{\phi_j^{n+1} - \phi_j^n}{\Delta t} + c\frac{\phi_j^n - \phi_{j-1}^n}{\Delta x} = 0. \tag{3.63}$$

Substitution of (3.57) into (3.63) gives

$$\mathrm{e}^{-\mathrm{i}\omega\Delta t} - 1 = \mu\left(\mathrm{e}^{-\mathrm{i}k\Delta x} - 1\right). \tag{3.64}$$

It follows that the exact dispersion relation and the exact solution are obtained in the special case when $\mu = 1$. Further analysis is facilitated by separating (3.64) into its real and imaginary parts

$$|A|\cos\omega_{\mathrm{r}}\Delta t - 1 = \mu(\cos k\Delta x - 1) \tag{3.65}$$

and

$$|A|\sin\omega_{\mathrm{r}}\Delta t = \mu\sin k\Delta x, \tag{3.66}$$

where, $\omega = \omega_{\mathrm{r}} + \mathrm{i}\omega_{\mathrm{i}}$, and $|A| \equiv \mathrm{e}^{\omega_{\mathrm{i}}\Delta t}$ is the modulus of the amplification factor. Squaring both sides of (3.65) and (3.66) and adding yields

$$|A|^2 = 1 - 2\mu(1-\mu)(1-\cos k\Delta x),$$

which implies that the donor-cell scheme is stable and damping for $0 \le \mu \le 1$, and that the maximum damping per time step occurs at $\mu = 1/2$.[9]

[9] This stability condition is identical to that obtained via the standard von Neumann stability analysis in Sect. 2.3.3.

The discrete-dispersion relation

$$\omega_{\mathrm{r}} = \frac{1}{\Delta t} \arctan\left(\frac{\mu \sin k\Delta x}{1 + \mu(\cos k\Delta x - 1)}\right)$$

may be obtained after dividing (3.66) by (3.65). The function $\arctan \omega_{\mathrm{r}}\Delta t$ is single-valued over the range of resolvable frequencies $0 \leq \omega_{\mathrm{r}} \leq \pi/\Delta t$, so as expected for a two-time-level scheme, there is no computational mode. In the limit of good numerical resolution,

$$c^* \equiv \frac{\omega_{\mathrm{r}}}{k} \approx c\left[1 - \frac{(k\Delta x)^2}{6}(1-\mu)(1-2\mu)\right],$$

showing that phase-speed error is minimized by choosing either $\mu = 1$ or $\mu = 1/2$. The donor-cell scheme is decelerating for $0 < \mu < 1/2$, and accelerating for $1/2 < \mu < 1$. The phase-speed error in the donor-cell scheme may be minimized by choosing a time step such that $\mu_{\mathrm{avg}} \approx 1/2$. Under such circumstances, the donor-cell method will generate less phase-speed error than the leapfrog centered-space scheme. Unfortunately, the good phase-speed characteristics of the donor-cell method are overshadowed by its large dissipation.

It is somewhat surprising that there are values of μ for which the donor-cell scheme is accelerating, since forward time differencing is decelerating and one-sided spatial differencing reduces the phase speed of solutions to the differential–difference advection equation. This example illustrates the danger of relying too heavily on results obtained through the independent analysis of space and time truncation error.

3.4.2 The Modified Equation

As an alternative to the discrete-dispersion equation, numerical dissipation and dispersion can be analyzed by examining a "modified" partial differential equation whose solution satisfies the finite-difference equation to a higher order of accuracy than the solution to the original partial differential equation. This technique is similar to that described in Sect. 3.3.2 except that since the truncation error includes derivatives with respect to both space and time, all the time derivatives must be expressed as spatial derivatives to isolate those terms responsible for numerical dissipation and dispersion. As an example, consider (3.63), the upstream approximation to the constant-wind-speed advection equation, which is a third-order-accurate approximation to the modified equation

$$\frac{\partial \psi}{\partial t} + c\frac{\partial \psi}{\partial x} = \frac{c\Delta x}{2}(1-\mu)\frac{\partial^2 \psi}{\partial x^2} - \frac{c(\Delta x)^2}{6}(1-\mu)(1-2\mu)\frac{\partial^3 \psi}{\partial x^3}. \tag{3.67}$$

Examination of this equation shows that upstream differencing generates numerical dissipation of $O(\Delta x)$ and numerical dispersion of $O\left[(\Delta x)^2\right]$. Both the dissipation and the dispersion are minimized as $\mu \to 1$, and the dispersion is also eliminated when $\mu = 1/2$.

In deriving the modified equation, one cannot use the original partial differential equation to express all the higher-order time derivatives as spatial derivatives because the finite-difference scheme must approximate the modified equation more accurately than the original partial differential equation (Warming and Hyett 1974). The upstream method (3.63) provides a first-order approximation to the advection equation (3.54), a second-order approximation to

$$\frac{\partial \psi}{\partial t} = -c\frac{\partial \psi}{\partial x} - \frac{\Delta t}{2}\frac{\partial^2 \psi}{\partial t^2} + c\frac{\Delta x}{2}\frac{\partial^2 \psi}{\partial x^2}, \tag{3.68}$$

and a third-order approximation to

$$\frac{\partial \psi}{\partial t} = -c\frac{\partial \psi}{\partial x} - \frac{\Delta t}{2}\frac{\partial^2 \psi}{\partial t^2} - \frac{(\Delta t)^2}{6}\frac{\partial^3 \psi}{\partial t^3} + c\frac{\Delta x}{2}\frac{\partial^2 \psi}{\partial x^2} - c\frac{(\Delta x)^2}{6}\frac{\partial^3 \psi}{\partial x^3}. \tag{3.69}$$

The third-order-accurate modified equation (3.67) is obtained by repeatedly substituting derivatives of (3.68) into (3.69) until all the first-order terms involving time derivatives are eliminated. The time derivatives in the remaining second-order terms can then be eliminated using the first-order-accurate relation (3.54).

3.4.3 Stable Schemes of Optimal Accuracy

The universe of possible finite-difference methods for the solution of the advection equation (3.54) is very large, and most of our analysis is focused on a few important methods. Here we briefly broaden our discussion to consider the family of two-time-level explicit schemes of the form

$$\phi_j^{n+1} = \sum_{k=-l}^{r} a_k \phi_{j+k}^n. \tag{3.70}$$

What properties must be satisfied by the a_k to yield stable solutions to (3.54) with the highest possible order of accuracy? As the stencil for the preceding scheme is widened by increasing l and r, one can obtain progressively higher-order accuracy. Yet in a situation reminiscent of the behavior of linear multistep methods for ordinary differential equations (see Sect. 2.4.3), the method with the highest order of accuracy for a given l and r may not be stable.

Suppose $c > 0$ and let us demand that the values of Δt for which the method is stable include all those satisfying $0 \le c\Delta t/\Delta x \le 1$. Then the maximum order of accuracy of the scheme is

$$p = \min(r + l, 2r + 2, 2l) \tag{3.71}$$

(Iserles and Strang 1983). If $l < r$, which corresponds to a downwind biased stencil, (3.71) gives $p = 2l$; this order of accuracy can be achieved using a centered scheme with $a_k = 0$ for $k > l$ and cannot be improved upon by adding more downstream points to the stencil. If $l > r + 1$, (3.71) gives $p = 2r + 2$, which may be achieved with $a_k = 0$ for $k < -r - 2$. Adding more than two upstream biased points to the stencil cannot both maintain stability and increase the order of accuracy.

3.4.4 The Lax–Wendroff Method

None of the schemes considered previously achieve $O\left[(\Delta t)^2\right]$ accuracy without multistage computation or implicitness or the use of data from two or more previous time levels. Lax and Wendroff (1960) proposed a general method for creating $O\left[(\Delta t)^2\right]$ schemes in which the time derivative is approximated by forward differencing and the $O(\Delta t)$ truncation error generated by that forward difference is canceled by terms involving finite-difference approximations to spatial derivatives. Needless to say, it is impossible to analyze the behavior of a Lax–Wendroff method properly without considering the combined effects of space and time differencing.

One important example of a Lax–Wendroff scheme is the following approximation to the advection equation (3.54):

$$\frac{\phi_j^{n+1} - \phi_j^n}{\Delta t} + c\left(\frac{\phi_{j+1}^n - \phi_{j-1}^n}{2\Delta x}\right) = \frac{c^2 \Delta t}{2}\left(\frac{\phi_{j+1}^n - 2\phi_j^n + \phi_{j-1}^n}{(\Delta x)^2}\right). \tag{3.72}$$

The lowest-order truncation error in the first term of (3.72), the forward time difference, is

$$\frac{\Delta t}{2}\frac{\partial^2 \psi}{\partial t^2}.$$

However, since ψ is the exact solution to the continuous problem (3.54),

$$\frac{\Delta t}{2}\frac{\partial^2 \psi}{\partial t^2} = \frac{\Delta t}{2}\frac{\partial}{\partial t}\left(-c\frac{\partial \psi}{\partial x}\right) = \frac{c^2 \Delta t}{2}\frac{\partial^2 \psi}{\partial x^2}.$$

The term on the right side of (3.72) will therefore cancel the $O(\Delta t)$ truncation error in the forward time difference to within $O\left[\Delta t(\Delta x)^2\right]$, and as a consequence, the entire scheme is $O\left[(\Delta t)^2\right] + O\left[(\Delta x)^2\right]$ accurate.

The second-order nature of (3.72) may also be demonstrated by expressing it as a two-step formula in which each individual step is centered in space and time. In the first step, intermediate values staggered in space and time are calculated from the relations

$$\frac{\phi_{j+\frac{1}{2}}^{n+\frac{1}{2}} - \frac{1}{2}\left(\phi_{j+1}^n + \phi_j^n\right)}{\frac{1}{2}\Delta t} = -c\left(\frac{\phi_{j+1}^n - \phi_j^n}{\Delta x}\right), \tag{3.73}$$

$$\frac{\phi_{j-\frac{1}{2}}^{n+\frac{1}{2}} - \frac{1}{2}\left(\phi_j^n + \phi_{j-1}^n\right)}{\frac{1}{2}\Delta t} = -c\left(\frac{\phi_j^n - \phi_{j-1}^n}{\Delta x}\right). \tag{3.74}$$

In the second step, ϕ_j^{n+1} is computed from

$$\frac{\phi_j^{n+1} - \phi_j^n}{\Delta t} = -c\left(\frac{\phi_{j+\frac{1}{2}}^{n+\frac{1}{2}} - \phi_{j-\frac{1}{2}}^{n+\frac{1}{2}}}{\Delta x}\right). \tag{3.75}$$

The single-step formula (3.72) may be recovered by using (3.73) and (3.74) to eliminate $\phi^{n+1/2}$ and $\phi^{n-1/2}$ from (3.75). One advantage of the two-step formulation is that its extension to more complex problems can be immediately apparent. For example, if the wind speed is a function of the spatial coordinate, c is replaced by $c_{j+1/2}$, $c_{j-1/2}$, and c_j in (3.73), (3.74), and (3.75), respectively. In contrast, the equivalent modification of the single-step formula (see (3.77)) is slightly less obvious.

The amplitude and phase-speed errors of the Lax–Wendroff approximation to the constant-wind-speed advection equation may be examined by substituting a solution of the form (3.58) into (3.72), which yields

$$|A|(\cos\omega_r\Delta t - i\sin\omega_r\Delta t) = 1 + \mu^2(\cos k\Delta x - 1) - i\mu\sin k\Delta x. \tag{3.76}$$

Equating the real and imaginary parts of the preceding equation, and then eliminating $|A|$, one obtains the discrete-dispersion relation

$$\omega_r = \frac{1}{\Delta t}\arctan\left(\frac{\mu\sin k\Delta x}{1 + \mu^2(\cos k\Delta x - 1)}\right).$$

In the limit $k\Delta x \ll 1$, ω_r/k reduces to (3.62), showing that for well-resolved waves the phase-speed error of the Lax–Wendroff method is identical to that of the leapfrog centered-space scheme. Eliminating ω_r from the real and imaginary parts of (3.76), one obtains

$$|A|^2 = 1 - \mu^2(1 - \mu^2)(1 - \cos k\Delta x)^2,$$

from which it follows that the Lax–Wendroff scheme is stable for $\mu^2 \leq 1$. Short wavelengths are damped most rapidly; the $2\Delta x$ wave is completely eliminated in a single time step if $|\mu| = 1/\sqrt{2}$. Since the shortest wavelengths are seriously in error – once again the phase speed of the $2\Delta x$ wave is zero – this scale-selective damping can be advantageous. Indeed, the scale-selectivity of the dissipation in the Lax–Wendroff scheme is the same as that of a fourth-order spatial filter. Unfortunately, the numerical analyst has little control over the actual magnitude of the dissipation because it is a function of the Courant number, and in most practical problems, μ will vary throughout the computational domain. The dependence of the damping on the Courant number is illustrated in Fig. 3.12, which compares solutions generated by the Lax–Wendroff method and the leapfrog scheme (3.59) using

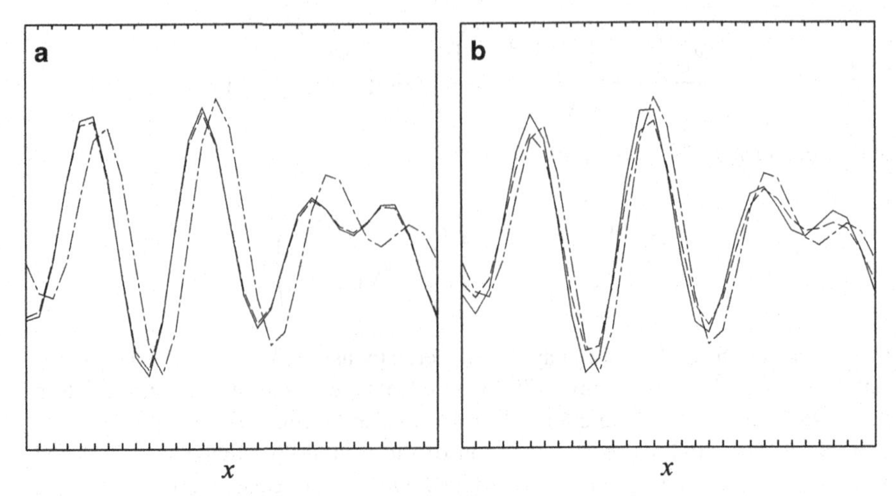

Fig. 3.12 Leapfrog, second-order space (*solid line*), Lax–Wendroff (*dashed line*), and exact (*dot-dashed line*) solutions for the advection of the sum of equal-amplitude $7.5\Delta x$ and $10\Delta x$ sine waves over a distance of 12 grid points using a Courant number of **a** 0.1 and **b** 0.75

Courant numbers of 0.75 and 0.1. When $\mu = 0.1$, the leapfrog and Lax–Wendroff schemes give essentially the same result, but when μ is increased to 0.75, the damping of the Lax–Wendroff solution relative to the leapfrog scheme is clearly evident. Figure 3.12 also demonstrates how the phase-speed error in both numerical solutions is reduced as the Courant number increases toward unity.

The term that cancels the $O(\Delta t)$ truncation error in a Lax–Wendroff scheme must be specifically reformulated for each new problem. The following three examples illustrate the general approach. If the flow velocity in (3.72) is a function of x, then

$$\frac{\partial^2 \psi}{\partial t^2} = c\frac{\partial}{\partial x}\left(c\frac{\partial \psi}{\partial x}\right),$$

and the right side of (3.72) becomes

$$\frac{c_j \Delta t}{2}\left(\frac{c_{j+\frac{1}{2}}(\phi_{j+1}^n - \phi_j^n) - c_{j-\frac{1}{2}}(\phi_j^n - \phi_{j-1}^n)}{(\Delta x)^2}\right). \tag{3.77}$$

If the flow is two-dimensional, the advection problem becomes

$$\frac{\partial \psi}{\partial t} + u\frac{\partial \psi}{\partial x} + v\frac{\partial \psi}{\partial y} = 0,$$

and if u and v are constant,

$$\frac{\partial^2 \psi}{\partial t^2} = u^2\frac{\partial^2 \psi}{\partial x^2} + v^2\frac{\partial^2 \psi}{\partial y^2} + 2uv\frac{\partial^2 \psi}{\partial x \partial y},$$

which must be approximated by spatial differences of at least first-order accuracy. Finally, consider a general system of "conservation laws" of the form

$$\frac{\partial \mathbf{v}}{\partial t} + \frac{\partial}{\partial x}\mathbf{F}(\mathbf{v}) = \mathbf{0},$$

where $\mathbf{v}$ and $\mathbf{F}$ are column vectors. Then

$$\frac{\partial^2 \mathbf{v}}{\partial t^2} = \frac{\partial}{\partial x}\left(\mathbf{J}\frac{\partial \mathbf{F}}{\partial x}\right), \tag{3.78}$$

where $\mathbf{J}$ is the Jacobian matrix whose ijth element is $\partial F_i/\partial v_j$. Once again, this matrix operator must be approximated by spatial differences of at least first-order accuracy.

In many applications the Lax–Wendroff method can be implemented more easily and more efficiently using the two-step method (3.73)–(3.75), or the following variant of the two-step method suggested by MacCormack (1969):

$$\tilde{\mathbf{v}}_j = \mathbf{v}_j^n - \frac{\Delta t}{\Delta x}\left[\mathbf{F}(\mathbf{v}_j^n) - \mathbf{F}(\mathbf{v}_{j-1}^n)\right],$$
$$\tilde{\tilde{\mathbf{v}}}_j = \tilde{\mathbf{v}}_j - \frac{\Delta t}{\Delta x}\left[\mathbf{F}(\tilde{\mathbf{v}}_{j+1}) - \mathbf{F}(\tilde{\mathbf{v}}_j)\right],$$
$$\mathbf{v}_j^{n+1} = \frac{1}{2}\left(\mathbf{v}_j^n + \tilde{\tilde{\mathbf{v}}}_j\right).$$

These two-step methods generate numerical approximations to the higher-order spatial derivatives required to cancel the $O(\Delta t)$ truncation error in the forward time difference without requiring the user to explicitly evaluate complex expressions like (3.78). The MacCormack method is particularly useful, since it easily generalizes to problems in two or more spatial dimensions.

In the classical Lax–Wendroff method, the spatial derivatives are approximated using centered differences, but other approximations are also possible. If the spatial dependence of ψ is not discretized, the Lax–Wendroff approximation to the advection equation (3.72) may be written

$$\frac{\phi^{n+1} - \phi^n}{\Delta t} + c\frac{d\phi}{dx} = \frac{c^2\Delta t}{2}\frac{d^2\phi}{dx^2}.$$

Warming and Beam (1976) proposed the following upwind approximation to the spatial derivatives in the preceding equation:

$$\phi_j^{n+1} = \phi_j^n - \mu\left(\phi_j^n - \phi_{j-1}^n\right) - \frac{\mu}{2}(1-\mu)\left(\phi_j^n - 2\phi_{j-1}^n + \phi_{j-2}^n\right), \tag{3.79}$$

which is accurate of order $O\left[(\Delta t)^2\right] + O(\Delta t\Delta x) + O\left[(\Delta x)^2\right]$ and is stable for $0 \le \mu \le 2$.

3.5 Diffusion, Sources, and Sinks

Molecular diffusion in one spatial dimension is governed by the diffusion equation

$$\frac{\partial \psi}{\partial t} = M \frac{\partial^2 \psi}{\partial x^2}, \tag{3.80}$$

where $M > 0$ is a molecular diffusivity. Suppose that (3.80) is approximated as

$$\frac{\phi_j^{n+1} - \phi_j^n}{\Delta t} = M \left(\frac{\phi_{j+1}^n - 2\phi_j^n + \phi_{j-1}^n}{(\Delta x)^2} \right). \tag{3.81}$$

The standard von Neumann stability analysis yields an amplification factor for this forward-time scheme of

$$A_\text{f} = 1 - 2\nu(1 - \cos k\Delta x), \tag{3.82}$$

where $\nu = M\Delta t/(\Delta x)^2$. The amplification factor is maximized for $k = \pi/\Delta x$ (the $2\Delta x$ mode), and thus $|A_\text{f}| \leq 1$ and the numerical solution decays with time, provided that $0 \leq \nu \leq 1/2$. Note that in contrast to the conditional stability criteria obtained for finite-difference approximations to the advection equation, this scheme does not remain stable as $\Delta t, \Delta x \to 0$ unless Δt decreases much more rapidly than Δx, i.e., unless $\Delta t/\Delta x \leq O(\Delta x)$. This makes the preceding scheme very inefficient at high spatial resolution.

Although it guarantees that the solution will not blow up, the criterion $0 \leq \nu \leq 1/2$ is not adequate to ensure a qualitatively correct simulation of the $2\Delta x$ mode. The amplitude b_k of the kth Fourier mode of the exact solution to (3.80) satisfies

$$\frac{db_k}{dt} = -Mk^2 b_k, \tag{3.83}$$

implying that the correct amplification factor for the kth mode is

$$\frac{b_k(t + \Delta t)}{b_k(t)} = \mathrm{e}^{-Mk^2\Delta t},$$

which (for $M > 0$) is a real number between 0 and 1. However, if $1/4 < \nu \leq 1/2$, the numerical amplification factor for the $2\Delta x$ mode lies in the interval $[-1, 0)$, and as a consequence, the sign of the $2\Delta x$ mode flips every time step as it gradually damps toward zero. If one wishes to avoid "overdamping" the poorly resolved modes, Δt must satisfy the more restrictive criterion that $0 \leq \nu \leq 1/4$.

Another approximation to (3.80) is provided by the Dufort–Frankel method:

$$\frac{\phi_j^{n+1} - \phi_j^{n-1}}{2\Delta t} = M \left(\frac{\phi_{j+1}^n - (\phi_j^{n+1} + \phi_j^{n-1}) + \phi_{j-1}^n}{\Delta x^2} \right). \tag{3.84}$$

The Dufort–Frankel method is sufficiently implicit to be stable for all values of Δt (see Problem 6); yet (3.84) can be easily solved for ϕ_j^{n+1} and stepped forward like an explicit method. The weakness of this approach is revealed by its truncation error, which is

$$O\left[(\Delta t)^2\right] + O\left[(\Delta x)^2\right] + O\left[(\Delta t/\Delta x)^2\right].$$

The same criterion required to preserve the stability of (3.81), that $\Delta t/\Delta t \le O(\Delta x)$ as $\Delta t, \Delta x \to 0$, is necessary to make (3.84) a consistent first-order method.

Although the Dufort–Frankel method is stable for all Δt, it may not even qualitatively model the effects of diffusion if the time step is large enough that $\nu = M\Delta t/(\Delta x)^2 > 1/2$. To understand why, note that the amplification factor for the Dufort–Frankel method is

$$A_{\text{df}} = \frac{2\nu\cos(k\Delta x) \pm \left(1 - 4\nu^2\sin^2(k\Delta x)\right)^{1/2}}{1 + 2\nu}. \tag{3.85}$$

If $\nu > 1/2$, the second term in the preceding equation becomes imaginary for some subset of the resolved waves, and individual modes change phase each time step, instead of simply damping in amplitude. Consider the $4\Delta x$ component of the solution ($k\Delta x = \pi/2$), for which this behavior is most pronounced. If $\nu > 1/2$, the first term in (3.85) is zero and both roots give pure imaginary values for A_{df} that produce decaying $4\Delta t$ oscillations in the $4\Delta x$ mode. If ν is large, the rate of decay becomes very small. Although its unconditional instability makes the Dufort–Frankel scheme notable from the theoretical standpoint, in many practical applications it has proved unable to correctly damp the short-wavelength components of the solution (Sun 1982).

As is the case with numerical approximations to the advection equation, the stability of an explicit finite-difference approximation to the diffusion equation is limited by the shortest-wavelength modes. The behavior of the shortest modes relative to the longer modes in the advection problem is, however, quite different from that in the diffusion problem. In the advection problem the shortest waves translate without loss of amplitude (and may even amplify as the result of deformation in the wind field or nonlinear processes), so any errors in the simulation of the short waves can have a serious impact on the accuracy of the overall solution. On the other hand, as implied by (3.83), diffusion preferentially damps the shortest modes, and after a brief time the amplitude in these modes becomes negligible relative to that of the total solution. Since the accuracy with which the short waves are simulated is irrelevant once those waves have dissipated, an acceptable approximation to the overall solution can often be obtained without accurately simulating the transient decay of the most poorly resolved initial perturbations. It can therefore be very advantageous to approximate the diffusion equation using A-stable schemes like the trapezoidal method, for which the time step is limited only by accuracy considerations. In particular, the time step can be chosen to accurately simulate the transient decay of the physical scales of primary interest, whereas any inaccuracies generated by this time step in the poorly resolved modes are hidden by their rapid decay.

If the time differencing in the finite-difference approximation to the diffusion equation is trapezoidal, the resulting scheme

$$\frac{\phi_j^{n+1} - \phi_j^n}{\Delta t} = \frac{M}{2}\left(\delta_x^2 \phi_j^{n+1} + \delta_x^2 \phi_j^n\right) \tag{3.86}$$

is known as the *Crank–Nicolson* method. The amplification factor for this scheme is

$$A_{\text{cn}} = \frac{1 - \nu(1 - \cos k\Delta x)}{1 + \nu(1 - \cos k\Delta x)},$$

and since $M > 0$, it follows that $|A_{\text{cn}}| \leq 1$ and the scheme is stable for all Δt. Assuming that boundary conditions are specified at the edges of the spatial domain, (3.86) constitutes a tridiagonal linear system for the unknown ϕ_j^{n+1}, which can be solved with minimal computational effort as discussed in the Appendix.

If M is constant, the Crank–Nicolson method can be extended to higher dimensions using operator splitting while preserving second-order accuracy because the spatial finite-difference operators will commute (see Sect. 4.3). In many problems the diffusivity is not uniform throughout the spatial domain, and in such cases the governing equation becomes

$$\frac{\partial \psi}{\partial t} = \frac{\partial}{\partial x}\left(M \frac{\partial \psi}{\partial x}\right).$$

The spatial derivative operator in the preceding equation may be discretized as

$$\frac{d\phi_j}{dt} = \frac{1}{\Delta x}\left(\frac{(M_{j+1} + M_j)}{2}\frac{(\phi_{j+1} - \phi_j)}{\Delta x} - \frac{(M_j + M_{j-1})}{2}\frac{(\phi_j - \phi_{j-1})}{\Delta x}\right).$$

In higher-dimensional problems where M is nonconstant, the spatial difference operators do not generally commute, so simple splitting methods will reduce the approximation to first-order accuracy. Second-order accuracy can nevertheless be efficiently obtained in two dimensions using the Peaceman–Rachford formulae (4.62) and (4.63).

3.5.1 Advection and Diffusion

In many applications it is necessary to consider the combined effects of both advection and diffusion. In one dimension this problem is governed by the equation

$$\frac{\partial \psi}{\partial t} + c\frac{\partial \psi}{\partial x} = M\frac{\partial^2 \psi}{\partial x^2}. \tag{3.87}$$

First we will examine accuracy issues that arise solely from the approximation of the spatial derivatives, then discuss choices for the time integration of advection–diffusion equations, and conclude by considering the combined effects of time and space differencing.

If the spatial first derivative is approximated by an upstream difference (with $c > 0$) and the second derivative is approximated by the standard three-point stencil, the resulting differential–difference approximation to (3.87) is

$$\frac{d\phi_j}{dt} + c\left(\frac{\phi_j - \phi_{j-1}}{\Delta x}\right) = M\left(\frac{\phi_{j+1} - 2\phi_j + \phi_{j-1}}{(\Delta x)^2}\right). \tag{3.88}$$

Evaluating the truncation error in the preceding equation shows that it is an $O\left[(\Delta x)^2\right]$-accurate approximation to the modified equation

$$\frac{\partial \psi}{\partial t} + c\frac{\partial \psi}{\partial x} = M\left(1 + \frac{Pe}{2}\right)\frac{\partial^2 \psi}{\partial x^2}, \tag{3.89}$$

where $Pe = c\Delta x/M$ is the numerical Péclet number. The Péclet number is a nondimensional parameter classically defined as the ratio of the strength of thermal advection to the strength of thermal diffusion.[10] Since the length scale in the numerical Péclet number is the grid spacing, Pe is a measure of the relative strengths of advection and diffusion at the smallest spatial scales resolved on the numerical mesh. A comparison of the modified equation (3.89) with the original advection–diffusion equation (3.87) shows that the differential–difference approximation (3.88) generates an inaccurate approximation to the diffusion term unless $Pe \ll 1$, i.e., unless diffusion dominates advective transport on the shortest resolvable scales. This difficulty arises because the total diffusion is dominated by numerical diffusion unless the molecular diffusivity is very large or the grid resolution is very fine.

As the horizontal resolution increases, the numerical Péclet number decreases, and, in principle, there is some grid size at which diffusive transport dominates advective transport in the shortest resolvable modes. Nevertheless, in many problems involving low-viscosity flow this grid size may be several orders of magnitude smaller than the physical scales of primary interest, so there is no possibility of resolving the scales at which molecular viscosity dominates numerical diffusion without exceeding the resources of the most advanced computers. Even when molecular diffusion has no direct influence on the resolved-scale fields, an essentially inviscid transport by sub-grid-scale eddies may produce turbulent mixing whose influence on the resolved-scale fields is often parameterized by a diffusion term in which the true molecular diffusivity M is replaced by an "eddy" diffusivity $\tilde{M}$ (Yih 1977, p. 572). The eddy diffusivity is generally parameterized such that $\tilde{M}$ is proportional to the mesh size, in which case the numerical Péclet number $c\Delta x/\tilde{M}$ does not decrease as the grid is refined, and the relative importance of *eddy* diffusion and advection remains constant as the mesh size is refined.

To accurately represent the diffusion term in low-viscosity flow, it is necessary to use a less diffusive approximation to $\partial\psi/\partial x$, such as a centered difference or a

[10] The Péclet number is completely analogous to the more familiar Reynolds number, which is the ratio of momentum advection to momentum diffusion. The difference between the Péclet and Reynolds numbers is due to the difference in the diffusivities of heat and momentum. In particular, the ratio of the Péclet number to the Reynolds number is equal to the Prandtl number, which is the ratio of the kinematic viscosity (or momentum diffusivity) to the thermal diffusivity.

higher-order one-sided difference. The truncation error for such differences is made up of terms containing high-order derivatives as discussed in Sect. 3.3.2. High even-order spatial derivatives are more scale-selective in their damping than the second-derivative operator $\partial^2\psi/\partial x^2$ (Fig. 3.8). As a consequence, the influence of physical diffusion on wavelengths longer than $6\Delta x$–$10\Delta x$ is easier to capture using higher-order difference schemes to approximate $\partial\psi/\partial x$.

Now consider the effects of time differencing on the stability of differential–difference approximations to the advection–diffusion equation in which the spatial derivatives are not discretized. Recall that if (3.87) is Fourier-transformed, the result is (2.18), which has the same form as the basic test problem for investigating absolute stability, (2.16). It is common to use different time differencing for the advection and diffusion terms, so to facilitate the analysis of methods which integrate the advection and diffusion terms differently it is convenient to express (2.16) in the equivalent form

$$\frac{d\psi}{dt} = \lambda\psi + i\omega\psi, \tag{3.90}$$

where, as introduced in Sect. 2.2.1, $\lambda = \Re\{\gamma\}$ and $\omega = \Im\{\gamma\}$. The first term on the right side of the preceding equation is the prototype for the integration of those terms representing diffusion; the second term is the prototype for advection.

As discussed in Sect. 2.4.3, the region of absolute stability for the leapfrog scheme is just a segment along the imaginary axis, and this method is not suitable for integrating terms involving diffusion. Perhaps the simplest way to retain leapfrog time differencing for the oscillatory forcing in (2.16) and obtain a scheme that is also stable for very weak dissipation is by evaluating the damping term at time level $n-1$ such that

$$\frac{\phi^{n+1} - \phi^{n-1}}{2\Delta t} = i\omega\phi^n + \lambda\phi^{n-1}. \tag{3.91}$$

Note that although the standard leapfrog method is accurate to $O\left[(\Delta t)^2\right]$, this approach introduces an $O(\Delta t)$ truncation error in the approximation of the damping. The amplification factor for the leapfrog-forward scheme (3.91) is

$$A_{\mathrm{lff}} = i\tilde{\omega} \pm (1 + 2\tilde{\lambda} - \tilde{\omega}^2)^{1/2},$$

where $\tilde{\omega} = \omega\Delta t$ and $\tilde{\lambda} = \lambda\Delta t$. When $1 + 2\tilde{\lambda} > \tilde{\omega}^2$, the square root is real; both amplification factors have the same magnitude, and

$$|A_{\mathrm{lff}}|^2 = 1 + 2\tilde{\lambda}.$$

Thus, the case $1 + 2\tilde{\lambda} > \tilde{\omega}^2$ will be obtained and the leapfrog-forward scheme will be stable if

$$\frac{\tilde{\omega}^2 - 1}{2} < \tilde{\lambda} \le 0.$$

The region of the $\tilde{\lambda}$–$\tilde{\omega}$ plane satisfying this inequality lies inside the curve labeled LFF in Fig. 3.13a. Consideration of the case $1 + 2\tilde{\lambda} \le \tilde{\omega}^2$ shows that the actual

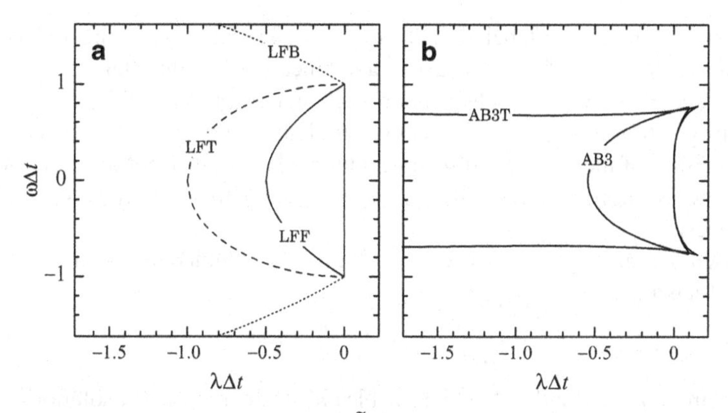

Fig. 3.13 **a** Regions of useful stability in the $\tilde{\lambda}$–$\tilde{\omega}$ plane if the oscillatory forcing is integrated using leapfrog differencing and the damping is integrated using the forward (*LFF*), trapezoidal (*LFT*), or backward (*LFB*) methods. **b** Region of absolute stability when the oscillatory forcing is integrated using the three-step Adams–Bashforth scheme and the damping is integrated using either the same three-step Adams–Bashforth formula (*AB3*, *solid line*) or the trapezoidal method (*AB3T*, *dashed line*). Note the difference in the horizontal and vertical scales

region of the $\tilde{\lambda}$–$\tilde{\omega}$ plane throughout which $|A_{\text{lff}}| \leq 1$ is a larger triangular region with vertices at $(\tilde{\lambda}, \tilde{\omega})$ equal to $(0, 1)$, $(-1, 0)$, and $(0, -1)$. This larger region of absolute stability is of no practical use, however, since A_{lff} is purely imaginary when $1 + 2\tilde{\lambda} \leq \tilde{\omega}^2$, and as a consequence, the numerical solution undergoes a $4\Delta t$ oscillation independent of the actual value of $\omega\Delta t$. Because the region of useful stability for the leapfrog-forward scheme is relatively small, this scheme is appropriate only for problems with very low viscosity. Even when the value of $\lambda\Delta t$ is as small as 0.3, stability considerations require a significant reduction in $\omega\Delta t$ relative to that which would be stable in the inviscid limit.

One might expect to do much better using an A-stable scheme to integrate the dissipation terms. A second-order method may be obtained if the damping term in (3.90) is approximated using trapezoidal time differencing such that

$$\frac{\phi^{n+1} - \phi^{n-1}}{2\Delta t} = \mathrm{i}\omega\phi^n + \frac{\lambda}{2}\left(\phi^{n+1} + \phi^{n-1}\right). \tag{3.92}$$

The amplification factor associated with the preceding equation is

$$A_{\text{lfb}} = \frac{\mathrm{i}\tilde{\omega} \pm (1 - \tilde{\lambda}^2 - \tilde{\omega}^2)^{1/2}}{1 - \tilde{\lambda}}.$$

If $\tilde{\omega}^2 + \tilde{\lambda}^2 \geq 1$, A_{lfb} is purely imaginary and the numerical solution undergoes spurious $4\Delta t$ oscillations. The combinations of $\tilde{\lambda}$ and $\tilde{\omega}$ that yield stable physically reasonable solutions satisfy

$$\tilde{\omega}^2 + \tilde{\lambda}^2 < 1 \quad \text{and} \quad \tilde{\lambda} \leq 0,$$

and lie within the curve labeled LFT in Fig. 3.13a. Since damping reduces the amplitude of the true solution to (2.16), and since the damping term in (3.92) is approximated by an A-stable scheme, one might hope that the stability condition for the leapfrog-differenced purely oscillatory problem would be a sufficient condition for the stability of the full hybrid scheme, but this is certainly not the case. Indeed, the increase in the region of stability relative to the leapfrog-backward scheme is quite modest.

A much larger region of absolute stability is obtained using the leapfrog-backward scheme

$$\frac{\phi^{n+1} - \phi^{n-1}}{2\Delta t} = i\omega\phi^n + \lambda\phi^{n+1}. \tag{3.93}$$

The combinations of $\tilde{\lambda}$ and $\tilde{\omega}$ that yield stable physically reasonable solutions satisfy

$$\tilde{\omega}^2 < 1 - 2\tilde{\lambda} \quad \text{and} \quad \tilde{\lambda} \leq 0. \tag{3.94}$$

The curve labeled LFB in Fig. 3.13a shows the region of useful stability for this scheme. Assuming $\lambda < 0$, the stability condition for the purely oscillatory problem ($|\omega\Delta t| < 1$) is sufficient to guarantee the stability of the leapfrog-backward approximation to (3.90).

The leapfrog-backward scheme is not a particularly attractive method because it is only first-order accurate, yet being implicit, it requires essentially the same computational overhead as the more accurate trapezoidal method. The stability and accuracy potentially available through a trapezoidal approximation to the damping term can be better realized if the leapfrog approximation to the oscillatory term is replaced by the three-step Adams–Bashforth method (2.80). The resulting Adams–Bashforth–trapezoidal approximation to (3.90) has the form

$$\frac{\phi^{n+1} - \phi^n}{\Delta t} = \frac{i\omega}{12}\left(23\phi^n - 16\phi^{n-1} + 5\phi^{n-2}\right) + \frac{\lambda}{2}\left(\phi^{n+1} + \phi^n\right). \tag{3.95}$$

The region of absolute stability for this method is plotted as the curve labeled AB3T in Fig. 3.13b. Also shown in that figure is the region of absolute stability for the standard three-step Adams–Bashforth scheme. As apparent in Fig. 3.13b, the region of absolute stability is dramatically improved when the dissipation term is approximated with the trapezoidal method instead of integrating the entire problem using the three-step Adams–Bashforth scheme. In particular, the region of absolute stability for (3.95) expands so that it is determined almost entirely by the value of $|\omega\Delta t|$. If Δt is small enough to yield stable solutions to the inviscid problem, then the full advection–diffusion problem will be stable for almost all values of $\lambda < 0$.

Since it remains stable when the diffusion is very strong, the Adams–Bashforth–trapezoidal approximation is well suited for simulating advection and diffusion in situations where there are both regions of zero viscosity and patches of high diffusivity. Patches of high eddy diffusivity may appear in a nominally inviscid fluid in localized regions where the flow is dynamically unstable to small-scale perturbations. The high eddy diffusion in these isolated regions can have a severe impact on the time step of the overall numerical integration unless an artificial cap is imposed on the maximum eddy diffusivity or the time differencing is approximated with a very stable method like the Adams–Bashforth–trapezoidal scheme.

The preceding analyses of the prototype ordinary differential equation (3.90) are sufficient to characterize the stability of the various time-differencing schemes qualitatively, but they do not yield the precise stability limits of the complete finite-difference equation obtained when the spatial derivatives in the advection–diffusion equation are approximated by finite differences. We conclude this section by examining two completely discrete approximations to the advection–diffusion equation.

First consider the forward-time centered-space approximation

$$\frac{\phi_j^{n+1} - \phi_j^n}{\Delta t} + c\delta_{2x}\phi_j^n = M\delta_x^2\phi_j^n,$$

which we expect will be unstable if M is too small relative to c. The standard von Neumann analysis shows that the numerical solution will be nonamplifying when

$$|A_k|^2 = [1 - 2\nu(1 - \cos k\Delta x)]^2 + \mu^2 \sin^2 k\Delta x \leq 1,$$

for all k in the interval $[0, \pi/\Delta x]$. Here, as before, $\mu = c\Delta t/\Delta x$ and $\nu = M\Delta t/(\Delta x)^2$. Necessary and sufficient conditions for the stability of this scheme are

$$0 \leq \nu \leq \frac{1}{2} \quad \text{and} \quad \mu^2 \leq 2\nu. \tag{3.96}$$

To establish the necessity of these conditions, note that the first condition is required for stability when $k\Delta x = \pi$, and the second condition is required for stability in the limit $k\Delta x \to 0$, in which case

$$|A_k|^2 \to 1 - (2\nu - \mu^2)(k\Delta x)^2.$$

To establish that (3.96) is sufficient for stability, suppose that $\mu^2 \leq 2\nu$. Then

$$\begin{aligned} |A_k|^2 &\leq (1 - 2\nu(1 - \cos k\Delta x))^2 + 2\nu \sin^2 k\Delta x \\ &= 1 - 2\nu(1 - 2\nu)(1 - \cos k\Delta x)^2, \end{aligned}$$

which is less than unity whenever $0 \leq \nu \leq 1/2$.

If leapfrog-forward differencing is used in the advection–diffusion equation to give the approximation

$$\delta_{2t}\phi_j^n + c\delta_{2x}\phi_j^n = M\delta_x^2\phi_j^{n-1}, \tag{3.97}$$

the amplification factor becomes

$$A = -i\mu \sin\theta \pm \left[-\mu^2 \sin^2\theta + 1 - 2\nu(1 - \cos\theta)\right]^{1/2}, \tag{3.98}$$

where μ and ν are defined above, and $\theta = k\Delta x$. One might imagine that the solution will be stable if the time step is small enough that both the advection and the diffusion problems considered in isolation are stable, which requires $|\mu| < 1$ and $\nu \leq 1/4$. In fact, the criteria required to ensure useful stability of all the resolved modes are more restrictive. For

$$1 - 2\nu(1 - \cos\theta) > \mu^2 \sin^2\theta, \tag{3.99}$$

the square root in (3.98) is real and

$$|A_k|^2 = 1 - 2\nu(1 - \cos(k\Delta x)).$$

The right side of the preceding equation is greater than zero by (3.99) and is less than unity because $\nu > 0$.

To determine the stability condition that ensures all resolved modes are stable, note that if (3.99) is not satisfied, A will be purely imaginary and the mode will oscillate with a nonphysical period of $4\Delta t$. This behavior is unacceptable (whether or not the numerical solution is actually amplifying), so we focus on determining the necessary and sufficient conditions for the satisfaction of (3.99), which may be expressed in the alternative form $F(\theta) > 0$, where

$$F(\theta) = 1 - 2\nu(1 - \cos\theta) - \mu^2(1 - \cos^2\theta).$$

Then

$$\frac{dF}{d\theta} = -2\sin\theta(\nu - \mu^2\cos\theta), \qquad \frac{d^2F}{d\theta^2} = -2(\nu\cos\theta + 2\cos^2\theta - 1).$$

It follows that the minimum of F with respect to θ occurs either (1) at $\cos\theta = -\nu/\mu^2$ for the case $\nu < \mu^2$, or (2) at the highest resolvable wavenumber, $\theta = \pi$ for the case $\nu \geq \mu^2$.

First consider the case $\nu < \mu^2$. Evaluating $F(\theta)$ at its minimum, $F(\theta) > 0$ becomes

$$1 - 2\nu - 2\nu^2/\mu^2 - \mu^2 + \nu^2/\mu^2 > 0,$$

or $\mu^2 > (\mu^2 + \nu)^2$. In the case $\nu \geq \mu^2$, the extremum is obtained by evaluating $F(\theta)$ at $\theta = \pi$, so $F(\theta) > 0$ requires $1 - 4\nu > 0$ or $\nu < 1/4$. The scheme is therefore stable when either of the following conditions is satisfied.

$$\mu^2 \leq \nu < \frac{1}{4} \qquad \text{(weak advection)},$$

$$\nu < \mu^2 \quad \text{and} \quad (\mu^2 + \nu)^2 < \mu^2 \qquad \text{(weak diffusion)}.$$

Note that $|\mu| \leq 1$ is a necessary condition for stability and that any nonzero value of M reduces the maximum stable time step relative to that in the inviscid limit $(M \to 0)$.

3.5.2 Advection with Sources and Sinks

Sources and sinks typically appear as functions of the temporal and spatial coordinates and the undifferentiated unknown variables. In problems such as chemically reacting flow, the source and sink terms couple many unknowns together, but only at a single point. In such cases the equations governing these reactions are often integrated as a separate fractional step (see Sect. 4.3) using one of the ordinary differential equation solvers described in Chap. 2. In this section we explore how the treatment of simple sources and sinks in unsplit integrations of the advection–source–sink equation influence the stability and accuracy of the numerical solution.

First, suppose that the source or sink is a function only of the coordinate variables, in which case the advection–source–sink equation is

$$\frac{\partial \psi}{\partial t} + c\frac{\partial \psi}{\partial x} = s(x,t).$$

Almost any finite-difference scheme suitable for the approximation of the pure advection problem can be trivially modified to approximate the preceding equation. The only subtlety involves the numerical specification of $s(x,t)$. It is natural to specify $s(x,t)$ at the finest spatial and temporal scales resolvable on the space–time grid, but this may generate excessive noise in the numerical solution. One should make a distinction between the shortest scale present on the numerical mesh and the shortest scale at which the finite-difference scheme can be expected to yield physically meaningful results. The accuracy of the solution is generally not improved by applying forcing at wavelengths too short to be adequately simulated by the numerical scheme. Since almost all numerical methods do a very poor job of simulating $2\Delta x$ waves and $2\Delta t$ oscillations, it is usually unwise to include spatial and temporal scales in $s(x,t)$ corresponding to wavelengths shorter than about $4\Delta x$ or periods shorter than about $4\Delta t$.[11] The optimal cutoff depends on the numerical scheme and the nature of the problem being approximated. See Lander and Hoskins (1997) for an example in which a cutoff wave number is determined for external forcing in a spectral model of the Earth's atmosphere.

Now suppose that the sink is a linear function of ψ, so that the advection–source–sink equation is

$$\frac{\partial \psi}{\partial t} + c\frac{\partial \psi}{\partial x} = -r\psi, \tag{3.100}$$

where positive values of r represent sinks and negative values represent sources. Confusion can arise in assessing the stability of numerical approximations to (3.100). Consider the stability of the upstream approximation

$$\frac{\phi_j^{n+1} - \phi_j^n}{\Delta t} + c\left(\frac{\phi_j^n - \phi_{j-1}^n}{\Delta x}\right) = -r\phi_j^n. \tag{3.101}$$

[11] For similar reasons, it is often unwise to include $2\Delta x$ features in the initial data.

The amplification factor for this scheme is

$$A = (1 - \mu - \lambda) + \mu e^{-ik\Delta x}, \tag{3.102}$$

where $\lambda = r\Delta t$ and $\mu = c\Delta t/\Delta x$. If $r < 0$, the true solution grows with time, and it is clearly inappropriate to require $|A| \leq 1$. In this case, all that is required is that the scheme be sufficiently stable to converge in the limit $\Delta t \to 0$, $\Delta x \to 0$, for which the von Neumann condition is

$$|A| \leq 1 + \eta\Delta t, \tag{3.103}$$

where η is a constant independent of Δt and Δx. To establish criteria guaranteeing satisfaction of (3.103) let

$$\tilde{A} = (1 - \mu) + \mu e^{-ik\Delta x},$$

which is just the amplification factor for upstream differencing. Since $A = \tilde{A} - r\Delta t$,

$$|A| \leq |\tilde{A}| + |r\Delta t|.$$

If $0 \leq \mu \leq 1$, then as demonstrated in Sect. 3.2.2, $|\tilde{A}| \leq 1$ and

$$|A| \leq 1 + |r|\Delta t,$$

which satisfies the stability criterion (3.103). The finite-difference approximation (3.101) to the advection–source equation is therefore stable whenever the associated approximation to the pure advection problem is stable.

Clearly, the method used in the preceding stability analysis can be generalized to a wider class of problems. Let $L(\psi)$ be a linear operator involving partial derivatives of ψ and consider the family of partial differential equations of the form

$$\frac{\partial \psi}{\partial t} + L(\psi) + r\psi = 0.$$

As demonstrated by Strang (1964), the range of Δt for which any explicit two-time-level approximation to the preceding partial differential equation satisfies the stability condition (3.103) is independent of the value of r. Unfortunately, it is rather easy to misinterpret this result. The Strang perturbation theorem guarantees only that the value of r has no influence on the ability of consistent finite-difference approximations to converge to the correct solution in the limit of Δt, $\Delta x \to 0$. The value of r does affect the boundedness of numerical solutions computed with finite values of Δt and Δx.

The solution to the advection–sink problem (3.100) is bounded whenever $r > 0$, and in such circumstances the numerical approximation obtained using finite Δx and Δt should satisfy the more strict stability condition $|A| \leq 1$. The conditions on $r\Delta t$ required to guarantee a nongrowing solution may be determined as follows. Using (3.102),

$$|A|^2 = (1 - \lambda)^2 - 2\mu(1 - \mu - \lambda)(1 - \cos k\Delta x).$$

Now consider two cases. First, if $\mu(1-\mu-\lambda) \geq 0$, the largest amplification factor occurs when $k = 0$ and

$$|A_{k=0}|^2 = (1-\lambda)^2.$$

In this case all $|A|$ are less than unity when $(1-\lambda)^2 \leq 1$ or $0 \leq \lambda \leq 2$. This inequality is always satisfied, because r, c, and therefore λ and μ are nonnegative, and by assumption $\mu(1-\mu-\lambda) \geq 0$, implying $1 \geq 1-\mu \geq \lambda$.

All the stability restrictions on Δt arise therefore from the second case, for which $\mu(1-\mu-\lambda) < 0$. In this case the largest amplification factor occurs for $k\Delta x = \pi$ and

$$|A_{k=\pi\Delta x}|^2 = (1-\lambda)^2 - 4\mu(1-\mu-\lambda) = (1-\lambda-2\mu)^2.$$

Thus, all $|A|$ are less than unity when $(1-\lambda-2\mu)^2 \leq 1$ or $0 \leq \lambda + 2\mu \leq 2$, or equivalently,

$$0 \leq \frac{c\Delta t}{\Delta x}\left(1 + \frac{r\Delta x}{2c}\right) \leq 1.$$

The last expression shows that the value of r ceases to restrict the maximum stable time step as $\Delta x \to 0$. This is consistent with the implication of the Strang perturbation theorem that the stability condition sufficient to guarantee convergence cannot depend on r. The value of r may, nevertheless, have a dramatic impact on the maximum stable time step when Δx is finite.

3.6 Summary

In this chapter we have investigated the performance of schemes for approximating advective and diffusive transport in one dimension. Let us now recapitulate the better methods discussed in this chapter and briefly summarize the conditions under which they might be expected to yield good results. Further analysis of the performance of these schemes in more complex situations will be presented in the following chapters.

First, consider the class of problems in which the solution is sufficiently smooth that it can always be properly resolved on the numerical mesh. This is typically the case in diffusion-dominated flow. Under these circumstances any stable method can be expected to converge to the correct result as the space–time grid is refined. Higher-order schemes will converge to *smooth* solutions more rapidly than low-order methods as the mesh size is decreased. Thus, even though higher-order methods require more computations per grid point per time step, genuinely high accuracy (i.e., several significant digits) can usually be achieved more efficiently by using a high-order scheme on a relatively coarse mesh than by using a low-order scheme on a finer mesh. One of the most efficient ways to achieve high-order accuracy in the representation of spatial derivatives for smooth flows is through spectral methods, which will be introduced in Chap. 6. Nevertheless, in practice it is common to model simple diffusion problems with the well-behaved Crank–Nicolson method, although this method is only second order.

Most low-viscosity flows do not remain completely smooth[12]. Instead, they develop at least some features with spatial scales shorter than or equal to that of an individual grid cell. Such small-scale features cannot be accurately captured by any numerical scheme, and the unavoidable errors in these small scales can feed back on the larger-scale flow and thereby exert a significant influence on the overall solution. In such circumstances there is no hope of computing an approximation to the correct solution that is accurate to several significant digits. Although the larger-scale features may be approximated with considerable quantitative accuracy, generally one must either be content with a qualitatively correct representation of the shortest-scale features or remove these features with some type of numerical smoothing. Since it is not realistic to expect convergence to the correct solution in such problems, it is not particularly important to use high-order methods. Instead, one generally employs the finest possible numerical grid, selects a method that captures the behavior of moderately resolved waves with reasonable fidelity, and ensures that any spurious poorly resolved waves are eliminated by either explicit or implicit numerical dissipation. The numerical dissipation associated with all the schemes considered in this chapter is applied throughout the entire numerical domain. An alternative approach will be considered in Chap. 5, in which the implicit dissipation is primarily limited to those regions where the approximate solution is discontinuous or very poorly resolved.

Given that some degree of dissipation must generally be included to generalize the methods described in this chapter to practical problems involving low-viscosity flow, the neutral amplification factors associated with leapfrog time differencing are less advantageous than they may first appear. As discussed in Chap. 2, the difficulties associated with time splitting that can arise in nonlinear problems make the leapfrog scheme relatively unattractive in comparison with the third-order Adams–Bashforth or third- or fourth-order Runge–Kutta methods. Even forward differencing is a possibility, provided that it is used in a Lax–Wendroff method and that the implicit diffusion in the Lax–Wendroff scheme is limited by using a sufficiently small time step.

The choice between centered and upstream-biased stencils for the approximation of spatial derivatives is less clear-cut. Approximations to the advection equation based on centered spatial differences typically require the use of an explicit fourth- or sixth-derivative dissipative filter and are therefore less efficient than a third-order upstream approximation. This lack of efficiency is compensated by two practical advantages. First, it is not necessary to determine the upstream direction at each grid point when formulating the computer algorithm to evaluate a centered spatial difference. The determination of the upstream direction is not difficult in advection problems where all signal propagation is directed along a clearly defined flow, but it can be far more difficult in problems admitting wave solutions that propagate both to the right and to the left. The second advantage of a centered difference

[12] An example of inviscid fluid motion that does remain smooth is provided by the barotropic vorticity equation, which will be discussed in Sect. 4.5.2.

used in conjunction with a spatial filter is that one can explicitly control the magnitude of the artificial dissipation, whereas the magnitude of the numerical dissipation associated with an upstream-biased difference is implicitly determined by the local wind speed. The compact schemes appear to provide particularly good formulae for the evaluation of centered spatial differences because they remain accurate at relatively short wavelengths ($3\Delta x$ or $4\Delta x$) and use information at a minimum number of spatial grid points, which reduces the amount of special coding required near the boundaries of the spatial domain.

Problems

1. Suppose that $f(x)$ is to be represented at discrete points x_j on an *uneven* mesh and that $\Delta_{j-1/2} = x_j - x_{j-1/2}$. Use Taylor series expansions to derive a second-order finite-difference approximation to $\mathrm{d}f/\mathrm{d}x$ using a three-point stencil of the form

$$\alpha f_{j+1} + \beta f_j + \gamma f_{j-1}.$$

Hint: The result may be written in the form

$$\left(\frac{\Delta_{j-\frac{1}{2}}}{\Delta_{j+\frac{1}{2}} + \Delta_{j-\frac{1}{2}}}\right)\left(\frac{f_{j+1} - f_j}{\Delta_{j+\frac{1}{2}}}\right) + \left(\frac{\Delta_{j+\frac{1}{2}}}{\Delta_{j+\frac{1}{2}} + \Delta_{j-\frac{1}{2}}}\right)\left(\frac{f_j - f_{j-1}}{\Delta_{j-\frac{1}{2}}}\right).$$

2. Determine an $O(\Delta x)^2$-accurate *one-sided* finite-difference approximation to the first derivative $\partial\psi/\partial x$. Use the minimum number of points. Suppose that the numerical solution ϕ_j is available at points x_j, and that the derivative will be calculated using points to the right of x_j (i.e., x_j, $x_{j+1}, \ldots$). Assume a constant grid spacing. How does the magnitude of the leading-order term in the truncation error of this one-sided approximation compare with that for the centered difference

$$\frac{\phi_{j+1} - \phi_{j-1}}{2\Delta x} \; ?$$

3. Determine those regions of the x–t plane in which the solution of

$$\left(\frac{\partial}{\partial t} + U\frac{\partial}{\partial x}\right)^2 \psi - c^2\frac{\partial^2\psi}{\partial x^2} = 0,$$

depends on ψ at some fixed point (x_0, t_0). Assuming that U and c are nonnegative constants, schematically plot these regions and label them as either the "domain of influence" or the "domain of dependence." Draw a plot for the case $U > c$ and a plot for $U < c$.

4. Explain how the unconditional stability of the trapezoidally time differenced one-dimensional advection equation

$$\frac{\phi_j^{n+1} - \phi_j^n}{\Delta t} + \frac{c}{2\Delta x}\left(\frac{\phi_{j+1}^{n+1} + \phi_{j+1}^n}{2} - \frac{\phi_{j-1}^{n+1} + \phi_{j-1}^n}{2}\right) = 0$$

is consistent with the CFL stability condition.

5. Consider the Lax–Friedrichs approximation to the scalar advection equation

$$\frac{\phi_j^{n+1} - \frac{1}{2}(\phi_{j+1}^n + \phi_{j-1}^n)}{\Delta t} + c\frac{\phi_{j+1}^n - \phi_{j-1}^n}{2\Delta x} = 0.$$

(a) Determine the truncation error for this scheme. Under what conditions does this scheme provide a consistent approximation to the advection equation? Would the condition required for consistency be difficult to satisfy in a series of simulations in which Δx is repeatedly halved?

(b) Determine the values of $c\Delta t/\Delta x$ for which this scheme is stable.

6. Show that the Dufort–Frankel method (3.84) does indeed produce stable, non-growing solutions to the one-dimensional diffusion equation (3.80) for any Δt.

7. When applied to the oscillation equation, Matsuno time differencing preferentially damps the higher frequencies (provided that $\kappa_{\max}\Delta t < 1/\sqrt{2}$). Yet, if we turn our attention to the constant-wind-speed advection equation, the Lax–Wendroff scheme (3.72) damps $2\Delta x$ waves much more rapidly than does the following combination of Matsuno time differencing and centered space differencing:

$$\tilde{\phi}^n = \phi^n - c\Delta t\, \delta_{2x}\phi^n, \qquad \phi^{n+1} = \phi^n - c\Delta t\, \delta_{2x}\tilde{\phi}^n.$$

Explain why. Consider only those time steps for which $c\Delta t$ times the effective horizontal wave number is less than $1/\sqrt{2}$.

8. Consider the shallow-water equations, linearized about a state at rest,

$$\frac{\partial u}{\partial t} + g\frac{\partial \eta}{\partial x} = 0, \qquad \frac{\partial \eta}{\partial t} + H\frac{\partial u}{\partial x} = 0.$$

Prove, *without* doing a von Neumann stability analysis, that the following finite-difference approximation to the preceding system must be unstable:

$$\frac{u_j^{n+1} - u_j^n}{\Delta t} + g\left(\frac{\eta_j^n - \eta_{j-1}^n}{\Delta x}\right) = 0,$$

$$\frac{\eta_j^{n+1} - \eta_j^n}{\Delta t} + H\left(\frac{u_j^n - u_{j-1}^n}{\Delta x}\right) = 0.$$

9. Determine the truncation error in the following approximation to the one-dimensional advection equation:

$$\frac{\phi_j^{n+1} - \phi_j^n}{\Delta t} + c\left(\frac{3\phi_j^n - 4\phi_{j-1}^n + \phi_{j-2}^n}{2\Delta x}\right) = 0.$$

Also determine the range of $\mu = c\Delta t/\Delta x$ over which the scheme is stable.

10. The method of Warming and Beam (3.79) uses the same numerical stencil as the formula in Problem 9. Show that it is more accurate and more stable. In particular show it is second-order accurate with truncation error $O(\Delta t^2 + \Delta t \Delta x + (\Delta x)^2)$, and that it is stable for $0 \leq \mu \leq 2$.

11. According to (3.71), what will be the maximum order of accuracy for a two-time-level explicit scheme that approximates the advection equation using r points on the downstream side of the numerical stencil and $r + 1$ points upstream? Give an example of one such scheme.

12. Suppose we try to reduce the phase-speed errors in the Lax–Wendroff scheme (3.72) by using the following approximation to the constant-wind-speed advection equation:

$$\frac{\phi_j^{n+1} - \phi_j^n}{\Delta t} + c\left(\frac{4}{3}\delta_{2x}\phi_j^n - \frac{1}{3}\delta_{4x}\phi_j^n\right) = \frac{c^2\Delta t}{2}\delta_x^2\phi_j^n.$$

Evaluate the truncation errors in this scheme to determine the leading-order dissipative and dispersive errors, and determine the condition (if any) under which the scheme is stable. Compare your results with those for (3.72) and for the leapfrog-time fourth-order-space scheme:

$$\delta_{2t}\phi_j^n + c\left(\frac{4}{3}\delta_{2x}\phi_j^n - \frac{1}{3}\delta_{4x}\phi_j^n\right) = 0.$$

13. Determine the order of accuracy and the stability properties of the "slant-derivative" approximation to the constant-wind-speed advection equation:

$$\frac{\phi_j^{n+1} - \phi_j^n}{\Delta t} + \frac{c}{2}\left(\frac{\phi_j^{n+1} - \phi_{j-1}^{n+1}}{\Delta x} + \frac{\phi_{j+1}^n - \phi_j^n}{\Delta x}\right) = 0.$$

14. Suppose that the advection equation is approximated by a second-order Runge–Kutta time difference and a centered second-order spatial difference. Show that the auxiliary condition $O(\Delta t) \leq O\left[(\Delta x)^{4/3}\right]$ is a necessary condition for this scheme to converge to the true solution in the limit $\Delta t, \Delta x \to 0$.

15. Suppose that the time derivative in the differential–difference equation (3.28) is approximated using a fourth-order Runge–Kutta scheme. Determine the maximum value of $c\Delta t/\Delta x$ for which this scheme will be stable using the stability criteria for the oscillation equation given in Table 2.2. Explain how this value can exceed unity without violating the CFL condition.

16. Derive the modified equation that is approximated through order 3 by the leapfrog-time centered-space scheme (3.59). Compare this with the modified equation for the Lax–Wendroff scheme

$$\frac{\partial \phi}{\partial t} + c\frac{\partial \phi}{\partial x} = -(1-\mu^2)\frac{c(\Delta x)^2}{6}\frac{\partial^3 \phi}{\partial x^3} - \mu(1-\mu^2)\frac{c(\Delta x)^3}{8}\frac{\partial^4 \phi}{\partial x^4},$$

where $\mu = c\Delta t/\Delta x$. Discuss whether the behavior of these two schemes, as illustrated in Fig. 3.12, is consistent with the leading-order error terms in each scheme's modified equation.

17. Derive an expression for the upstream approximation to the constant-wind-speed advection equation that remains upstream independent of the sign of the velocity field. Express the upstream spatial derivative as the combination of a centered-space derivative and a diffusive smoother in a manner similar to that in (3.46).

18. Under what condition will solutions to the leapfrog-backward approximation (3.93) produce spurious oscillations of period $4\Delta t$? Let the region of useful stability be the values of $\omega\Delta t$ and $\lambda\Delta t$ for which the solution is nonamplifying and not spuriously oscillating with a period of $4\Delta t$. Derive the conditions (3.94) defining the region of useful stability for the leapfrog-backward approximation to (3.90).

19. *Compute solutions to the constant-wind-speed advection equation on the periodic domain $0 \le x \le 1$ subject to the initial condition $\psi(x,0) = \sin^6(2\pi x) + R(x)$, where $R(x)$ is a small random number in the interval $[-5 \times 10^{-7}, 5 \times 10^{-7}]$. Use centered-space differencing $\partial\psi/\partial x \approx \delta_{2x}\phi$ and set $c = 0.1$.

(a) Compare the exact solution with numerical solutions obtained using forward and leapfrog differencing. Use a Courant number $c\Delta t/\Delta x = 0.1$ and plot your solutions at $t = 50$ using a vertical scale that includes $-40 \le \phi \le 40$. Compare and explain the results obtained using $\Delta x = 1/20$, 1/40, and 1/80. Use a single forward time step to initialize the leapfrog integration.

(b) Compare the exact solution with numerical solutions obtained using Heun (second-order Runge–Kutta) and leapfrog differencing. Use a Courant number $c\Delta t/\Delta x = 0.5$ and plot your solutions at $t = 60$ using a vertical scale that includes $-2 \le \phi \le 2$. Compare and explain the results obtained using $\Delta x = 1/20$, 1/40, 1/80, and 1/160.

(c) Repeat the simulation in (b) with $\Delta x = 1/160$ but use a Courant number of 1.2 and integrate to $t = 2.1$. Compare and contrast the nature of the instabilities exhibited by the forward, leapfrog, and Heun methods in the simulations in (a), (b), and (c).

20. *Find solutions to the advection equation

$$\frac{\partial \psi}{\partial t} + c\frac{\partial \psi}{\partial x} = 0$$

in a periodic domain $0 \leq x \leq 1$. Suppose that $c = 0.2\ \mathrm{ms}^{-1}$ and

$$\psi(x,0) = \begin{cases} 9^4\left[(x-\frac{5}{6})^2 - (\frac{1}{9})^2\right]^2, & \text{if } |x-\frac{5}{6}| \leq \frac{1}{9}; \\ 0, & \text{otherwise}. \end{cases}$$

Obtain solutions using (1) leapfrog time differencing and centered second-order spatial differencing, (2) upstream (or donor-cell) differencing, and (3) the Lax–Wendroff method. Choose $\Delta x = 1/36$. Examine the sensitivity of the numerical solutions to the Courant number $(c\Delta t/\Delta x)$. Try Courant numbers of 0.1, 0.5, and 0.9. For each Courant number, submit a plot of the three numerical solutions and the exact solution at time $t = 5$. Scale the vertical axis on the plot to the range $-0.6 \leq \psi \leq 1.6$. Discuss the relative quality of the solutions and their dependence on the Courant number. Is the dependence of the solutions on the Courant number consistent with the modified equation (3.67) and the results obtained in Problem 16?

21. *Consider the leapfrog-time fourth-order-space approximation to the constant-wind-speed advection equation

$$\delta_{2t}\phi_j^n + c\left(\frac{4}{3}\delta_{2x}\phi_j^n - \frac{1}{3}\delta_{4x}\phi_j^n\right) = 0.$$

(a) Determine the maximum Courant number $(c\Delta t/\Delta x)$ for which this scheme is stable.

(b) Repeat the comparison in Problem 20 including results from this fourth-order scheme and the leapfrog-time second-order-space scheme on each plot. Use Courant numbers 0.1, 0.5, and 0.72. Does the accuracy of both approximate solutions improve as the Courant number approaches its maximum stable value? Why or why not?

22. *Compute solutions to the one-dimensional diffusion equation (3.80) over the time interval $0 \leq t \leq 1/4$ on the periodic domain $0 \leq x \leq 1$, subject to the initial condition

$$\psi(x,0) = \sin(2\pi x) + \cos(6\pi x)/2 + \sin(20\pi x)/5 + R,$$

where R is a randomly distributed number in the interval $[0, 4 \times 10^{-7}]$ and $M = 0.01$.

(a) Derive an expression for the exact solution to this problem for the special case $R = 0$.

(b) Compute three approximate solutions using the Crank–Nicolson scheme (3.86). Compute a coarse-resolution solution using $\Delta x = 1/20$ and setting $M\Delta t/(\Delta x)^2 = 1/2$. Then halve Δt and Δx and compute another approximate solution with $\Delta x = 40$ and $M\Delta t/(\Delta x)^2 = 1$. Finally halve Δt and Δx one additional time and submit plots for all three resolutions at $t = 0.25$.

(c) Obtain a second set of approximate solutions using forward time differencing (3.81) at the same space and time resolutions considered in (b) and submit those plots.

(d) Finally, repeat the same series of integrations using the DuFort–Frankel method (3.84). Initialize the Dufort–Frankel method with a single forward step. Submit plots at $t = 0.25$ for the same three resolutions considered in (b).

(e) Discuss the differences between the solutions produced by the three methods and their probable source. In this instance, which appear to be the more critical constraints on the values of Δt and Δx: those associated with ensuring the stability of the forward scheme or those required for the consistency of the DuFort–Frankel method?

Chapter 4
Beyond One-Dimensional Transport

The basic properties of finite-difference methods were explored in Chap. 3 by applying each scheme to simple prototype problems involving advection and diffusion in one spatial dimension. The equations governing wavelike geophysical flows include additional complexities. In particular, the flow may depend on several unknown functions that are related by a system of partial differential equations, the unknowns may be functions of more than two independent variables, and the equations may be nonlinear. In this chapter we will examine some of the additional considerations that arise in the design and analysis of finite-difference schemes for the approximation of these more general problems.

4.1 Systems of Equations

Suppose that the problem of interest involves several unknown functions of x and t and that the governing equations for the system are linear with constant coefficients. An example of this type is the linearized one-dimensional shallow-water system

$$\frac{\partial u}{\partial t} + U\frac{\partial u}{\partial x} + g\frac{\partial h}{\partial x} = 0, \tag{4.1}$$

$$\frac{\partial h}{\partial t} + U\frac{\partial h}{\partial x} + H\frac{\partial u}{\partial x} = 0, \tag{4.2}$$

in which U and $u(x,t)$ represent the mean and perturbation fluid velocity, H and $h(x,t)$ are the mean and perturbation fluid depth, and g is the gravitational acceleration. The procedure for determining the truncation error, consistency, and order of accuracy of finite-difference approximations to a system such as (4.1) and (4.2) is identical to that discussed in Sect. 3.1. Taylor series expansions for the exact solution at the various grid points $(x_0, x_0 \pm \Delta x, \dots)$ are substituted into each finite difference, and the order of accuracy of the overall scheme is determined by the

D.R. Durran, *Numerical Methods for Fluid Dynamics: With Applications to Geophysics*, Texts in Applied Mathematics 32, DOI 10.1007/978-1-4419-6412-0_4,

lowest powers of Δx and Δt appearing in the truncation error. The stability analysis for finite-difference approximations to systems of partial differential equations is, on the other hand, more complex than that for a single equation.

4.1.1 Stability

Recall that a von Neumann stability analysis of the finite-difference approximation to a linear constant-coefficient scalar equation is performed by determining the magnitude of the amplification factor A_k. Here, as in Sect. 3.2.2, the amplification factor for a two-time-level scheme is defined such that a single step of the finite-difference integration maps the Fourier component e^{ikx} to $A_k e^{ikx}$. However, when the governing equations are approximated by a linear constant-coefficient system of finite-difference equations, the kth Fourier component of the solution is represented by the vector $\mathbf{v}_k$, and the amplification factor becomes an *amplification matrix*[1] $\mathbf{A}_k$. For example, in the shallow-water system (4.1) and (4.2), the vector representing the kth Fourier mode is

$$\mathbf{v}_k = \begin{pmatrix} u_k \\ h_k \end{pmatrix} e^{ikx}.$$

If the true solution does not grow with time, an appropriate stability condition is that

$$\|\mathbf{v}_k^n\| = \|\mathbf{A}_k^n \mathbf{v}_k^0\| \leq \|\mathbf{v}_k^0\|,$$

for all n and all wave numbers k resolved on the numerical mesh. For a single scalar equation, this condition reduces to (3.13). If the true solution grows with time, or if one is interested only in establishing sufficient conditions for the convergence of a consistent finite-difference scheme, the preceding condition should be relaxed to

$$\|\mathbf{v}_k^n\| = \|\mathbf{A}_k^n \mathbf{v}_k^0\| \leq C_T \|\mathbf{v}_k^0\| \quad \text{for all} \quad n\Delta t \leq T$$

and all sufficiently small values of Δt and Δx. Here C_T may depend on T, the time period over which the equations are integrated, but not on Δt or Δx. In the case of a single scalar equation, the preceding condition reduces to (3.12). Possible vector norms for use in these inequalities include $\|\;\|_\infty$ and $\|\;\|_2$ (defined by (3.10) and (3.11)).

4.1.1.1 Power Bounds on Matrices

The preceding stability conditions may be expressed in terms of the amplification matrix after introducing the concept of matrix norms. The norm of a matrix

[1] Except in this section, the amplification matrix will be written without the subscript k for conciseness.

is defined in terms of the more familiar vector norm such that if $\mathbf{B}$ is an $M \times N$ matrix and $\mathbf{z}$ a column vector of length N, then

$$\|\mathbf{B}\| = \sup_{\|\mathbf{z}\|=1} \|\mathbf{Bz}\| = \sup_{\|\mathbf{z}\|\neq \mathbf{0}} \frac{\|\mathbf{Bz}\|}{\|\mathbf{z}\|}.$$

In particular, if b_{ij} is the element in the ith row and jth column of $\mathbf{B}$, then

$$\|B\|_\infty = \max_{1\le i \le M} \sum_{j=1}^{N} |b_{ij}|$$

and

$$\|B\|_2 = \rho_\mathrm{m}(\mathbf{B}^*\mathbf{B})^{1/2},$$

where $\mathbf{B}^*$ is the conjugate transpose of $\mathbf{B}$, and ρ_m is the *spectral radius*, defined as the maximum in absolute value of the eigenvalues of a square matrix.[2]

Necessary and sufficient conditions for the stability of a constant-coefficient linear system may be expressed using this norm notation as

$$\|\mathbf{A}_k^n\| \le 1 \tag{4.3}$$

for nongrowing solutions, and as

$$\|\mathbf{A}_k^n\| \le C_T \tag{4.4}$$

in cases where the true solution grows with time or where the interest is only in ensuring that the numerical solution will be sufficiently stable to converge in the limit of $\Delta x, \Delta t \to 0$. (Once again, C_T depends on time, but not on Δx and Δt.)

Up to this point, the stability analysis for the single scalar equation and that for the system are essentially the same. The difference between the two arises when one attempts to reduce the preceding conditions on $\|\mathbf{A}_k^n\|$ to a constraint on $\|\mathbf{A}_k\|$. In the scalar case the necessary and sufficient condition that $|A_k^n| \le 1$ is just $|A_k| \le 1$. On the other hand, when the amplification factor is a matrix, the necessary and sufficient conditions for $\mathbf{A}_k$ to be "power-bounded" are rather complex. Since $\|\mathbf{A}_k^n\| \le \|\mathbf{A}_k\|^n$ (by the fundamental properties of any norm), the condition $\|\mathbf{A}_k\| \le 1$ will ensure stability. This condition is not, however, necessary for stability, as may be seen by considering the matrix

$$\mathbf{E} = \begin{pmatrix} 1 & -1 \\ 0 & -1 \end{pmatrix},$$

for which $\|\mathbf{E}\|_\infty = 2$, $(\|\mathbf{E}\|_2)^2 = (3 + \sqrt{5})/2$, and yet for all positive integers m, $\mathbf{E}^{2m}$ is the identity matrix, whose norm is unity.

---

[2] The standard mathematical notation for the spectral radius of a matrix $\mathbf{A}$ is $\rho(\mathbf{A})$. We use $\rho_\mathrm{m}(\mathbf{A})$ to distinguish the spectral radius from the density.

The precise necessary and sufficient conditions for an arbitrary matrix to be power-bounded are given by the Kreiss matrix theorem (Kreiss 1962; Strikwerda 1989, p. 188) and are relatively complicated. Necessary conditions for the boundedness of $\|\mathbf{A}_k^n\|$ can, however, be expressed quite simply. To have $\|\mathbf{A}_k^n\| \leq 1$, i.e., to have a nongrowing numerical solution, it is necessary that

$$\rho_{\mathrm{m}}(\mathbf{A}_k) \leq 1. \tag{4.5}$$

To satisfy the bound on the amplification matrix for growing solutions (4.4), it is necessary that

$$\rho_{\mathrm{m}}(\mathbf{A}_k) \leq 1 + \eta \Delta t, \tag{4.6}$$

where η is a constant independent of Δx and Δt. The fact that the preceding expressions are not sufficient conditions for stability is illustrated by the matrix

$$\mathbf{F} = \begin{pmatrix} 1 & 1 \\ 0 & 1 \end{pmatrix},$$

for which $\rho_{\mathrm{m}}(\mathbf{F}) = 1$, but

$$\mathbf{F}^n = \begin{pmatrix} 1 & n \\ 0 & 1 \end{pmatrix},$$

so $\|\mathbf{F}^n\|$ grows linearly with n. This linear growth is, however, much weaker than the geometric growth in $\|\mathbf{F}^n\|$ that would occur if the spectral radius of $\mathbf{F}$ were bigger than 1. The fact that the condition $\rho_{\mathrm{m}}(\mathbf{A}_k) \leq 1$ is capable of eliminating all highly unstable cases with geometrically growing solutions is an indication that the spectral radius criteria (4.5) and (4.6) are "almost" strong enough to ensure stability. Indeed, if $\mathbf{A}_k$ can be transformed to a diagonal matrix, which is frequently the situation when hyperbolic partial differential equations are approximated by finite differences, (4.5) and (4.6) are both necessary and sufficient conditions for stability. Even if $\mathbf{A}_k$ cannot be transformed to a diagonal matrix, (4.5) will be sufficient to ensure nongrowing solutions, provided that the moduli of all but one of the eigenvalues of $\mathbf{A}_k$ are strictly less than unity.

In most practical applications, the governing equations will contain either nonlinear terms or linear terms with variable coefficients, and to perform a von Neumann stability analysis, one must first approximate the full equations with a frozen-coefficient linearized system. Subsequent analysis of the frozen-coefficient linearized system yields necessary, but not sufficient, conditions for the stability of the numerical solution to the original problem. It is therefore often not profitable to exert great effort to determine sufficient conditions for the stability of the frozen-coefficient linearized system. Instead, it is common practice to evaluate (4.5) or (4.6) with the understanding that they provide necessary conditions for the stability of both the original problem and the associated frozen-coefficient linear system, but do not guarantee stability in either case.

4.1.1.2 Reanalysis of Leapfrog Time Differencing

A simple system of finite-difference equations illustrating the preceding concepts can be obtained by writing the leapfrog time-differenced approximation to the oscillation equation (2.19) as

$$\phi^{n+1} = \chi^n + 2i\kappa\Delta t\phi^n, \tag{4.7}$$

$$\chi^{n+1} = \phi^n. \tag{4.8}$$

Although in this example the original problem is governed by a single ordinary differential equation rather than a system of partial differential equations, the stability analysis of the finite-difference system proceeds as if (4.7) and (4.8) had been obtained directly from a more complicated problem. Let $\mathbf{A}$ denote the amplification matrix obtained by writing the system in the matrix form

$$\begin{pmatrix} 2i\kappa\Delta t & 1 \\ 1 & 0 \end{pmatrix} \begin{pmatrix} \phi \\ \chi \end{pmatrix}^n = \begin{pmatrix} \phi \\ \chi \end{pmatrix}^{n+1}.$$

The eigenvalues of the amplification matrix satisfy

$$\lambda^2 - 2i\kappa\Delta t\lambda - 1 = 0,$$

which is the same quadratic equation as obtained for the amplification factor in the analysis of the leapfrog scheme in Sect. 2.3.4, where it was shown that $|\lambda_\pm| = 1$ if and only if $|\kappa\Delta t| \leq 1$. Thus, the necessary condition for stability $\rho_m(\mathbf{A}) \leq 1$ is satisfied when $|\kappa\Delta t| \leq 1$.

Sufficient conditions for stability are easy to obtain if the amplification matrix is diagonalizable, since $\rho_m(\mathbf{A}) \leq 1$ will then be both a necessary and a sufficient condition for stability. Any matrix can be transformed to a diagonal matrix if it has a complete set of linearly independent eigenvectors. The eigenvectors of the leapfrog amplification matrix

$$\begin{pmatrix} i\kappa\Delta t + [1 - (\kappa\Delta t)^2]^{1/2} \\ 1 \end{pmatrix} \quad \text{and} \quad \begin{pmatrix} i\kappa\Delta t - [1 - (\kappa\Delta t)^2]^{1/2} \\ 1 \end{pmatrix}$$

are linearly independent for $\kappa\Delta t \neq \pm 1$. The leapfrog scheme must therefore be stable when $|\kappa\Delta t| < 1$. When $\kappa\Delta t = 1$, however, the eigenvectors of the leapfrog amplification matrix are not linearly independent, and the matrix is not diagonalizable. In this case,

$$\mathbf{A} = \begin{pmatrix} 2i & 1 \\ 1 & 0 \end{pmatrix} \quad \text{and} \quad \mathbf{A}^n = i^n \begin{pmatrix} n+1 & -in \\ -in & 1-n \end{pmatrix}.$$

Since $\|\mathbf{A}^n\|$ grows linearly with n, the leapfrog scheme is not stable for $\kappa\Delta t = 1$. Similar reasoning shows that the choice $\kappa\Delta t = -1$ is also unstable. The overall conclusion, that the leapfrog differencing is stable for $|\kappa\Delta t| < 1$, is identical to that obtained in Sect. 2.3.4.

4.1.1.3 The Discrete-Dispersion Relation

Another useful way to obtain necessary conditions for the stability of wavelike solutions of systems of partial differential equations is to evaluate the discrete-dispersion relation. The discrete-dispersion relation is particularly simple in instances where the original system is approximated by finite differences that are centered in space and time. For example, suppose the one-dimensional shallow-water system (4.1) and (4.2) is approximated using leapfrog-time centered-second-order space differencing as

$$\delta_{2t} u + U \delta_{2x} u + g \delta_{2x} h = 0, \tag{4.9}$$

$$\delta_{2t} h + U \delta_{2x} h + H \delta_{2x} u = 0. \tag{4.10}$$

(The finite-difference operator δ_{nx} is defined in the Appendix by (A.1).) As in the case of the scalar advection equation discussed in Sect. 3.4.1, the construction of the discrete-dispersion relation mimics the procedure used to obtain the dispersion relation for the continuous system. Wave solutions to the discretized shallow-water equations are sought in the form

$$u_j^n = u_0 \mathrm{e}^{\mathrm{i}(kj\Delta x - \omega n \Delta t)}, \qquad h_j^n = h_0 \mathrm{e}^{\mathrm{i}(kj\Delta x - \omega n \Delta t)}, \tag{4.11}$$

where u_0 and h_0 are complex constants determining the wave amplitude, and the physically relevant portion of the solution is the real part of u_j^n and h_j^n. Substitution of (4.11) into the finite-differenced governing equations (4.9) and (4.10) yields

$$\left(-\frac{\sin \omega \Delta t}{\Delta t} + U \frac{\sin k \Delta x}{\Delta x}\right) u_0 + g \frac{\sin k \Delta x}{\Delta x} h_0 = 0,$$
$$\left(-\frac{\sin \omega \Delta t}{\Delta t} + U \frac{\sin k \Delta x}{\Delta x}\right) h_0 + H \frac{\sin k \Delta x}{\Delta x} u_0 = 0.$$

Nontrivial values of u_0 and h_0 will satisfy the preceding pair of homogeneous equations when the determinant of the coefficients of u_0 and h_0 is zero, which requires

$$\left(\frac{\sin \omega \Delta t}{\Delta t} - U \frac{\sin k \Delta x}{\Delta x}\right)^2 = gH \left(\frac{\sin k \Delta x}{\Delta x}\right)^2,$$

or defining $c = \sqrt{gH}$,

$$\sin \omega \Delta t = \frac{\Delta t}{\Delta x} (U \pm c) \sin k \Delta x. \tag{4.12}$$

In the limit of Δt, $\Delta x \to 0$, this discrete-dispersion relation approaches the dispersion relation for the continuous problem $\omega = (U \pm c)k$.

The discrete-dispersion relation for the linearized shallow-water system is identical to that for the scalar advection equation (3.60) except that it supports two physical modes moving at velocities $U + c$ and $U - c$. The amplitude and phase-speed error in each wave may be analyzed in the same manner as that in the scalar advection problem (see Sect. 3.4.1). The analysis of amplitude error, for example, proceeds by examining the amplification factor $e^{\Im(\omega)\Delta t}$ by which the waves (4.11) grow or decay during each time step. Since the horizontal wave number (k) of periodic waves is real, the imaginary part of ω will be zero unless the right side of (4.12) exceeds unity. When the right side is greater than unity, one of the ω satisfying (4.12) has a positive imaginary part, and the numerical solution grows at each time step. Thus, a necessary condition for the stability of the finite-difference scheme is that

$$\left| \frac{\Delta t}{\Delta x}(U \pm c) \sin k\Delta x \right| \leq 1$$

for all k resolvable on the numerical mesh, or since $k = \pi/(2\Delta x)$ is a resolvable wave,

$$(|U| + c) \frac{\Delta t}{\Delta x} \leq 1. \tag{4.13}$$

This condition is not quite sufficient to guarantee stability; since the time differencing is leapfrog, sufficient conditions for stability require strict inequality in (4.13).

4.1.2 Staggered Meshes

When simulating a system of equations with several unknowns, it is not necessary to define all the unknown variables at the same grid points. Significant improvements in the accuracy of the short-wavelength components of the solution can sometimes be obtained by the use of *staggered* meshes.

4.1.2.1 Spatial Staggering

Consider, once again, the linearized shallow-water system (4.1) and (4.2) and to reveal the benefits of staggering more clearly, suppose that $U = 0$. The finite-difference approximations (4.9) and (4.10) assume that the perturbation velocity (u) and depth (h) are defined at the same grid points, as shown schematically in Fig. 4.1a. An alternative arrangement is shown in Fig. 4.1b, in which the grid points

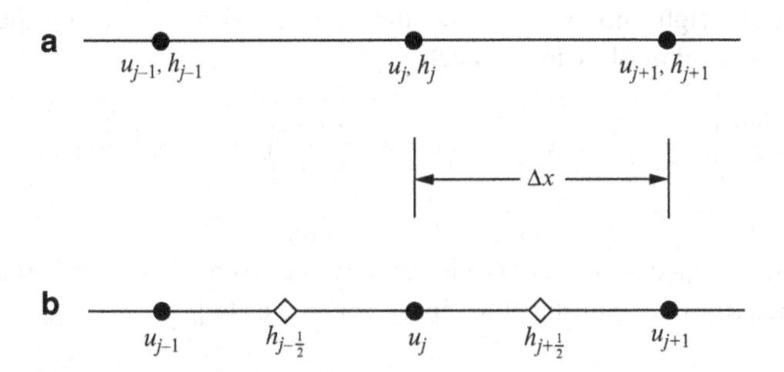

Fig. 4.1 Distribution of u and h on **a** an unstaggered and **b** a staggered mesh

where h is defined are shifted $\Delta x/2$ to the right (or left) of u grid points. A centered-difference approximation to the linearized shallow-water system with $U = 0$ may be written for the staggered mesh as

$$\delta_{2t} u_j + g\delta_x h_j = 0, \tag{4.14}$$

$$\delta_{2t} h_{j+\frac{1}{2}} + H\delta_x u_{j+\frac{1}{2}} = 0. \tag{4.15}$$

The discrete-dispersion relation for the solution to these equations is

$$\sin \omega \Delta t = \pm \frac{2c\Delta t}{\Delta x} \sin\left(\frac{k\Delta x}{2}\right), \tag{4.16}$$

from which it follows that (4.14) and (4.15) are stable when $|c\Delta t/\Delta x| < 1/2$. The maximum time step available for integrations on the staggered mesh is only half that which may be used on the unstaggered mesh. The more stringent restriction on the time step is, however, not entirely bad, because shorter time steps are generally required by spatial differencing schemes that more faithfully capture the high-frequency components of the solution. Analysis of the truncation error shows that both the staggered and the unstaggered schemes are $O\left[(\Delta x)^2\right]$ and that the leading-order truncation error is smaller for the staggered scheme.

A more revealing comparison of the accuracy of each scheme is provided by examining their discrete-dispersion relations in the limit of good temporal resolution ($\omega\Delta t \to 0$). Let c_{u} and c_{s} denote the phase speeds of the numerical solutions on the unstaggered and staggered meshes, respectively; then

$$c_{\mathrm{u}} = \frac{c}{k\Delta x} \sin k\Delta x \qquad \text{and} \qquad c_{\mathrm{s}} = \frac{2c}{k\Delta x} \sin\left(\frac{k\Delta x}{2}\right).$$

Curves showing c_{u} and c_{s} are plotted as a function of spatial resolution in Fig. 4.2. Also plotted in Fig. 4.2 is the phase speed obtained when the explicit fourth-order difference (3.2) is used to approximate the spatial derivatives on the unstaggered mesh. As evident in Fig. 4.2, the error in the phase speed of the poorly resolved waves is much smaller on the staggered mesh. In particular, the $2\Delta x$ wave propagates at 64% of the correct speed on the staggered mesh but remains stationary on the unstaggered mesh.

Substantial improvements in the group velocity of the shortest waves are also achieved using the staggered mesh. Assuming good temporal resolution, the group velocities of the right-moving wave for the second-order schemes on the unstaggered and staggered meshes are, respectively,

$$\left(\frac{\partial \omega}{\partial k}\right)_{\mathrm{u}} = c\cos k\Delta x \qquad \text{and} \qquad \left(\frac{\partial \omega}{\partial k}\right)_{\mathrm{s}} = c\cos\left(\frac{k\Delta x}{2}\right).$$

The group velocity of a $2\Delta x$ wave is $-c$ on the unstaggered mesh and zero on the staggered mesh. Since the correct group velocity is c, both schemes generate serious error, but the error on the unstaggered mesh is twice as large.

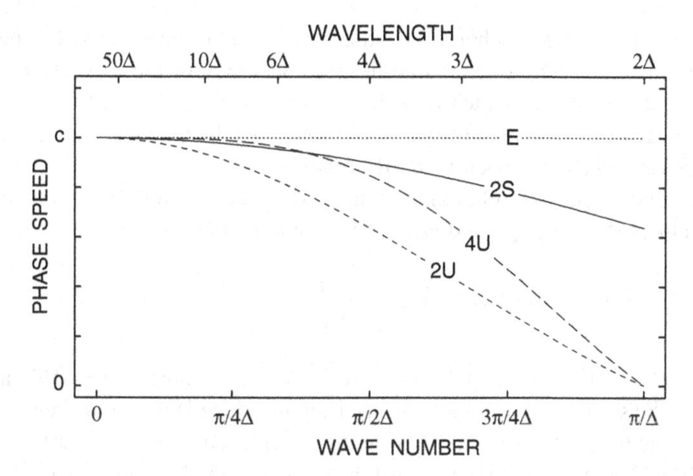

Fig. 4.2 Phase speed in the limit of good temporal resolution as a function of spatial resolution for the exact solution (*E*), for second-order approximations on a staggered mesh c_s (*2S*) and an unstaggered mesh c_u (*2U*), and for centered fourth-order spatial derivatives on an unstaggered mesh (*4U*)

4.1.2.2 Temporal Staggering and Forward–Backward Differencing

Just as every unknown variable need not be defined at every spatial grid point, it is also not necessary to define all the unknown variables at each time level. For example, the finite-difference scheme (4.14) and (4.15) could be stepped forward in time using only the values of u at the odd time levels ($[2n+1]\Delta t$) and h at the even time levels ($2n\Delta t$). Staggering u and h in time could, therefore, halve the total computation required for a given simulation. Unfortunately, time staggering can be difficult to program and may be incompatible with the time differencing used to integrate other terms in the governing equations, such as those representing advection by the mean wind in (4.1) and (4.2).

Sometimes the advantages of time staggering can be achieved without actually staggering the unknowns in time by evaluating the various terms in the governing equations at different time levels. In the case of the shallow-water system, the benefits of true time staggering can be obtained using forward–backward differencing. One possible forward–backward formulation of the spatially staggered finite-difference approximation to the linearized shallow-water system (4.14) and (4.15) is

$$\delta_t u_j^{n+\frac{1}{2}} + g\delta_x h_j^n = 0, \tag{4.17}$$

$$\delta_t h_{j+\frac{1}{2}}^{n+\frac{1}{2}} + H\delta_x u_{j+\frac{1}{2}}^{n+1} = 0. \tag{4.18}$$

The momentum equation (4.17) is first updated using forward differencing, and then the continuity equation (4.18) is integrated using backward differencing. The backward difference does not introduce an implicit coupling between the unknowns in the forward–backward scheme because u^{n+1} is computed in (4.17) before it is required in (4.18). The overall stability and accuracy of the forward–backward scheme is independent of which equation is updated first; the continuity equation could be integrated first with a forward difference and then the momentum equation could be advanced using a backward difference.

The discrete-dispersion relation associated with the forward–backward approximation on the spatially staggered mesh, (4.17) and (4.18), is

$$\sin\left(\frac{\omega\Delta t}{2}\right) = \pm\frac{c\Delta t}{\Delta x}\sin\left(\frac{k\Delta x}{2}\right). \tag{4.19}$$

If $|c\Delta t/\Delta x| < 1$, there will be one[3] real-valued ω satisfying (4.19), and the scheme will be stable. The time-step restriction introduced by spatial staggering can therefore be avoided if leapfrog differencing is replaced by the forward–backward scheme. In addition, forward–backward differencing involves only two time levels and thereby avoids the introduction of computational modes.

In the case of the linearized shallow-water system with $U = 0$, forward–backward differencing on a spatially staggered mesh is clearly superior to the leapfrog spatially unstaggered scheme. However, in applications where several different terms appear in each governing equation, it is often impossible to choose a single staggering that improves the accuracy of every term. In such situations the advantages of staggering can be substantially reduced. As an example, suppose that the preceding forward–backward spatially staggered approximation is to be extended to shallow-water problems with nonzero mean flow. The simplest $O\left[(\Delta x)^2\right]$ approximation to the spatial derivatives in the advection terms is the same centered difference used in the unstaggered equations (4.9) and (4.10). The staggering of u with respect to h does not interfere with the construction of these centered differences, but it does not improve their accuracy either. The incorporation of the advection terms in the forward–backward time difference poses more of a problem, since a forward-difference approximation to the advection equation (see Sect. 2.3.2) is unstable. One possible approach is to perform the forward–backward differencing over an interval of $2\Delta t$ and to use the intermediate time level to evaluate the advection terms with leapfrog differencing as follows:

$$\delta_{2t}u_j^n + U\delta_{2x}u_j^n + g\delta_x h_j^{n-1} = 0,$$

$$\delta_{2t}h_{j+\frac{1}{2}}^n + U\delta_{2x}h_{j+\frac{1}{2}}^n + H\delta_x u_{j+\frac{1}{2}}^{n+1} = 0.$$

The discrete-dispersion relation for this system is

$$\sin\omega\Delta t = \frac{\Delta t}{\Delta x}\left(U\sin k\Delta x \pm 2c\sin\left(\frac{k\Delta x}{2}\right)\right),$$

[3] Recall that the range of resolvable frequencies is $0 \le \omega \le \pi/\Delta t$, which correspond to wave periods between ∞ and $2\Delta t$.

and is identical to that which would be obtained if leapfrog time differencing were used to integrate every term. The benefits of forward–backward time differencing have been lost. Once again, the numerical scheme includes computational modes, and for $c \gg |U|$ the maximum stable time step is half that allowed in the spatially unstaggered scheme. The benefits of spatial staggering are retained, but apply only to that portion of the total velocity of propagation that is produced by the pressure gradient and divergence terms (i.e., by the mechanisms that remain active in the limit $U \to 0$). In situations where $c \gg |U|$, spatial staggering yields substantial improvement, but in those cases where $|U| \gg c$, the errors in the $2\Delta x$ waves introduced by the advection terms dominate the total solution and mask the benefits of spatial staggering. One way to improve accuracy when $|U| \geq c$ is to use fourth-order centered differencing for the advection terms. Fourth-order differencing *is not used to obtain highly accurate approximations to the well-resolved waves, but rather to reduce the phase-speed error in the moderately resolved waves* to a value comparable to that generated by second-order staggered differencing (compare curves 2S and 4U over wavelengths ranging from $4\Delta x$ to $10\Delta x$ in Fig. 4.2). Reinecke and Durran (2009) give an example where fourth-order approximations for horizontal advection together with second-order approximations to other derivatives on a staggered mesh give much better results than those obtained using second-order derivatives for all terms.

4.2 Three or More Independent Variables

In most time-dependent problems of practical interest, the unknowns are functions of three or four independent variables (i.e., time and two or three spatial coordinates). The accuracy, consistency, and stability of finite-difference approximations to higher-dimensional equations are determined using essentially the same procedures described in Chap. 3. Two specific examples will be considered in the following sections: scalar advection in two dimensions and the Boussinesq equations.

4.2.1 Scalar Advection in Two Dimensions

The advection equation for two-dimensional flow can be approximated using leapfrog-time centered-space schemes that are obvious generalizations of the finite-difference approximations employed in the one-dimensional problem. New considerations involving the incorporation of mixed spatial derivatives do, however, arise in designing accurate and efficient forward-in-time approximations. These considerations will be explored after first examining schemes that are centered in space and time.

4.2.1.1 Centered-In-Time Schemes

When explicit finite-difference schemes for the integration of one-dimensional problems are extended to two or more spatial dimensions, the stability criteria for the multidimensional problems are often more stringent than those for the one-dimensional formulation. As an example, consider the two-dimensional advection equation

$$\frac{\partial \psi}{\partial t} + U\frac{\partial \psi}{\partial x} + V\frac{\partial \psi}{\partial y} = 0, \tag{4.20}$$

which may be approximated by leapfrog-time, centered-second-order-space differencing as

$$\delta_{2t}\phi + U\delta_{2x}\phi + V\delta_{2y}\phi = 0. \tag{4.21}$$

Let $\mu = U\Delta t/\Delta x$ and $\nu = V\Delta t/\Delta y$ be the Courant numbers for flow parallel to the x- and y-axes. The finite-difference equation (4.21) has discrete solutions of the form

$$\phi^j_{m,n} = \mathrm{e}^{\mathrm{i}(km\Delta x + \ell n\Delta y - \omega j\Delta t)},$$

provided that ω, k, and ℓ satisfy the discrete dispersion relation

$$\sin(\omega\Delta t) = \mu\sin(k\Delta x) + \nu\sin(\ell\Delta y). \tag{4.22}$$

A necessary condition for stability is that ω be real, or equivalently that

$$|\mu| + |\nu| \le 1. \tag{4.23}$$

As discussed in connection with (3.60), the sufficient condition for stability actually requires strict inequality in (4.23) to avoid weakly growing modes such as

$$\phi^j_{m,n} = j\cos\left[\pi(m+n-j)/2\right],$$

which is a solution to (4.21) when $\mu = \nu = \frac{1}{2}$. The distinction between strict inequality and the condition given in (4.23) is, however, of little practical significance.

To better compare this stability condition with that for one-dimensional advection, suppose that $\Delta x = \Delta y = \Delta s$ and express the wind components in terms of wind speed c and direction θ such that $U = c\cos\theta$ and $V = c\sin\theta$. Then the stability condition may be written

$$c\Big(|\cos\theta| + |\sin\theta|\Big)\frac{\Delta t}{\Delta s} < 1.$$

The left side of the preceding inequality is maximized when the wind blows diagonally across the mesh. If C denotes a bound on the magnitude of the two-dimensional velocity vector, the stability condition becomes $C\Delta t/\Delta s < 1/\sqrt{2}$. Comparing this with the corresponding result for one-dimensional flow, it is apparent that the maximum stable time step in the two-dimensional case is decreased by a factor of $1/\sqrt{2}$.

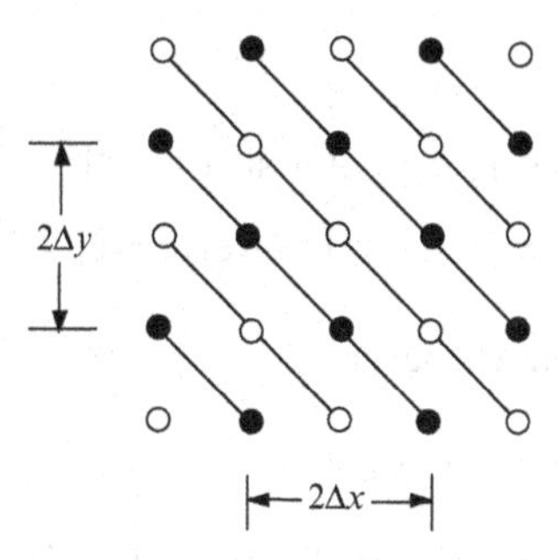

Fig. 4.3 Distribution of wave crest (*solid circles*) and wave troughs (*open circles*) in the shortest-wavelength disturbance resolvable on a square mesh in which $\Delta x = \Delta y$

The stability condition for two-dimensional flow is more restrictive than that for the one-dimensional case because shorter-wavelength disturbances are present on the two-dimensional mesh. The manner in which two-dimensional grids can support wavelengths shorter than $2\Delta x$ is illustrated in Fig. 4.3. In the case shown in Fig. 4.3, $\Delta x = \Delta y = \Delta s$. Grid points beneath a wave crest are indicated by solid circles; those beneath a trough are indicated by open circles. The apparent wavelength parallel to both the x-axis and the y-axis is $2\Delta s$; however, the true wavelength measured along the line $x = y$ is $\sqrt{2}\Delta s$. The maximum stable time step is inversely proportional to the highest frequency resolvable by the numerical scheme, and in the case of the advection equation, the frequency is proportional to the wave number times the wind speed. Since the wave number of a diagonally propagating wave exceeds the apparent wave numbers in the x and y directions by a factor of $\sqrt{2}$, the maximum resolvable frequency is increased by the same factor, and the maximum stable time step is reduced by $1/\sqrt{2}$.

One way to avoid this restriction on the maximum stable time step is to average each spatial derivative as follows (Abarbanel and Gottlieb 1976):

$$\delta_{2t}\phi + U\left\langle\delta_{2x}\phi\right\rangle^{2y} + V\left\langle\delta_{2y}\phi\right\rangle^{2x} = 0, \tag{4.24}$$

where $\langle\,\rangle^x$ is an averaging operator defined by

$$\langle f(x)\rangle^{nx} = \left[\frac{f(x + n\Delta x/2) + f(x - n\Delta x/2)}{2}\right]. \tag{4.25}$$

The discrete dispersion relation for this "averaging" scheme is

$$\sin(\omega\Delta t) = \mu\sin(k\Delta x)\cos(\ell\Delta y) + \nu\sin(\ell\Delta y)\cos(k\Delta x). \tag{4.26}$$

Let $\xi = k\Delta x$ and $\zeta = \ell\Delta y$, and note that Schwarz's inequality[4] implies

$$|\sin\xi||\cos\zeta| + |\sin\zeta||\cos\xi| \le \left(\sin^2\xi + \cos^2\xi\right)^{1/2}\left(\sin^2\zeta + \cos^2\zeta\right)^{1/2}.$$

[4] $\sum_j a_j b_j \le \left(\sum_j a_j^2\right)^{1/2}\left(\sum_j b_j^2\right)^{1/2}$.

Since

$$
\begin{aligned}
|\sin(\omega \Delta t)| &= |\mu \sin \xi \cos \zeta + \nu \sin \zeta \cos \xi| \\
&\leq \max \{|\mu|, |\nu|\} \Big(|\sin \xi| |\cos \zeta| + |\sin \zeta| |\cos \xi| \Big),
\end{aligned}
$$

real values of ω are obtained whenever

$$
\max \{|\mu|, |\nu|\} \leq 1. \tag{4.27}
$$

As before, suppose that $U = c \cos \theta$, $V = c \sin \theta$, $\Delta x = \Delta y = \Delta s$, and that C is a bound on $|c|$; then requiring strict inequality in (4.27) to guarantee that the leapfrog time difference does not admit weakly unstable modes, the stability condition becomes $C\Delta t/\Delta s < 1$, which is identical to that for the one-dimensional case.

Although the averaging scheme is potentially more efficient because it permits longer time steps, it is also less accurate. This loss of accuracy is not clearly reflected in the truncation error, which is $O\left[(\Delta x)^2\right] + O\left[(\Delta y)^2\right]$ for both the nonaveraged method (4.21) and the averaging scheme (4.24). The problems with the averaging scheme appear in the representation of the poorly resolved waves. As discussed in connection with Fig. 4.3, shorter waves are resolvable on a two-dimensional grid, and if properly represented by the spatial differencing, they *should* generate higher-frequency oscillations and reduce the maximum stable time step. The averaging scheme avoids such time-step reduction by artificially reducing the phase speeds of the diagonally propagating waves.

The phase-speed errors introduced by the spatial differencing in both methods may be examined by a generalization of the one-dimensional approach discussed in Sect. 3.3.1. First consider the propagation of two-dimensional waves in the nondiscretized problem. Waves of the form

$$
\psi(x, y, t) = e^{i(kx + \ell y - \omega t)}
$$

satisfy the two-dimensional advection equation (4.20), provided that

$$
\omega = \mathbf{v} \cdot \mathbf{k}, \tag{4.28}
$$

where $\mathbf{v}$ is the velocity vector and $\mathbf{k}$ is the wave number vector with components (k, ℓ). If K denotes the magnitude of $\mathbf{k}$, then the x and y components of the wave number vector may be expressed as

$$
k = K \cos \theta \quad \text{and} \quad \ell = K \sin \theta, \tag{4.29}
$$

where θ is the angle between the wave number vector and the x-axis. The dispersion relation (4.28) implies that all apparent wave propagation is parallel to the wave number vector. Consider, therefore, the case in which the velocity vector is parallel to the wave number vector. Then if c is the wind speed,

$$
U = c \cos \theta \quad \text{and} \quad V = c \sin \theta. \tag{4.30}
$$

Substituting (4.29) and (4.30) into the dispersion relation demonstrates that $c = \omega/K$, implying that the phase speed is equal to the wind speed.

As in the one-dimensional case, the phase speeds of the waves generated by the finite-difference approximations (4.21) and (4.24) are defined as $c^* = \omega^*/K$, where ω^* is the frequency satisfying the appropriate discrete-dispersion relation: either (4.22) or (4.26). In the limit of good time resolution, the phase speed for the nonaveraging scheme is

$$c^*_{\mathrm{na}} = \frac{1}{K}\left(U\,\frac{\sin(k\Delta x)}{\Delta x} + V\,\frac{\sin(\ell\Delta y)}{\Delta y}\right),$$

and that for the averaging scheme is

$$c^*_{\mathrm{a}} = \frac{1}{K}\left(U\,\frac{\sin(k\Delta x)}{\Delta x}\cos(\ell\Delta y) + V\,\frac{\sin(\ell\Delta y)}{\Delta y}\cos(k\Delta x)\right).$$

Suppose $\Delta x = \Delta y = \Delta s$ and define $\beta = K\Delta s$; then using (4.29) and (4.30) to evaluate the velocity and wave number components in the preceding expressions, the relative phase speed for each scheme becomes

$$\frac{c^*_{\mathrm{na}}}{c} = \frac{\cos\theta\sin(\beta\cos\theta) + \sin\theta\sin(\beta\sin\theta)}{\beta}$$

and

$$\frac{c^*_{\mathrm{a}}}{c} = \frac{\cos\theta\sin(\beta\cos\theta)\cos(\beta\sin\theta) + \sin\theta\sin(\beta\sin\theta)\cos(\beta\cos\theta)}{\beta}.$$

These expressions for the relative phase speed were evaluated for wavelengths $2\pi/K = 2\Delta s$, $3\Delta s$, $4\Delta s$, and $6\Delta s$ and are plotted as a function of θ in Fig. 4.4. Figure 4.4 shows a polar plot in which the relative phase speed of a $3\Delta s$ wave propagating along a ray extending outward from the origin at an angle θ with respect to the x-axis is plotted as the radial distance between the origin and the point where that ray intersects the dashed curve labeled "3." As indicated in Fig. 4.4a, the nonaveraging scheme does not resolve the propagation of $2\Delta s$ waves parallel to either the x-axis or the y-axis, but $2\Delta s$ waves can move at greatly reduced speed along the diagonal line $x = y$ ($\theta = \pi/4$). The phase-speed error diminishes as the wavelength increases, with the maximum error in $6\Delta s$ waves being no larger than 20%. These results can be compared with the relative phase speed curves for the averaging scheme plotted in Fig. 4.4b. The averaging scheme generates substantial errors in the phase speed of waves moving diagonally along the line $x = y$; the $2\Delta s$ wave does not propagate at all, and even the $6\Delta s$ wave is significantly retarded. These reduced phase speeds allow the averaging scheme to remain stable for large time steps, but as is apparent in Fig. 4.4, the enhanced stability is obtained at the cost of increased phase-speed errors in the poorly and moderately resolved waves.

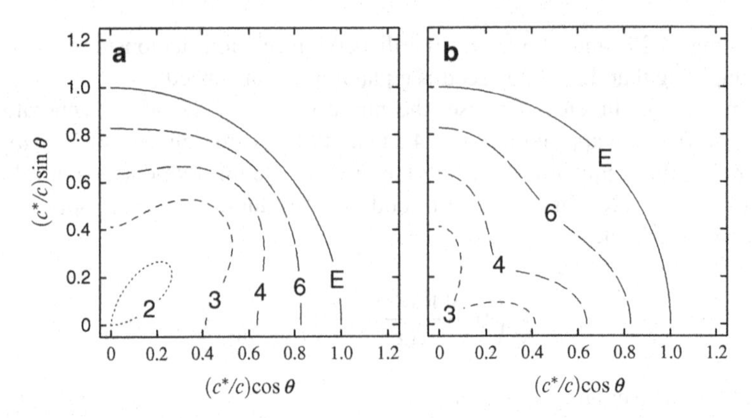

Fig. 4.4 Polar plot of the relative phase speeds of $2\Delta s$ (*shortest dashed line*), $3\Delta s$, $4\Delta s$, and $6\Delta s$ (*longest dashed line*) waves generated by **a** the nonaveraged finite-difference formula and **b** the averaging scheme. Also plotted is the curve for perfect propagation (*E*), which is independent of the wavelength and appears as a circular arc of radius unity

4.2.1.2 Forward-In-Time Schemes

Assuming that $U \geq 0$ and $V \geq 0$, one generalization of the upstream method to the two-dimensional advection equation (4.20) is

$$\delta_t \phi_{m,n}^{j+\frac{1}{2}} + U\delta_x \phi_{m-\frac{1}{2},n}^{j} + V\delta_y \phi_{m,n-\frac{1}{2}}^{j} = 0. \tag{4.31}$$

The stability of this scheme may be investigated using the standard von Neumann method. Let

$$\phi_{m,n}^{j} = A^j \mathrm{e}^{\mathrm{i}(km\Delta x + \ell n\Delta y)}. \tag{4.32}$$

Then

$$A = 1 - \mu(1 - \mathrm{e}^{-\mathrm{i}\xi}) - \nu(1 - \mathrm{e}^{-\mathrm{i}\zeta}), \tag{4.33}$$

where as before, $\mu = U\Delta t/\Delta x$, $\nu = V\Delta t/\Delta y$, $\xi = k\Delta x$, and $\zeta = \ell\Delta y$. Necessary and sufficient conditions for stability are

$$0 \leq \mu, \qquad 0 \leq \nu, \qquad \text{and} \qquad \mu + \nu \leq 1. \tag{4.34}$$

The necessity of the preceding conditions may be established by considering the three cases $\xi = 0$, $\zeta = 0$, and $\xi = \zeta$, for each of which the dependence of the amplification factor on the wave number reduces to an expression of the same form as in the one-dimensional case (3.22). The sufficiency of (4.34) follows from

$$\begin{aligned} |A| &\leq |1 - \mu - \nu| + |\mu \mathrm{e}^{-\mathrm{i}\xi}| + |\nu \mathrm{e}^{-\mathrm{i}\zeta}| \\ &= |1 - \mu - \nu| + |\mu| + |\nu|, \end{aligned}$$

which implies that $|A| \leq 1$ whenever μ and ν satisfy (4.34).

Suppose that $\Delta x = \Delta y = \Delta s$ and that C is a bound on the magnitude of the two-dimensional velocity vector; then provided that the spatial differences are evaluated in the upstream direction, the stability condition is $C\Delta t/\Delta s \leq 1/\sqrt{2}$. As was the case with the leapfrog approximation (4.21), the maximum stable time step is approximately 30% less than that in the analogous one-dimensional problem. In contrast to the situation with centered-in-time schemes, it is, however, possible to improve the stability of forward-in-time approximations to the two-dimensional advection equation while simultaneously improving at least some aspects of their accuracy.

One natural way to derive the upstream approximation to the one-dimensional advection equation is through the method of characteristics (Courant et al. 1952). The true solution of the one-dimensional advection equation is constant along characteristic curves whose slopes are $\mathrm{d}x/\mathrm{d}t = U$. The characteristic curve passing through the point $[m\Delta x, (j+1)\Delta t]$ also passes through $[(m-\mu)\Delta x, j\Delta t]$, and as discussed in Sect. 7.1.1, the upstream scheme

$$\phi_m^{j+1} = (1-\mu)\phi_m^j + \mu\phi_{m-1}^j$$

is obtained if the value of ϕ^j at $(m-\mu)\Delta x$ is estimated from ϕ_m^j and ϕ_{m-1}^j by linear interpolation. The method of characteristics is naturally extended to the two-dimensional advection problem using bilinear interpolation, in which case

$$\begin{aligned}\phi_{m,n}^{j+1} = (1-\mu)&\left[(1-\nu)\phi_{m,n}^j + \nu\phi_{m,n-1}^j\right]\\ &+\mu\left[(1-\nu)\phi_{m-1,n}^j + \nu\phi_{m-1,n-1}^j\right]\end{aligned} \tag{4.35}$$

(Bates and McDonald 1982). Colella (1990), who derived the same scheme using a finite-volume argument (see Sect. 5.9.2), has referred to this scheme as the corner transport upstream (CTU) method.

The CTU method may be expressed in the alternative form

$$\delta_t\phi_{m,n}^{j+\frac{1}{2}} + U\delta_x\phi_{m-\frac{1}{2},n}^j + V\delta_y\phi_{m,n-\frac{1}{2}}^j = UV\Delta t\delta_x\delta_y\phi_{m-\frac{1}{2},n-\frac{1}{2}}^j, \tag{4.36}$$

which shows that it differs from (4.31) by a term that is a finite-difference approximation to

$$UV\Delta t\frac{\partial^2\psi}{\partial x\partial y}.$$

The addition of this cross-derivative term improves the stability of the CTU scheme relative to (4.31). Substituting (4.32) into (4.36) yields

$$A = \left(1-\mu+\mu e^{-i\xi}\right)\left(1-\nu+\nu e^{-i\zeta}\right).$$

Each factor in the preceding equation has the same form as (3.22), so the magnitude of each factor will be less than 1, and the CTU scheme will be stable if $0 \leq \mu \leq 1$

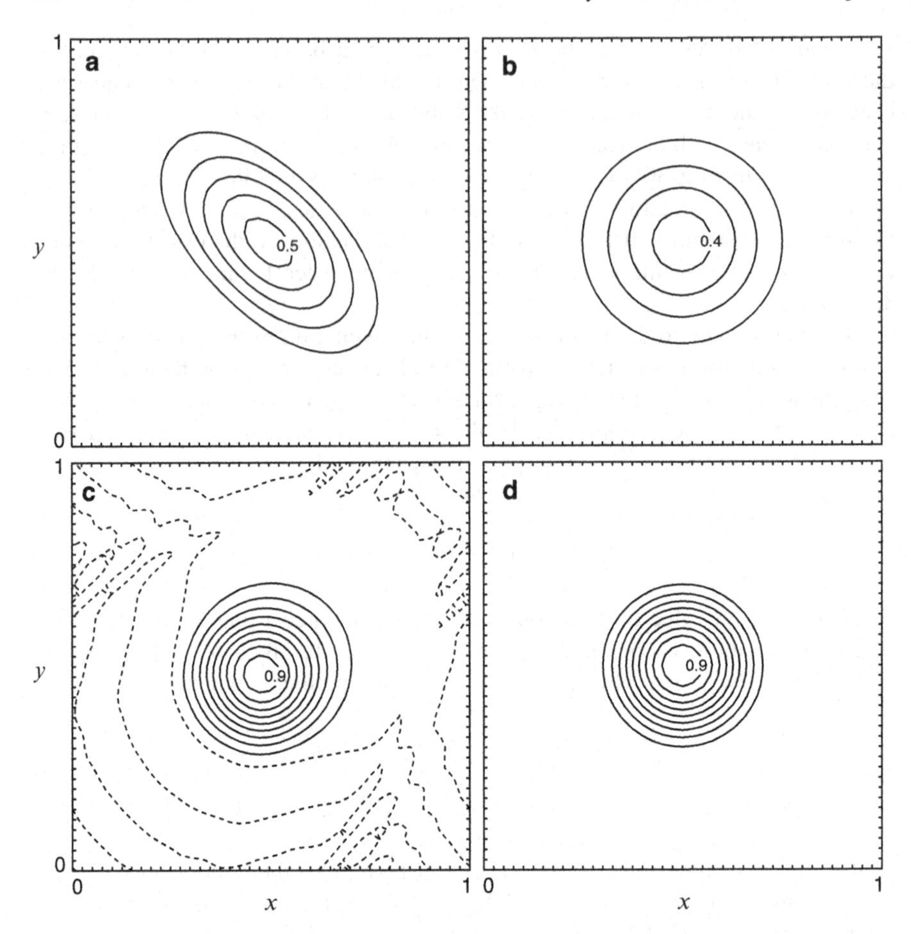

Fig. 4.5 Contours of the solution to the constant-wind-speed advection equation on a two-dimensional periodic mesh obtained using **a** the two-dimensional upstream method, (**b**) the corner transport upstream method, **c** the upstream-biased Lax-Wendroff method, and **d** the true solution. The contour interval is 0.1, and the zero contour is *dashed*

and $0 \leq \nu \leq 1$. If the computational mesh is uniform and the wind speed is bounded by C, the stability condition becomes $C\Delta t/\Delta s \leq 1$, which is identical to that for the upstream approximation to the one-dimensional problem.

Solutions generated by the CTU method are compared with those obtained using the two-dimensional upstream scheme (4.31) in Fig. 4.5. The spatial domain is $0 \leq x \leq 1, 0 \leq y \leq 1$ and is discretized using a square mesh with $\Delta x = \Delta y = 0.025$. The lateral boundary conditions are periodic, and the initial condition is

$$\phi(x, y, 0) = \frac{1}{2}[1 + \cos(\pi r)],$$

where

$$r(x, y) = \min\left(1,\ 4\sqrt{(x - 1/2)^2 + (y - 1/2)^2}\,\right).$$

The wind is directed diagonally across the mesh with $U = V = 1$. The time step is chosen such that $\mu = \nu = 0.5$. The results are displayed at $t = 1$, at which time the flow has made exactly one circuit around the domain. The solution obtained using (4.31) is shown in Fig. 4.5a; that obtained using the CTU method appears in Fig. 4.5b, and the true solution is plotted in Fig. 4.5d. Since they are first-order methods, both upstream solutions are heavily damped. The solution generated by the two-dimensional upstream method has also developed a pronounced asymmetry, whereas that produced by the CTU method appears axisymmetric. The CTU solution is, however, damped slightly more than the two-dimensional upstream solution.

The tendency of the two-dimensional upstream scheme to distort the solution as shown in Fig. 4.5a can be understood by noting that (4.31) is a second-order approximation to the modified equation

$$\frac{\partial \psi}{\partial t} + U\frac{\partial \psi}{\partial x} + V\frac{\partial \psi}{\partial y} = \frac{U\Delta x}{2}(1-\mu)\frac{\partial^2 \psi}{\partial x^2} + \frac{V\Delta y}{2}(1-\nu)\frac{\partial^2 \psi}{\partial y^2} - UV\Delta t\frac{\partial^2 \psi}{\partial x \partial y}.$$

The CTU method, on the other hand, is a second-order approximation to

$$\frac{\partial \psi}{\partial t} + U\frac{\partial \psi}{\partial x} + V\frac{\partial \psi}{\partial y} = \frac{U\Delta x}{2}(1-\mu)\frac{\partial^2 \psi}{\partial x^2} + \frac{V\Delta y}{2}(1-\nu)\frac{\partial^2 \psi}{\partial y^2};$$

the mixed spatial derivative does not appear because it is canceled (to second order) by the finite difference on the right side of (4.36). The influence of the mixed spatial derivative on the error in the two-dimensional upstream scheme can be isolated by considering the simplified equation

$$\frac{\partial \varphi}{\partial t} = -\frac{\partial^2 \varphi}{\partial x \partial y}.$$

Expressing the preceding equation in a coordinate system rotated by 45°, so that the new independent variables are $r = x + y$ and $s = x - y$, yields

$$\frac{\partial \varphi}{\partial t} = \frac{\partial^2 \varphi}{\partial s^2} - \frac{\partial^2 \varphi}{\partial r^2}.$$

Thus, perturbations in ψ diffuse along lines of constant r and "antidiffuse" along lines of constant s. Whenever $UV > 0$, this process of diffusion and antidiffusion tends to distort the solution as shown in Fig. 4.5a. In contrast, the leading-order error in the CTU method is purely isotropic when $U = V$ and $\Delta x = \Delta y$.

Second-order forward-in-time approximations can be obtained using the Lax–Wendroff method. The scheme

$$\delta_t \phi^{j+\frac{1}{2}} + U\delta_{2x}\phi^j + V\delta_{2y}\phi^j = \frac{U^2\Delta t}{2}\delta_x^2\phi^j + \frac{V^2\Delta t}{2}\delta_y^2\phi^j \tag{4.37}$$

has sometimes been proposed as a generalization of the one-dimensional Lax–Wendroff method for constant-wind-speed advection in two dimensions, but this scheme is not second-order accurate because the right side is not a second-order approximation to the leading-order truncation error in the forward time difference,

$$\frac{\Delta t}{2}\frac{\partial^2 \psi}{\partial t^2} = \frac{\Delta t}{2}\left(u^2\frac{\partial^2 \psi}{\partial x^2} + v^2\frac{\partial^2 \psi}{\partial y^2} + 2uv\frac{\partial^2 \psi}{\partial x \partial y}\right).$$

In addition to its lack of accuracy, (4.37) is unstable for all Δt.

A stable second-order Lax–Wendroff approximation to the two-dimensional constant-wind-speed advection equation may be written in the form

$$\delta_t \phi^{j+\frac{1}{2}} + U\delta_{2x}\phi^j + V\delta_{2y}\phi^j = \frac{U^2\Delta t}{2}\delta_x^2\phi^j + \frac{V^2\Delta t}{2}\delta_y^2\phi^j + UV\Delta t\delta_{2x}\delta_{2y}\phi^j.$$

Necessary and sufficient conditions for the stability of this method are that

$$\mu^{2/3} + \nu^{2/3} \le 1$$

(Turkel 1977). If C is a bound on the magnitude of the two-dimensional wind vector and $\Delta x = \Delta y = \Delta s$, the stability condition becomes $C\Delta t/\Delta s \le 1/2$, which is more restrictive than that for the two-dimensional leapfrog and upstream schemes, and much more restrictive than the stability condition for the CTU method.

The stability of the two-dimensional Lax–Wendroff approximation can be greatly improved using an upstream finite-difference approximation to the mixed spatial derivative (Leonard et al. 1993). If $U \ge 0$ and $V \ge 0$, the resulting scheme is

$$\delta_t \phi_{m,n}^{j+\frac{1}{2}} + U\delta_{2x}\phi_{m,n}^j + V\delta_{2y}\phi_{m,n}^j$$
$$= \frac{U^2\Delta t}{2}\delta_x^2\phi_{m,n}^j + \frac{V^2\Delta t}{2}\delta_y^2\phi_{m,n}^j + UV\Delta t\delta_x\delta_y\phi_{m-\frac{1}{2},n-\frac{1}{2}}^j. \quad (4.38)$$

The stability condition for this scheme is identical to that for the CTU method, $0 \le \mu \le 1$ and $0 \le \nu \le 1$ (Hong et al. 1997). If the mixed spatial derivative is calculated in the upstream direction and the magnitude of the two-dimensional wind vector is bounded by C, the stability condition for an isotropic mesh may be expressed as $C\Delta t/\Delta s \le 1$.

Numerical solutions computed using (4.38) appear in Fig. 4.5c. The initial condition and the physical and numerical parameters are identical to those used to obtain the two-dimensional upstream and CTU solutions. As might be expected in problems where there is adequate numerical resolution, the amplitude error in the second-order solution is far less than that in either first-order solution. The leading-order dispersive error in the second-order method does, however, generate regions where ϕ is slightly negative. Techniques for minimizing or eliminating these spurious negative values will be discussed in Chap. 5.

4.2.2 Systems of Equations in Several Dimensions

Although the stability analysis of systems of linear finite-difference equations in several spatial dimensions is conceptually straightforward, in practice it can be somewhat tedious. The easiest way to obtain necessary conditions for the stability of linear centered-difference approximations to problems involving wavelike flow is to examine the discrete-dispersion relation. As an example, consider the linearized Boussinesq equations governing the incompressible flow of a continuously stratified fluid in the x–z plane. If U and 0 are the constant basic-state horizontal and vertical wind speeds, the linearized versions of (1.61)–(1.63) become

$$\frac{\partial u}{\partial t} + U\frac{\partial u}{\partial x} + \frac{\partial P}{\partial x} = 0, \tag{4.39}$$

$$\frac{\partial w}{\partial t} + U\frac{\partial w}{\partial x} + \frac{\partial P}{\partial z} = b, \tag{4.40}$$

$$\frac{\partial b}{\partial t} + U\frac{\partial b}{\partial x} + N^2 w = 0, \tag{4.41}$$

$$\frac{\partial u}{\partial x} + \frac{\partial w}{\partial z} = 0, \tag{4.42}$$

where, as before, P is the perturbation pressure divided by ρ_0, b is the buoyancy, and N^2 is the Boussinesq approximation to the Brunt–Väisälä frequency.

This system is often discretized using the staggered mesh shown in Fig. 4.6, which is sometimes referred to as the Arakawa "C" grid (Arakawa and Lamb 1977). One important property of the C grid is that it allows an accurate computation of the pressure gradient and velocity divergence using a compact stencil on the staggered mesh, as in the following finite-difference approximation to (4.39)–(4.42)

$$\delta_{2t} u_{m-\frac{1}{2},n} + U\delta_{2x} u_{m-\frac{1}{2},n} + \delta_x P_{m-\frac{1}{2},n} = 0, \tag{4.43}$$

$$\delta_{2t} w_{m,n-\frac{1}{2}} + U\delta_{2x} w_{m,n-\frac{1}{2}} + \delta_z P_{m,n-\frac{1}{2}} = \left\langle b_{m,n-\frac{1}{2}} \right\rangle^z, \tag{4.44}$$

$$\delta_{2t} b_{m,n} + U\delta_{2x} b_{m,n} + N^2 \langle w_{m,n} \rangle^z = 0, \tag{4.45}$$

$$\delta_x u_{m,n} + \delta_z w_{m,n} = 0. \tag{4.46}$$

This system of finite-difference equations does not provide a complete algorithm for the time integration of the Boussinesq system because it does not include an equation for P^{j+1}. There is no equation for P^{j+1} because the Boussinesq system does not have a prognostic pressure equation. Techniques for determining the time tendency of the pressure field will be discussed in Sect. 8.1. For the present, it is assumed that the pressure is determined in some unspecified way that guarantees satisfaction of the finite-difference system (4.43)–(4.46).

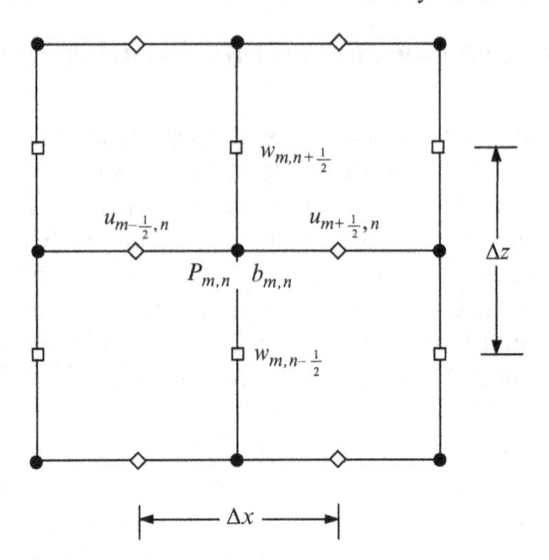

Fig. 4.6 Distribution of the dependent variables on a staggered mesh for the finite-difference approximation of the two-dimensional Boussinesq system

Substituting solutions of the form

$$\begin{pmatrix} u \\ w \\ b \\ P \end{pmatrix}_{m,n}^{j} = \begin{pmatrix} u_0 \\ w_0 \\ b_0 \\ P_0 \end{pmatrix} \mathrm{e}^{\mathrm{i}(km\Delta x + \ell n \Delta z - \omega j \Delta t)}$$

into the finite-difference system (4.43)–(4.46) and setting the determinant of the coefficients of u_0, w_0, b_0, and P_0 to zero, one obtains the discrete-dispersion relation

$$\tilde{\omega} = U\tilde{k}_2 \pm \frac{\tilde{N}\tilde{k}_1}{(\tilde{k}_1^2 + \tilde{\ell}^2)^{1/2}}, \tag{4.47}$$

in which

$$\tilde{k}_1 = \frac{\sin(k\Delta x/2)}{\Delta x/2}, \qquad \tilde{k}_2 = \frac{\sin(k\Delta x)}{\Delta x},$$

$$\tilde{\omega} = \frac{\sin \omega\Delta t}{\Delta t}, \qquad \tilde{\ell} = \frac{\sin(\ell\Delta z/2)}{\Delta z/2}, \qquad \tilde{N} = N\cos(\ell\Delta z/2).$$

The preceding relation is identical to the dispersion relation for the continuous problem except that the true frequencies and wave numbers are replaced by their numerical approximations. Note that there are two different approximations to the horizontal wave number: $\tilde{k}_1$ arises from a centered finite-difference approximation to the horizontal derivative on a Δx-wide stencil, and $\tilde{k}_2$ is associated with finite differences on a $2\Delta x$-wide stencil. The factor $\tilde{k}_1$ is associated with the discretized pressure-gradient and divergence operators, whereas $\tilde{k}_2$ is associated with the advection operator.

The numerical solution should not grow with time because linear-wave solutions to Boussinesq equations are nonamplifying. A necessary condition for the absence of amplifying waves in the discretized solution is that $\tilde{\omega}$ be real or, equivalently, that the magnitude of Δt times the right side of (4.47) be less than 1. Since the factor multiplying N in (4.47) is bounded by unity, this stability condition is

$$\left(\frac{|U|}{\Delta x} + N\right)\Delta t \leq 1.$$

4.3 Splitting into Fractional Steps

More efficient integration schemes can often be obtained by splitting complex finite-difference formulae into a series of fractional steps. As an example, the two-dimensional advection equation (4.20) might be approximated by the scheme

$$\phi^s = \phi^n - \frac{U\Delta t}{2}\delta_{2x}(\phi^n + \phi^s), \tag{4.48}$$

$$\phi^{n+1} = \phi^s - \frac{V\Delta t}{2}\delta_{2y}(\phi^s + \phi^{n+1}), \tag{4.49}$$

where ϕ^s is a temporary quantity computed during the first fractional step. (Note that ϕ^s is not a consistent approximation to the true solution at any particular time level, which complicates the specification of boundary conditions for ϕ^s and makes it difficult to use multilevel time differencing in time-split methods.) If the computational domain contains $N_x \times N_y$ grid points, each integration step of (4.48) and (4.49) requires the solution of $N_x + N_y$ tridiagonal systems. In contrast, a single integration step of the corresponding unsplit formula,

$$\phi^{n+1} = \phi^n - \frac{U\Delta t}{2}\delta_{2x}(\phi^n + \phi^{n+1}) - \frac{V\Delta t}{2}\delta_{2y}(\phi^n + \phi^{n+1}), \tag{4.50}$$

requires the solution of a linear system with an $N_xN_y \times N_xN_y$ coefficient matrix whose bandwidth is the smaller of $2N_x + 1$ and $2N_y + 1$. The fractional-step approach is more efficient because fewer computations are required to solve the $N_x + N_y$ tridiagonal problems than to solve the single linear system associated with the band matrix. Some loss of accuracy may, however, be introduced when the problem is split into fractional steps.

To examine the accuracy and stability of fractional-step splittings in a general context, consider the class of partial differential equations of the form

$$\frac{\partial \psi}{\partial t} = \mathscr{L}\psi = \mathscr{L}_1\psi + \mathscr{L}_2\psi, \tag{4.51}$$

where $\mathscr{L}$ is the linear operator formed by the sum of two linear operators $\mathscr{L}_1$ and $\mathscr{L}_2$. In the preceding case of two-dimensional advection,

$$\mathscr{L}_1 = -U\frac{\partial}{\partial x} \quad \text{and} \quad \mathscr{L}_2 = -V\frac{\partial}{\partial y}, \tag{4.52}$$

and $\mathscr{L}$ is split into operators involving derivatives parallel to each spatial coordinate. In other applications, the governing equations might be split into subproblems representing different physical processes. In a simulation of chemically reacting flow, for example, the terms representing advection might be grouped together into $\mathscr{L}_1$, whereas terms describing chemistry might appear in $\mathscr{L}_2$.

Assuming for simplicity in the following analysis that $\mathscr{L}$ is time independent, one may write the exact solution to (4.51) in the form $\psi(t) = \exp(t\mathscr{L})\psi(0)$, where the exponential of the operator $\mathscr{L}$ is defined by the infinite series

$$\exp(t\mathscr{L}) = I + t\mathscr{L} + \frac{t^2}{2}\mathscr{L}^2 + \frac{t^3}{6}\mathscr{L}^3 + \cdots,$$

and I is the identity operator. The change in ψ over one time step is therefore

$$\psi(t+\Delta t) = \exp[(\Delta t+t)\mathscr{L}]\psi(0) = \exp(\Delta t\mathscr{L})\exp(t\mathscr{L})\psi(0) = \exp(\Delta t\mathscr{L})\psi(t).$$

Suppose that a numerical approximation to the preceding expression has the form

$$\phi^{n+1} = \mathscr{F}(\Delta t)\phi^n. \tag{4.53}$$

If the global truncation error in this approximation is $O\left[(\Delta t)^n\right]$, the local truncation error[5] is $O\left[(\Delta t)^{n+1}\right]$ and

$$\mathscr{F}(\Delta t) = \exp(\Delta t\mathscr{L}) + O\left[(\Delta t)^{n+1}\right]. \tag{4.54}$$

In practice, $\mathscr{F}$ may involve approximations to spatial derivatives, but the fundamental properties of the fractional-step method can be explored without explicitly considering the discretization of the spatial derivatives.

4.3.1 Split Explicit Schemes

The unsplit forward-difference approximation

$$\mathscr{F}(\Delta t) = (I + \Delta t\mathscr{L}_1 + \Delta t\mathscr{L}_2)$$

satisfies (4.54) with $n = 1$, as would be expected, since forward differencing is $O(\Delta t)$ accurate. It is easy to achieve the same level of accuracy using

[5] See Sect. 2.2.3 for the definition of local and global truncation error.

$O(\Delta t)$-accurate fractional steps. For example, the split scheme consisting of the two forward steps

$$\phi^s = (I + \Delta t \mathscr{L}_1)\phi^n, \tag{4.55}$$

$$\phi^{n+1} = (I + \Delta t \mathscr{L}_2)\phi^s \tag{4.56}$$

generates the approximate finite-difference operator

$$(I + \Delta t \mathscr{L}_2)(I + \Delta t \mathscr{L}_1) = I + \Delta t \mathscr{L}_1 + \Delta t \mathscr{L}_2 + (\Delta t)^2 \mathscr{L}_2 \mathscr{L}_1,$$

and is therefore $O(\Delta t)$ accurate.

It is more difficult to design split schemes that are $O\left[(\Delta t)^2\right]$ accurate unless the operators $\mathscr{L}_1$ and $\mathscr{L}_2$ commute. Suppose the forward differences in (4.55) and (4.56) are replaced by second-order numerical operators $\mathscr{F}_1$ and $\mathscr{F}_2$. One possible choice for $\mathscr{F}_1$ and $\mathscr{F}_2$ is the second-order Runge–Kutta method, in which (4.53) would be evaluated in the two stages

$$\phi^* = \phi^n + \frac{\Delta t}{2\beta} \mathscr{L} \phi^n,$$

$$\phi^{n+1} = \phi^n + \beta \Delta t \mathscr{L} \phi^* + (1 - \beta) \Delta t \mathscr{L} \phi^n,$$

where β is a free parameter (see Sect. 2.3). A second possibility is the Lax–Wendroff method, in which $\mathscr{L}$ and $\mathscr{L}^2$ are both evaluated during a single forward time step such that

$$\phi^{n+1} = \phi^n + \Delta t \mathscr{L} \phi^n + \frac{(\Delta t)^2}{2} \mathscr{L}^2 \phi^n.$$

Whatever the exact formulation of $\mathscr{F}_1$ and $\mathscr{F}_2$, since they are of second order,

$$\mathscr{F}_1(\Delta t) = I + \Delta t \mathscr{L}_1 + \frac{(\Delta t)^2}{2} \mathscr{L}_1^2 + O\left[(\Delta t)^3\right],$$

$$\mathscr{F}_2(\Delta t) = I + \Delta t \mathscr{L}_2 + \frac{(\Delta t)^2}{2} \mathscr{L}_2^2 + O\left[(\Delta t)^3\right],$$

and the composite operator for a complete integration step is

$$[\mathscr{F}_2(\Delta t)][\mathscr{F}_1(\Delta t)] = \\ I + \Delta t(\mathscr{L}_2 + \mathscr{L}_1) + \frac{(\Delta t)^2}{2}(\mathscr{L}_2^2 + 2\mathscr{L}_2\mathscr{L}_1 + \mathscr{L}_1^2) + O\left[(\Delta t)^3\right].$$

The preceding expression will not be a second-order approximation to the exact operator

$$\exp(\Delta t \mathscr{L}) = I + \Delta t(\mathscr{L}_1 + \mathscr{L}_2) + \frac{(\Delta t)^2}{2}(\mathscr{L}_1^2 + \mathscr{L}_1\mathscr{L}_2 + \mathscr{L}_2\mathscr{L}_1 + \mathscr{L}_2^2) + O\left[(\Delta t)^3\right]$$

unless $\mathscr{L}_1\mathscr{L}_2 = \mathscr{L}_2\mathscr{L}_1$. Unfortunately, in many practical applications $\mathscr{L}_1$ and $\mathscr{L}_2$ do not commute. For example, if $\mathscr{L}_1$ and $\mathscr{L}_2$ are the one-dimensional advection operators defined by (4.52) and U and V are functions of x and y, then

$$\mathscr{L}_1\mathscr{L}_2 = U\frac{\partial V}{\partial x}\frac{\partial}{\partial y} + UV\frac{\partial^2}{\partial x \partial y},$$

$$\mathscr{L}_2\mathscr{L}_1 = V\frac{\partial U}{\partial y}\frac{\partial}{\partial x} + UV\frac{\partial^2}{\partial x \partial y},$$

and $\mathscr{L}_1\mathscr{L}_2 \neq \mathscr{L}_2\mathscr{L}_1$ unless

$$U\frac{\partial V}{\partial x} = V\frac{\partial U}{\partial y} = 0.$$

Strang (1968) noted that even if $\mathscr{L}_1$ and $\mathscr{L}_2$ do not commute, $\mathscr{F}_1$ and $\mathscr{F}_2$ can still be used to construct the following $O\left[(\Delta t)^2\right]$ operator:

$$[\mathscr{F}_1(\Delta t/2)][\mathscr{F}_2(\Delta t)][\mathscr{F}_1(\Delta t/2)]. \tag{4.57}$$

It might appear that this splitting requires 50% more computation than the binary products considered previously. However, since

$$[\mathscr{F}_2(\Delta t)] = [\mathscr{F}_2(\Delta t/2)][\mathscr{F}_2(\Delta t/2)] + O\left[(\Delta t)^3\right],$$

the scheme

$$\{[\mathscr{F}_1(\Delta t/2)][\mathscr{F}_2(\Delta t/2)]\}\{[\mathscr{F}_2(\Delta t/2)][\mathscr{F}_1(\Delta t/2)]\} \tag{4.58}$$

could be evaluated in lieu of (4.57) while preserving second-order accuracy. Replacing $\Delta t/2$ by Δt in (4.58), it becomes apparent that all that is required to maintain $O\left[(\Delta t)^2\right]$ accuracy is to reverse the order of the individual operators on alternate time steps.

Problems can be split into more than two subproblems in a similar manner. If the original problem is approximated by a series of numerical operators, $\mathscr{F}_1$, $\mathscr{F}_2$, ..., $\mathscr{F}_N$ and the least accurate of the $\mathscr{F}_j$ is $O\left[(\Delta t)^n\right]$, the simplest fractional step splitting

$$[\mathscr{F}_1(\Delta t)][\mathscr{F}_2(\Delta t)]\ldots[\mathscr{F}_N(\Delta t)]$$

is $O(\Delta t)$ unless all the individual operators commute, in which case the accuracy is $O\left[(\Delta t)^n\right]$. When $\mathscr{F}_1, \mathscr{F}_2, \ldots, \mathscr{F}_N$ do not commute, but are all at least $O\left[(\Delta t)^2\right]$, second-order accuracy can be obtained using Strang splitting, which again simply requires that the order of the individual operators be reversed on alternate steps. Two adjacent time steps then have the form

$$\{[\mathscr{F}_1(\Delta t)][\mathscr{F}_2(\Delta t)]\ldots[\mathscr{F}_N(\Delta t)]\}\{[\mathscr{F}_N(\Delta t)][\mathscr{F}_{N-1}(\Delta t)]\ldots[\mathscr{F}_1(\Delta t)]\}. \tag{4.59}$$

The second-order accuracy of the preceding expression follows from the two-operator case. Let $[\mathscr{F}_{N-1,N}(\Delta t)]$ be a second-order approximation to the *unsplit* sum of the exact operators $\exp[\Delta t(\mathscr{L}_{N-1} + \mathscr{L}_N)]$. Using the second-order accuracy of (4.58),

$$[\mathscr{F}_{N-2}(\Delta t)][\mathscr{F}_{N-1}(\Delta t)][\mathscr{F}_N(\Delta t)][\mathscr{F}_N(\Delta t)][\mathscr{F}_{N-1}(\Delta t)][\mathscr{F}_{N-2}(\Delta t)] = \\ [\mathscr{F}_{N-2}(\Delta t)][\mathscr{F}_{N-1,N}(2\Delta t)][\mathscr{F}_{N-2}(\Delta t)] + O\,[\![(\Delta t)^3.$$

The right side of the preceding expression is the basic Strang-split operator (4.57) with the time step doubled, and is therefore a second-order approximation to $\exp[2\Delta t(\mathscr{L}_{N-2} + \mathscr{L}_{N-1} + \mathscr{L}_N)]$ that may be premultiplied and postmultiplied by $[\mathscr{F}_{N-3}(\Delta t)]$ to continue the argument by induction.

4.3.2 Split Implicit Schemes

Although Strang splitting can be used to obtain second-order accuracy when explicit time differencing is used in the individual fractional steps, other techniques are required when the time differencing is implicit. The trapezoidal scheme is the most important second-order implicit time difference used in split schemes. The trapezoidal approximation to the general partial differential equation (4.51) may be expressed using the preceding operator notation as

$$\left[I - \frac{\Delta t}{2}\mathscr{L}\right]\phi^{n+1} = \left[I + \frac{\Delta t}{2}\mathscr{L}\right]\phi^n,$$

or

$$\phi^{n+1} = \left[I - \frac{\Delta t}{2}\mathscr{L}\right]^{-1}\left[I + \frac{\Delta t}{2}\mathscr{L}\right]\phi^n.$$

If trapezoidal time differencing is employed in two successive fractional steps, as in (4.48) and (4.49), the composite operator is

$$\mathscr{F}(\Delta t) = \left[I - \frac{\Delta t}{2}\mathscr{L}_2\right]^{-1}\left[I + \frac{\Delta t}{2}\mathscr{L}_2\right]\left[I - \frac{\Delta t}{2}\mathscr{L}_1\right]^{-1}\left[I + \frac{\Delta t}{2}\mathscr{L}_1\right]. \quad (4.60)$$

This operator may be expanded using the formula for the sum of a geometric series,

$$(1 - x)^{-1} = 1 + x + x^2 + x^3 + \cdots,$$

to yield

$$\mathscr{F}(\Delta t) = I + \Delta t(\mathscr{L}_2 + \mathscr{L}_1) + \frac{(\Delta t)^2}{2}(\mathscr{L}_2^2 + 2\mathscr{L}_2\mathscr{L}_1 + \mathscr{L}_1^2) + O\left[(\Delta t)^3\right].$$

This is the same expression as obtained using second-order explicit differences in each fractional step, and as before it will not agree with $\exp(\Delta t\mathscr{L}_1 + \Delta t\mathscr{L}_2)$ through $O\left[(\Delta t)^2\right]$ unless $\mathscr{L}_1$ and $\mathscr{L}_2$ commute.

Even if $\mathscr{L}_1$ and $\mathscr{L}_2$ do not commute, an $O\left[(\Delta t)^2\right]$ approximation can be achieved by the following permutation of the operators in (4.60):

$$\mathscr{F}(\Delta t) = \left[I - \frac{\Delta t}{2}\mathscr{L}_2\right]^{-1}\left[I - \frac{\Delta t}{2}\mathscr{L}_1\right]^{-1}\left[I + \frac{\Delta t}{2}\mathscr{L}_1\right]\left[I + \frac{\Delta t}{2}\mathscr{L}_2\right].$$

The resulting scheme,

$$\left[I - \frac{\Delta t}{2}\mathscr{L}_1\right]\left[I - \frac{\Delta t}{2}\mathscr{L}_2\right]\phi^{n+1} = \left[I + \frac{\Delta t}{2}\mathscr{L}_1\right]\left[I + \frac{\Delta t}{2}\mathscr{L}_2\right]\phi^n, \tag{4.61}$$

may be efficiently implemented using the Peaceman–Rachford *alternating direction* algorithm

$$\left[I - \frac{\Delta t}{2}\mathscr{L}_1\right]\phi^s = \left[I + \frac{\Delta t}{2}\mathscr{L}_2\right]\phi^n, \tag{4.62}$$

$$\left[I - \frac{\Delta t}{2}\mathscr{L}_2\right]\phi^{n+1} = \left[I + \frac{\Delta t}{2}\mathscr{L}_1\right]\phi^s. \tag{4.63}$$

To demonstrate the equivalence of (4.61) and the Peaceman–Rachford formulation, apply $I - \frac{\Delta t}{2}\mathscr{L}_1$ to each side of (4.63) and observe that

$$\begin{aligned}
\left[I - \frac{\Delta t}{2}\mathscr{L}_1\right]\left[I - \frac{\Delta t}{2}\mathscr{L}_2\right]\phi^{n+1} &= \left[I - \frac{\Delta t}{2}\mathscr{L}_1\right]\left[I + \frac{\Delta t}{2}\mathscr{L}_1\right]\phi^s \\
&= \left[I + \frac{\Delta t}{2}\mathscr{L}_1\right]\left[I - \frac{\Delta t}{2}\mathscr{L}_1\right]\phi^s \\
&= \left[I + \frac{\Delta t}{2}\mathscr{L}_1\right]\left[I + \frac{\Delta t}{2}\mathscr{L}_2\right]\phi^n,
\end{aligned}$$

where the second equality is obtained because $\mathscr{L}_1$ commutes with itself, and substitution from (4.62) is used to form the final equality.

4.3.3 Stability of Split Schemes

When the numerical operators $\mathscr{F}_1$ and $\mathscr{F}_2$ commute, the stability of the split scheme $\mathscr{F}_1\mathscr{F}_2$ is guaranteed by the stability of the individual operators. To demonstrate this, suppose $\mathbf{A}_1$ and $\mathbf{A}_2$ are the amplification matrices associated with $\mathscr{F}_1$ and $\mathscr{F}_2$, and note that if $\mathscr{F}_1$ and $\mathscr{F}_2$ commute, their amplification matrices also commute. The amplification matrix for the split scheme is $\mathbf{A}_1\mathbf{A}_2$, and

$$\begin{aligned}
\|(\mathbf{A}_1\mathbf{A}_2)^n\| &= \|\mathbf{A}_1\mathbf{A}_2\mathbf{A}_1\mathbf{A}_2\cdots\mathbf{A}_1\mathbf{A}_2\| \\
&= \|(\mathbf{A}_1)^n(\mathbf{A}_2)^n\|
\end{aligned} \tag{4.64}$$

$$\leq \|(\mathbf{A}_1)^n\| \, \|(\mathbf{A}_2)^n\|, \tag{4.65}$$

and it is apparent that the split scheme inherits the stability properties of the individual operators.

If $\mathscr{F}_1$ and $\mathscr{F}_2$ do not commute, equality (4.64) does not hold, and

$$\|(\mathbf{A}_1\mathbf{A}_2)^n\| \leq \|\mathbf{A}_1\|^n \, \|\mathbf{A}_2\|^n$$

is the best bound that can be obtained without specific knowledge of $\mathbf{A}_1$ and $\mathbf{A}_2$. The preceding expression guarantees the stability of the split scheme when $\|\mathbf{A}_1\|$ and $\|\mathbf{A}_2\|$ are less than or equal to unity; however, as discussed in Sect. 4.1.1.1, $\|\mathbf{A}_1\| \leq 1$ is not necessary for the stability of $\mathscr{F}_1$.

As an illustration of the preceding discussion, consider the system of ordinary differential equations

$$\frac{du}{dt} = \mathrm{i}cv + \mathrm{i}bu, \tag{4.66}$$

$$\frac{dv}{dt} = \mathrm{i}cu, \tag{4.67}$$

where b and c are real constants. Oscillatory solutions to the preceding problem exist of the form

$$\begin{pmatrix} u \\ v \end{pmatrix} = \begin{pmatrix} 1 \\ -c/\omega \end{pmatrix} A\mathrm{e}^{-\mathrm{i}\omega t},$$

where A is an arbitrary amplitude and ω is one of the two real roots to the dispersion relation

$$\omega^2 + b\omega - c^2 = 0.$$

A split scheme in which each step is stable but the composite scheme is unconditionally unstable can be obtained by constructing the following finite-difference approximation to (4.66) and (4.67). In the first step, integrate the terms involving c using forward–backward differencing,

$$\frac{u^s - u^n}{\Delta t} = \mathrm{i}cv^n,$$

$$\frac{v^s - v^n}{\Delta t} = \mathrm{i}cu^s,$$

and then integrate the term involving b using trapezoidal differencing:

$$\frac{u^{n+1} - u^s}{\Delta t} = \mathrm{i}b\left(\frac{u^{n+1} + u^s}{2}\right),$$

$$v^{n+1} = v^s.$$

Letting $\hat{c} = c\Delta t$, one may write the first step in matrix form as

$$\begin{pmatrix} 1 & 0 \\ -\mathrm{i}\hat{c} & 1 \end{pmatrix} \begin{pmatrix} u^s \\ v^s \end{pmatrix} = \begin{pmatrix} 1 & \mathrm{i}\hat{c} \\ 0 & 1 \end{pmatrix} \begin{pmatrix} u^n \\ v^n \end{pmatrix},$$

or

$$\begin{pmatrix} u^s \\ v^s \end{pmatrix} = \begin{pmatrix} 1 & \mathrm{i}\hat{c} \\ \mathrm{i}\hat{c} & 1-\hat{c}^2 \end{pmatrix} \begin{pmatrix} u^n \\ v^n \end{pmatrix}. \tag{4.68}$$

Denote the matrix in (4.68) by $\mathbf{A}_1$. The eigenvalues of $\mathbf{A}_1$ are

$$1 - \frac{\hat{c}^2}{2} \pm \frac{\mathrm{i}\hat{c}}{2}(4-\hat{c}^2)^{1/2}.$$

For $|\hat{c}| \leq 2$, the magnitude of both eigenvalues is unity, and since $\mathbf{A}_1$ is symmetric, the scheme is stable. Observe, however, that the norm of $\mathbf{A}_1$,

$$\|\mathbf{A}_1\|_2 = \left(1 + \frac{\hat{c}^4}{2} + \frac{\hat{c}^2}{2}(4+\hat{c}^4)^{1/2}\right)^{1/2},$$

exceeds unity for all nonzero Δt.

If $\hat{b} = b\Delta t$, the second fractional step may be written as

$$\begin{pmatrix} u^{n+1} \\ v^{n+1} \end{pmatrix} = \begin{pmatrix} \dfrac{2+\mathrm{i}\hat{b}}{2-\mathrm{i}\hat{b}} & 0 \\ 0 & 1 \end{pmatrix} \begin{pmatrix} u^s \\ v^s \end{pmatrix}. \tag{4.69}$$

Let $\mathbf{A}_2$ represent the amplification matrix in (4.69). One can easily show that $\|\mathbf{A}_2\| = 1$, so this scheme is also stable.

The amplification matrix for the composite scheme is

$$\mathbf{A}_2\mathbf{A}_1 = \begin{pmatrix} \dfrac{2+\mathrm{i}\hat{b}}{2-\mathrm{i}\hat{b}} & \dfrac{2+\mathrm{i}\hat{b}}{2-\mathrm{i}\hat{b}}\mathrm{i}\hat{c} \\ \mathrm{i}\hat{c} & 1-\hat{c}^2 \end{pmatrix}.$$

Since $\mathbf{A}_2\mathbf{A}_1 \neq \mathbf{A}_1\mathbf{A}_2$, the stability of the individual steps does not guarantee the stability of the composite scheme. Moreover, the inequality

$$\|\mathbf{A}_2\mathbf{A}_1\| \leq \|\mathbf{A}_2\|\|\mathbf{A}_1\|$$

cannot be used to show stability, since $\|\mathbf{A}_1\| > 1$. In fact, numerical calculations show that the magnitude of the largest eigenvalue of $\mathbf{A}_2\mathbf{A}_1$ is greater than unity for all $|\hat{b}|, |\hat{c}| > 0$, so the composite scheme is unconditionally unstable.

4.4 Linear Equations with Variable Coefficients

Some of the simplest equations of practical interest are linear equations with variable coefficients. Consider, for example, one-dimensional advection by a spatially varying wind speed, which is governed by the partial differential equation

$$\frac{\partial \psi}{\partial t} + c(x)\frac{\partial \psi}{\partial x} = 0. \tag{4.70}$$

Suppose $\phi_j(t)$ is the numerical approximation to $\psi(t, j\Delta x)$ and that c is available at the same set of spatial grid points. One obvious differential–difference approximation to the preceding equation is

$$\frac{d\phi_j}{dt} + c_j \delta_{2x}\phi_j = 0. \tag{4.71}$$

The stability of this scheme is often assessed by "freezing" $c(x)$ at some constant value c_0 and studying the stability of the family of frozen-coefficient problems obtained by varying c_0 over the range of all possible $c(x)$. It should be noted, however, that in some pathological examples there is no relation between the stability of the variable-coefficient problem and the corresponding family of frozen-coefficient problems (see Problem 10).

Suppose that the time derivative in (4.71) is replaced by leapfrog time differencing; then a necessary condition for the stability of the resulting scheme is

$$\max_x |c(x)| \frac{\Delta t}{\Delta x} < 1.$$

If this stability condition is violated in some small region of the flow, the instability will initially be confined to the same region and will appear as a packet of rapidly amplifying short waves typically having wavelengths between $2\Delta x$ and $4\Delta x$. If the numerical solution and the variable coefficients remain smooth and well resolved, the frozen-coefficient analysis can also yield sufficient conditions for stability. To guarantee stability via a frozen-coefficient analysis, the numerical scheme must include some dissipative smoothing (Gustafsson et al. 1995, p. 235). The stability of some completely nondissipative methods can, nevertheless, be established by the energy method (see Sect. 4.4.2).

A second reasonable differential–difference approximation to (4.70) may be written in the form

$$\frac{d\phi_j}{dt} + \left\langle \langle c_j \rangle^x \delta_x \phi_j \right\rangle^x = 0, \tag{4.72}$$

where the averaging operator $\langle\, \rangle^x$ is defined by (A.2) in the Appendix. When c is a constant, the preceding expression is identical to (4.71). If identical time differences are employed to solve (4.71) and (4.72), a frozen-coefficient stability analysis will yield the same stability condition for each scheme. An analysis of truncation error, performed by substituting Taylor series expansions for c and ψ into the differential–difference equations, shows that both schemes are accurate to $O\left[(\Delta x)^2\right]$. Is there any practical difference between (4.71) and (4.72)? There is, but the difference is not obvious unless one considers problems in which c and ϕ are poorly resolved on the numerical mesh or situations where the true solution has additional conservation properties (such as advection in a nondivergent flow) that are not automatically retained by finite-difference approximations.

First consider the problems that can arise when there are large-amplitude poorly resolved perturbations in ϕ or c. Under such circumstances, both of the preceding numerical approximations can exhibit serious instabilities. The structure and growth rates of the unstable perturbations generated by each scheme can, however, be very different. Perhaps the most useful way to explore these instabilities is to examine the aliasing error produced by (4.71) and (4.72).

4.4.1 Aliasing Error

Aliasing error occurs when a short-wavelength fluctuation is sampled at discrete intervals and misinterpreted as a longer-wavelength oscillation. The shortest wavelength that can be represented on a numerical grid is twice the grid interval; all shorter wavelengths will be aliased. Figure 4.7 illustrates the aliasing of a $4\Delta x/3$ wave into a $4\Delta x$ wave. The apparent equivalence of the $4\Delta x/3$ and $4\Delta x$ waves follows from the fact that for all integers n the relation

$$e^{ikj\Delta x} = \left[e^{i2n\pi}\right]^j e^{ikj\Delta x} = e^{i(k+2n\pi/\Delta x)j\Delta x} \tag{4.73}$$

is satisfied at all spatial locations $j\Delta x$ on the discrete mesh. In the case shown in Fig. 4.7, the wave number of the aliased wave is $k = (2\pi)/(4\Delta x/3) = 3\pi/(2\Delta x)$, and the wave number of the resolved wave is $-\pi/(2\Delta x)$, so (4.73) applies with $n = -1$. The change in the sign of the wave number during aliasing is visible in Fig. 4.7 as the 180° phase shift between the original and the aliased wave.

Aliasing error may occur when the initial data are represented on a discrete grid or projected onto a truncated series of Fourier expansion functions. Aliasing error can also occur during the computation of the product $c\partial\psi/\partial x$ on a finite-resolution numerical grid. To illustrate how the product of two spatially varying functions may introduce aliasing error, suppose that the product $\phi(x)\chi(x)$ is computed at a discrete set of grid points. If $\phi = e^{ik_1x}$ and $\chi = e^{ik_2x}$, then $\phi\chi = e^{i(k_1+k_2)x}$. Since ϕ and χ were representable on the numerical grid, $|k_1|, |k_2| \leq \pi/\Delta x$. It is possible, however, that the wave number of their product lies in the range $\pi/\Delta x < |k_1+k_2| \leq 2\pi/\Delta x$, in which case the product cannot be resolved on the numerical mesh and will be misrepresented as a longer wave. The wave number $\tilde{k}$ into which a binary product is aliased is determined by the

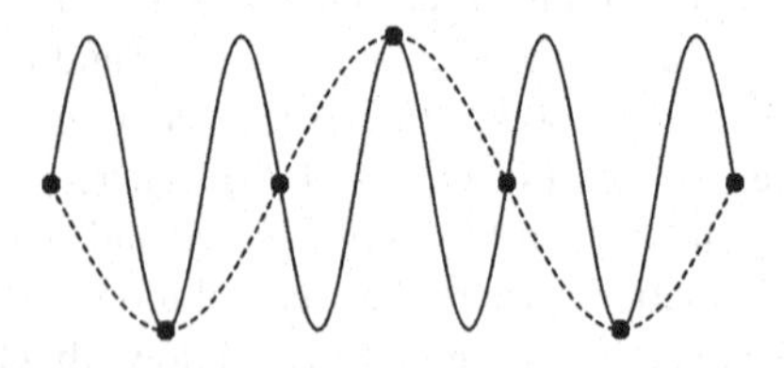

Fig. 4.7 Misrepresentation of a $4\Delta x/3$ wave as a $4\Delta x$ wave when sampled on a discrete mesh

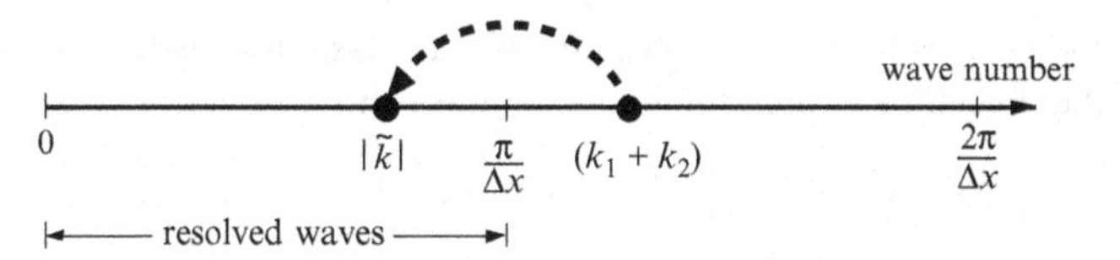

Fig. 4.8 Aliasing of $k_1 + k_2$ into $\tilde{k}$ such that $|\tilde{k}| = 2\pi/\Delta x - (k_1 + k_2)$ appears to be the symmetric reflection of $k_1 + k_2$ about the cutoff wave number $\pi/\Delta x$

relation

$$\tilde{k} = \begin{cases} k_1 + k_2 - 2\pi/\Delta x, & \text{if } k_1 + k_2 > \pi/\Delta x, \\ k_1 + k_2 + 2\pi/\Delta x, & \text{if } k_1 + k_2 < -\pi/\Delta x. \end{cases} \tag{4.74}$$

There is no aliasing when $|k_1 + k_2| \leq \pi/\Delta x$. In particular, if both ϕ and χ are $4\Delta x$ waves or longer, their product will not be aliased. A graphical diagram of the aliasing process may be created by plotting $k_1 + k_2$ and the cutoff wave number $\pi/\Delta x$ on a number line extending from zero to $2\pi/\Delta x$. Since $(k_1 + k_2 + |\tilde{k}|)/2 = \pi/\Delta x$, $|\tilde{k}|$ appears as the reflection of $k_1 + k_2$ about the cutoff wave number, as illustrated in Fig. 4.8. Note that the product of two extremely short waves is aliased into a relatively long wave. For example, the product of a $2\Delta x$ wave and a $2.5\Delta x$ wave appears as a $10\Delta x$ wave.

4.4.1.1 Unstable Growth Through Aliasing Error

Let us now consider the effects of aliasing error on stability.[6] Suppose that (4.70) is approximated as the differential–difference equation (4.71) and that a solution is sought over the periodic domain $[-\pi, \pi]$. Let the spatial domain be discretized into the $2N + 1$ points

$$x_j = \frac{\pi j}{N}, \quad j = -N, \ldots, N,$$

and suppose that the initial data are representable as the sum of the Fourier modes in the set $\{e^0, e^{iNx/2}, e^{-iNx/2}, e^{iNx}\}$. This set of four modes is closed under multiplication on the discrete mesh owing to aliasing error; for example,

$$e^{iNx_j/2} \times e^{iNx_j} = e^{i\pi j/2} \times e^{i\pi j} = e^{i2\pi j} e^{-i\pi j/2} = e^{-iNx_j/2}.$$

Let c and ϕ be arbitrary combinations of these four Fourier modes. Under the assumption that c is real, the velocity

$$c(x_j) = c_0 + c_{n/2} e^{i\pi j/2} + c_{-n/2} e^{-i\pi j/2} + c_n e^{i\pi j}$$

may be alternatively expressed as

$$c(x_j) = c_0 + (c_r + ic_i) e^{i\pi j/2} + (c_r - ic_i) e^{-i\pi j/2} + c_n e^{i\pi j}, \tag{4.75}$$

[6] The following example was anticipated by Miyakoda (1962) and Gary (1979).

where the coefficients c_0, c_r, c_i, and c_n are all real. Assuming that ϕ is also real, we may write the preceding expression in the similar form

$$\phi(x_j) = a_0 + (a_r + ia_i)e^{i\pi j/2} + (a_r - ia_i)e^{-i\pi j/2} + a_n e^{i\pi j}. \tag{4.76}$$

Substituting (4.75) and (4.76) into the nonaveraging scheme (4.71) and requiring the linearly dependent terms to sum to zero yields

$$\begin{aligned}
\dot{a}_0 &= 2(a_i c_r - a_r c_i)/\Delta x, \\
\dot{a}_n &= 2(a_i c_r + a_r c_i)/\Delta x, \\
\dot{a}_r &= a_i(c_n + c_0)/\Delta x, &(4.77) \\
\dot{a}_i &= a_r(c_n - c_0)/\Delta x, &(4.78)
\end{aligned}$$

where the dot denotes differentiation with respect to time. Eliminating a_i between (4.77) and (4.78), one obtains

$$\ddot{a}_r = \frac{c_n^2 - c_0^2}{(\Delta x)^2} a_r. \tag{4.79}$$

A similar equation holds for a_i. According to (4.79), the behavior of the $4\Delta x$ wave in ϕ is determined by the relative magnitudes of the mean wind speed and the $2\Delta x$ wind-speed perturbation. If the mean wind is stronger than the $2\Delta x$ perturbation, a_r oscillates sinusoidally. On the other hand, if c_n exceeds c_0, the $4\Delta x$ component in ϕ grows exponentially. The growth criterion $c_n > c_0$ is particularly simple in the special case $c_r = c_i = 0$. Then growth will occur whenever the velocity changes sign. This exponential growth is clearly a nonphysical instability, since the true solution is constant along the characteristic curves $dx/dt = c(x)$ and therefore bounded between the maximum and minimum initial values of ϕ.

The preceding discussion demonstrates that the nonaveraging scheme (4.71) can be unstable in a rather pathological case. Both (4.71) and (4.72) can produce unstable growth in less pathological examples, provided that the wind field forces the development of unresolvable short-wavelength perturbations in ϕ. An example of this type is illustrated in Fig. 4.9, in which the initial distributions of c and ϕ are smooth and well resolved but ϕ develops unresolvable perturbations as a result of 180° changes in the wind direction. In this example the wind speed and the initial condition on ψ are

$$c(x) = 0.5\cos(4\pi x) \qquad \text{and} \qquad \psi(x,0) = \sin(2\pi x).$$

The computational domain is $0 \le x \le 1$ and $\Delta x = 1/80$. The time derivatives in (4.71) and (4.72) were integrated using a fourth-order Runge–Kutta method and a small Courant number. The velocity and the initial condition ϕ^0 are plotted together in Fig. 4.9a. The velocity field is convergent in the portions of the spatial domain labeled "Con" along the bottom of Fig. 4.9a and is divergent in the regions labeled "Div." The character of the true solution is illustrated in Fig. 4.9b, which shows a very high resolution simulation at $t = 0.5$ s. Observe that the true solution is tending toward a square wave of amplitude $\sqrt{2}/2$ with a unit-amplitude spike at the left edge

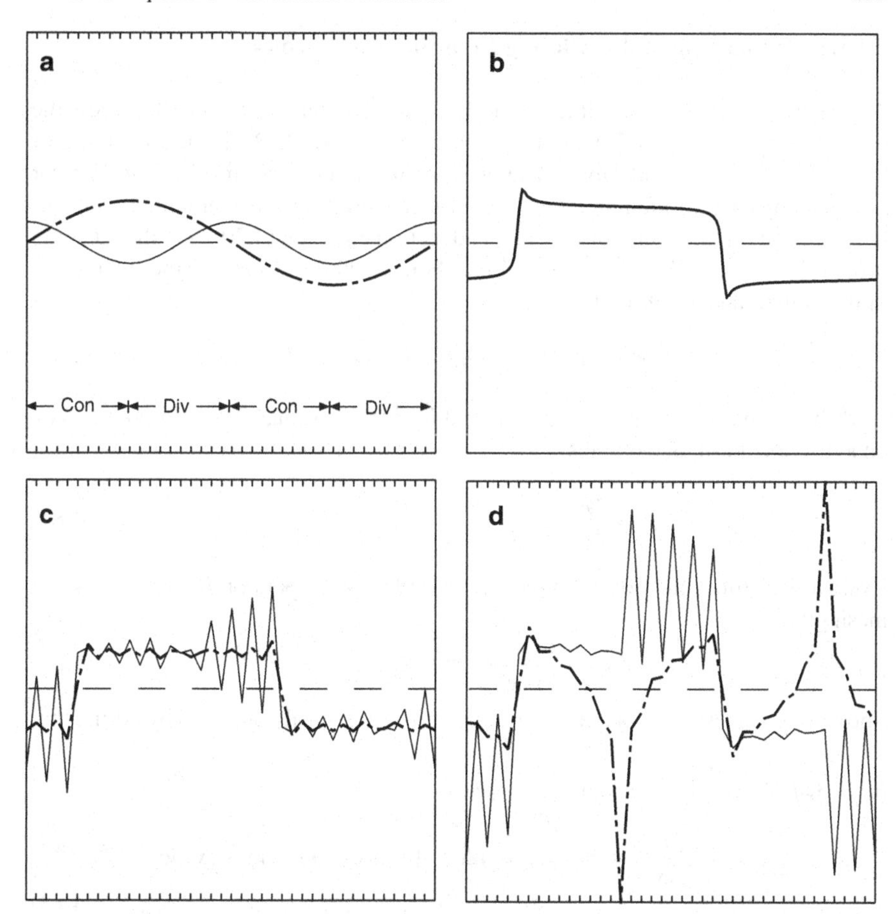

Fig. 4.9 Comparison of two differential–difference solutions to the one-dimensional advection equation in a spatially reversing flow. **a** Initial condition (*dot-dashed line*) and the time-invariant velocity (*solid line*). Also indicated are the regions where the flow is convergent (*Con*) or divergent (*Div*). **b** Solution to a high-resolution simulation at $t = 0.5$ s. Averaging (*dot-dashed line*) and nonaveraging (*thin solid line*) differential–difference solutions at **c** $t = 1.0$ s and **d** 2.2 s. The *long dashed line* is the zero line

of each plateau. The spikes are located at the nodal points in the velocity field where the flow is convergent. Away from the spike the solution is tending toward the initial value of ψ at the divergent node. A comparison of the two differential–difference solutions is shown at $t = 1.0$ s in Fig. 4.9c and at $t = 2.2$s in Fig. 4.9d. In Fig. 4.9c and d the solution obtained with the nonaveraging scheme (4.71) is dominated by a growing $2\Delta x$ component. In contrast, the solution calculated with the averaging scheme (4.72) never develops a large-amplitude $2\Delta x$ component. At the earlier time (Fig. 4.9c), the averaging scheme generates a solution that is reasonably accurate and far superior to the nonaveraging result. This superiority is lost, however, by the later time (Fig. 4.9d), at which the averaging scheme has generated a longer-wavelength disturbance that rapidly amplifies and dominates the solution.

4.4.1.2 Comparison of the Aliasing Error in Two Schemes

As suggested by the preceding example, one important difference between the nonaveraging scheme (4.71) and the averaging method (4.72) lies in the nature of the aliasing error generated by each approximation. Let us examine the aliasing error produced by each formula in a more general context. Suppose that numerical solutions are sought to the one-dimensional advection equation (4.70), and that at some instant in time, c and ϕ are expanded into Fourier modes. Consider the interaction of the individual pair of modes

$$c = c_{k_1} \mathrm{e}^{\mathrm{i}k_1 x} \quad \text{and} \quad \phi = a_{k_2} \mathrm{e}^{\mathrm{i}k_2 x}. \tag{4.80}$$

If the unapproximated product of c and $\partial\phi/\partial x$ is evaluated at grid points $j\Delta x$ on a discrete mesh, one obtains

$$c\frac{\partial \phi}{\partial x} = \mathrm{i}c_{k_1} a_{k_2} k_2 \mathrm{e}^{\mathrm{i}(k_1+k_2)j\Delta x}. \tag{4.81}$$

Evaluating c times the nonaveraging spatial-difference operator $\delta_{2x}\phi$ on the same mesh gives

$$c_j \delta_{2x}\phi_j = \frac{\mathrm{i}c_{k_1} a_{k_2}}{\Delta x}(\sin k_2 \Delta x)\mathrm{e}^{\mathrm{i}(k_1+k_2)j\Delta x}. \tag{4.82}$$

The analogous result for the averaging scheme is most easily obtained by noting that

$$\begin{aligned}\left\langle \langle c_j \rangle^x \delta_x \phi_j \right\rangle^x &= \frac{1}{2}\left[\delta_{2x}(c_j\phi_j) + c_j\delta_{2x}\phi_j - \phi_j\delta_{2x}c_j\right] \\ &= \frac{\mathrm{i}c_{k_1} a_{k_2}}{2\Delta x}\big(\sin[(k_1+k_2)\Delta x] + \sin k_2 \Delta x - \sin k_1 \Delta x\big)\mathrm{e}^{\mathrm{i}(k_1+k_2)j\Delta x}.\end{aligned} \tag{4.83}$$

In the limit of good numerical resolution, $k_1\Delta x,\ k_2\Delta x \to 0$, and each of the above expressions is equivalent, which is to be expected, since (4.82) and (4.83) are both second-order approximations to (4.81). As $(k_1+k_2)\Delta x$ approaches π, the numerical formulae may become inaccurate, but the most serious problems develop when $\pi < |(k_1+k_2)\Delta x| \leq 2\pi$ and the product term is aliased into a longer wavelength.

Suppose that a wave of wave number k_2 in the ϕ field is interacting with some disturbance in the velocity field to force $\mathrm{d}\phi/\mathrm{d}t$ at an aliased wave number $\tilde{k}$. According to (4.74), there is only one resolvable wave number in the velocity field that could alias into $\tilde{k}$ through interaction with k_2 during the approximation of the product $c\partial\psi/\partial x$. The rate at which this aliasing occurs can be computed as follows. Without loss of generality, assume that the unresolvable wave number is positive (i.e., $k_1 + k_2 > \pi/\Delta x$), in which case $\tilde{k}$ is negative and

$$\tilde{k} = k_1 + k_2 - 2\pi/\Delta x. \tag{4.84}$$

Suppose that at a given instant both interacting waves have unit amplitude, i.e., $|c_{k_1}| = |a_{k_2}| = 1$. Let $C_{k_2 \to \tilde{k}} = \mathrm{d}\,|a_{\tilde{k}}|/\mathrm{d}t$ denote the rate at which interactions

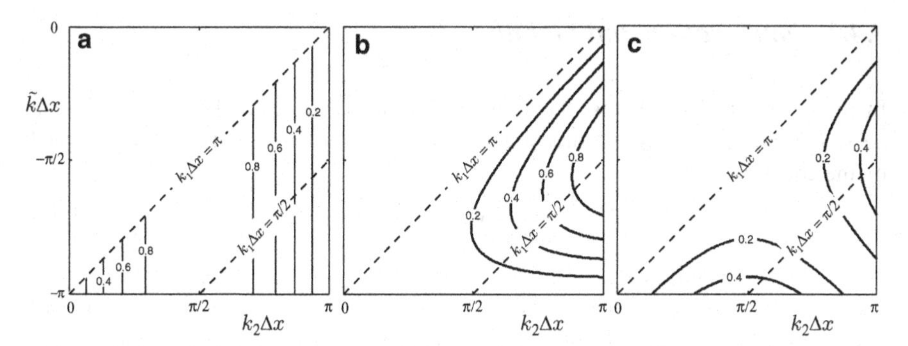

Fig. 4.10 Contours of the spurious growth rate at wave number $\tilde{k}$ due to interactions between wave numbers k_1 and k_2 as a function of $k_2\Delta x$ and $\tilde{k}\Delta x$ for **a** the nonaveraging scheme $C^{\mathrm{N}}_{k_2\to\tilde{k}}$, **b** the averaging scheme $C^{\mathrm{A}}_{k_2\to\tilde{k}}$, and **c** the averaging scheme for nondivergent flow $C^{\mathrm{ndiv}}_{k_2\to\tilde{k}}$. Contours of the wave number k_1 involved in these interactions are plotted as *diagonal dashed lines*. The contour plot in **c** is discussed in Sect. 4.4.2

between the wave numbers k_1 and k_2 force the growth at the aliased wave number. Using (4.84) to eliminate k_1 from (4.83), one obtains a growth rate for the averaging scheme of

$$C^{\mathrm{A}}_{k_2\to\tilde{k}} = \frac{|\sin\tilde{k}\Delta x + \sin k_2\Delta x - \sin(\tilde{k} - k_2)\Delta x|}{2\Delta x}.$$

The growth rate for the nonaveraging scheme,

$$C^{\mathrm{N}}_{k_2\to\tilde{k}} = \frac{|\sin k_2\Delta x|}{\Delta x},$$

can be obtained directly from (4.82).

A contour plot of $C^{\mathrm{N}}_{k_2\to\tilde{k}}$ as a function of k_2 and $\tilde{k}$ appears in Fig. 4.10a. Contours of the wave number k_1 involved in these interactions (computed from (4.84)) also appear plotted as a function of $(k_2, \tilde{k})$ as the dashed diagonal lines in Fig. 4.10. Since $k_1 \leq \pi/\Delta x$ (because every wave must be resolved on the numerical mesh), no aliasing can occur for the $(k_2, \tilde{k})$ combinations above the diagonal in Fig. 4.10a, and this region of the plot is left blank. Equivalent data, showing contours of $C^{\mathrm{A}}_{k_2\to\tilde{k}}$ for the averaging scheme, appear in Fig. 4.10b.

As indicated in Fig. 4.10, the rate at which wavelengths in the concentration field longer than $4\Delta x$ ($k_2\Delta x < \pi/2$) undergo aliasing is much weaker in the averaging scheme than in the nonaveraging approach. In addition, the nonaveraging scheme allows every wave number on the resolvable mesh ($0 \leq k_2 \leq \pi/\Delta x$) to combine with some disturbance in the velocity field to produce aliasing into $2\Delta x$ waves ($\tilde{k}\Delta x = -\pi$). On the other hand, the averaging scheme does not allow any aliasing into the $2\Delta x$ wave, although aliasing is permitted into the longer wavelengths. The practical impact of this difference in aliasing is evident in the numerical comparisons shown in Fig. 4.9, in which the aliasing error in the nonaveraging scheme appears primarily at $2\Delta x$, whereas the errors that eventually develop in the averaging scheme appear at longer wavelengths.

4.4.2 Conservation and Stability

In many practical problems, the flow field is nondivergent. The advection of a passive scalar ψ by a two-dimensional nondivergent flow $\mathbf{v} = (u, v)$ is governed by the equation

$$\frac{\partial \psi}{\partial t} + \mathbf{v} \cdot \nabla \psi = 0, \tag{4.85}$$

subject to the constraint

$$\nabla \cdot \mathbf{v} = 0. \tag{4.86}$$

Assuming the spatial domain is periodic or that there is no flow through its boundary, the domain averages of both ψ and ψ^2 are conserved. The conservation of ψ is easily shown by first combining the *advective form* (4.85) with (4.86) to obtain the *flux* (or divergence) form

$$\frac{\partial \psi}{\partial t} + \nabla \cdot (\psi \mathbf{v}) = 0. \tag{4.87}$$

Then, denoting the domain integral by an overbar,

$$\frac{d}{dt}\overline{\psi} = \overline{\frac{\partial \psi}{\partial t}} = -\overline{\nabla \cdot (\psi \mathbf{v})} = 0,$$

where the last equality follows from the assumed conditions on the boundary. Conservation of $\overline{\psi^2}$, or equivalently $\|\psi\|_2^2$, may be demonstrated using the advective form (4.85)

$$\frac{d}{dt}\overline{\psi^2} = 2\overline{\psi\frac{\partial \psi}{\partial t}} = -2\overline{\psi\,(\mathbf{v} \cdot \nabla \psi)} = -\overline{(\nabla \psi^2 \cdot \mathbf{v})} = -\overline{\nabla \cdot (\psi^2 \mathbf{v})} + \overline{\psi^2 \nabla \cdot \mathbf{v}} = 0.$$

The first term in the final integrand is zero because of the assumed boundary conditions; the second term is zero because the flow is nondivergent.

In most applications it is helpful if numerical methods for the simulation of advection by nondivergent flow satisfy the discrete equivalents of these conservation laws. This is particularly true if one wants to analyze the mass or energy budgets of the numerical solution ϕ. In addition, conservation of $\|\phi\|_2$ guarantees that even completely inviscid methods are stable. Nevertheless, it should be noted that methods can still be accurate and converge to the correct *smooth* solution without exactly satisfying either of these conservation relations.

4.4.2.1 Conservation of Mass and Flux Form

The flux form (4.87) is an example of a *conservation law*. Conservation laws can be expressed by equations of the general form

$$\frac{\partial \psi}{\partial t} + \nabla \cdot \mathbf{f} = 0, \tag{4.88}$$

which states that the local rate of change of ψ is determined by the convergence of the flux $\mathbf{f}$. It follows from the divergence theorem that the integral of ψ over the entire domain is determined by the net flux through the boundaries, and thus $\overline{\psi}$ is conserved if $\mathbf{f}$ is periodic over the domain or if the component of $\mathbf{f}$ normal to the boundary vanishes at the boundary.

Numerical approximations to conservation laws will conserve the sum of the grid-point values if the fluxes are specified such that the flux out of each cell is identical to the flux into the adjacent cell. Semidiscrete approximations to (4.88) satisfying this constraint include those that may be expressed in the particular flux form

$$\frac{d\phi_{i,j}}{dt} + \frac{F_{i+\frac{1}{2},j} - F_{i-\frac{1}{2},j}}{\Delta x} + \frac{G_{i,j+\frac{1}{2}} - G_{i,j-\frac{1}{2}}}{\Delta y} = 0, \tag{4.89}$$

where F and G are the numerical approximations to the x and y components of $\mathbf{f}$ at the cell interfaces. Two possible ways to arrive at a finite-difference approximation to the flux form of the transport equation (4.87) are

$$\frac{d\phi}{dt} + \delta_x\Big(\langle u\phi\rangle^x \Big) + \delta_y\Big(\langle v\phi\rangle^y \Big) = 0$$

and

$$\frac{d\phi}{dt} + \delta_x\Big(\langle u\rangle^x \langle \phi\rangle^x \Big) + \delta_y\Big(\langle v\rangle^y \langle \phi\rangle^y \Big) = 0. \tag{4.90}$$

Both of the preceding schemes are in the general form (4.89), and therefore both conserve the sum of ϕ over the domain.

4.4.2.2 Stability and Skew Symmetry

The sum of ϕ^2 over the domain will remain constant with time in the semidiscrete case if the spatial finite-difference operator is skew symmetric. Let $\mathbf{u}$ be a column vector containing the approximate solution at each spatial grid point and $\mathbf{A}$ a matrix containing the finite-difference operators approximating the terms involving spatial derivatives in a linear partial differential equation. Then the approximate solution satisfies a set of linear differential–difference equations of the form

$$\frac{d\mathbf{u}}{dt} = \mathbf{Au}. \tag{4.91}$$

This system will conserve $\|\mathbf{u}\|_2$ if the matrix $\mathbf{A}$ is skew symmetric, as may be verified by noting that if $\mathbf{A} = -\mathbf{A}^{\mathrm{T}}$; then

$$\frac{d\,\|\mathbf{u}\|_2^2}{dt} = \frac{d}{dt}(\mathbf{u}^{\mathrm{T}}\mathbf{u}) = (\mathbf{Au})^{\mathrm{T}}\mathbf{u} + \mathbf{u}^{\mathrm{T}}(\mathbf{Au}) = \mathbf{u}^{\mathrm{T}}\mathbf{A}^{\mathrm{T}}\mathbf{u} + \mathbf{u}^{\mathrm{T}}\mathbf{Au} = 0.$$

As a simple example suppose that advection by a constant wind speed c_0 in a periodic domain is approximated as

$$\frac{d\phi}{dt} + c_0\delta_{2x}\phi = 0 \tag{4.92}$$

and let $\mathbf{u} = (\phi_1, \phi_2, \ldots, \phi_N)^{\mathrm{T}}$. Then $\mathbf{A} = -c_0\mathbf{S}$, where $\mathbf{S}$ is the $N \times N$ matrix

$$\frac{1}{2\Delta x}\begin{pmatrix} 0 & 1 & & & -1 \\ -1 & 0 & 1 & & \\ 0 & -1 & 0 & 1 & \\ & & \ddots & \ddots & \ddots \\ 1 & & & -1 & 0 \end{pmatrix}.$$

Clearly $c_0\mathbf{S}$ is skew symmetric. Now suppose the wind speed is a function of x and (4.92) is replaced by

$$\frac{d\phi_j}{dt} + c_j\delta_{2x}\phi_j = 0. \tag{4.93}$$

Let $\mathbf{C}$ be a diagonal matrix such that $c_{j,j} = c(j\Delta x)$, in which case (4.91) takes the form

$$\frac{d\mathbf{u}}{dt} + \mathbf{CSu} = \mathbf{0}.$$

The matrix $\mathbf{CS}$ is not skew symmetric, and (4.93) will not conserve $\sum \phi^2$. Another alternative, the simple flux form

$$\frac{d\phi_j}{dt} + \delta_{2x}(c_j\phi_j) = 0,$$

leads to

$$\frac{d\mathbf{u}}{dt} + \mathbf{SCu} = \mathbf{0},$$

but $\mathbf{SC}$ is also not skew symmetric.

The sum $\mathbf{SC} + \mathbf{CS}$ is, however, skew symmetric, since $(\mathbf{SC})^{\mathrm{T}} = \mathbf{C}^{\mathrm{T}}\mathbf{S}^{\mathrm{T}} = -\mathbf{CS}$ and similarly $(\mathbf{CS})^{\mathrm{T}} = -\mathbf{SC}$. One might therefore simulate advection in a spatially variable wind field using the formula

$$\frac{d\phi_j}{dt} + \frac{1}{2}\left[\delta_{2x}(c_j\phi_j) + c_j\delta_{2x}\phi_j\right] - \frac{1}{2}\phi_j\delta_{2x}c_j = 0. \tag{4.94}$$

The expression in square brackets is skew symmetric and the final term is included to give a consistent approximation to the advective form (4.70).

The conservation properties of the preceding skew-symmetric form are most apparent if we return to the problem of advection in a two-dimensional nondivergent flow. Suppose that the nondivergence condition (4.86) is enforced numerically as

$$\delta_{2x}u + \delta_{2y}v = 0. \tag{4.95}$$

If the advective operators in both x and y in (4.85) are expressed in the skew-symmetric form (4.94), (4.95) may be used to simply the result to

$$\frac{\partial \phi}{\partial t} + \frac{1}{2}\left[\delta_{2x}(u\phi) + u\delta_{2x}\phi\right] + \frac{1}{2}\left[\delta_{2y}(v\phi) + y\delta_{2y}\phi\right] = 0. \tag{4.96}$$

Since the operators in (4.96) are skew symmetric, $\sum \phi^2$ will be conserved and the numerical solution will be stable. This conservation property mirrors that for $\overline{\psi^2}$ satisfied by the continuous problem. Using (4.95), one may show that (4.96) is algebraically equivalent to both the flux form (4.90) and the averaging-operator advective form

$$\frac{d\phi}{dt} + \left\langle \langle u \rangle^x \delta_x \phi \right\rangle^x + \left\langle \langle v \rangle^y \delta_y \phi \right\rangle^y = 0, \tag{4.97}$$

which is the same advective operator introduced in (4.72). The equivalence to (4.90) immediately implies that all three methods will conserve the sum of ϕ_j over the domain. Although it is less helpful for immediately assessing the conservation of $\sum \phi$ and $\sum \phi^2$, (4.97) is useful since it will exactly preserve horizontally uniform fields of ϕ, whereas if the velocity is spatially varying, both of the other formulae can introduce small fluctuations in ϕ due to roundoff errors. Both (4.90) and (4.97) are also more natural choices than (4.96) for staggered meshes in which the velocity normal to the interface between each pair of grid cells is located at the center of that interface (as in Fig. 4.6).

If the averaging-operator form (4.97) is used to simulate advection and the velocities satisfy the numerical nondivergence relation (4.95), the aliasing error is reduced relative to that for the same operator in the divergent one-dimensional case. In particular, if only the x-structure of the perturbations is considered for simplicity, the last term in (4.83) is zero and the normalized rate at which perturbations of wave number k_2 in the wind field interact with fluctuations at wave number k_1 in the concentration field to generate aliasing errors at wave number $\tilde{k}$ becomes

$$C^{\text{ndiv}}_{k_2 \to \tilde{k}} = \frac{|\sin \tilde{k}\Delta x + \sin k_2 \Delta x|}{2\Delta x}.$$

As in Sect. 4.4.1, the unresolvable wave number is assumed to be positive (i.e., $k_1 + k_2 > \pi/\Delta x$), in which case $\tilde{k}$ is negative and given by (4.84).

Figure 4.10c shows contours of $C^{\text{ndiv}}_{k_2 \to \tilde{k}}$ as a function of k_2 and $\tilde{k}$. The maximum forcing for aliasing associated with (4.97) is half that of both the simple advective operator

$$\frac{d\phi}{dt} + u\delta_{2x}\phi + v\delta_{2y}\phi = 0 \tag{4.98}$$

(Fig. 4.10a) and the averaging-operator advective in divergent flow (Fig. 4.10b). The contrast between (4.97) and (4.98) is particularly pronounced, with (4.98) generating much stronger aliasing at almost all wave number combinations except those involving the $2\Delta x$ wave in the velocity field. Figure 4.10c also characterizes the aliasing generated by the flux form (4.90) and the skew-symmetric form (4.96), since all three are algebraically equivalent.

4.4.2.3 The Effect of Time Differencing on Conservation

Differential–difference equations that conserve $\|\phi\|_2$, such as (4.97), generally cease to be conservative when the time derivative is approximated by finite differences. Nevertheless, one type of time differencing that does preserve the conservation properties of linear differential–difference equations is trapezoidal differencing. The conservation properties of trapezoidal time differencing may be demonstrated by writing the differential–difference equation in the general form

$$\frac{d\phi_j}{dt} + L(\phi_j) = 0, \tag{4.99}$$

where L is a linear finite-difference operator including all the spatial differences. As a preliminary step, note that for the differential–difference equation (4.99) to conserve $\|\phi\|_2$, the linear operator L must have the algebraic property

$$\sum_j \varphi_j L(\varphi_j) = 0, \tag{4.100}$$

where φ_j is any discrete function defined on the numerical mesh and the summation is taken over all the grid points.

Approximating (4.99) with trapezoidal time differences yields

$$\frac{\phi_j^{n+1} - \phi_j^n}{\Delta t} + \frac{L(\phi_j^{n+1}) + L(\phi_j^n)}{2} = 0.$$

Multiplying the preceding equation by $(\phi_j^{n+1} + \phi_j^n)$, using the linearity of L, summing over the discrete mesh, and using (4.100), one obtains

$$\sum_j \left[(\phi_j^{n+1})^2 - (\phi_j^n)^2 \right] = -\frac{\Delta t}{2} \sum_j \left[\left(\phi_j^{n+1} + \phi_j^n\right) L \left(\phi_j^{n+1} + \phi_j^n\right) \right] = 0,$$

which implies that $\|\phi^{n+1}\|_2 = \|\phi^n\|_2$.

4.5 Nonlinear Instability

As discussed in the preceding section, the stability of finite-difference approximations to linear equations with variable coefficients can be determined by examining the stability of the associated family of frozen-coefficient problems – provided that the solution and the variable coefficients are sufficiently smooth and well resolved on the numerical mesh. One may attempt to analyze the stability of nonlinear equations through a similar procedure. First, the nonlinear equations are linearized; then a frozen-coefficient analysis is performed to determine stability conditions for the linearized problem. This approach gives necessary conditions for stability, but as was the case with variable-coefficient linear equations, it may give misleading results in situations where the solution is dominated by poorly resolved short-wavelength perturbations. Unfortunately, the caveat that the solution must remain smooth and well resolved is a much more serious impediment to the analysis of nonlinear finite-difference equations because such equations can rapidly generate unresolvable short-wave perturbations from very smooth initial data.

Rigorous demonstrations of nonlinear stability typically rely on the energy method. The stability of schemes for the simulation of the Euler equations for incompressible flow is often easier to establish if the nonlinear advective terms are written in skew-symmetric form (Morinishi et al. 1998). Using tensor notation in which repeated indices are summed, the nonlinear generalization of the skew-symmetric operator in (4.96) approximates

$$\frac{1}{2}\frac{\partial}{\partial x_j}(u_j u_i) + \frac{1}{2}u_j\frac{\partial u_i}{\partial x_j}, \tag{4.101}$$

As in (4.96), the preceding expression is an equally weighted combination of the flux form

$$\frac{\partial}{\partial x_j}(u_j u_i)$$

and the advective form

$$u_j\frac{\partial u_i}{\partial x_j}.$$

In addition to facilitating the construction of stable approximations, methods based on the skew-symmetric form (4.101) may also generate less aliasing error (Blaisdell et al. 1996), which would be consistent with the analysis of the simpler scalar advection equation considered in the previous section.

In the following we will focus on the construction of stable finite-difference approximations to two simpler nonlinear problems: Burgers's equation and the barotropic vorticity equation. Solutions to Burgers's equation often develop shocks and discontinuities whose accurate approximation requires the use of methods that will be presented in Chap. 5. The schemes that will be considered in this section provide very simple examples illustrating the stabilization of numerical approximations to a nonlinear problem by a judicious choice of finite-difference formulae. These schemes are not, however, recommended for practical applications involving the simulation of problems with shocks or discontinuous solutions. The opportunity for practical application of the ideas illustrated using Burgers's equation arises in attempting to avoid nonlinear instabilities in numerical solutions to the barotropic vorticity equation. Solutions to the barotropic vorticity equation never develop shocks and remain essentially as smooth as the initial data.

4.5.1 Burgers's Equation

The inviscid Burgers's equation,

$$\frac{\partial \psi}{\partial t} + \psi\frac{\partial \psi}{\partial x} = 0, \tag{4.102}$$

is an example of a nonlinear partial differential equation whose solution rapidly develops unresolvable short-wavelength components. If $\psi(x,0) = f(x)$ at some initial time $t = 0$, solutions to this problem can be written in the implicit form

$$\psi(x,t) = f\left(x - \psi(x,t)t\right),$$

which implies that f is constant along the characteristic curves

$$x - \psi(x,t)t = x_0.$$

Here x_0 is the x-intercept of the curve at $t = 0$. Since $\psi = f$, ψ must also be constant along each characteristic curve, and every characteristic is therefore a straight line. In any region where $\partial\psi/\partial x$ is negative, the characteristics will converge, and for some sufficiently large value of t, these converging characteristics must cross. At those points where two (or more) characteristics intersect, the solution is multivalued and exhibits a discontinuity, or shock. If the initial condition is smooth, the time when the solution first develops a shock t_c can be determined by examining the rate at which gradients of ψ steepen. Define $S(x,t) = \partial\psi/\partial x$ and note that

$$\frac{dS}{dt} = \frac{\partial S}{\partial t} + \psi\frac{\partial S}{\partial x} = \frac{\partial}{\partial x}\left(\frac{\partial \psi}{\partial t} + \psi\frac{\partial \psi}{\partial x}\right) - \left(\frac{\partial \psi}{\partial x}\right)^2 = -S^2.$$

Integration of the preceding expression yields

$$S = \left(t + S(x_0,0)^{-1}\right)^{-1}. \tag{4.103}$$

The first discontinuity, or shock, develops when S becomes infinite at a time $t_c = -S(x_0,0)^{-1}$ determined by the most negative initial value of $\partial\psi/\partial x$. This behavior may be compared with that for the linear problem with variable coefficients shown in Fig. 4.9a and b, in which the characteristic curves never cross (but rather approach the lines $x = 1/8$ and $x = 5/8$ asymptotically) and true discontinuities do not develop over any finite time interval.

Suppose that solutions to Burgers's equation are sought on the periodic domain $0 \le x \le 1$ subject to the initial condition $\psi(x,0) = \sin(2\pi x)$. When (4.102) is approximated by the advective-form differential–difference equation

$$\frac{d\phi_j}{dt} + \phi_j\delta_{2x}\phi_j = 0, \tag{4.104}$$

with $\Delta x = 1/50$, the numerical solution appears as shown in Fig. 4.11.[7] The numerical solution provides a good approximation to the true solution at $t = 0.13$, at which time the true solution is still smooth and easy to resolve on the discrete grid. But by $t = 0.22$ the true solution has developed a shock, and the numerical solution misrepresents the shock as a steep gradient bounded by a series of large-amplitude

[7] The solution shown in Fig. 4.11 was obtained using a fourth-order Runge–Kutta method and a very small time step to accurately approximate the time derivative in (4.104).

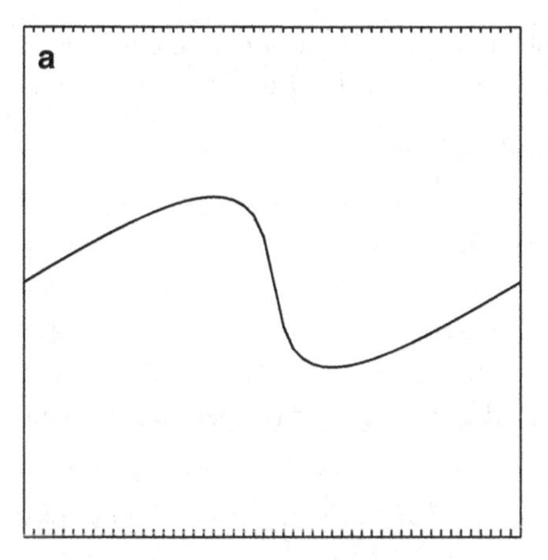

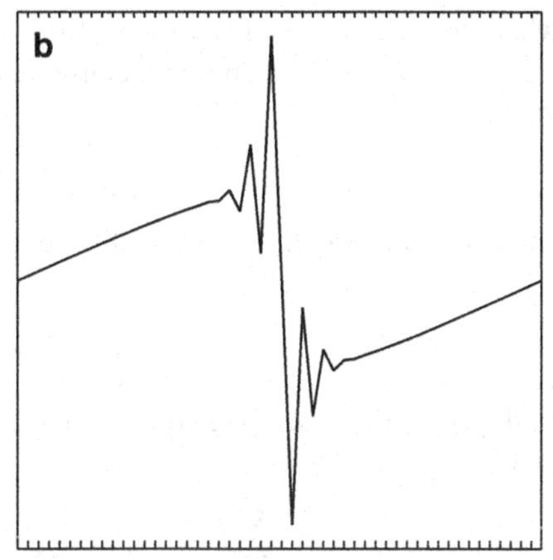

Fig. 4.11 Differential–difference solution to Burgers's equation obtained using (4.104) at **a** $t = 0.13$ and **b** $t = 0.22$

short-wavelength perturbations. These short-wavelength perturbations are amplifying rapidly, and as a consequence, $\|\phi\|_2$ is growing without bound. The growth in $\|\phi\|_2$ is an instability, since the L_2 norm of the true solution does not increase with time. If the solution is smooth, $\|\psi\|_2$ is conserved along with all other moments, i.e.,

$$\int_0^1 [\psi(x)]^p \, dx = 0 \tag{4.105}$$

for any positive integer p. The property (4.105) may be derived by multiplying (4.102) by $p\psi^{p-1}$, which yields

$$0 = p\psi^{p-1}\left(\frac{\partial \psi}{\partial t} + \psi\frac{\partial \psi}{\partial x}\right) = \frac{\partial \psi^p}{\partial t} + \frac{p}{p+1}\frac{\partial \psi^{p+1}}{\partial x},$$

and then integrating this equation over the periodic domain. If the solution contains discontinuities, the preceding manipulations are not valid, but one can show that $\partial\|\psi\|_2/\partial t$ is never positive (see Sect. 5.1.2).

The inability of the advective-form differential–difference scheme to conserve $\|\phi\|_2$ can be demonstrated by multiplying (4.104) by ϕ_j and summing over the domain to obtain

$$\begin{aligned}
\frac{d}{dt}\sum_j \phi_j^2 &= -\sum_j \left(\frac{\phi_j^2\phi_{j+1} - \phi_j^2\phi_{j-1}}{\Delta x}\right) \\
&= -\frac{1}{\Delta x}\left(\sum_j \phi_j^2\phi_{j+1} - \sum_j \phi_{j+1}^2\phi_j\right) \\
&= \sum_j \phi_j\phi_{j+1}\left(\frac{\phi_{j+1} - \phi_j)}{\Delta x}\right),
\end{aligned} \tag{4.106}$$

where the second equality follows from the periodicity of the solution. One might attempt to obtain a scheme that conserves $\|\phi\|_2$ by rewriting Burgers's equation in the flux form

$$\frac{\partial \psi}{\partial t} + \frac{1}{2}\frac{\partial \psi^2}{\partial x} = 0$$

and approximating this with the differential–difference equation

$$\frac{d\phi_j}{dt} + \frac{1}{2}\delta_{2x}\phi_j^2 = 0. \tag{4.107}$$

Multiplying the preceding equation by ϕ_j and summing over the periodic domain yields

$$\frac{d}{dt}\sum_j \phi_j^2 = -\frac{1}{2}\sum_j \phi_j\phi_{j+1}\left(\frac{\phi_{j+1}-\phi_j}{\Delta x}\right), \tag{4.108}$$

which demonstrates that the flux form also fails to conserve $\|\phi\|_2$. Since the terms representing the nonconservative forcing in (4.106) and (4.108) differ only by a factor of $-1/2$, it is possible to obtain a scheme that does conserve $\|\phi\|_2$ using a weighted average of the advective- and flux-form schemes. The resulting "conservative form" is

$$\frac{d\phi_j}{dt} + \frac{1}{3}\phi_j\delta_{2x}\phi_j + \frac{1}{3}\delta_{2x}\phi_j^2 = 0. \tag{4.109}$$

Figure 4.12 shows a comparison of the solutions to (4.104) and (4.109). The test problem is the same test considered previously in connection with Fig. 4.11, except that the vertical scale of the plotting domain shown in Fig. 4.12 has been reduced, and Fig. 4.12b now shows solutions at $t = 0.28$. The unstable growth of the short-wavelength oscillations generated by advective-form differencing can be observed by comparing the solution at $t = 0.22$ (Fig. 4.11b) and that at $t = 0.28$ (Fig. 4.12b). As illustrated in Fig. 4.12b, short-wavelength oscillations also develop in the conservative-form solution, but these oscillations do not continue to amplify.[8] The flux form (4.107) yields a solution (not shown) to this test problem that looks qualitatively similar to the conservative-form solution shown in Fig. 4.12b, although the spurious oscillations in the flux-form result are actually somewhat weaker. It is perhaps surprising that the short-wavelength oscillations are smaller in the flux-form solution than in the conservative-form solution and that the flux-form solution does not show a tendency toward instability. In fact, practical experience suggests that the flux-form difference (4.107) is not particularly susceptible to nonlinear instability.

[8] Even though they do not lead to instability, the short-wavelength oscillations in the conservative-form solution to Burgers's equation are nonphysical and are not present in the correct generalized solution to Burgers's equation, which satisfies the Rankine–Hugoniot condition (5.11) at the shock and is smooth away from the shock. After the formation of the shock, the correct generalized solution ceases to conserve $\|\phi\|^2$, so it can no longer be well approximated by the numerical solution obtained using the conservative-form difference. To obtain good numerical approximations to discontinuous solutions to Burgers's equation it is necessary to use the methods discussed in Chap. 5.

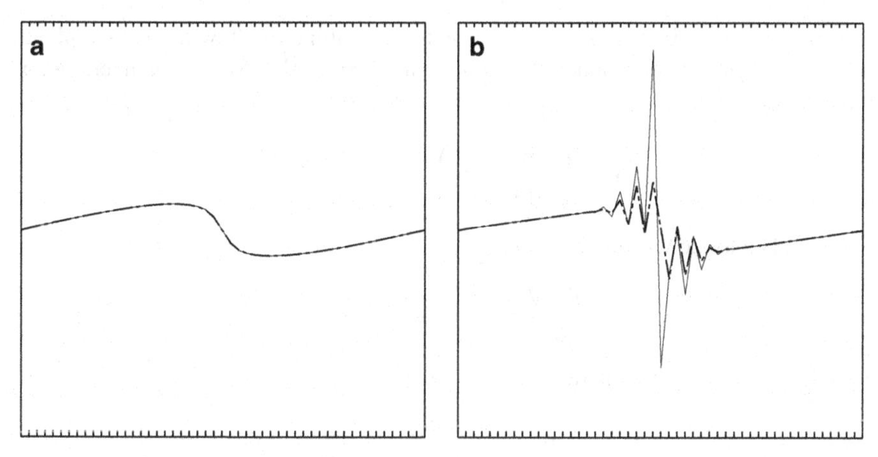

Fig. 4.12 Differential–difference solution to Burgers's equation at **a** $t = 0.13$ and **b** $t = 0.28$ obtained using the advective form ((4.104), *thin solid line*) and the conservative form ((4.109), *dot-dashed line*)

Fornberg (1973), nevertheless, demonstrated that both the advective form and the flux form are unstable (and that the conservative form is stable) when the discretized initial condition has the special form $\ldots, 0, -1, 1, 0, -1, 1, 0, \ldots$.

The instabilities that develop in the preceding solutions to Burgers's equation appear to be associated with the formation of the shock. The development of a shock is not, however, a prerequisite for the onset of nonlinear instability, and such instabilities may occur in numerical simulations of very smooth flow. One example in which nonlinear instability develops in a smooth flow is provided by the viscous Burgers's equation

$$\frac{\partial \psi}{\partial t} + \psi \frac{\partial \psi}{\partial x} = \nu \frac{\partial^2 \psi}{\partial x^2}, \tag{4.110}$$

where ν is a coefficient of viscosity. The true solution to the viscous Burgers's equation never develops a shock, but the advective-form differential–difference approximation to (4.110) becomes unstable for sufficiently small values of ν.

4.5.2 The Barotropic Vorticity Equation

A second example involving the development of nonlinear instability in very smooth flow is provided by the equation governing the vorticity in a two-dimensional incompressible homogeneous fluid,

$$\frac{\partial \zeta}{\partial t} + \mathbf{u} \cdot \nabla \zeta = 0. \tag{4.111}$$

Here **u** is the two-dimensional velocity vector describing the flow in the x–y plane and ζ is the vorticity component along the z-axis. Since the flow is nondivergent, **u** and ζ may be expressed in terms of a stream function ψ such that

$$\mathbf{u} = \left(-\frac{\partial \psi}{\partial y}, \frac{\partial \psi}{\partial x}\right) \qquad \zeta = \nabla^2 \psi,$$

and (4.111) may be written as

$$\frac{\partial \nabla^2 \psi}{\partial t} + J(\psi, \nabla^2 \psi) = 0, \tag{4.112}$$

where J is the Jacobian operator

$$J(p,q) = \frac{\partial p}{\partial x}\frac{\partial q}{\partial y} - \frac{\partial p}{\partial y}\frac{\partial q}{\partial x}.$$

In atmospheric science, (4.112) is known as the barotropic vorticity equation.

Fjørtoft (1953) demonstrated that if the initial conditions are smooth, solutions to (4.112) must remain smooth in the sense that there can be no net transfer of energy from the larger spatial scales into the smaller scales. Fjørtoft's conclusion follows from the properties of the domain integral of the Jacobian operator. Let $\overline{p}$ denote the domain integral of p, and suppose, for simplicity, that the domain is periodic in x and y. Then by the assumed periodicity of the spatial domain[9]

$$\overline{J(p,q)} = \overline{\frac{\partial}{\partial x}\left(p\frac{\partial q}{\partial y}\right)} - \overline{\frac{\partial}{\partial y}\left(p\frac{\partial q}{\partial x}\right)} = 0.$$

As a consequence,

$$\overline{pJ(p,q)} = \overline{J(p^2/2, q)} = 0, \tag{4.113}$$

and

$$\overline{qJ(p,q)} = \overline{J(p, q^2/2)} = 0. \tag{4.114}$$

The preceding relations may be used to demonstrate that the domain-integrated kinetic energy and the domain-integrated enstrophy (half the vorticity squared) are both conserved. First consider the enstrophy, $\zeta^2/2 = (\nabla^2\psi)^2/2$. Multiplying (4.112) by $\nabla^2\psi$ and integrating over the spatial domain yields

$$\frac{\partial}{\partial t}\left(\overline{\frac{(\nabla^2\psi)^2}{2}}\right) + \overline{(\nabla^2\psi) J(\psi, \nabla^2\psi)} = 0,$$

which, using (4.114), reduces to

$$\frac{\partial}{\partial t}\left(\overline{\frac{(\nabla^2\psi)^2}{2}}\right) = 0.$$

[9] Equivalent conservation properties hold in a rectangular domain in which the normal velocity is zero at all points along the boundary.

The conservation of the domain-integrated kinetic energy, $\mathbf{u}\cdot\mathbf{u}/2 = \nabla\psi\cdot\nabla\psi/2$, may be demonstrated by multiplying (4.112) by ψ, applying the vector identity

$$\psi\frac{\partial\nabla^2\psi}{\partial t} = \nabla\cdot\left(\psi\frac{\partial\nabla\psi}{\partial t}\right) - \frac{\partial}{\partial t}\left(\frac{\nabla\psi\cdot\nabla\psi}{2}\right),$$

integrating over the periodic spatial domain, and using (4.113) to obtain

$$\frac{\partial}{\partial t}\left(\overline{\frac{\nabla\psi\cdot\nabla\psi}{2}}\right) = 0.$$

Now suppose that the stream function is expanded in a Fourier series along the x and y coordinates

$$\psi = \sum_k\sum_\ell a_{k,\ell}e^{i(kx+\ell y)} = \sum_{k,\ell}\psi_{k,\ell},$$

and define the total wave number κ such that $\kappa^2 = k^2 + \ell^2$. By the periodicity of the domain and the orthogonality of the Fourier modes,

$$\overline{\mathbf{u}\cdot\mathbf{u}} = \overline{\nabla\psi\cdot\nabla\psi} = \overline{\nabla\cdot(\psi\nabla\psi)} - \overline{\psi\nabla^2\psi} = -\overline{\psi\nabla^2\psi} = \sum_{k,\ell}\kappa^2\overline{\psi_{k,\ell}^2}$$

and

$$\overline{\zeta^2} = \overline{(\nabla^2\psi)^2} = \sum_{k,\ell}\kappa^4\overline{\psi_{k,\ell}^2}.$$

The two preceding relations may be used to evaluate an average wave number, κ_{avg}, given by the square root of the ratio of the domain-integrated enstrophy to the domain-integrated kinetic energy,

$$\kappa_{\text{avg}} = \left(\frac{\overline{\zeta^2}}{\overline{\mathbf{u}\cdot\mathbf{u}}}\right)^{1/2}.$$

Since the domain-integrated enstrophy and the domain-integrated kinetic energy are both conserved, κ_{avg} does not change with time. Any energy transfers that take place from larger to smaller scales must be accompanied by a second energy transfer from smaller to larger scales to conserve κ – there can be no systematic energy cascade into the short-wavelength components of the solution.

Suppose that the barotropic vorticity equation (4.112) is approximated using centered second-order differences in space and time such that

$$\delta_{2t}(\tilde{\nabla}^2\phi) + \tilde{J}(\phi, \tilde{\nabla}^2\phi) = 0,$$

where the numerical approximation to the horizontal Laplacian operator is

$$\tilde{\nabla}^2\phi = (\delta_x^2 + \delta_y^2)\phi$$

and the numerical approximation to the Jacobian operator is

$$\tilde{J}(p,q) = (\delta_{2x}p)(\delta_{2y}q) - (\delta_{2y}p)(\delta_{2x}q).$$

Phillips (1959) showed that solutions obtained using the preceding scheme are subject to an instability in which short-wavelength perturbations suddenly amplify without bound. This instability cannot be controlled by reducing the time step, and it occurs using values of Δt that are well below the threshold required to maintain the stability of equivalent numerical approximations to the linearized constant-coefficient problem. Phillips demonstrated that this instability could be controlled by removing all waves with wavelengths shorter than four grid intervals, thereby eliminating the possibility of aliasing error.

A more elegant method of stabilizing the solution was proposed by Arakawa (1966), who suggested reformulating the numerical approximation to the Jacobian to preserve the discrete analogue of the relations (4.113) and (4.114) and thereby obtain a numerical scheme that conserves both the domain-integrated enstrophy and the domain-integrated kinetic energy. In particular, Arakawa proposed the following approximation to the Jacobian:

$$\begin{aligned}\tilde{J}_a(p,q) = \frac{1}{3}\left[(\delta_{2x}p)(\delta_{2y}q) - (\delta_{2y}p)(\delta_{2x}q)\right] \\ + \frac{1}{3}\left[\delta_{2x}(p\,\delta_{2y}q) - \delta_{2y}(p\,\delta_{2x}q)\right] + \frac{1}{3}\left[\delta_{2y}(q\,\delta_{2x}p) - \delta_{2x}(q\,\delta_{2y}p)\right].\end{aligned}$$

The Arakawa Jacobian satisfies the numerical analogue of (4.113) and (4.114),

$$\sum_{m,n} p_{m,n}\,\tilde{J}_a(p_{m,n},q_{m,n}) = \sum_{m,n} q_{m,n}\,\tilde{J}_a(p_{m,n},q_{m,n}) = 0, \tag{4.115}$$

where the summation is taken over all grid points in the computational domain. As a consequence of (4.115), solutions to

$$\frac{\partial}{\partial t}\left(\tilde{\nabla}^2\phi\right) + \tilde{J}_a(\phi,\tilde{\nabla}^2\phi) = 0 \tag{4.116}$$

conserve their domain-integrated enstrophy and kinetic energy and must therefore also conserve the discretized equivalent of the average wave number κ_{avg}. Since the average wave number is conserved, there can be no net amplification of the short-wavelength components in the numerical solution. The numerical solution is not only stable, it also remains smooth.

Any numerical approximation to the barotropic vorticity equation will be stable if it conserves the domain-integrated kinetic energy, since that is equivalent to the conservation of $\|\mathbf{u}\|_2$. The enstrophy-conservation property of the Arakawa Jacobian does more, however, than guarantee stability; it prevents a systematic cascade of energy into the shortest waves resolvable on the discrete mesh. In designing a numerical approximation to the barotropic vorticity equation it is clearly appropriate to chose a finite-difference scheme like the Arakawa Jacobian that inhibits the downscale cascade of energy. On the other hand, it is not clear that schemes that limit the cascade of energy to small scales are appropriate in those fluid-dynamical applications where there actually is a systematic transfer of kinetic energy from large to small scale. Indeed, any accurate numerical approximation to the equations governing such flows must replicate this downscale energy transfer.

One natural approach to the elimination of nonlinear instability in systems that support a downscale energy cascade is through the parameterization of unresolved turbulent dissipation. In high-Reynolds-number (nearly inviscid) flow, kinetic energy is ultimately transferred to very small scales before being converted to internal energy by viscous dissipation, yet the storage limitations of digital computers do not allow most numerical simulations to be conducted with sufficient spatial resolution to resolve all the small-scale eddies involved in this energy cascade. Under such circumstances the kinetic energy transferred downscale during the numerical simulation will tend to accumulate in the smallest scales resolvable on the numerical mesh, and it is generally necessary to remove this energy by some type of scale-selective dissipation. The scale-selective dissipation constitutes a parameterization of the influence of the unresolved eddies on the resolved-scale flow and should be designed to represent the true behavior of the physical system as closely as possible. Regardless of the exact formulation of the energy-removal scheme, it will tend to stabilize the solution and prevent nonlinear instability.

Many fluid flows contain limited regions of active small-scale turbulence and relatively larger patches of dynamically stable laminar flow. Since eddy diffusion will not be active outside the regions of parameterized turbulence, a scale-selective background dissipation, similar to Phillips's (1959) technique of removing all wavelengths shorter than four grid intervals, is often required to avoid nonlinear instability. This dissipation may be implicitly included in the time differencing or in an upwind-biased spatial difference, or it may be explicitly added to an otherwise non-damping method using formulae such as those discussed in Sect. 3.3.3. Although it is not required for stability, a small amount of background dissipation may also be incorporated in numerical approximations to linear partial differential equations to damp those short-wavelength components of the numerical solution whose phase speed and group velocity are most seriously in error.

Problems

1. Verify that the leapfrog time-differenced shallow-water equations (4.14) and (4.15) support a computational mode and that the forward–backward-differenced system (4.17) and (4.18) does not, by solving their respective discrete-dispersion relations for ω.

2. Eliminate h from the finite-difference equations for the leapfrog staggered scheme (4.14) and (4.15) and compare the resulting higher-order finite-difference approximation to the second-order partial differential equation

$$\frac{\partial^2 u}{\partial t^2} - c^2 \frac{\partial^2 u}{\partial x^2} = 0$$

with the expression that arises when h is eliminated from the forward–backward approximation on the staggered mesh (4.17) and (4.18). What does this comparison suggest about the number of computational modes admitted by each numerical approximation?

3. Suppose that numerical solutions to the two-dimensional Boussinesq system (4.39)–(4.42) are obtained using the staggered grid shown in Fig. 4.6 except that the distribution of the variables is modified so that b is colocated with the w rather than the P points.

(a) Write down appropriate modifications for the discretized vertical momentum and buoyancy equations (4.44) and (4.45), and derive the discrete dispersion relation for this system.

(b) Assume that all resolved modes are quasi-hydrostatic, so that $\tilde{k}_1^2$ can be neglected with respect to $\tilde{\ell}^2$ in the denominator of (4.47) and in the result derived in (a). Also suppose the mean wind is zero. Compare the horizontal and vertical group velocities for the numerical solutions on each staggered grid with the exact expression from the nondiscretized quasi-hydrostatic system.

4. Derive the amplification factor and stability condition given in the text for the CTU method (4.36). Show that including the cross-derivative term in the CTU method always decreases the amplification factor relative to that obtained with the standard two-dimensional upstream scheme (4.31).

5. Show that, at least for some combinations of U and V, the false two-dimensional Lax–Wendroff scheme (4.37) is unstable for all Δt.

6. Determine the range of Δt (if any) over which the backward, forward, and leapfrog schemes give a stable approximation to

$$\frac{d\psi}{dt} = r\psi.$$

Consider the cases $r > 0$ and $r < 0$. The true solution to this equation preserves the sign of $\psi(t = 0)$. What, if any, additional restrictions must be placed on Δt to ensure that the numerical solution for each method is both stable and sign preserving.

7. Suppose the Lax–Wendroff method is used to obtain an $O\left[(\Delta t)^2\right]$-accurate approximation to the advection–diffusion equation (3.87). Show that before discretizing the spatial derivatives, the scheme has the form

$$\frac{\phi^{n+1} - \phi^n}{\Delta t} + c\frac{\partial \phi^n}{\partial x} - M\frac{\partial^2 \phi^n}{\partial x^2} = \frac{\Delta t}{2}\left(c^2\frac{\partial^2 \phi^n}{\partial x^2} - 2cM\frac{\partial^3 \phi^n}{\partial x^3} + M^2\frac{\partial^4 \phi^n}{\partial x^4}\right).$$

Comment on the probable utility of this approach.

8. Derive expressions for the boundaries for the regions of useful stability for the leapfrog-backward scheme (3.93) and the leapfrog–trapezoidal method (3.92) shown in Fig. 3.13a.

9. The following approximation to the advection–diffusion equation (3.87) is unstable:

$$\delta_{2t}\phi_j^n + c\delta_{2x}\phi_j^n = M\delta_x^2\phi_j^n.$$

Modify the right side of the above equation to stabilize the method, at least for sufficiently small Δt, but *do not make the scheme implicit.* Prove that your modified scheme is indeed stable for sufficiently small values of Δt. It is not necessary to work out the exact range of Δt over which the scheme is stable.

10. The analysis of the frozen-coefficient problem does not always correctly indicate the behavior of solutions to partial differential equations with variable coefficients. Consider the initial-value problem

$$\frac{\partial \psi}{\partial t} - \mathrm{i}\frac{\partial}{\partial x}\left(\sin x \frac{\partial \psi}{\partial x}\right) = 0, \tag{4.117}$$

$\psi(x,0) = f(x)$ on the interval $-\infty < x < \infty$.

(a) Show that the L_2 norm of the solution to this problem does not grow with time.

(b) Freeze the coefficients at $x = 0$ and show that the resulting problem is ill posed because its solution does not depend continuously on the initial data. (*Hint:* Consider

$$\psi_1(x,0) = \mathrm{e}^{ik_1 x} \quad \text{and} \quad \psi_2(x,0) = \mathrm{e}^{ik_2 x}$$

and show that $\|\psi_1(x,0) - \psi_2(x,0)\|$ is bounded, whereas $\|\psi_1(x,t) - \psi_2(x,t)\|$ can be arbitrarily large for any finite t.)
Since stable numerical solutions cannot be obtained for ill-posed problems, the stability of a numerical approximation to (4.117) cannot be determined by examining the stability of the family of all frozen-coefficient problems.

11. Suppose that (4.97) and (4.98) are applied to model tracer advection in a closed rectangular domain with no velocity normal to the boundaries and that the boundaries are located at the edges (as opposed to the centers) of the outermost grid cells. Let the differential–difference equations generated by each scheme be expressed as a linear system of the form (4.91). Write down the coefficient matrix **A** for each scheme, and show that the matrix associated with (4.97) is skew symmetric, whereas that associated with (4.98) is not.

12. *The linearized one-dimensional Rossby adjustment problem for an atmosphere with no mean wind is governed by the equations

$$\frac{\partial u}{\partial t} - fv + g\frac{\partial h}{\partial x} = 0, \qquad \frac{\partial v}{\partial t} + fu = 0, \qquad \frac{\partial h}{\partial t} + H\frac{\partial u}{\partial x} = 0.$$

Compare the approximate solution to these equations obtained using leapfrog differencing on an unstaggered mesh

$$\begin{aligned} \delta_{2t}u - fv + g\delta_{2x}h &= 0,\\ \delta_{2t}v + fu &= 0,\\ \delta_{2t}h + H\delta_{2x}u &= 0 \end{aligned}$$

with those obtained using forward–backward time differencing on the staggered mesh shown in Fig. 4.1:

$$\begin{aligned} \delta_t u_j^{n+\frac{1}{2}} - fv_j^n + g\delta_x h_j^n &= 0,\\ \delta_t v_j^{n+\frac{1}{2}} + fu_j^{n+1} &= 0,\\ \delta_t h_{j+\frac{1}{2}}^{n+\frac{1}{2}} + H\delta_x u_{j+\frac{1}{2}}^{n+1} &= 0. \end{aligned}$$

Assume that v, which is not shown in Fig. 4.1, is defined at the same points as u. Let the spatial domain be periodic on the interval $0 \le x \le 2{,}000$ km, but show your solutions only in the domain $600 \le x \le 14{,}000$ km. Let $f = 10^{-4}\,\mathrm{s}^{-1}$ and $c = \sqrt{gH} = 10\,\mathrm{ms}^{-1}$. For initial conditions choose $u(x,0) = v(x,0) = 0$, and let the height field be given by a slightly smoothed unit-amplitude square wave with nodes at $x = 0$ and $x = 1{,}000$ km. Obtain this smoothed square wave by three iterative applications of the filter

$$\phi_j^f = \frac{1}{4}(\phi_{j+1} + 2\phi_j + \phi_{j-1})$$

to a pure square wave. Let $\Delta x = 3\frac{1}{3}$ km.

(a) Show solutions for all three fields at the time step closest to $t = 21{,}000$ s. Use Courant numbers $(c\Delta t/\Delta x)$ of 0.9 and 0.1. Discuss the quality of the two solutions. Explain the source of the difference between the two solutions.

(b) Eliminate the smoothing step from the initialization and discuss the impact on the solution.

(Note that analytic solutions to this problem are given in Gill 1982, Sects. 7.2, 7.3.)

13. *Compute numerical solutions to the variable-wind-speed advection equation (4.70) in a periodic domain $0 \le x \le 2$. Choose

$$c(x) = \begin{cases} 0.3 - 1.5(x - \frac{1}{3})\sin(3\pi x)\sin(12\pi x), & \text{if } \frac{1}{3} \le x \le \frac{2}{3},\\ 0.3, & \text{otherwise,} \end{cases}$$

and use the initial condition

$$\psi(x,0) = \begin{cases} \frac{1}{4}(\cos(8\pi(x - \frac{1}{8})) + 1)^2 & \text{if } |x - \frac{1}{8}| \le \frac{1}{8};\\ 0, & \text{otherwise}. \end{cases}$$

(a) Given that

$$\int_{1/3}^{2/3} \frac{\mathrm{d}x}{c(x)} = 1.391,$$

find the correct x-location of the peak of the initial distribution at time $t = 3$. Describe the shape and location of the true solution at $t = 3$.

(b) Compare numerical solutions obtained using the second-order approximations

$$\delta_{2t}\phi + c\delta_{2x}\phi = 0 \tag{4.118}$$

and

$$\delta_{2t}\phi + \left\langle \langle c\rangle^x \delta_x \phi\right\rangle^x = 0$$

with the fourth-order space schemes

$$\delta_{2t}\phi + c\left[\frac{4}{3}\delta_{2x}\phi - \frac{1}{3}\delta_{4x}\phi\right] = 0$$

and

$$\delta_{2t}\phi + \frac{4}{3}\left\langle \langle c\rangle^x \delta_x \phi\right\rangle^x - \frac{1}{3}\left\langle \langle c\rangle^{2x} \delta_{2x} \phi\right\rangle^{2x} = 0.$$

Assume that c and ϕ are located at the same points. Use $\Delta x = 1/32$ and a Courant number of 0.6 based on the maximum wind speed. Take a single forward step to initialize the leapfrog integration. Plot the left half of the domain and show the solutions at $t = 0$, $t = 1.5$, and $t = 3$. Also plot the wind speed. Compare the numerical solutions with the exact solution determined in (a) at time $t = 3$.

(c) Retry the previous simulations using $\Delta x = 1/64$ and discuss the degree of improvement.

(d) Now try adding a fourth-order spatial filter to each scheme in the $\Delta x = 1/32$ case. Lag the filter in time. For example, (4.118) becomes

$$\delta_{2t}\phi_j^n + c\delta_{2x}\phi_j^n = -\gamma\left(\phi_{j+2}^{n-1} - 4\phi_{j+1}^{n-1} + 6\phi_j^{n-1} - 4\phi_{j-1}^{n-1} + \phi_{j-2}^{n-1}\right).$$

Discuss the dependence of the solution on the parameter $\gamma\Delta t$. As a start, set $\gamma\Delta t = 0.01$.

Chapter 5
Conservation Laws and Finite-Volume Methods

As demonstrated in the preceding chapters, the errors in most numerical solutions increase dramatically as the physical scale of the simulated disturbance approaches the minimum scale resolvable on the numerical mesh. When solving equations for which smooth initial data guarantee a smooth solution at all later times, such as the barotropic vorticity equation (4.112), one can avoid any difficulties associated with poor numerical resolution by using a sufficiently fine computational mesh. But if the governing equations allow an initially smooth field to develop shocks or discontinuities, as is the case with Burgers's equation (4.102), there is no hope of maintaining adequate numerical resolution throughout the simulation, and special numerical techniques must be used to control the development of overshoots and undershoots in the vicinity of the shock. Numerical approximations to equations with discontinuous solutions must also satisfy additional conditions beyond the stability and consistency requirements discussed in Chap. 3 to guarantee that the numerical solution converges to the correct solution as the spatial grid interval and the time step approach zero.

The possibility of erroneous convergence to a function that does not approximate the true discontinuous solution can be demonstrated by comparing numerical solutions to the generalized Burgers equation in *advective form*

$$\frac{\partial \psi}{\partial t} + \psi^2 \frac{\partial \psi}{\partial x} = 0 \tag{5.1}$$

with those generated by analogous solutions to the same equation in *flux form*

$$\frac{\partial \psi}{\partial t} + \frac{\partial}{\partial x}\left(\frac{\psi^3}{3}\right) = 0. \tag{5.2}$$

As will be explained in Sect. 5.1, if the initial conditions are specified by the step function

$$\psi(x,0) = \begin{cases} 1, & \text{if } x \leq 0, \\ 0 & \text{otherwise,} \end{cases} \tag{5.3}$$

D.R. Durran, *Numerical Methods for Fluid Dynamics: With Applications to Geophysics*, Texts in Applied Mathematics 32, DOI 10.1007/978-1-4419-6412-0_5,

the correct solution consists of a unit-amplitude step propagating to the right at speed $1/3$. An upstream finite-difference approximation to the advective form (5.1) was calculated using

$$\frac{\phi_j^{n+1} - \phi_j^n}{\Delta t} + \left(\frac{\phi_j^n + \phi_{j-1}^n}{2}\right)^2 \left(\frac{\phi_j^n - \phi_{j-1}^n}{\Delta x}\right) = 0, \tag{5.4}$$

and an upstream approximation to the flux form (5.2) was obtained using

$$\frac{\phi_j^{n+1} - \phi_j^n}{\Delta t} + \frac{(\phi_j^n)^3 - (\phi_{j-1}^n)^3}{3\Delta x} = 0. \tag{5.5}$$

Figure 5.1a shows a comparison of the exact and the numerical solutions at $t = 2.4$ on the subdomain $0.5 \leq x \leq 1.0$. The computations were performed using $\Delta x = 0.02$ and a time step such that $\max[\psi(x,0)]\Delta t/\Delta x = 0.5$. Both schemes yield plausible-looking approximations to the correct solution (shown by the dot-dashed line), but the numerical solution obtained using advective-form differencing (shown by the solid line) moves at the wrong speed. As illustrated in Fig. 5.1b, in which the numerical solutions are recalculated after reducing Δx and Δt by a factor of 4, the speed of the solution generated by the advective-form approximation is not significantly improved by decreasing Δx and Δt. The advective-form approximation simply does not converge to the correct solution in the limit $\Delta x \to 0$ and $\Delta t \to 0$. The difficulties that can be associated with advective-form finite differencing are even more apparent if (5.1) is approximated using the scheme

$$\frac{\phi_j^{n+1} - \phi_j^n}{\Delta t} + (\phi_j^n)^2 \frac{\phi_j^n - \phi_{j-1}^n}{\Delta x} = 0$$

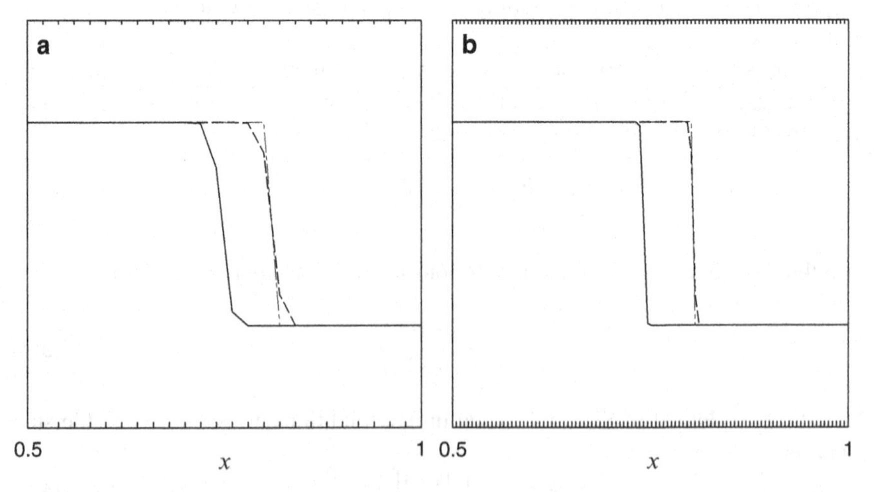

Fig. 5.1 Exact (*dot-dashed line*), upstream advective-form (*solid line*) and upstream flux-form (*dashed line*) solutions to the generalized Burgers equation at $t = 2.4$ on the subdomain $0.5 \leq x \leq 1$: **a** $\Delta x = 0.02$, $\Delta t = 0.01$; **b** $\Delta x = 0.005$, $\Delta t = 0.0025$

and the initial data

$$\phi_j^0 = \begin{cases} 1, & \text{if } j \le j_0, \\ 0, & \text{otherwise.} \end{cases}$$

In this case, the finite-difference approximation to $\psi^2 \partial\psi/\partial x$ is zero at every grid point, and the numerical solution is stationary. To understand how advective-form finite-difference approximations can converge to invalid solutions to the generalized Burgers equation, it is helpful to review the sense in which discontinuous functions constitute solutions to partial differential equations.

5.1 Conservation Laws and Weak Solutions

Many of the partial differential equations arising in fluid dynamics can be expressed as a system of *conservation laws* of the form

$$\frac{\partial \mathbf{u}}{\partial t} + \sum_j \frac{\partial}{\partial x_j} \mathbf{f}_j(\mathbf{u}) = \mathbf{0}, \tag{5.6}$$

which states that the rate of change of $\mathbf{u}$ at each point is determined by the convergence of the fluxes $\mathbf{f}_j$ at that point. An example of this type is provided by the one-dimensional shallow-water equations. Let u denote the velocity and h the fluid depth, and suppose that there is no bottom topography; then conservation of mass requires

$$\frac{\partial h}{\partial t} + \frac{\partial}{\partial x}(hu) = 0,$$

and conservation of momentum implies

$$\frac{\partial}{\partial t}(hu) + \frac{\partial}{\partial x}\left(hu^2 + g\frac{h^2}{2}\right) = 0.$$

If a function contains a discontinuity, it cannot be the solution to a partial differential equation in the conventional sense, because derivatives are not defined at the discontinuity. Instead, the solution is required to satisfy a family of related integral equations. Consider solutions to the scalar conservation law

$$\frac{\partial \psi}{\partial t} + \frac{\partial}{\partial x} f(\psi) = 0 \tag{5.7}$$

on the unbounded domain $-\infty < x < \infty$. Integrating this conservation law over the intervals $[x_1, x_2]$ and $[t_1, t_2]$, one obtains

$$\int_{x_1}^{x_2} \psi(x, t_2)\, \mathrm{d}x = \int_{x_1}^{x_2} \psi(x, t_1)\, \mathrm{d}x + \int_{t_1}^{t_2} f(\psi(x_1, t))\, \mathrm{d}t - \int_{t_1}^{t_2} f(\psi(x_2, t))\, \mathrm{d}t, \tag{5.8}$$

which states that the total change in ψ over the region $x_1 \le x \le x_2$ is determined by the time-integrated fluxes through the boundary of that region. This integral form of the conservation law can usually be derived from first physical principles as easily as the differential form (5.7), and unlike the differential form, the integral form can be satisfied by piecewise-continuous functions. If ψ satisfies the integral equation (5.8) on every subdomain $[x_1, x_2] \times [t_1, t_2]$, then ψ is a *weak solution* of the conservation law. Differentiable weak solutions are also solutions to the partial differential equation (5.7) and are uniquely determined by the initial data. Nondifferentiable weak solutions may, however, be nonunique.

5.1.1 The Riemann Problem

Weak solutions to the conservation law (5.7) are particularly easy to obtain when the initial data are constant on each side of a single discontinuity. This combination of a scalar conservation law and piecewise-constant initial data containing a single discontinuity is known as a *Riemann problem.* Riemann problems have solutions in which the initial discontinuity propagates at a constant speed s, as indicated schematically in Fig. 5.2. Assuming for notational convenience that at $t = 0$ the discontinuity is at $x = 0$, this solution has the form

$$\psi(x,t) = \begin{cases} \psi_{\mathrm{L}} & \text{if } x - st < 0, \\ \psi_{\mathrm{R}} & \text{otherwise,} \end{cases} \tag{5.9}$$

where $\psi_{\mathrm{L}} = \psi(x_{\mathrm{L}})$, $\psi_{\mathrm{R}} = \psi(x_{\mathrm{R}})$, and it has been assumed that x_{L} and x_{R} are located sufficiently far upstream and downstream that the jump does not propagate past these points during the time interval of interest. The speed of the shock may be determined as follows. From (5.9),

$$\int_{x_{\mathrm{L}}}^{x_{\mathrm{R}}} \psi(x,t)\,\mathrm{d}x = (st - x_{\mathrm{L}})\psi_{\mathrm{L}} + (x_{\mathrm{R}} - st)\psi_{\mathrm{R}},$$

and thus

$$\frac{d}{dt}\int_{x_{\mathrm{L}}}^{x_{\mathrm{R}}} \psi(x,t)\,\mathrm{d}x = s(\psi_{\mathrm{L}} - \psi_{\mathrm{R}}). \tag{5.10}$$

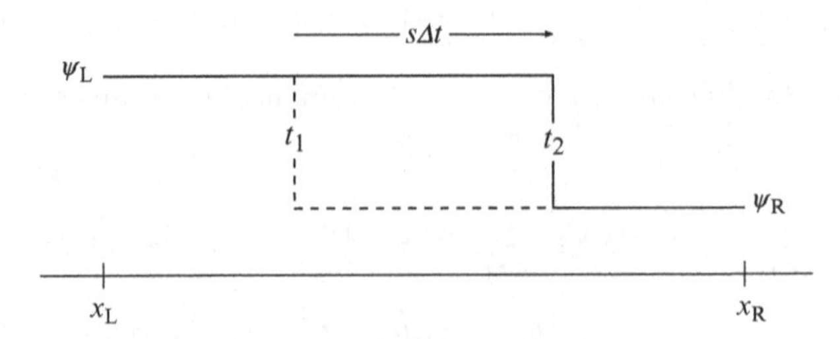

Fig. 5.2 The displacement of a jump propagating to the right at speed s over time $\Delta t = t_2 - t_1$

Integrating (5.7) over the interval $[x_\mathrm{L}, x_\mathrm{R}]$, one obtains

$$\frac{d}{dt}\int_{x_\mathrm{L}}^{x_\mathrm{R}} \psi(x,t)\,\mathrm{d}x = f(\psi_\mathrm{L}) - f(\psi_\mathrm{R}),$$

which together with (5.10) implies that

$$s = \frac{f(\psi_\mathrm{L}) - f(\psi_\mathrm{R})}{\psi_\mathrm{L} - \psi_\mathrm{R}}. \tag{5.11}$$

This equation for the speed of the jump is known as the *Rankine–Hugoniot condition*. Note that the Rankine–Hugoniot condition requires the jump in the weak solutions plotted in Fig. 5.1 to propagate at a speed of $1/3$. The Rankine–Hugoniot condition is frequently derived from first principles in various physical applications. For example, Stoker (1957, Sects. 10.6.6, 10.7.7) derives the Rankine–Hugoniot condition for the one-dimensional shallow-water system by constructing mass and momentum budgets for a control volume containing the shock.

As previously mentioned, nondifferentiable weak solutions need not be uniquely determined by the initial data, and if more than one weak solution exists, it is necessary to select the physically relevant solution. When the solutions to equations representing real physical systems develop discontinuities, one of the physical assumptions used to derive those equations is often violated. Solutions to the inviscid Euler equations may suggest that discontinuities develop in supersonic flow around an airfoil, but the velocity and thermodynamic fields around an airfoil never actually become discontinuous. The discontinuities predicted by the Euler equations actually appear as narrow regions of steep gradients that are stabilized against further scale collapse by viscous dissipation and diffusion. The discontinuous inviscid solution may be considered to be the limit of a series of viscous solutions in which the viscosity is progressively reduced to zero. Thus, one strategy for selecting the physically significant weak solution would be to conduct a series of viscous simulations with progressively smaller viscosities and choose the weak solution toward which the viscous solutions converge. This, of course, is a highly inefficient strategy, and it may be impossible to implement in actual simulations of high-Reynolds-number flow, where any realistic value for the molecular viscosity may be too low to significantly influence the solution on the spatial scales resolvable on the numerical grid. In addition, any attempt to include realistic viscosities in the numerical solution reintroduces precisely those mathematical complications that were eliminated when the full physical system was originally approximated by the simpler inviscid model.

5.1.2 Entropy-Consistent Solutions

It is therefore preferable to obtain alternative criteria for selecting the physically relevant weak solution. These criteria may be derived directly from physical principles. Stoker (1957) eliminated nonphysical shocks in shallow-water flow by requiring that "the water particles do not gain energy upon crossing a shock front." In gas dynamics, thermodynamic principles require that entropy be nondecreasing at the shock. Generalized entropy conditions can also be derived for any system of one or two scalar conservation laws of the form (5.7) by considering the limiting behavior of the corresponding viscous system as the viscosity approaches zero (Lax 1971).

For example, a generalized entropy function for the inviscid Burgers equation

$$\frac{\partial \psi}{\partial t} + \frac{\partial}{\partial x}\left(\frac{\psi^2}{2}\right) = 0 \tag{5.12}$$

is ψ^2. When ψ is a weak solution to Burgers's equation, ψ^2 is a weak solution to the inequality

$$\frac{\partial \psi^2}{\partial t} + \frac{\partial}{\partial x}\left(\frac{2\psi^3}{3}\right) \leq 0. \tag{5.13}$$

If the solutions to Burgers's equation are differentiable, the left side of (5.13) is identically zero and the time rate of change of the integral of ψ^2 over any spatial domain is equal to the divergence of the entropy flux, $2\psi^3/3$, through the edges of the domain. But if the solution of Burgers's equation is discontinuous, (5.13) can no longer be satisfied by an equality. The sense of the inequality demanded by (5.13) is that which matches the limiting behavior of ψ^2 for solutions to the viscous Burgers equation

$$\frac{\partial \psi}{\partial t} + \frac{\partial}{\partial x}\left(\frac{\psi^2}{2}\right) = \epsilon \frac{\partial^2 \psi}{\partial x^2}$$

as $\epsilon \to 0$ (LeVeque 1992, p. 37).

Consider two possible weak solutions to the inviscid Burgers equation (5.12), both of which are consistent with the initial condition

$$\psi(x,0) = \begin{cases} 0, & \text{if } x \leq 0, \\ 1, & \text{otherwise.} \end{cases}$$

The first solution, shown in Fig. 5.3a, consists of a unit-amplitude downward[1] jump moving to the right at the speed given by the Rankine–Hugoniot condition, which is a speed of $1/2$. The second solution, shown in Fig. 5.3b, is the *rarefaction wave*, or expansion fan, given by

$$\psi(x,t) = \begin{cases} 0, & \text{if } x \leq 0, \\ x/t, & \text{if } 0 < x < t, \\ 1, & \text{otherwise.} \end{cases}$$

Note that the central point in the rarefaction wave moves at the same speed as the shock. The validity of the rarefaction-wave solution in the interval $0 < x < t$ can be confirmed by substituting $\psi = x/t$ into (5.12). The validity of the solution in

[1] The jump is "downward" in the sense that the fluid level drops during the passage of the discontinuity.

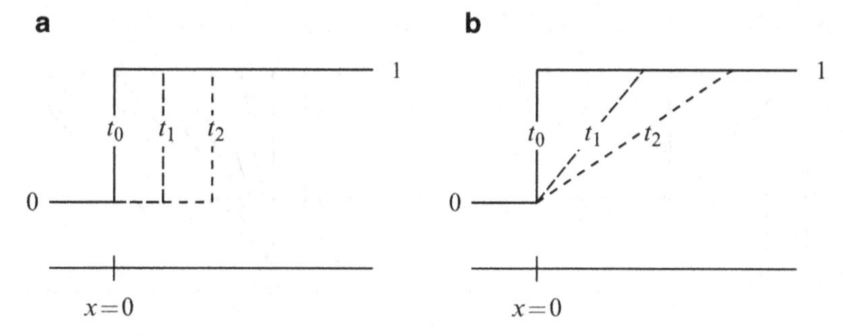

Fig. 5.3 **a** An entropy-violating shock, and **b** the entropy-consistent rarefaction wave

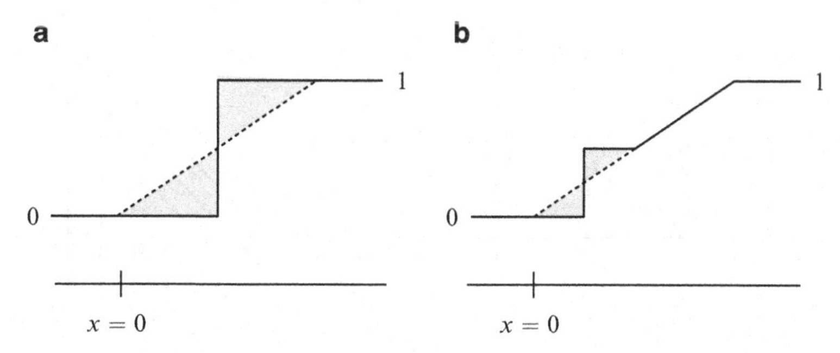

Fig. 5.4 Spatial distribution of ψ in a rarefaction wave compared with that in **a** an entropy-violating shock, and **b** the combination of a small entropy-violating shock and a rarefaction wave

any larger domain follows from the fact that the shock is a weak solution, since it moves at the speed determined by the Rankine–Hugoniot condition, and as indicated in Fig. 5.4a, the rate of change of $\int \psi \mathrm{d}x$ over any domain including the interval $0 \leq x \leq t$ is the same for the shock and the rarefaction wave. An infinite number of other weak solutions also exist, such as the small shock following a rarefaction wave shown in Fig. 5.4b.

Characteristic curves for the shock and rarefaction-wave solutions are plotted in Fig. 5.5. Those characteristics that intersect the trajectory of the shock are directed away from the shock, i.e., they originate at some point along the trajectory of the shock and do not continue back to the line $t = 0$ along which the initial data are specified. As a consequence, the initial data do not determine the value of $\psi(x,t)$ throughout the entire $t > 0$ half-plane, which is clearly a nonphysical situation. In contrast, all the characteristics associated with the rarefaction wave originate from the line $t = 0$, and the solution is everywhere determined by the initial data. The rarefaction wave is therefore the physically relevant weak solution.

If the initial data in the preceding example are reflected about the point $x = 0$, the unique and physically relevant solution consists of a unit-amplitude *upward* jump propagating to the right at speed $1/2$. This solution is shown in Fig. 5.6, together with a representative set of its characteristic curves. In this case, the characteristic curves are directed into the jump, so the initial data do determine the solution.

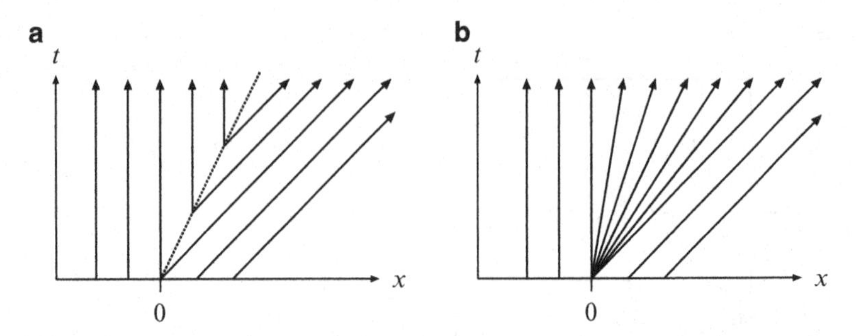

Fig. 5.5 Characteristic curves for **a** an entropy-violating shock, and **b** the entropy-consistent rarefaction wave. The trajectory of the shock is indicated by the *heavy dashed line*

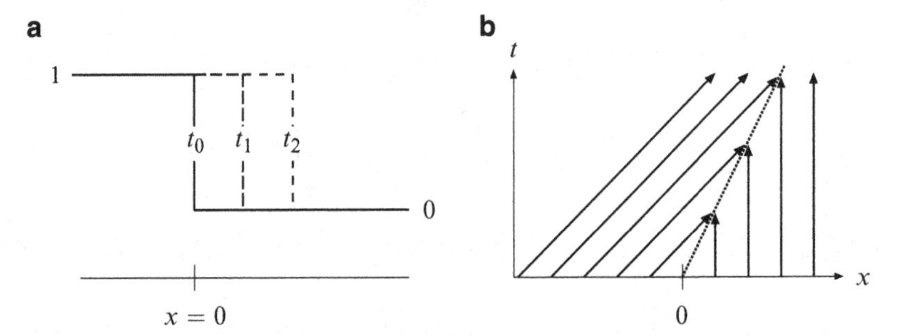

Fig. 5.6 **a** An entropy-consistent shock, and **b** characteristic curves associated with that shock. The trajectory of the jump is indicated by the *heavy dashed line*

Indeed, the intersecting characteristics indicate the need for a discontinuity in the solution, because otherwise the solution would have to be double-valued at the point where two different characteristics meet.

These results are consistent with the entropy condition (5.13), which may be evaluated for jump solutions to Burgers's equation as follows. As in Fig. 5.2, let $\psi_L = \psi(x_L)$, $\psi_R = \psi(x_R)$, and assume that x_L and x_R are located sufficiently far upstream and downstream that the jump does not pass these points during the time interval $[t_1, t_2]$. Following the same derivation that led to (5.10),

$$\frac{d}{dt}\int_{x_L}^{x_R} \psi^2(x,t)\,dx = s(\psi_L^2 - \psi_R^2), \tag{5.14}$$

where s is the speed of the jump. Integrating the left side of (5.13) over the domain $[x_L, x_R] \times [t_1, t_2]$, one obtains

$$\int_{t_1}^{t_2}\int_{x_L}^{x_R}\left[\frac{\partial \psi^2}{\partial t} + \frac{\partial}{\partial x}\left(\frac{2\psi^3}{3}\right)\right] dx\,dt =$$

$$\int_{x_L}^{x_R} \psi^2(x,t_2)\,dx - \int_{x_L}^{x_R} \psi^2(x,t_1)\,dx + \int_{t_1}^{t_2} \frac{2}{3}\left(\psi_R^3 - \psi_L^3\right) dt.$$

Defining $\Delta t = t_2 - t_1$, expanding the first term on the right side in a Taylor series about time t_1 using (5.14), and substituting $s = (\psi_L + \psi_R)/2$,

$$\int_{t_1}^{t_2}\int_{x_L}^{x_R}\left[\frac{\partial\psi^2}{\partial t} + \frac{\partial}{\partial x}\left(\frac{2\psi^3}{3}\right)\right] \mathrm{d}x\,\mathrm{d}t$$

$$= s\Delta t(\psi_L^2 - \psi_R^2) + O\left[(\Delta t)^2\right] + \frac{2}{3}\Delta t(\psi_R^3 - \psi_L^3)$$

$$= \frac{\Delta t}{6}(\psi_R - \psi_L)^3 + O\left[(\Delta t)^2\right]. \tag{5.15}$$

Taking the limit $\Delta t \to 0$, it follows that the only jumps that can satisfy the entropy condition (5.13) are those for which ψ_L exceeds ψ_R.

As suggested by the preceding examples, one criterion for determining the entropy-consistent solution is to demand that all characteristic curves intersecting the shock be directed in toward the trajectory of the shock. For scalar conservation laws, this condition is satisfied, provided that for all ψ between ψ_L and ψ_R

$$\frac{f(\psi_L) - f(\psi)}{\psi_L - \psi} \geq s \geq \frac{f(\psi) - f(\psi_R)}{\psi - \psi_R} \tag{5.16}$$

(Oleinik 1957). Since s is given by the Rankine–Hugoniot condition (5.11), this inequality reduces to the simple condition that $\psi_L > \psi_R$ when $f(\psi)$ is convex, i.e., when the chord connecting any two points $(\psi_1, f(\psi_1))$ and $(\psi_2, f(\psi_2))$ lies entirely above the graph of f. Because the flux appearing in Burgers's equation is convex, the entropy-consistent shock shown in Fig. 5.6 can be distinguished from the entropy-violating shock shown in Fig. 5.3 by the criterion $\psi_L > \psi_R$, which agrees with (5.15).

5.2 Finite-Volume Methods and Convergence

There are two special difficulties that can arise in attempting to compute discontinuous solutions to partial differential equations. First, as suggested by the spurious solution generated by the advective-form upstream finite-difference approximation (5.1), the numerical scheme might converge to a function that is not a weak solution of the conservation law. Second, the numerical method may converge to a genuine weak solution, but it may fail to converge to the physically relevant entropy-consistent solution.

The possibility of numerical solutions converging to a function that is not a weak solution to the governing equation can be avoided by using a method that can be expressed in *conservation form*.[2] An approximation to the scalar conservation

[2] The scheme need only be algebraically equivalent to a method of the form (5.17). For example the advective-form approximation (4.97) is algebraically equivalent to the conservation form (4.90) when the velocity field satisfies (4.95).

law (5.7) is in conservation form if

$$\frac{\phi_j^{n+1} - \phi_j^n}{\Delta t} + \left(\frac{F_{j+\frac{1}{2}} - F_{j-\frac{1}{2}}}{\Delta x} \right) = 0, \tag{5.17}$$

where $F_{j\pm 1/2}$ are numerical approximations to the fluxes $f[\psi((j \pm 1/2)\Delta x)]$ such that

$$F_{j+\frac{1}{2}} = F(\phi_{j-p}^n, \phi_{j-p+1}^n, \ldots, \phi_{j+q+1}^n),$$
$$F_{j-\frac{1}{2}} = F(\phi_{j-p-1}^n, \phi_{j-p}^n, \ldots, \phi_{j+q}^n),$$

and p and q are integers.

If the true solution and the fluxes are differentiable, ϕ_j can approximate $\psi(x_j, t^n)$ and each term in (5.17) may be interpreted as a finite difference. The construction of high-order finite-difference approximations to the flux divergence in (5.17) is considered in Sect. 5.7.1. If, on the other hand, the true solution is not differentiable, numerical approximations to weak solutions of the conservation law may be obtained using *finite-volume methods*. In the finite-volume approach, ϕ_j approximates the spatial average of ψ over grid cell j,

$$\phi_j^n \approx \frac{1}{\Delta x} \int_{x_j - \Delta x/2}^{x_j + \Delta x/2} \psi(x, t^n)\, \mathrm{d}x,$$

and $F_{j+1/2}$ approximates the time-averaged flux through the interface between grid cells j and $j+1$,

$$F_{j+\frac{1}{2}} \approx \frac{1}{\Delta t} \int_{t^n}^{t^{n+1}} f(\psi(x_j + \Delta x/2, t))\, \mathrm{d}t\,.$$

Averaging the exact conservation law (5.7) over the jth grid cell and integrating over a single time step gives

$$\frac{1}{\Delta x} \int_{x_j - \Delta x/2}^{x_j + \Delta x/2} \psi(x, t^{n+1})\, \mathrm{d}x = \frac{1}{\Delta x} \int_{x_j - \Delta x/2}^{x_j + \Delta x/2} \psi(x, t^n)\, \mathrm{d}x$$
$$- \frac{1}{\Delta x} \left(\int_{t^n}^{t^{n+1}} f(\psi(x_j + \Delta x/2, t))\, \mathrm{d}t - \int_{t^n}^{t^{n+1}} f(\psi(x_j - \Delta x/2, t))\, \mathrm{d}t \right).$$

Finite-volume methods approximate the preceding expression as

$$\phi_j^{n+1} = \phi_j^n - \frac{\Delta t}{\Delta x} \left(F_{j+\frac{1}{2}} - F_{j+\frac{1}{2}} \right).$$

Solutions to finite-volume methods automatically satisfy

$$\sum_{j=j_1}^{j_2} \phi_j^{n+1} \Delta x = \sum_{j=j_1}^{j_2} \phi_j^n \Delta x + \Delta t F_{j_1 - \frac{1}{2}} - \Delta t F_{j_2 + \frac{1}{2}}, \tag{5.18}$$

which is a discrete approximation to an arbitrary member of the family of integral equations (5.8) satisfied by any weak solution to the exact conservation law.

Thus, if a set of finite-volume solutions converges to some function as Δx, $\Delta t \to 0$, all that is required to ensure that function is a weak solution to (5.7) is a consistency condition. Suppose that the numerical fluxes are smooth functions of the grid-point values (at a minimum, F must be Lipschitz continuous[3]) and that these fluxes are consistent with the conservation law (5.7) in the sense that

$$F(\psi_0, \psi_0, \ldots, \psi_0) = f(\psi_0),$$

i.e., that the numerical fluxes generated by a spatially and temporally uniform ψ_0 are identical to the true flux generated by the same constant value of ψ_0. Then, if the numerical solutions converge to some function as Δx, $\Delta t \to 0$, that function must be a weak solution of (5.7) (Lax and Wendroff 1960; LeVeque 2002). Note that the results presented in Fig. 5.1 are consistent with this theorem because (5.5) is in conservation form with $F_{j+1/2} = \phi_j^3/3$ but (5.4) is not algebraically equivalent to any scheme in conservation form.

Although finite-volume methods are particularly appropriate for those problems with discontinuous solutions, they can also generate excellent approximations to conservation laws with smooth solutions. One important advantage of the finite-volume approach is that it provides a general method for creating two-time-level approximations algebraically equivalent to (5.17) in which the time stepping does *not* rely on Euler time differencing. Two-time-level finite-volume methods are explicitly constructed using this approach in Sect. 5.6.

5.2.1 Monotone Schemes

There is no guarantee that a consistent method in conservation form will generate results that actually converge to a weak solution. The theorem of Lax and Wendroff ensures only that if the numerical solution does converge, it will converge to a weak solution. Convergence to the entropy-consistent weak solution is, however, guaranteed whenever a consistent method in conservation form is *monotone* (Kuznecov and Vološin 1976; Harten et al. 1976; Crandall and Majda 1980b). Recall that a real-valued function is "monotone increasing" if $g(x) \leq g(y)$ whenever $x \leq y$. A finite-volume (or finite-difference) method is *monotone* if ϕ_j^{n+1} is a monotone-increasing function of each grid-cell value of ϕ appearing in the finite-volume formula. If the scheme is expressed in the functional form

$$\phi_j^{n+1} = H(\phi_{j-p}^n, \ldots, \phi_{j+q+1}^n),$$

the condition that the method be monotone is

$$\frac{\partial\, H(\phi_{j-p}, \ldots, \phi_{j+q+1})}{\partial \phi_i} \geq 0 \tag{5.19}$$

[3] Any differentiable function is Lipschitz continuous.

for each integer i in the interval $[j-p,\ j+q+1]$. If the method is linear in the ϕ_i^n, the method will be monotone if and only if the coefficients of all the ϕ_i^n are nonnegative.

The upstream approximation to the flux form of the constant-wind-speed advection equation

$$\frac{\partial \psi}{\partial t} + \frac{\partial}{\partial x}(c\psi) = 0 \tag{5.20}$$

is

$$\phi_j^{n+1} = (1-\mu)\phi_j^n + \mu\phi_{j-1}^n, \tag{5.21}$$

where $\mu = c\Delta t/\Delta x$. Thus, according to (5.19), the preceding method is monotone for $0 \le \mu \le 1$, which is identical to the standard stability condition for the upstream scheme. As suggested by this example, the range of Δt for which a consistent method in conservation form is monotone is a subset of the range of Δt for which the same scheme is stable when used to approximate problems with smooth solutions. The class of monotone methods is, however, far more restrictive than the class of stable finite-volume methods because *any monotone method is at most first-order accurate* (Godunov 1959; Harten et al. 1976). The only exceptions occur in special cases of no practical significance such as when perfect results are obtained using (5.21) with $\mu = 1$. The leading-order truncation error in any monotone first-order approximation to (5.7) is diffusive (e.g., Sect. 3.4.2), which makes the scheme a higher-order approximation to a viscous problem and ensures that the numerical solution converges to the entropy-consistent solution.

5.2.2 Total Variation Diminishing Methods

First-order methods do not provide a particularly efficient way to obtain accurate numerical solutions; better results can often be obtained using higher-order schemes. Although they are not monotone, many of these schemes satisfy the weaker stability condition that they are *total variation nonincreasing*. The total variation of a one-dimensional grid-point function is defined as

$$\mathrm{TV}(\phi) = \sum_{j=1}^{N-1} |\phi_{j+1} - \phi_j|,$$

where N is the total number of grid points in the numerical domain. The total variation of a continuous function on the interval $[a,b]$ may be defined in an analogous manner as the supremum over all possible subdivisions of the domain $a = x_1 < x_2 < \cdots < x_N = b$ of

$$\sum_{j=1}^{N-1} |\psi(x_{j+1}) - \psi(x_j)|,$$

or equivalently as

$$\mathrm{TV}(\psi) = \limsup_{\epsilon \to 0} \frac{1}{\epsilon} \int_{-\infty}^{\infty} |\psi(x+\epsilon) - \psi(x)|\, \mathrm{d}x.$$

A numerical method is total variation nonincreasing if

$$\mathrm{TV}(\phi^{n+1}) \leq \mathrm{TV}(\phi^n). \tag{5.22}$$

Although slightly imprecise, it is common practice and easier on the tongue to refer to a method that is total variation nonincreasing as *total variation diminishing*, or TVD. This convention will be followed in the remainder of this book, so (5.22) is the working definition of a TVD method.

Solutions to a consistent finite-volume method in conservation form are guaranteed to converge to weak solutions of the exact conservation law whenever the scheme is TVD. The nature of this convergence is, however, complicated by the fact that there may be several nonunique weak solutions to a given conservation law. If the scheme is TVD, the infimum, over the set of all possible weak solutions, of the difference between the numerical solution and each weak solution is guaranteed to go to zero as $\Delta x \rightarrow 0$ and $\Delta t \rightarrow 0$, but a sequence of numerical solutions computed with successively smaller values of Δx and Δt need not smoothly converge to any particular weak solution (LeVeque 1992, p. 164). Of course, the goal is to obtain an approximation that converges to the entropy-consistent solution, and this is generally accomplished by demanding that every approximate solution also satisfy a discrete form of the entropy condition.

The family of monotone finite-volume schemes is a subset of the family of TVD schemes, which are in turn a subset of an even more general class of monotonicity-preserving methods (Harten 1983). A method is *monotonicity preserving* if it will preserve monotone-increasing or monotone-decreasing initial data. For example, if $\phi_j^0 \geq \phi_{j+1}^0$ for all j, then the numerical solution generated by a monotonicity-preserving method has the property that $\phi_j^n \geq \phi_{j+1}^n$ for all n and j. Monotonicity-preserving methods generate approximate solutions that are free from spurious ripples. In particular, no new local extrema are generated in the numerical solution.

One might hope to create a second-order TVD or monotonicity-preserving method by adding enough spatial smoothing to a second-order nondissipative scheme to prevent the development of spurious ripples near the discontinuity. Suppose that numerical solutions to the constant-wind-speed advection equation (5.20) are computed for the case $c = 1/3$ using the second-order scheme

$$\delta_{2t}\phi_j^n + \delta_{2x}\left(c\phi_j^n\right) = \delta_x^4\left(\gamma_4\phi_j^{n-1}\right), \tag{5.23}$$

which is a leapfrog centered-space approximation plus a fourth-derivative filter. Let the strength of the fourth-derivative filter be maximized by setting $\gamma_4\Delta t = 1/32$, which removes all amplitude from the $2\Delta x$ wave in a single leapfrog time step. Solutions computed subject to the initial condition given by the step function (5.3) are shown in Fig. 5.7a at time $t = 2.4$ on the subdomain $0.5 \leq x \leq 1$. This solution was

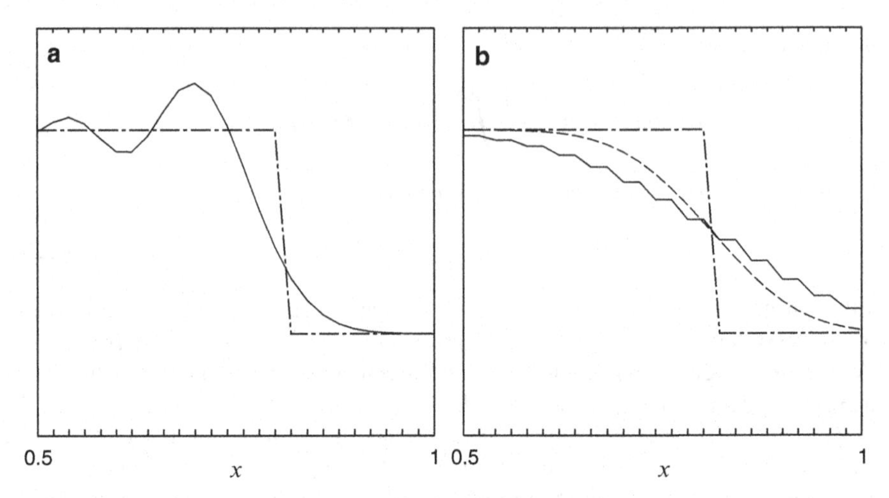

Fig. 5.7 Numerical and exact solutions to the constant-wind-speed advection equation at $t = 2.4$ for the subdomain $0.5 \leq x \leq 1$: **a** exact (*dash-dotted line*) and filtered leapfrog (*solid line*) solutions; **b** exact (*dash-dotted* line), Lax–Friedrichs (*solid line*), and upstream (*dashed line*) solutions

calculated using a Courant number of 0.5 and $\Delta x = 0.02$. Although the strength of the fourth-derivative filter is maximized, spurious ripples still appear behind the leading edge of the jump, implying that (5.23) is neither TVD nor monotonicity preserving. The failure of this attempt to create a second-order monotonicity-preserving method could have been predicted on the basis of the theorem of Godunov (1959), who showed that any linear monotonicity-preserving method is at most first-order accurate.

Godunov's theorem implies the only way to construct higher-order TVD schemes is through the use of nonlinear finite-difference or finite-volume formulae. Several such nonlinear schemes will be considered in the following sections. In most cases these methods combine some information from a higher-order approximation with the smooth solution from a monotone first-order scheme in an attempt to maintain the sharpness of the numerically simulated discontinuity without developing spurious ripples.

When simulating advection of passive tracers in one dimension, one typically uses upstream differencing as the monotone scheme because it is less diffusive than other monotone methods.[4] Figure 5.7b illustrates the superiority of upstream solutions to the advection equation (5.20) over those obtained using another monotone scheme, the Lax–Friedrichs method

$$\frac{\phi_j^{n+1} - \frac{1}{2}(\phi_{j+1}^n + \phi_{j-1}^n)}{\Delta t} + \left(\frac{c\phi_{j+1}^n - c\phi_{j-1}^n}{2\Delta x}\right) = 0. \tag{5.24}$$

[4] In two dimensions the corner transport upstream (CTU) method (4.35) would be a good alternative to coordinate-parallel upstream differences.

As in the leapfrog simulation shown in Fig. 5.7a, the initial condition was specified by the step function (5.3), and both solutions were calculated using a Courant number of 0.5 and $\Delta x = 0.02$. The numerical diffusion generated by the Lax–Friedrichs scheme is larger than that produced by the upstream method (see Problem 5). Thus, despite the obvious stair step in the Lax–Friedrichs solution, its long-wavelength components are more strongly damped than those in the upstream solution. This $2\Delta x$ stair step arises from the discontinuity in the initial data and disappears if the initial width of the jump is increased from a single grid interval to $2\Delta x$.

The Lax–Friedrichs method does have one important advantage relative to the upstream method: it is a central scheme that can be applied independently of the direction of signal propagation. It is trivial to determine the direction of signal propagation for the advection equation, but more general systems of conservation laws such as (5.6) support waves moving in different directions, and the unknown variables must be reexpressed as Riemann invariants[5] before applying the upstream method to the solution of such problems. Those invariants moving toward the right are stepped forward using an upstream stencil which is the mirror image of that applied to the leftward-moving invariants. Central schemes avoid this complication, and in addition they avoid the excessive numerical diffusion associated with (5.24) when formulated as higher-order TVD methods (Nessyahu and Tadmor 1990; Kurganov and Tadmor 2000).

5.3 Discontinuities in Geophysical Fluid Dynamics

Although hydraulic jumps can develop from smooth initial conditions in shallow-water flow and fronts can form in association with mid-latitude low-pressure systems, true dynamical discontinuities do not develop from smooth initial data in most other geophysical problems. Geophysically significant motions in a continuously stratified fluid can be well described by filtered sets of equations, such as the Boussinesq system (see Sect. 1.2). In contrast to the shallow-water system, these filtered equations do not form a hyperbolic system, their linear wave solutions are dispersive, and their nonlinear solutions do not form strong shocks.

Nevertheless, scale contraction does frequently occur in geophysical flows as the result of stretching and shearing deformation by the velocity field. The kinematic effects of flow deformation on an initially circular distribution of a passive tracer are illustrated in Fig. 5.8. As the scale of the tracer distribution shrinks in the direction perpendicular to the axis of dilatation, the concentration field will eventually become difficult to resolve adequately on a given numerical mesh, but a true discontinuity never develops in any finite time. The only discontinuities supported by the advection equation are a special type of shock known as a contact discontinuity, in which a preexisting discontinuity is simply carried along by the moving fluid.

[5] See Sect. 9.1.1 for an example.

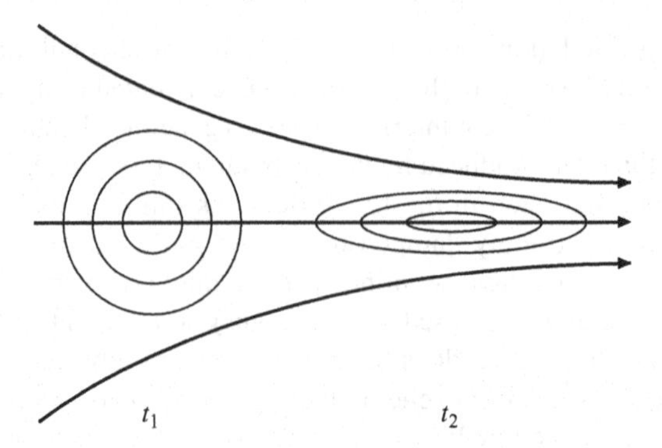

Fig. 5.8 Deformation of a tracer field in a confluent flow field from a circular pattern at t_1 into a cigar shape at t_2

True discontinuities can be generated from initially smooth data at atmospheric fronts (Hoskins and Bretherton 1972), but even in this case the processes producing the scale collapse in the frontal zone are primarily advective.

One might suppose that tracer transport in the scale-contracting flow illustrated in Fig. 5.8 is described by a conservation law of the form

$$\frac{\partial \psi}{\partial t} + \frac{\partial}{\partial x} f(\psi) + \frac{\partial}{\partial y} g(\psi) = 0, \tag{5.25}$$

which is the generalization of (5.7) to two dimensions. In fact, the local rate of change in the mass of a tracer transported by a two-dimensional flow is described by a slightly different conservation law,

$$\frac{\partial \psi}{\partial t} + \frac{\partial}{\partial x}(u\psi) + \frac{\partial}{\partial y}(v\psi) = 0. \tag{5.26}$$

In contrast to (5.25), the fluxes in (5.26) are not determined solely by ψ, but depend on velocity components that are functions of the independent variables x, y, and t.[6]

The conservation laws (5.25) and (5.26) do, nevertheless, have a number of common properties. In particular, pairs of entropy-consistent solutions ψ and ϑ to either (5.25) or (5.26) share the property that if $\psi(x, y, 0) \geq \vartheta(x, y, 0)$ for all x and y at some initial time 0, then $\psi(x, y, t) \geq \vartheta(x, y, t)$ for all x and y, and all $t \geq 0$. If approximate numerical solutions to either (5.25) or (5.26) are computed with a monotone scheme, those solutions have same property, i.e., if $\phi^0_{i,j} \geq \theta^0_{i,j}$ for all i and j, then $\phi^n_{i,j} \geq \theta^n_{i,j}$ for all i, j, and n. The special case $\theta^0_{i,j} = 0$ is particularly important, since then $\phi^n_{i,j} \geq \theta^n_{i,j}$ implies that monotone schemes will not generate spurious negative values from

[6] If the velocity and other dynamical fields are being predicted along with ψ, the full system may be expressed in the form (5.6), which is autonomous (having no explicit dependence on $\mathbf{x}$ and t). Nevertheless, in many practical transport problems the velocity is treated as an externally specified field.

nonnegative initial data. More generally, monotone schemes yield numerical solutions to either (5.25) or (5.26) that are free from spurious ripples in the vicinity of discontinuities and poorly resolved gradients. This is perhaps the most useful property of monotone approximations to the tracer transport equation, since unlike (5.25), (5.26) is a linear partial differential equation whose weak solutions are uniquely determined by the initial and boundary data. There is therefore no need to employ monotone schemes (or to demand satisfaction of some entropy condition) to ensure that consistent, conservation-form approximations to (5.26) converge to the correct solution as Δx, Δy, and Δt approach zero.

As discussed previously, monotone methods are not actually used in most practical applications because they are only first-order accurate and highly diffusive. In regions where the solution is smooth, more accurate approximations to the one-dimensional conservation law (5.7) can be obtained using TVD methods. One might hope to pursue the same strategy in designing approximations to the two-dimensional tracer transport equation, but there are difficulties. The first problem is that except for special cases of no practical importance, all TVD approximations to the two-dimensional nonlinear conservation law (5.25) are at most first-order accurate (Goodman and LeVeque 1985). Thus, in contrast to the one-dimensional case, there are no second-order-accurate TVD approximations to (5.25).

The second and more fundamental problem is that although the entropy-consistent solution to the nonlinear conservation law (5.25) is TVD (or, more precisely, total variation nonincreasing), the total variation in the true solution to the tracer transport equation can increase with time – even when the velocity field is nondivergent! The non-TVD nature of the solutions to (5.26) follows from the circumstance that the total variation of $\psi(x, y)$ is not invariant under coordinate rotations.[7] The total variation of a two-dimensional function is conventionally defined as

$$TV(\psi) = \limsup_{\epsilon \to 0} \frac{1}{\epsilon} \int_{-\infty}^{\infty} \int_{-\infty}^{\infty} |\psi(x+\epsilon, y) - \psi(x, y)| \, dx \, dy$$
$$+ \limsup_{\epsilon \to 0} \frac{1}{\epsilon} \int_{-\infty}^{\infty} \int_{-\infty}^{\infty} |\psi(x, y+\epsilon) - \psi(x, y)| \, dx \, dy.$$

Suppose that the initial conditions for (5.26) are

$$\psi(x, y, 0) = \begin{cases} 1 & \text{if } |x| \leq 1 \text{ and } |y| \leq 1, \\ 0 & \text{otherwise,} \end{cases}$$

and that the flow is in solid-body rotation with $u = -y$ and $v = x$. After the distribution of ψ rotates through an angle of 45°, its total variation will increase by

[7] The nonconservation of the total variation under coordinate rotations, which was pointed out to this author by Joe Tenerelli, appears to be a weakness in the mathematical definition of the total variation of a two-dimensional function. A second weakness appears in the physical units that are associated with total variation. If ψ and φ are variables with arbitrary physical units Q and x and y are spatial coordinates in meters, then $TV(\psi(x))$ has units of Q, whereas $TV(\varphi(x, y))$ has units of meters times Q.

a factor of $\sqrt{2}$. Accurate finite-volume approximations to (5.26) cannot therefore be TVD. Useful schemes for the simulation of tracer transport can nevertheless be obtained by borrowing techniques used to generate TVD approximations to the one-dimensional conservation law (5.7).

Instead of demanding that the scheme be TVD, it is possible to control the development of spurious oscillations by regulating the behavior of the local maxima and minima in the solution. Smooth solutions to the nonlinear conservation law (5.25) also satisfy the advective-form equation

$$\frac{\partial \psi}{\partial t} + \frac{df}{d\psi}\frac{\partial \psi}{\partial x} + \frac{dg}{d\psi}\frac{\partial \psi}{\partial y} = 0. \tag{5.27}$$

Similarly, *if the velocity field is nondivergent,*

$$\frac{\partial u}{\partial x} + \frac{\partial v}{\partial y} = 0, \tag{5.28}$$

and if ψ satisfies (5.26), then it must also satisfy the advective form[8]

$$\frac{\partial \psi}{\partial t} + u\frac{\partial \psi}{\partial x} + v\frac{\partial \psi}{\partial y} = 0. \tag{5.29}$$

Solutions to both (5.27) and (5.29) conserve the amplitude of all local maxima and minima in the initial data. Flux-corrected transport (FCT) algorithms, which will be considered in the next section, exploit this property of the true solution to control the development of ripples near a discontinuity.

The remainder of this chapter will be primarily devoted to the examination of methods for the simulation of discontinuities or poorly resolved gradients in nondivergent flow. The first topic considered is one-dimensional nondivergent flow, which can occur only if the velocity is constant. The one-dimensional constant-wind-speed advection equation is also a member of the family of autonomous conservation laws of the form (5.7). As a consequence, the study of the constant-wind-speed advection equation serves as an introduction to both fluid transport problems of the form (5.26) and nonlinear hyperbolic conservation laws of the form (5.25). The extension of these results to nonuniform two-dimensional flow is discussed in Sect. 5.9. The extension of the one-dimensional constant-wind-speed problem to nonlinear systems of conservation laws, which is beyond the scope of this text, is discussed in Godlewski and Raviart (1996) and LeVeque (2002).

[8] Equation (5.29) states that the tracer concentration (typically expressed as a dimensionless ratio, such as grams per kilogram or parts per billion) is conserved following the motion of each fluid parcel. In contrast, (5.26) states that the local rate of change of the mass of the tracer at a fixed point in space is determined by the divergence of the tracer mass flux at that point.

5.4 Flux-Corrected Transport

Flux-corrected transport (FCT) was proposed by Boris and Book (1973) as a way of approximating a conservation law with a high-order scheme in regions where the solution is smooth while using a low-order monotone scheme where the solution is poorly resolved or discontinuous. The concept of FCT and the algorithms for its implementation were further generalized by Zalesak (1979). Zalesak suggested approximating the scalar conservation law (5.7) with a finite-difference formula in the conservation form

$$\frac{\phi_j^{n+1} - \phi_j^n}{\Delta t} + \left(\frac{F_{j+\frac{1}{2}} - F_{j-\frac{1}{2}}}{\Delta x} \right) = 0 \tag{5.30}$$

and then computing the fluxes $F_{j\pm1/2}$ in several steps as follows:

1. Compute a set of low-order fluxes $F^{\mathrm{l}}_{j+1/2}$ using a monotone scheme.
2. Compute a set of high-order fluxes $F^{\mathrm{h}}_{j+1/2}$ using a high-order scheme.
3. Compute the *antidiffusive fluxes*

$$A_{j+\frac{1}{2}} = F^{\mathrm{h}}_{j+\frac{1}{2}} - F^{\mathrm{l}}_{j+\frac{1}{2}}.$$

4. Compute a monotone estimate of the solution at $(n+1)\Delta t$ (also known as the "transported and diffused" solution),

$$\phi_j^{\mathrm{td}} = \phi_j^n - \frac{\Delta t}{\Delta x}\left(F^{\mathrm{l}}_{j+\frac{1}{2}} - F^{\mathrm{l}}_{j-\frac{1}{2}} \right).$$

5. Correct the $A_{j+1/2}$ so that the final "antidiffusion" step does not generate new maxima or minima. The correction procedure may be expressed mathematically by defining

$$A^{\mathrm{c}}_{j+\frac{1}{2}} = C_{j+\frac{1}{2}}\, A_{j+\frac{1}{2}}, \qquad 0 \le C_{j+\frac{1}{2}} \le 1.$$

   The procedure for computing $C_{j+1/2}$ will be discussed shortly.
6. Perform the "antidiffusion" step

$$\phi_j^{n+1} = \phi_j^{\mathrm{td}} - \frac{\Delta t}{\Delta x}\left(A^{\mathrm{c}}_{j+\frac{1}{2}} - A^{\mathrm{c}}_{j-\frac{1}{2}} \right).$$

If all the $C_{j+1/2}$ were unity, the preceding algorithm would give results identical to the higher-order scheme, and if all the $C_{j+1/2}$ were zero, the solution would be identical to that obtained with the monotone scheme. Criteria for determining $C_{j+1/2}$ are usually designed to prevent the development of new maxima and minima and to prohibit the amplification of existing extrema.

5.4.1 Flux Correction: The Original Proposal

Boris and Book (1973) did not actually give a formula for $C_{j+1/2}$, but offered the following expression for the corrected fluxes:

$$A^{\mathrm{c}}_{j+\frac{1}{2}} = S_{j+\frac{1}{2}} \max\left\{0,\ \min\left[|A_{j+\frac{1}{2}}|,\right.\right.$$

$$\left.\left. S_{j+\frac{1}{2}}\left(\phi^{\mathrm{td}}_{j+2} - \phi^{\mathrm{td}}_{j+1}\right)\frac{\Delta x}{\Delta t},\ S_{j+\frac{1}{2}}\left(\phi^{\mathrm{td}}_{j} - \phi^{\mathrm{td}}_{j-1}\right)\frac{\Delta x}{\Delta t}\right]\right\},$$

where

$$S_{j+\frac{1}{2}} = \operatorname{sgn}\left(A_{j+\frac{1}{2}}\right).$$

The logic behind this formula can be understood by considering the case where $A_{j+1/2} \geq 0$, so the preceding may be written as

$$\frac{\Delta t}{\Delta x} A^{\mathrm{c}}_{j+\frac{1}{2}} = \max\left\{0,\ \min\left[A_{j+\frac{1}{2}}\frac{\Delta t}{\Delta x},\ \left(\phi^{\mathrm{td}}_{j+2} - \phi^{\mathrm{td}}_{j+1}\right),\ \left(\phi^{\mathrm{td}}_{j} - \phi^{\mathrm{td}}_{j-1}\right)\right]\right\}. \tag{5.31}$$

Two typical configurations for ϕ_j are shown in Fig. 5.9; observe that in both cases the antidiffusive flux is directed up-gradient, which is the most common situation. The increase in ϕ^n_{j+1}, and the decrease in ϕ^n_j, that would be produced by the uncorrected antidiffusive flux $A_{j+1/2}$ is given by the first argument of the "min" function in (5.31). The second argument of the min function ensures that the corrected flux is not large enough to generate a new maximum by rendering $\phi_{j+1} > \phi_{j+2}$. This type of limitation on the antidiffusive flux would apply in the case shown in the left panel in Fig. 5.9. If ϕ_{j+1} is already greater than ϕ_{j+2}, the antidiffusive flux is zeroed to prevent the amplification of the preexisting extrema at ϕ_{j+1}; this situation is illustrated in the right panel in Fig. 5.9. The third argument of the min function ensures that no new minima are created and that no preexisting minima are amplified at ϕ_j.

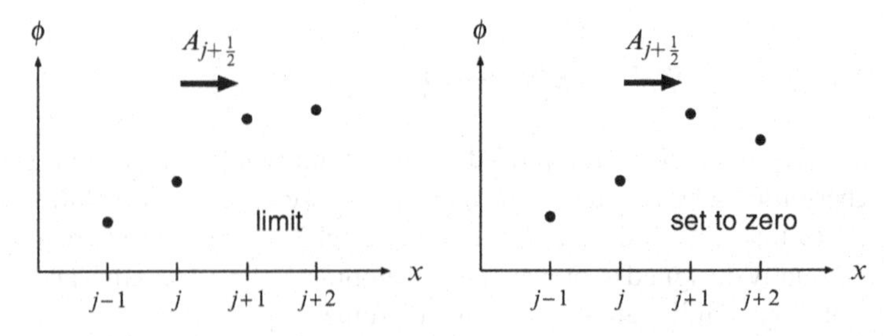

Fig. 5.9 Two possible configurations in which an antidiffusive flux, indicated by the *heavy arrow*, may be modified by a flux limiter

5.4.2 The Zalesak Corrector

Zalesak (1979) noted that the preceding algorithm limits each antidiffusive flux without considering the action of the antidiffusive fluxes at neighboring grid points and that this can lead to an unnecessarily large reduction in the antidiffusive flux. For example, although the antidiffusive flux shown in the left panel in Fig. 5.9 will tend to decrease ϕ_j, this decrease is likely to be partly compensated by an up-gradient antidiffusive flux directed from grid point $j-1$ into grid point j. Zalesak proposed the following algorithm, which considers the net effect of both antidiffusive fluxes to minimize the correction to those fluxes and thereby keep the algorithm as close as possible to that which would be obtained using the higher-order scheme:

1. As an optional preliminary step, set certain down-gradient antidiffusive fluxes to zero, such that

$$\begin{aligned} A_{j+\frac{1}{2}} = 0, \qquad & \text{if} \qquad A_{j+\frac{1}{2}}(\phi^{\text{td}}_{j+1} - \phi^{\text{td}}_{j}) < 0 \\ & \text{and either} \quad A_{j+\frac{1}{2}}(\phi^{\text{td}}_{j+2} - \phi^{\text{td}}_{j+1}) < 0 \\ & \text{or} \qquad A_{j+\frac{1}{2}}(\phi^{\text{td}}_{j} - \phi^{\text{td}}_{j-1}) < 0. \end{aligned} \tag{5.32}$$

Zalesak refers to this as a cosmetic correction, and it is usually omitted. This cosmetic correction has, nevertheless, been used in the FCT computations shown in this chapter. It has no effect on the solution shown in Fig. 5.10a, makes a minor improvement in the solution shown in Fig. 5.10b, and makes a major improvement in the solution shown in Fig. 5.13b.

2. Evaluate the range of permissible values for ϕ^{n+1}_j:

$$\phi^{\max}_j = \max\left(\phi^n_{j-1}, \phi^n_j, \phi^n_{j+1}, \phi^{\text{td}}_{j-1}, \phi^{\text{td}}_j, \phi^{\text{td}}_{j+1}\right),$$
$$\phi^{\min}_j = \min\left(\phi^n_{j-1}, \phi^n_j, \phi^n_{j+1}, \phi^{\text{td}}_{j-1}, \phi^{\text{td}}_j, \phi^{\text{td}}_{j+1}\right).$$

If the flow is nondivergent, the ϕ^{td} are not needed in the preceding formulae because the extrema predicted by the monotone scheme will be of lower amplitude than those at the beginning of the time step. If, however, the flow is divergent, then the local minima and maxima in the true solution may be increasing, and the increase predicted by the monotone scheme should be considered in determining $\phi^{\max}$ and $\phi^{\min}$.

3. Compute the sum of all antidiffusive fluxes into grid cell j,

$$P^+_j = \max\left(0, A_{j-\frac{1}{2}}\right) - \min\left(0, A_{j+\frac{1}{2}}\right).$$

4. Compute the maximum net antidiffusive flux divergence that will preserve $\phi^{n+1}_j \le \phi^{\max}_j$,

$$Q^+_j = \left(\phi^{\max}_j - \phi^{\text{td}}_j\right)\frac{\Delta x}{\Delta t}. \tag{5.33}$$

5. Compute the required limitation on the net antidiffusive flux into grid cell j,

$$R_j^+ = \begin{cases} \min\left(1, Q_j^+/P_j^+\right) & \text{if } P_j^+ > 0, \\ 0 & \text{if } P_j^+ = 0. \end{cases}$$

6. Compute the corresponding quantities involving the net antidiffusive flux out of grid cell j,

$$\begin{aligned} P_j^- &= \max\left(0, A_{j+\frac{1}{2}}\right) - \min\left(0, A_{j-\frac{1}{2}}\right). \\ Q_j^- &= \left(\phi_j^{\text{td}} - \phi_j^{\min}\right)\frac{\Delta x}{\Delta t}. \\ R_j^- &= \begin{cases} \min\left(1, Q_j^-/P_j^-\right) & \text{if } P_j^- > 0, \\ 0 & \text{if } P_j^- = 0. \end{cases} \end{aligned} \tag{5.34}$$

7. Limit the antidiffusive flux so that it neither produces an overshoot in the grid cell into which it is directed nor generates an undershoot in the grid cell out of which it flows:

$$C_{j+\frac{1}{2}} = \begin{cases} \min\left(R_{j+1}^+, R_j^-\right) & \text{if } A_{j+\frac{1}{2}} \geq 0, \\ \min\left(R_j^+, R_{j+1}^-\right) & \text{if } A_{j+\frac{1}{2}} < 0. \end{cases}$$

Two examples illustrating the performance of the Zalesak FCT algorithm on the constant-wind-speed one-dimensional advection equation are shown in Fig. 5.10. In these examples, the monotone flux is computed using the upstream method with

$$F_{j+\frac{1}{2}}^{\text{l}} = \frac{c}{2}(\phi_j + \phi_{j+1}) - \frac{|c|}{2}(\phi_{j+1} - \phi_j), \tag{5.35}$$

and the high-order flux is computed using the flux form of the Lax–Wendroff method such that

$$F_{j+\frac{1}{2}}^{\text{h}} = \frac{c}{2}(\phi_j + \phi_{j+1}) - \frac{c^2 \Delta t}{2\Delta x}(\phi_{j+1} - \phi_j). \tag{5.36}$$

The calculations were performed in a wide periodic domain, only the center portion of which is shown in the figure. In each case the wind speed is constant, and the Courant number is 0.5.

The curves shown in Fig. 5.10a are solutions to the same traveling-jump problem considered in connection with Fig. 5.7, except that the solutions are plotted at $t = 1.8$. The solution computed using FCT is shown by the solid line. Also shown are the exact solution and the approximate solutions obtained using upstream differencing and using the Lax–Wendroff method without FCT. The FCT scheme is almost identical to the uncorrected Lax–Wendroff method except near the top of the step, where the flux-correction procedure completely eliminates the dispersive

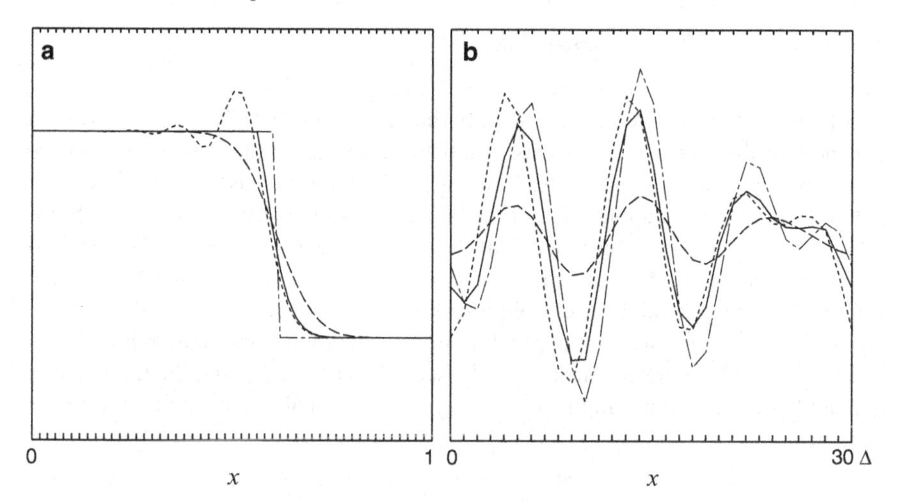

Fig. 5.10 Results from two constant-wind-speed advection tests: exact solution (*thin dash-dotted* line) and numerical solutions obtained with the Zalesak flux-corrected transport (FCT) combination of upstream and Lax–Wendroff differencing (*solid* line), upstream differencing (*long dashed line*), and the Lax–Wendroff scheme (*short-dashed line*)

ripples apparent in the uncorrected Lax–Wendroff solution. The FCT scheme is not only superior to the higher-order scheme, it also captures the steepness of the jump much better than the upstream method.

The curves in Fig. 5.10b show solutions to the test problem considered in Chap. 3 in which the sum of equal-amplitude $7.5\Delta x$ and $10\Delta x$ waves is advected over a distance of twelve grid points (cf. Fig. 3.6). Once again the FCT solution is clearly superior to that obtained using upstream differencing. Although this test case does not involve shocks or discontinuities, the FCT solution remains roughly comparable in quality to that obtained with the uncorrected Lax–Wendroff method. In particular, the FCT solution exhibits more damping but less phase-speed error than that obtained with the Lax–Wendroff method.

As suggested by the preceding tests, the FCT approach allows one to obtain ripple-free solutions that are far superior to those computed by simple upstream differencing. One might attempt to obtain further improvements by computing the high-order flux using an extremely accurate method. Zalesak (1979), for example, gives a formula for an eighth-order-accurate method. Such very high order formulae are seldom used in practical applications. In part, this may be due to the unattractive compromises that may arise using high-order formulae with forward-in-time differencing. The more fundamental problem, however, is that the scheme reduces to a monotone method near any maxima and minima and is therefore only first-order accurate near extrema. The empirically estimated order of accuracy of the preceding FCT scheme is less than 2 and is unlikely to be substantially improved by using a more accurate scheme to compute the high-order flux (see Table 5.1 in Sect. 5.5.2).

5.4.3 Iterative Flux Correction

Substantial improvements in the neighborhood of smooth well-resolved extrema can, nevertheless, be achieved by using a better estimate for the low-order solution. One strategy for obtaining a better low-order solution is to reuse the standard FCT solution in an iterative application of the flux-correction procedure (Schär and Smolarkiewicz 1996). As a consequence of the flux-correction algorithm, the standard FCT solution is free from spurious ripples and can serve as an improved estimate for the "transported and diffused solution" in a second iteration. That portion of the antidiffusive flux that was not applied in the first iteration is the maximum antidiffusive flux available for application in the second iteration. Letting the tilde denote a quantity defined for use in the second iteration, the final step of the first iteration becomes

$$\tilde{\phi}_j^{\mathrm{td}} = \phi_j^{\mathrm{td}} - \frac{\Delta t}{\Delta x}\left(A^{\mathrm{c}}_{j+\frac{1}{2}} - A^{\mathrm{c}}_{j-\frac{1}{2}}\right),$$

and the new antidiffusive flux becomes

$$\tilde{A}_{j+\frac{1}{2}} = A_{j+\frac{1}{2}} - A^{\mathrm{c}}_{j+\frac{1}{2}}.$$

The antidiffusive flux is limited using precisely the same flux-correction algorithm as used in the first iteration, and the final estimate for ϕ^{n+1} is obtained using

$$\phi_j^{n+1} = \tilde{\phi}_j^{\mathrm{td}} - \frac{\Delta t}{\Delta x}\left(\tilde{A}^{\mathrm{c}}_{j+\frac{1}{2}} - \tilde{A}^{\mathrm{c}}_{j-\frac{1}{2}}\right).$$

This iteration can be very effective in improving the solution near well-resolved extrema such as the crest of a sine wave, but it does not noticeably improve the solution near a discontinuous step.

5.5 Flux-Limiter Methods

The strategy behind flux-limiter methods is similar to that underlying FCT in that the numerical fluxes used in both methods are a weighted sum of the fluxes computed by a monotone first-order scheme and a higher-order method. In flux-limiter methods, however, the limiter that apportions the flux between the high- and low-order schemes is determined without actually computing a low-order solution (ϕ^{td}). This limiter is expressed as a function of the local solution at the previous time level in a manner guaranteeing that the scheme generates TVD approximations to the one-dimensional scalar conservation law (5.7) and that the scheme is second-order accurate except in the vicinity of the extrema of ϕ.

Flux-limiter methods approximate (5.7) with a finite-difference scheme in the conservation form (5.30) using the flux

$$F_{j+\frac{1}{2}} = F^{\mathrm{l}}_{j+\frac{1}{2}} + C_{j+\frac{1}{2}}\left(F^{\mathrm{h}}_{j+\frac{1}{2}} - F^{\mathrm{l}}_{j+\frac{1}{2}}\right),$$

where F^{l} and F^{h} denote the fluxes obtained using monotone and high-order schemes, and $C_{j+1/2}$ is a multiplicative limiter. As in the FCT algorithm discussed previously, the high-order flux is recovered when $C_{j+1/2} = 1$, and the performance of the scheme is highly dependent on the algorithm for specifying $C_{j+1/2}$. We again demand that $C_{j+1/2} \geq 0$, but as will become evident, it is advantageous to allow $C_{j+1/2}$ to exceed unity. In scalar problems in which the phase speed of the disturbance is greater than zero,[9] $C_{j+1/2}$ is calculated as a nonlinear function of the local solution $C(r_{j+1/2})$, where

$$r_{j+\frac{1}{2}} = \frac{\phi_j - \phi_{j-1}}{\phi_{j+1} - \phi_j}$$

is the ratio of the slope of the solution across the cell interface upstream of $j + 1/2$ to the slope of the solution across the interface at $j + 1/2$. The parameter $r_{j+1/2}$ is approximately unity where the numerical solution is smooth and is negative when there is a local maximum or minimum immediately upstream of the cell interface at $j + 1/2$.

5.5.1 Ensuring That the Scheme Is TVD

Criteria guaranteeing that a flux-limiter method is TVD may be obtained by noting that a finite-difference scheme of the form

$$\phi_j^{n+1} = \phi_j^n - G_{j-\frac{1}{2}} \left(\phi_j^n - \phi_{j-1}^n\right) + H_{j+\frac{1}{2}} \left(\phi_{j+1}^n - \phi_j^n\right) \tag{5.37}$$

will be TVD provided that for all j

$$0 \leq G_{j+\frac{1}{2}}, \quad 0 \leq H_{j+\frac{1}{2}}, \quad \text{and} \quad G_{j+\frac{1}{2}} + H_{j+\frac{1}{2}} \leq 1 \tag{5.38}$$

(Harten 1983). This may be verified by observing that (5.37) and (5.38) imply

$$\begin{aligned}\left|\phi_{j+1}^{n+1} - \phi_j^{n+1}\right| \leq {} & \left(1 - G_{j+\frac{1}{2}} - H_{j+\frac{1}{2}}\right) \left|\phi_{j+1}^n - \phi_j^n\right| \\ & + G_{j-\frac{1}{2}} \left|\phi_j^n - \phi_{j-1}^n\right| + H_{j+\frac{3}{2}} \left|\phi_{j+2}^n - \phi_{j+1}^n\right|.\end{aligned}$$

Summing over all j and shifting the dummy index in the last two summations yields

$$\begin{aligned}\sum_j \left|\phi_{j+1}^{n+1} - \phi_j^{n+1}\right| \leq {} & \sum_j \left(1 - G_{j+\frac{1}{2}} - H_{j+\frac{1}{2}}\right) \left|\phi_{j+1}^n - \phi_j^n\right| \\ & + \sum_j G_{j+\frac{1}{2}} \left|\phi_{j+1}^n - \phi_j^n\right| + \sum_j H_{j+\frac{1}{2}} \left|\phi_{j+1}^n - \phi_j^n\right| \\ = {} & \sum_j \left|\phi_{j+1}^n - \phi_j^n\right|.\end{aligned}$$

[9] The general case, in which the phase speed is either positive or negative, is discussed in Sect. 5.5.3.

Sweby (1984) presented a systematic derivation of the possible functional forms for $C(r)$ that yield TVD flux-limited methods when the monotone scheme is upstream differencing and the high-order scheme is a member of a family of second-order methods that includes the Lax–Wendroff and Warming–Beam methods. Suppose that the constant-wind-speed advection equation (5.20) is approximated using the flux form of the Lax–Wendroff method and that $c > 0$. The Lax–Wendroff flux (5.36) can be expressed as

$$F^{\mathrm{LW}}_{j+\frac{1}{2}} = c\phi_j + \frac{c}{2}(1-\mu)(\phi_{j+1} - \phi_j),$$

where $\mu = c\Delta t/\Delta x$. The first term of the preceding expression is the numerical flux for upstream differencing (in a flow with $c > 0$). The second term is an increment to the upstream flux that can be multiplied by $C_{j+1/2}$ to obtain the "limited" flux

$$F_{j+\frac{1}{2}} = c\phi_j + \frac{c}{2}(1-\mu)(\phi_{j+1} - \phi_j)C_{j+\frac{1}{2}}. \tag{5.39}$$

The finite-difference scheme obtained after evaluating the divergence of these limited fluxes may be written

$$\begin{aligned} \phi_j^{n+1} = \phi_j^n &- \left[\mu - \frac{\mu}{2}(1-\mu)C_{j-\frac{1}{2}}\right]\left(\phi_j^n - \phi_{j-1}^n\right) \\ &- \frac{\mu}{2}(1-\mu)C_{j+\frac{1}{2}}\left(\phi_{j+1}^n - \phi_j^n\right). \end{aligned} \tag{5.40}$$

To arrive at a scheme that is TVD, one natural approach would be to choose

$$\begin{aligned} G_{j-\frac{1}{2}} &= \mu - \frac{\mu}{2}(1-\mu)C_{j-\frac{1}{2}}, \\ H_{j+\frac{1}{2}} &= -\frac{\mu}{2}(1-\mu)C_{j+\frac{1}{2}} \end{aligned}$$

and attempt to determine a function $C(r_{j+1/2}) \equiv C_{j+1/2}$ that will guarantee satisfaction of (5.38). Unfortunately, this approach fails, since by assumption, $C(r_{j+1/2}) \geq 0$, and thus $H_{j+1/2} < 0$ whenever the Courant number falls in the range $0 \leq \mu \leq 1$.

As an alternative, Sweby suggested setting

$$\begin{aligned} G_{j-\frac{1}{2}} &= \mu + \frac{\mu}{2}(1-\mu)\left[C_{j+\frac{1}{2}}\left(\frac{\phi_{j+1}^n - \phi_j^n}{\phi_j^n - \phi_{j-1}^n}\right) - C_{j-\frac{1}{2}}\right], \\ H_{j+\frac{1}{2}} &= 0. \end{aligned}$$

Then the TVD criteria (5.38) will be satisfied if

$$0 \leq G_{j-\frac{1}{2}} \leq 1$$

for all j, or equivalently, if

$$0 \leq \mu\left[1 + \frac{1}{2}(1-\mu)\left(\frac{C_{j+\frac{1}{2}}}{r_{j+\frac{1}{2}}} - C_{j-\frac{1}{2}}\right)\right] \leq 1.$$

If the CFL condition ($0 \leq \mu \leq 1$) holds for the upstream scheme, the criteria for the method to be TVD reduce to

$$\frac{-2}{1-\mu} \leq \frac{C_{j+\frac{1}{2}}}{r_{j+\frac{1}{2}}} - C_{j-\frac{1}{2}} \leq \frac{2}{\mu},$$

or

$$\left| \frac{C_{j+\frac{1}{2}}}{r_{j+\frac{1}{2}}} - C_{j-\frac{1}{2}} \right| \leq 2.$$

Suppose that $r_{j+1/2} > 0$. Then since $C(r)$ is assumed to be nonnegative, the preceding inequality is satisfied when

$$0 \leq \frac{C(r)}{r} \leq 2 \quad \text{and} \quad 0 \leq C(r) \leq 2. \tag{5.41}$$

Now consider the case $r_{j+1/2} \leq 0$. Negative values of r occur at the local maxima and minima of ϕ_j, where the flux must be completely determined by the monotone upstream method to avoid increasing the total variation; it is therefore necessary[10] to choose $C(r) = 0$ when $r < 0$. Note that the condition $C(r) = 0$ when $r < 0$ is implicitly included in the inequalities (5.41).

The inequalities (5.41) define the shaded region of the (r, C)-plane shown in Fig. 5.11a, which is the locus of all curves $C(r)$ that make the flux-limited method TVD. The range of possible choices for $C(r)$ can be further restricted if it is required that the method be second-order accurate whenever $r > 0$. Noting that $C_{j-1/2}$ depends on the value of ϕ_{j-2}, the flux-limited scheme (5.40) has the form

$$\phi_j^{n+1} = H(\phi_{j-2}^n, \phi_{j-1}^n, \phi_j^n, \phi_{j+1}^n).$$

All second-order approximations to the advection problem that have the preceding form are weighted averages of the Lax–Wendroff method and the method of Warming and Beam. As discussed previously, the flux-limited scheme becomes the Lax–Wendroff method if $C(r) = 1$. In a similar way, specifying $C(r) = r$ converts the scheme to the method of Warming and Beam (3.79). Curves corresponding to

$$C^{\mathrm{LW}}(r) = 1 \quad \text{and} \quad C^{\mathrm{WB}}(r) = r$$

are also plotted in Fig. 5.11a. Of course, $C^{\mathrm{LW}}(r)$ and $C^{\mathrm{WB}}(r)$ do not lie entirely within the shaded TVD region because neither the Lax–Wendroff method nor that of Warming and Beam is TVD. Nevertheless, to make the flux-limited scheme second-order accurate away from local maxima and minima (i.e., for $r > 0$), $C(r)$ must be

[10] Although it is necessary to choose $C(r) = 0$ when $r < 0$ to keep the scheme TVD, this is actually a poor choice if the solution is smooth and well resolved in the vicinity of the extremum. Well-resolved extrema would be captured more accurately using the higher-order scheme. See Sect. 5.8.

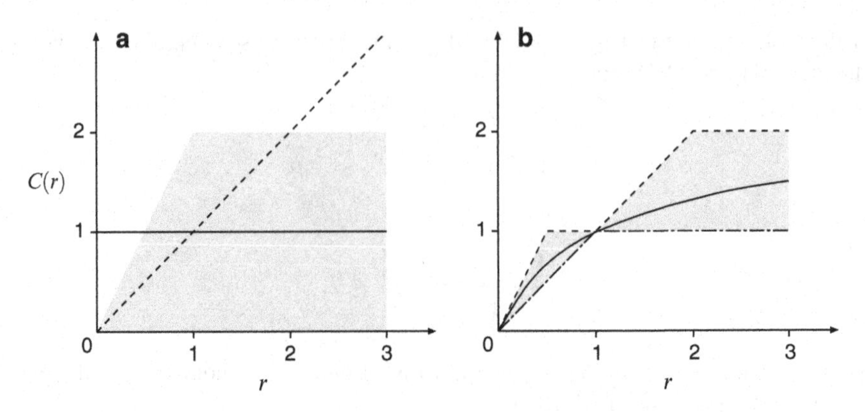

Fig. 5.11 **a** *Shading* indicates the region in which $C(r)$ must lie to give a total variation diminishing (TVD) method. *Heavy lines* indicate $C^{\mathrm{LW}}(r)$ (*solid line*) and $C^{\mathrm{WB}}(r)$ (*dashed line*). **b** *Shading* indicates the region in which $C(r)$ must lie to give a TVD scheme that is an internal average of the Lax–Wendroff method and the method of Warming and Beam. Three possible limiters are also indicated: the "superbee" (*dashed line*), minmod (*dot-dashed line*), and Van Leer (*solid line*)

a weighted average of $C^{\mathrm{LW}}(r)$ and $C^{\mathrm{WB}}(r)$. Sweby suggests that the best results are obtained if this weighted average is an internal average such that

$$C(r) = [1 - \theta(r)]C^{\mathrm{LW}}(r) + \theta(r)C^{\mathrm{WB}}(r), \tag{5.42}$$

where $0 \leq \theta(r) \leq 1$. This portion of the total TVD region is indicated by the shaded area in Fig. 5.11b, which will be referred to as the "second-order" TVD region, although the true second-order TVD region includes external averages of the Lax–Wendroff and Warming–Beam methods and is larger than the shaded area in Fig. 5.11b.

5.5.2 Possible Flux Limiters

Possible choices for the specific functional form of $C(r)$ that yield a TVD method satisfying (5.42) include the "minmod" limiter

$$C(r) = \max[0, \min(1, r)], \tag{5.43}$$

which is the dot-dashed curve following the lower boundary of the second-order TVD region in Fig. 5.11b; the "superbee" limiter (Roe 1985)

$$C(r) = \max[0, \min(1, 2r), \min(2, r)], \tag{5.44}$$

which lies along the upper boundary of the second-order TVD region; and the van Leer limiter (van Leer 1974)

$$C(r) = \frac{r + |r|}{1 + |r|}, \tag{5.45}$$

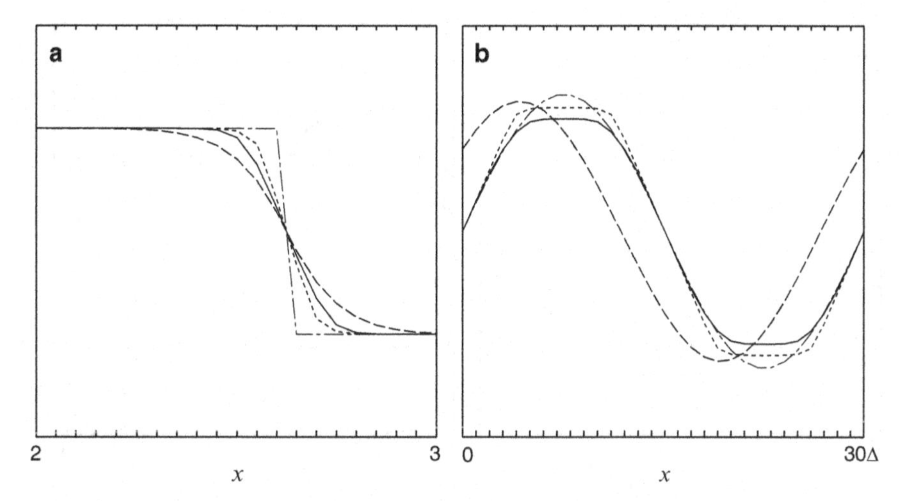

Fig. 5.12 Comparison of flux-limited approximations using the superbee (*short-dashed line*) and monotonized centered (MS) (*solid line*) limiters with **a** the minmod limiter (*long-dashed line*) in a case with a propagating step, and **b** the Lax–Wendroff solution (*dashed line*) in a case with a well-resolved sinusoidal distribution. The exact solution is shown by the *thin dot-dashed line*

which is the smooth curve in Fig. 5.11b. Also of note, but not plotted, is the monotonized centered (MC) limiter (van Leer 1977)

$$C(r) = \max\left[0, \min\left(2r, \frac{1+r}{2}, 2\right)\right]. \tag{5.46}$$

The performance of several different limiters is compared in Fig. 5.12. Figure 5.12a shows results from the same test problem as considered in Fig. 5.10a except that the horizontal grid size is increased from $1/50$ to $1/20$ and the solution is displayed at time 7.8 to better reveal small differences between the various solutions. Inspection of Fig. 5.12a shows that the minmod limiter allows the most numerical diffusion, the superbee limiter allows the least, and the MC limiter performs almost as well as the superbee. Although the superbee limiter works best on the example shown in Fig. 5.12a, the MC limiter may be the best choice for general applications. The weakness of the superbee limiter is illustrated in Fig. 5.12b, which shows flux-limited and Lax–Wendroff approximations to a problem whose correct solution is a unit-amplitude sine wave propagating to the right at speed 1/10 on the periodic domain $0 \leq x \leq 1$. In this example $\Delta x = 1/30$, the Courant number is 0.5 and the solution is shown at $t = 200$, at which point the initial distribution has made 20 circuits around the periodic domain. The superbee and MC limiters clearly flatten the crests and troughs in the flux-limited approximation to this well-resolved sine wave. As the superbee limiter flattens the crests and troughs it incorrectly amplifies the solution near the edges of the flattened extrema, but no such spurious amplification is generated by the MC limiter; the MC-limited solution remains

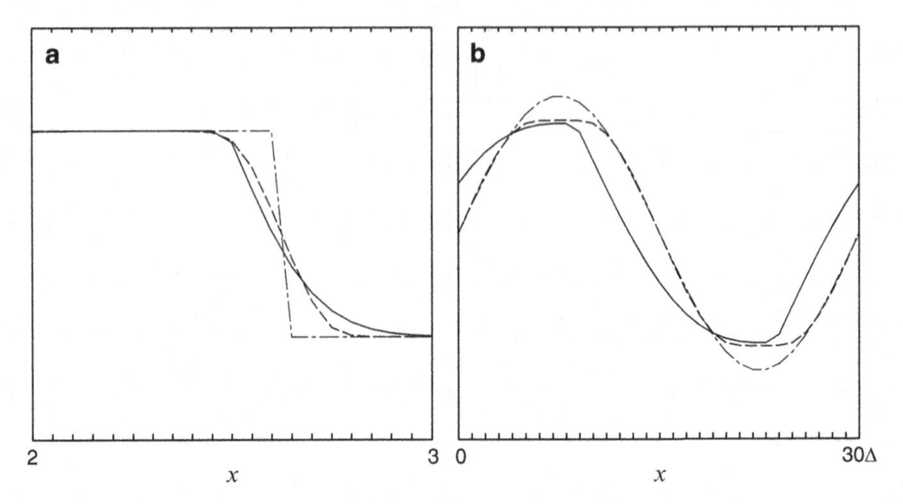

Fig. 5.13 Comparison of MC flux-limited (*long-dashed line*) and FCT (*solid line*) approximate solutions with the exact solution (*thin dashed-dotted line*) for the two test cases shown in Fig. 5.12

within the envelope of the true solution. Although the flux-limited solutions show distortion in the peaks and troughs, they are almost completely free from phase-speed error, and as a consequence, the overall character of the flux-limited solution is superior to that obtained with the Lax–Wendroff method (i.e., with no limiter), which exhibits a substantial phase lag and very modest damping.

Figure 5.13 shows a comparison of the MC flux-limited scheme with the Zalesak FCT algorithm discussed in Sect. 5.4.3. The two test problems are identical to those just considered in Fig. 5.12. Both methods were implemented using the same methods to evaluate the monotone and high-order fluxes (specifically upstream differencing and the Lax–Wendroff method). The solutions obtained with the MC flux-limited method are clearly superior to those obtained using the FCT scheme. The tendency of the FCT scheme to deform the sine wave into a sawtooth can, however, be eliminated using a second iterative pass of the FCT algorithm as discussed at the end of Sect. 5.4.2 (Schär and Smolarkiewicz 1996, Fig. 4).

Since FCT and flux-limiter methods both revert to first-order schemes in the vicinity of minima and maxima, they do not give fully second order approximations in problems like the sine-wave-advection test shown in Figs. 5.12b and 5.13b. The effective order of accuracy of these schemes can be empirically determined for the sine-wave-advection test by performing a series of simulations in which both Δx and Δt are repeatedly halved (so that all simulations are performed with the same Courant number of 0.5). Fitting a function of the form $\alpha(\Delta x)^p$ to the error as Δx decreases from $1/40$ to $1/320$ yields the approximate values for p listed in Table 5.1. As a check on the quality of this calculation, the empirically determined orders of accuracy for the upstream and Lax–Wendroff schemes are also listed in Table 5.1. The result for the upstream scheme is slightly in error, and could be improved by continuing the computation using double precision on still finer

Table 5.1 Empirically determined order of accuracy for constant-wind-speed advection of a sine wave. (FCT denotes flux-corrected transport, MC monotonized centered)

Scheme	Estimated order of accuracy
Upstream	0.9
Minmod flux limiter	1.6
Superbee flux limiter	1.6
Zalesak FCT	1.7
MC flux limiter	1.9
Lax–Wendroff	2.0

grids. The effective order of accuracy of the MC flux-limited scheme is higher than that of the other flux-limited and FCT methods. In fact, at all resolutions between $\Delta x = 1/40$ and $\Delta x = 1/320$ the actual error computed with the MC flux-limited scheme is lower than that obtained using any of the other methods (including the Lax–Wendroff scheme).

A final example is provided by the test problem from Chap. 3 in which the initial condition is the superposition of equal-amplitude $7.5\Delta x$ and $10\Delta x$ waves. The numerical parameters for this test are identical to those described in connection with Fig. 5.10b. Figure 5.14a compares the minmod, MC, and superbee flux-limited solutions to these problems. As was the case with the propagating step considered in Fig. 5.12a, the superbee limiter gives the best results, the minmod limiter is too diffusive, and the MC limiter is almost as good as the superbee. Figure 5.14b shows a comparison of the MC flux-limited and FCT solutions to the same problem. The superiority of the MC flux-limited solution over the FCT solution is not as clear-cut as in the examples shown in Fig. 5.13. The flux-limited solution is more heavily damped but exhibits less phase-speed error than that obtained with FCT.

The performance of all the schemes shown in Figs. 5.12–5.14 is improved by increasing the Courant number toward unity, since the upstream and Lax–Wendroff methods both give perfect results when $\mu = 1$. In practical applications with a temporally and spatially varying wind field there is, however, no hope of stepping the solution forward at each grid point using a local Courant number of unity. The solutions shown in Figs. 5.10–5.14 were obtained using $\mu = 0.5$ and are similar to those obtained at smaller Courant numbers. Note that neither the FCT nor the flux-limited methods approach the accuracy obtained on the test problems in Figs. 5.13b and 5.14 using a simple explicit fourth-order spatial difference and an accurate time difference. Although FCT and flux-limiter methods are highly useful in problems with discontinuities and unresolved gradients, conventional finite-difference schemes may perform much better when the solution is at least moderately resolved. The resolution required to make a high-order finite-difference method attractive need not be particularly high; the $7.5\Delta x$-wavelength component in the solution shown in Fig. 5.14 is certainly not well resolved.

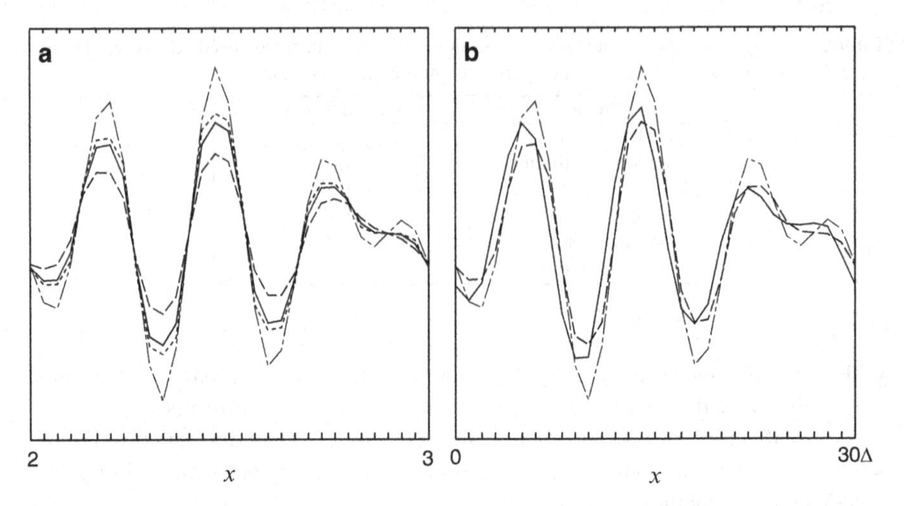

Fig. 5.14 Comparison of numerical approximations to an advection problem whose exact solution (shown by the *thin dot-dashed line*) consists of equal-amplitude $7.5\Delta x$ and $10\Delta x$ waves: **a** flux-limited approximations using the superbee (*short-dashed line*), MC (*solid line*), and minmod (*long-dashed line*) limiters; **b** MC flux-limited solution (*long-dashed* line) and the FCT solution (*solid line*)

5.5.3 Flow Velocities of Arbitrary Sign

To accommodate velocities of arbitrary sign, the definitions of $F_{j+1/2}$ and $r_{j+1/2}$ must be modified as follows. The Lax–Wendroff flux (5.36) may be expressed in terms of the upstream flux for advection by a velocity of arbitrary sign (5.35) as

$$F^{\mathrm{h}}_{j+\frac{1}{2}} = F^{\mathrm{l}}_{j+\frac{1}{2}} + \frac{|c|}{2}\left(1 - \frac{|c|\Delta t}{\Delta x}\right)\left(\phi^n_{j+1} - \phi^n_j\right). \tag{5.47}$$

The total corrected flux may therefore be expressed as

$$F^n_{j+\frac{1}{2}} = \frac{c}{2}\left(\phi^n_{j+1} + \phi^n_j\right) - \frac{1}{2}\left[(1 - C^n_{j+\frac{1}{2}})|c| + \frac{c^2\Delta t}{\Delta x}C^n_{j+\frac{1}{2}}\right]\left(\phi^n_{j+1} - \phi^n_j\right). \tag{5.48}$$

The value of $r_{j+1/2}$ used in the evaluation of the flux limiter $C^n_{j+1/2}$ should be computed as the ratio of the slope of the solution across the cell interface upstream of $j + 1/2$ to the slope of the solution across the interface at $j + 1/2$. Defining $\gamma = -\mathrm{sgn}(c_j)$, this ratio becomes

$$r^n_{j+\frac{1}{2}} = \frac{\phi^n_{j+\gamma+1} - \phi^n_{j+\gamma}}{\phi^n_{j+1} - \phi^n_j}.$$

If the velocity varies as a function of time, an $O\left[(\Delta t)^2\right]$ approximation to the velocity at $(n + 1/2)\Delta t$ should be used in (5.48) to preserve second-order accuracy (at least at locations away from the extrema of ϕ). Suitable approximations can be obtained by averaging the velocities between time levels n and $n+1$, or by extrapolating forward from time levels n and $n-1$ (see Problem 11). A formula for the approximation of advective fluxes in spatially varying nondivergent velocity fields is presented in Sect. 5.9.3.

5.6 Subcell Polynomial Reconstruction

Formulae very similar to those obtained with the flux-limiter approach can be derived by approximating the solution within each grid volume as a piecewise-linear function that has been reconstructed from the cell-averaged values. The evolution of these relatively simple piecewise-linear functions over a time interval Δt is then computed (or in more difficult problems, approximated). In the final step, the updated piecewise-linear solution is averaged over each grid cell. Van Leer (1974, 1977) developed the piecewise-linear approach in a series of five papers building on a similar approach introduced by Godunov (1959), who used piecewise-constant functions. Colella and Woodward (1984) subsequently extended the method to use piecewise parabolas.

The following sections present piecewise-constant approximations to nonlinear one-dimensional scalar conservation laws of the form (5.7) and both piecewise-linear and piecewise-parabolic approximations to the constant-wind-speed advection equation. The discontinuous Galerkin method, discussed in Sect. 6.6, may be considered an extension of this approach to even higher order polynomials.

5.6.1 Godunov's Method

In Godunov's method, the cell-averaged values at each individual time step are used to define a piecewise-constant function such that

$$\tilde{\phi}(x,t^n) = \phi_j^n \quad \text{for} \quad x_{j-\frac{1}{2}} \le x \le x_{j+\frac{1}{2}},$$

where $t^n = n\Delta t$ and $x_{j+1/2} = x_j + \Delta x/2$. Using the function $\tilde{\phi}(x,t^n)$ as the initial condition, one may obtain an approximate solution to the original conservation law at t^{n+1} by solving the Riemann problems associated with the discontinuities in $\tilde{\phi}$ at the interface of each grid cell. The exact solution to these Riemann problems can be easily obtained for a scalar conservation law or for linear systems of conservation laws, at least until the signals emanating from each interface begin to interact.[11]

[11] See LeVeque (2002) for a discussion of approximate techniques for the solution of Riemann problems involving nonlinear systems of conservation laws.

As the final step in the integration cycle, ϕ_j^{n+1} is obtained by averaging the individual Riemann solutions over the jth grid cell,

$$\phi_j^{n+1} = \frac{1}{\Delta x} \int_{x_{j-\frac{1}{2}}}^{x_{j+\frac{1}{2}}} \tilde{\phi}(x, t^{n+1})\, \mathrm{d}x.$$

In fact, it is not necessary to actually compute the solutions to each Riemann problem, since the integral form of the conservation law (5.8) implies that

$$\phi_j^{n+1} = \frac{1}{\Delta x} \int_{x_{j-\frac{1}{2}}}^{x_{j+\frac{1}{2}}} \tilde{\phi}(x, t^n)\, \mathrm{d}x - \frac{1}{\Delta x} \int_{t^n}^{t^{n+1}} f\left[\tilde{\phi}(x_{j+\frac{1}{2}}, t)\right] \mathrm{d}t$$
$$+ \frac{1}{\Delta x} \int_{t^n}^{t^{n+1}} f\left[\tilde{\phi}(x_{j-\frac{1}{2}}, t)\right] \mathrm{d}t. \tag{5.49}$$

The first integral in the preceding equation is simply ϕ_j^n. The other two integrals may be trivially evaluated, provided that the integrand is constant over the time $t^n < t < t^{n+1}$, which will be the case if the Courant number, $|c\Delta t/\Delta x|$, is less than unity (where c is the speed of the fastest-moving wave or shock). Note that the maximum time step permitted by this condition allows the Riemann solutions to interact within each grid cell, but these interactions can be ignored, since they do not change the value of $\tilde{\phi}$ at the cell interfaces and therefore do not complicate the evaluation of the integrals in (5.49).

The solution of the Riemann problem at each cell interface is determined by the initial values of $\tilde{\phi}$ on each side of the interface. In most cases, disturbances in the form of waves or shocks will propagate either rightward or leftward from the cell interface, and the fluxes in (5.49) will be correctly evaluated if $\tilde{\phi}$ is replaced by the value of ϕ^n that is upstream of the interface with respect to the propagation of the wave or shock. For smooth ψ, (5.7) may be expressed in the advective form

$$\frac{\partial \psi}{\partial t} + \frac{df}{d\psi}\frac{\partial \psi}{\partial x} = 0, \tag{5.50}$$

which shows that $\mathrm{d}f/\mathrm{d}\psi$ is the speed at which smooth perturbations in ψ propagate along the x-axis. Thus, one might approximate the solution with a finite-volume method

$$\phi_j^{n+1} = \phi_j^n - \frac{\Delta t}{\Delta x}\left[F(\phi_{j+}^n) - F(\phi_{j-}^n)\right] \tag{5.51}$$

in which the upstream direction is estimated using a numerical approximation to $\mathrm{d}f/\mathrm{d}\psi$ such that

$$F(\phi_{j+}) = F(\phi_{(j+1)-}) = \begin{cases} f(\phi_j) & \text{if } [f(\phi_{j+1}) - f(\phi_j)]/[\phi_{j+1} - \phi_j] \geq 0, \\ f(\phi_{j+1}) & \text{otherwise.} \end{cases} \tag{5.52}$$

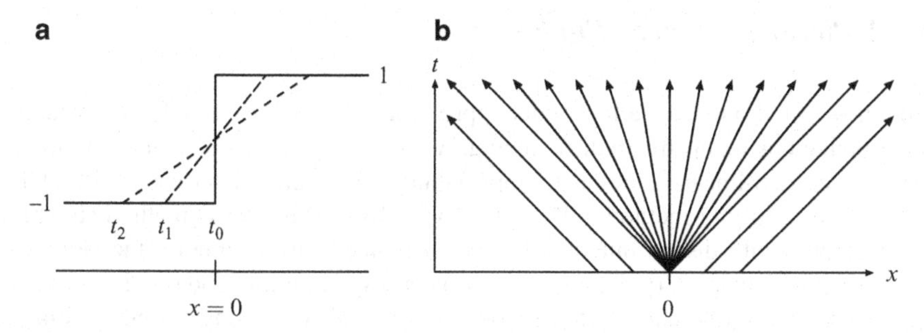

Fig. 5.15 **a** A transonic rarefaction wave; the position of the left edge of the wave front is indicated at three consecutive time intervals. **b** Characteristic curves associated with this wave

According to the Rankine–Hugoniot condition (5.11), the upstream flux is also correctly selected when the solution contains a discontinuity in the form of a propagating jump.

An erroneous result can, however, be generated if the entropy-consistent solution to the Riemann problem at a cell interface is a rarefaction wave in which the disturbance spreads both to the right and to left of the interface. For example, if the solution to Burgers's equation (5.12) is approximated using this scheme with initial data

$$\phi_j^0 = \begin{cases} 1 & \text{if } j \geq 0, \\ -1 & \text{if } j < 0, \end{cases}$$

the numerical solution will be a steady entropy-violating shock, since $F(\phi_{j+}^n) = F(\phi_{j-}^n) = 1/2$ for all j and n. The correct entropy-consistent solution is the rarefaction wave, or expansion fan, illustrated in Fig. 5.15a.

Rarefaction waves in which $\mathrm{d}f/\mathrm{d}\psi$ passes through zero at some point within the wave are known as *transonic rarefaction* waves. As a result of the transonic rarefaction, $\tilde{\phi}(x_{j+1/2}, t)$ assumes the value of ϕ for which the phase speed of the wave is zero (i.e., the value of ϕ for which the characteristics are parallel to the t-axis in the x–t plane – see Fig. 5.15b). Entropy-consistent solutions to the Riemann problem at each interface will be obtained if the upstream fluxes are determined according to the prescription

$$F(\phi_{j+}) = F(\phi_{(j+1)-}) = \begin{cases} \displaystyle\min_{\phi_j \leq \phi \leq \phi_{j+1}} f(\phi) & \text{if } \phi_j \leq \phi_{j+1}, \\ \displaystyle\max_{\phi_{j+1} \leq \phi \leq \phi_j} f(\phi) & \text{if } \phi_j > \phi_{j+1} \end{cases} \tag{5.53}$$

(LeVeque 1992, p. 145). Let ϕ_s be the value of ϕ for which the phase speed of the wave is zero in the transonic rarefaction. In the case shown in Fig. 5.15, the flux obtained from (5.53) will be $f(\phi_s)$ because the minimum value of $f(\phi)$ occurs when the local phase speed, $\mathrm{d}f/\mathrm{d}\phi$, is zero.

5.6.2 Piecewise-Linear Functions

Godunov's method yields a first-order approximation that is essentially identical to that obtained using upstream differencing. A second-order method can be obtained using piecewise-linear functions to approximate the solution over each grid interval, but the resulting method will not be TVD. To obtain a TVD method suitable for problems with discontinuous solutions, it is necessary to modify the slope of the piecewise-linear interpolating functions near discontinuities and poorly resolved gradients. This modification of the slope is accomplished using "slope-limiting" algorithms that are closely related to the flux-limiting procedures discussed in Sect. 5.5.2.

Although the actual computations may be organized in a more efficient manner, the procedure for advancing the numerical solution one time step is equivalent to the following three-step process. In the first step, the cell-averaged values are used to "reconstruct" a piecewise-linear function within each grid cell of the form

$$\tilde{\phi}(x,t^n) = \phi_j^n + \sigma_j^n(x - x_j) \quad \text{for} \quad x_{j-\frac{1}{2}} \leq x \leq x_{j+\frac{1}{2}}.$$

In the second step, the conservation law is integrated over a time Δt using $\tilde{\phi}(x,t^n)$ as the initial condition. In the third and final step, ϕ_j^{n+1} is obtained by averaging $\tilde{\phi}(x,t^{n+1})$ over each grid cell. The special considerations required to keep this method TVD are primarily connected with the first step, since if the conservation law is of the form (5.7), the solution obtained in the second step is TVD and the averaging in the third step does not increase, and often decreases, the total variation. The increase in total variation permitted in the first step must be kept small enough that the averaging in the third step yields a TVD scheme. The maximum increase in total variation during the first step is controlled by imposing limits on the slopes σ_j^n; the most severe limitation would be to set $\sigma_j^n = 0$, in which case the scheme reduces to Godunov's method.

In comparison with Godunov's method, in which exact solutions of the conservation law can be obtained relatively easily at each cell interface by solving a series of Riemann problems, the problems to be solved at each interface in the second step of the piecewise-linear method are more difficult, because $\tilde{\phi}$ is not constant on each side of the initial discontinuity. General techniques for obtaining acceptable approximations to the solution required in the second step are discussed in LeVeque (2002). In the following we will once again focus on the special case of the constant-wind-speed advection equation (5.20), for which the solution required in the second step is simply

$$\tilde{\phi}(x,t^{n+1}) = \tilde{\phi}(x - c\Delta t, t^n).$$

Assuming that $c > 0$ and averaging $\tilde{\phi}(x,t^{n+1})$ over $(x_{j-1/2}, x_{j+1/2})$, the second and third steps yield

$$\phi_j^{n+1} = \phi_j^n - \mu\left(\phi_j^n - \phi_{j-1}^n\right) - \frac{\mu}{2}(1-\mu)\Delta x(\sigma_j^n - \sigma_{j-1}^n), \tag{5.54}$$

where $\mu = c\Delta t/\Delta x$. The preceding equation can be recast in the conservation form (5.17) by setting

$$F_{j+\frac{1}{2}} = c\left(\phi_j^n + \tfrac{1}{2}(1-\mu)\Delta x\, \sigma_j^n\right). \tag{5.55}$$

The same expression for $F_{j+1/2}$ can be obtained by evaluating the flux swept out through the cell boundary at $x_{j+1/2}$ in time Δt using the relation

$$\Delta t\, F_{j+\frac{1}{2}} = \int_{x_{j+\frac{1}{2}}-c\Delta t}^{x_{j+\frac{1}{2}}} \tilde{\phi}_j(x, t^n)\, \mathrm{d}x.$$

If the slopes of the piecewise-linear functions are defined such that

$$\sigma_j = \frac{\phi_{j+1} - \phi_j}{\Delta x}, \tag{5.56}$$

then (5.54) reduces to the Lax–Wendroff method, which is not TVD. To make (5.54) TVD, the slope can be limited by a multiplicative constant $C_{j+1/2}$ such that

$$\sigma_j = \left(\frac{\phi_{j+1} - \phi_j}{\Delta x}\right) C_{j+\frac{1}{2}}, \tag{5.57}$$

in which case (5.55) becomes (5.39) and the scheme is identical to the flux-limited Lax–Wendroff method. As with the family of flux-limiter methods, there are a variety of reasonable choices for $C_{j+1/2}$, and every flux limiter defined in Sect. 5.5.2 can be reinterpreted as a slope limiter via (5.57). Indeed, the behavior of some flux limiters is easier to understand when they are interpreted as slope limiters. For example, let

$$a = \frac{\phi_{j+1} - \phi_j}{\Delta x}, \qquad b = \begin{cases} \dfrac{\phi_j - \phi_{j-1}}{\Delta x} & \text{if } c \geq 0, \\[2ex] \dfrac{\phi_{j+2} - \phi_{j+1}}{\Delta x} & \text{if } c < 0, \end{cases}$$

and define the minmod function MM such that

$$\mathrm{MM}(v_1, v_2, \ldots) = \begin{cases} \min(v_1, v_2, \ldots) & \text{if } v_j > 0 \text{ for all } j, \\ \max(v_1, v_2, \ldots) & \text{if } v_j < 0 \text{ for all } j, \\ 0 & \text{otherwise.} \end{cases} \tag{5.58}$$

Then the minmod limiter defined by (5.43) implies that $\sigma_j = \mathrm{MM}(a, b)$, which guarantees that the magnitude of the slope of ϕ between grid points j and $j+1$ is no larger than the slope over the next interval upstream, except that the slope is set to zero at an extremum of ϕ.

The effect of the minmod limiter is illustrated in Fig. 5.16. Note that the minmod limiter ensures that the total variation associated with the grid average values themselves is not increased when the piecewise-linear approximation $\tilde{\phi}(x, t^n)$ is created. Nevertheless, flux-limiter methods can be TVD even if $TV[\tilde{\phi}(x, t^n)] > TV(\phi^n)$

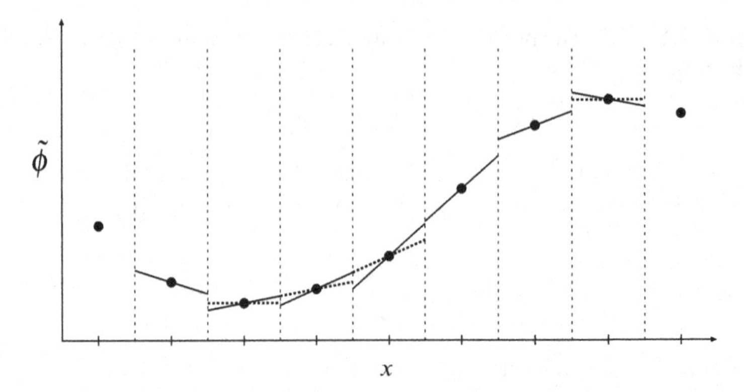

Fig. 5.16 A piecewise-linear finite-volume approximation. Cell-average values are plotted as *heavy points*. The *solid line segments* show the $\tilde{\phi}(x)$ obtained using the Lax–Wendroff slopes (5.56). *Dashed line segments* show the modification to $\tilde{\phi}(x)$ introduced by the minmod slope limiter assuming $c > 0$. Slopes in the rightmost and leftmost grid cells are omitted

because the total variation of the cell averages at time t^{n+1} is reduced when $\tilde{\phi}(x, t^{n+1})$ is averaged over each cell to determine ϕ^{n+1}. Indeed, the most effective slope limiters allow the total variation of the piecewise-linear approximation to exceed $TV(\phi^n)$. For example, the MC limiter (5.46) implies that $\sigma_j = \text{MM}(2a, 2b, (a+b)/2)$. Whenever $ab > 0$, the slope allowed by the MC limiter is steeper than that allowed by the minmod limiter. Assuming the adjacent slopes have the same sign, the MC limiter sets σ_j to the average of a and b, unless one of these slopes exceeds the other by more than a factor of 3, in which case σ_j is set to twice the value for the gentler slope.

5.6.3 The Piecewise-Parabolic Method

Third-order accuracy can be obtained using piecewise-quadratic functions, without limiting, to define the structure within each cell. The piecewise-parabolic method not only gives higher-order asymptotic accuracy than piecewise-linear approximations, it also generates less damping at coarsely resolved wavelengths such as $4\Delta x$ (Lauritzen 2007). Extremely short wavelengths, such as $2\Delta x$, are, however, not improved by switching from piecewise-linear to piecewise-quadratic polynomials.

Consider again the problem of advection of a scalar tracer at a constant speed c. A formula for the flux through the cell interface at $x_{j+1/2}$ that is valid for both positive and negative fluid velocities can be obtained as follows. Let the values of the scalar at the right and left edges of the jth cell be ϕ_{j+} and ϕ_{j-}, respectively. These values are interpolated using a four-point stencil of cell-average values centered around each edge such that

$$\phi_{j+} = \phi_{(j+1)-} = \left[7(\phi_j + \phi_{j+1}) - (\phi_{j-1} + \phi_{j+2})\right]/12. \tag{5.59}$$

The parabolic reconstruction within the finite volume is initially set to match the values at the right and left edges of each cell and to have same cell-average concentration, ϕ_j. As will be discussed shortly, the coefficients defining each parabola may then be adjusted to ensure the resulting scheme does not generate unphysical ripples or new local extrema.

Suppose $c \geq 0$, and define a local coordinate within the jth cell as $\xi = (x_{j+1/2} - x)/\Delta x$. This choice of ξ facilitates the evaluation of the flux by placing the origin at the edge of the cell where the flux will be evaluated; note that ξ decreases linearly from 1 at $x_{j-1/2}$ to 0 at $x_{j+1/2}$. The parabolic interpolant within cell j may be expressed in the form

$$\tilde{\phi}_j(\xi) = a_0 + a_1\xi + a_2\xi^2,$$

where

$$a_0 = \phi_{\xi_0}, \tag{5.60}$$

$$a_1 = -4\phi_{\xi_0} - 2\phi_{\xi_1} + 6\overline{\phi}, \tag{5.61}$$

$$a_2 = 3\phi_{\xi_0} + 3\phi_{\xi_1} - 6\overline{\phi}, \tag{5.62}$$

and

$$\phi_{\xi_0} = \tilde{\phi}_j(0) = \phi_{j+}, \quad \phi_{\xi_1} = \tilde{\phi}_j(1) = \phi_{j-}, \quad \overline{\phi} = \phi_j.$$

Again let $\mu = c\Delta t/\Delta x$. As illustrated in Fig. 5.17a, the flux swept across $x_{j+1/2}$ in time Δt satisfies

$$\Delta t F_{j+\frac{1}{2}} = \int_{x_{j+\frac{1}{2}}-c\Delta t}^{x_{j+\frac{1}{2}}} \tilde{\phi}_j(x)\,\mathrm{d}x = \int_{\mu}^{0} \tilde{\phi}_j(\xi)\frac{dx}{d\xi}\,\mathrm{d}\xi = \Delta x \int_0^{\mu} a_0 + a_1\xi + a_2\xi^2\,\mathrm{d}\xi,$$

which implies

$$F_{j+\frac{1}{2}} = c\left(a_0 + \frac{a_1}{2}|\mu| + \frac{a_2}{3}\mu^2\right). \tag{5.63}$$

Since $\mu \geq 0$ when $c \geq 0$, it is not necessary to specify the absolute value of μ in the preceding equation; however, the absolute value must be used in the $c < 0$ case.

The flux across $x_{j+1/2}$ must be computed in an upwind direction to satisfy the Courant–Friedrichs–Lewy condition that the numerical domain of dependence include the domain of dependence of the true solution. Thus, if $c < 0$, the relevant parabola is that within the cell centered at x_{j+1}. This parabola is defined with respect to $\xi = (x - x_{j+1/2})/\Delta x$ such that

$$\tilde{\phi}_{j+1}(\xi) = a_0 + a_1\xi + a_2\xi^2,$$

and

$$\phi_{\xi_0} = \phi_{(j+1)-}, \quad \phi_{\xi_1} = \phi_{(j+1)+}, \quad \overline{\phi} = \phi_{j+1}.$$

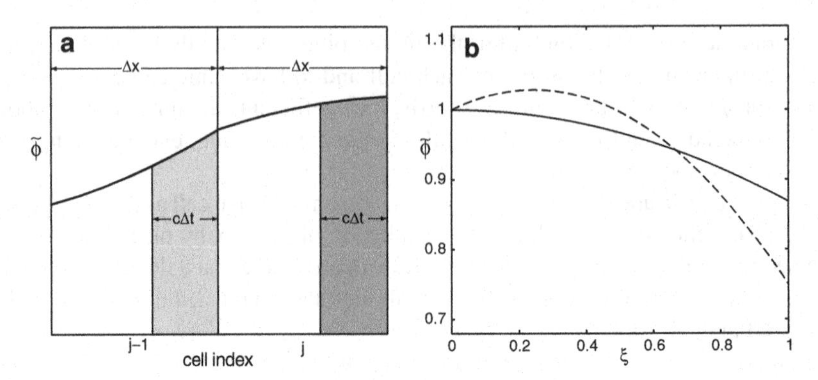

Fig. 5.17 **a** A piecewise-parabolic approximation for the two cells spanning the interval $[x_{j-3/2}, x_{j+1/2}]$. Assuming $c > 0$, the flux swept across $x_{j+1/2}$ in time Δt has the same numerical value as the area of the *dark shaded region* of width $c\Delta t$. The flux swept across $x_{j-1/2}$ has the same value as the area as the *lightly shaded region*. **b** A parabola with a local maximum inside the cell (*dashed line*) and the modified monotonic parabola (*solid line*)

Note that in the $c < 0$ case, ξ increases from 0 at $x_{j+1/2}$ to 1 at $x_{j+3/2}$. The definitions (5.60)–(5.62) are unchanged, and $\Delta t F_{j+1/2}$ is now given by

$$\int_{x_{j+\frac{1}{2}}-c\Delta t}^{x_{j+\frac{1}{2}}} \tilde{\phi}_{j+1}(x)\,\mathrm{d}x = \int_{-\mu}^{0} \tilde{\phi}_{j+1}(\xi)\frac{dx}{d\xi}\,\mathrm{d}\xi = \Delta x \int_{-\mu}^{0} a_0 + a_1\xi + a_2\xi^2\,\mathrm{d}\xi.$$

It follows that the flux $F_{j+1/2}$ may again be expressed in the form (5.63).

The preceding method can be modified to avoid the generation of spurious ripples near steep gradients by ensuring that no new local maxima or minima appear when the parabolic distribution of $\tilde{\phi}$ within each cell is reconstructed from the cell-average values. There are two opportunities for the generation of new maxima and minima. The first occurs if values of ϕ at the cell interfaces, estimated from (5.59), do not lie within the range of the ϕ_j in the neighboring cells. The second way new extrema can be generated is if the reconstructed parabola has a maximum or minimum within the cell itself. A variety of algorithms have been proposed to accomplish this limiting, beginning with the seminal paper by Colella and Woodward (1984). The strategy presented below largely follows Zerroukat et al. (2006), except that no attempt is made to avoid limiting smooth extrema. Methods for identifying and avoiding the spurious limiting of smooth extrema will be discussed in Sect. 5.8.

To avoid generating new extrema at the cell interfaces, compute

$$\phi_{j\pm}^{\mathrm{mon}} = \min(\phi_{j\pm\frac{1}{2}}^{\max}, \max(\phi_{j\pm\frac{1}{2}}^{\min}, \phi_{j\pm})),$$

where $\phi_{j\pm 1/2}^{\max,\min} = \max, \min(\phi_j, \phi_{j\pm 1})$. The values of $\phi_{j\pm}^{\mathrm{mon}}$ will lie in the interval $[\phi_j, \phi_{j\pm 1}]$ and should be used in lieu of $\phi_{j\pm}$ in the evaluation of ϕ_{ξ_0} and ϕ_{ξ_1}.

To avoid the generation of extrema inside a cell, check if $\tilde{\phi}'(\xi) = 0$ for $0 < \xi < 1$, or equivalently, if $0 < -a_1/(2a_2) < 1$. If this condition is met, the parabola must be modified to remove a local extremum. There are two possible cases:

1. If $(\overline{\phi} - \phi_{\xi_0})(\overline{\phi} - \phi_{\xi_1}) > 0$, the parabola is replaced by the piecewise-constant function $\tilde{\phi}(\xi) = \overline{\phi}$.
2. Otherwise a new monotonic parabola can be constructed with the correct cell-average value, whose values at the edge of the cell lie within the range of the ϕ_j in the neighboring cells.

   (a) If $|\overline{\phi} - \phi_{\xi_1}| > |\overline{\phi} - \phi_{\xi_0}|$, the extremum is moved to $\xi = 0$ by replacing (5.60)–(5.62) with

   $$a_0 = \phi_{\xi_0}, \quad a_1 = 0, \quad a_2 = -3\phi_{\xi_0} + 3\overline{\phi}.$$

   (b) If $|\overline{\phi} - \phi_{\xi_1}| < |\overline{\phi} - \phi_{\xi_0}|$, the extremum is moved to $\xi = 1$ by setting

   $$a_0 = -2\phi_{\xi_1} + 3\overline{\phi}, \quad a_1 = 6\phi_{\xi_1} - 6\overline{\phi}, \quad a_2 = -3\phi_{\xi_1} + 3\overline{\phi}.$$

   Both the original and the modified parabolas for case 2a are plotted in Fig. 5.17b. Note that value of $\tilde{\phi}(1)$ for the modified parabola lies between $\overline{\phi}$ and the value of ϕ_{ξ_1} for the original parabola, and as such remains bounded by the values of ϕ_j on each side of the interface.

Examples showing the behavior of the piecewise-parabolic method will be presented in Sect. 5.9.5.

5.7 Essentially Nonoscillatory and Weighted Essentially Nonoscillatory Methods

As noted in Sect. 5.6.2, when the minmod limiter is used to limit the Lax–Wendroff flux at the cell interface $x_{j+1/2}$, it compares the slope of the piecewise-linear function interpolating the solution at the endpoints of the interval $[x_j, x_{j+1}]$ with the slope over the next interval upstream and, provided both slopes have the same sign, it selects the gentler of the two. This strategy of using an adaptive stencil to choose the "smoothest" polynomial interpolant can be extended to higher-order schemes using essentially nonoscillatory (ENO) and weighted essentially nonoscillatory (WENO) methods. The primary advantage of ENO and WENO methods in comparison with flux-limiter methods is not in their treatment of discontinuities, but rather in their ability to maintain genuinely high order accuracy in the vicinity of smooth maxima and minima.

Both ENO and WENO methods may be applied in either a finite-volume or a finite-difference context. Here we will consider only the finite-difference approach because it generalizes very easily to higher-dimensional problems. Indeed, one significant advantage of the finite-difference ENO and WENO methods in comparison with FCT and flux-limiter methods is the ease with which they may be applied in higher dimensions.

5.7.1 Accurate Approximation of the Flux Divergence

The finite-difference approximation to the scalar conservation law (5.7) is

$$\frac{\phi_j^{n+1} - \phi_j^n}{\Delta t} + \left(\frac{F_{j+\frac{1}{2}} - F_{j-\frac{1}{2}}}{\Delta x} \right) = 0, \tag{5.64}$$

where $F_{j\pm1/2}$ are numerical approximations to $f[\psi(x_{j\pm1/2})]$ and the grid spacing Δx is uniform. The preceding expression is in conservation form, but in contrast to the finite-volume formalism, ϕ_j approximates the grid-point value $\psi(x_j)$ rather than a cell-average value.

As a first step we defer matters related to the time discretization and the distinction between ENO and WENO methods, and simply seek approximations for the numerical fluxes such that

$$\frac{F_{j+\frac{1}{2}} - F_{j-\frac{1}{2}}}{\Delta x} = \frac{d f(\psi)}{dx}(x_j) + O\left[(\Delta x)^k\right]. \tag{5.65}$$

An error-free expression for the flux divergence could be obtained if we knew a function $h(x)$ for which

$$f[\psi(x)] = \frac{1}{\Delta x} \int_{x-\Delta x/2}^{x+\Delta x/2} h(\xi)\, d\xi, \tag{5.66}$$

since differentiating the preceding equation yields

$$\frac{d f(\psi)}{dx}(x) = \frac{h(x + \Delta x/2) - h(x - \Delta x/2)}{\Delta x},$$

implying that (5.65) could be satisfied exactly by choosing $F_{j+1/2} = h(x_{j+1/2})$. The next goal, therefore, will be to determine a polynomial $p(x)$ that approximates $h(x)$ with sufficient accuracy that (5.65) may be satisfied by setting $F_{j+1/2} = p(x_{j+1/2})$.

Provided the leading-order term in the truncation error is smooth, sufficient accuracy may be achieved if

$$F_{j+\frac{1}{2}} \equiv p(x_{j+\frac{1}{2}}) = h(x_{j+\frac{1}{2}}) + O\left[(\Delta x)^k\right]. \tag{5.67}$$

One might suppose that $O\left[(\Delta x)^{k+1}\right]$ accuracy would be required because of the division by Δx in (5.65), but letting $E(x)$ be the smooth leading-order error, and noting that both E and $\partial E/\partial x$ will be $O\left[(\Delta x)^k\right]$; the error at $x_{j+1/2}$ may be expanded in a Taylor series about $x_{j-1/2}$ such that

$$F_{j+\frac{1}{2}} = h(x_{j+\frac{1}{2}}) + E(x_{j-\frac{1}{2}}) + \Delta x \frac{\partial E}{\partial x}(x_{j-\frac{1}{2}}) + O\left[(\Delta x)^{k+1}\right].$$

It follows that $F_{j+1/2} - F_{j-1/2}$ is $O\left[(\Delta x)^{k+1}\right]$, and that (5.67) provides sufficient accuracy.

The connection between the grid-point values ϕ_j and the polynomial that approximates $h(x)$ is realized through the *primitive* of h,

$$H(x) = \int_{-\infty}^{x} h(\xi)\,\mathrm{d}\xi. \tag{5.68}$$

Although the lower limit of the preceding integral is nominally $-\infty$, divided differences of H are all that is ultimately required, so alternatively, the lower limit of this integral could be specified at any fixed point well to the "left" of the polynomial stencil. Using (5.68) and then (5.66),

$$H(x_{j+\frac{1}{2}}) = \sum_{m=-\infty}^{j} \int_{x_{m-\frac{1}{2}}}^{x_{m+\frac{1}{2}}} h(\xi)\,\mathrm{d}\xi = \Delta x \sum_{m=-\infty}^{j} f[\psi(x_j)], \tag{5.69}$$

which provides a formula for determining $H_{j+1/2} \equiv H(x_{j+1/2})$ from the grid-point data.

Let $P(x)$ be the unique polynomial of degree k that interpolates H at $k+1$ consecutive points on the cell boundaries

$$x_{j-\ell-\frac{1}{2}}, \ldots, x_{j-\ell+k-\frac{1}{2}}, \tag{5.70}$$

with integer ℓ satisfying $0 \leq \ell < k$, so that $x_{j-1/2}$ and $x_{j+1/2}$ are included in the interpolation stencil. The desired polynomial approximation to $h(x)$ is the derivative of this function, $p(x) \equiv \mathrm{d}P/\mathrm{d}x$. For one thing, the average of p over the cell at x_j is identical to the average of h over the same cell since

$$\begin{aligned}
\int_{x_{j-\frac{1}{2}}}^{x_{j+\frac{1}{2}}} p(\xi)\,\mathrm{d}\xi = \int_{x_{j-\frac{1}{2}}}^{x_{j+\frac{1}{2}}} \frac{dP}{dx}(\xi)\,\mathrm{d}\xi &= P(x_{j+\frac{1}{2}}) - P(x_{j-\frac{1}{2}}) \\
&= H(x_{j+\frac{1}{2}}) - H(x_{j-\frac{1}{2}}) \\
&= \int_{-\infty}^{x_{j+\frac{1}{2}}} h(\xi)\,\mathrm{d}\xi - \int_{-\infty}^{x_{j-\frac{1}{2}}} h(\xi)\,\mathrm{d}\xi \\
&= \int_{x_{j-\frac{1}{2}}}^{x_{j+\frac{1}{2}}} h(\xi)\,\mathrm{d}\xi;
\end{aligned}$$

here the third equality follows because $P(x)$ interpolates $H(x)$ at the cell boundaries. The accuracy requirement (5.67) is also satisfied because, from basic interpolation theory, $P(x)$ is an $O\left[(\Delta x)^{k+1}\right]$ approximation to $H(x)$, so their derivatives $p(x)$ and $h(x)$ are equal to $O\left[(\Delta x)^{k}\right]$.

Given initial data $\phi_j = \psi(x_j)$, one could therefore obtain approximations to the fluxes satisfying (5.65) in the following manner: (1) evaluate the $H_{j+1/2}$ from the $f(\phi_j)$; (2) choose the stencil over which the polynomial is defined by choosing ℓ, the left shift of the stencil with respect to $x_{j-1/2}$; (3) form the polynomial P that interpolates the $H_{j+1/2}$ and differentiate it to obtain $p(x)$; and (4) set $F_{j+1/2} = p(x_{j+1/2})$. Since Δx is uniform it is never necessary to actually follow this conceptual procedure; once the stencil of points over which the interpolation is to be performed has been specified, the weighting coefficients for the data at each grid point may be obtained from a precomputed table. In general, one arrives at a relation of the form

$$F_{j+\frac{1}{2}} = \sum_{s=0}^{k-1} c_{\ell,s} f(\phi_{j-\ell+s}), \tag{5.71}$$

where the constants $c_{\ell,s}$ are determined from the table by the order of the interpolation k and by the left shift of the stencil ℓ relative to grid point x_j.

5.7.2 ENO Methods

Standard finite-difference approximations can be obtained via the procedure just described by choosing a fixed interpolation stencil, i.e., by choosing a fixed value for the left shift ℓ. When designing schemes that are at least second-order accurate, there is more than one possible choice for ℓ. For example, if a second-order scheme is desired ($k = 2$), there are two sets of consecutive points defined by (5.70) containing the interval $[x_{j-1/2}, x_{j+1/2}]$. Using the stencil $\{x_{j-3/2}, x_{j-1/2}, x_{j+1/2}\}$, corresponding to the choice $\ell = 1$, one obtains

$$F_{j+\frac{1}{2}} = -\frac{1}{2} f(\phi_{j-1}) + \frac{3}{2} f(\phi_j), \tag{5.72}$$

whereas the stencil $\{x_{j-1/2}, x_{j+1/2}, x_{j+3/2}\}$, which is obtained if $\ell = 0$, yields

$$F_{j+\frac{1}{2}} = \frac{1}{2} f(\phi_j) + \frac{1}{2} f(\phi_{j+1}). \tag{5.73}$$

The key idea in the ENO method is to adaptively choose the interpolation stencil to obtain the "smoothest" approximation over the interval $[x_{j-1/2}, x_{j+1/2}]$. If, for example, there is an isolated discontinuity in the interval $[x_{j-3/2}, x_{j-1/2}]$, one would avoid the stencil that includes this interval and compute the flux according to (5.73) in preference to (5.72). The smoothest stencil is determined by successively examining each potential higher-order contribution to $P(x)$ as expressed in the form of a Newton polynomial.

A Newton polynomial interpolant is constructed from *divided differences* of the gridded data. Divided differences of a function $\alpha(x)$ are denoted by square brackets and are defined such that $\alpha[x_0] = \alpha(x_0)$,

$$\alpha[x_0, x] = \frac{\alpha(x) - \alpha(x_0)}{x - x_0},$$

and through the recursion

$$\alpha[x_0, \ldots, x_{m-1}, x_m, x] = \frac{\alpha[x_0, \ldots, x_{m-1}, x] - \alpha[x_0, \ldots, x_{m-1}, x_m]}{x - x_m}.$$

If $P(x)$ is a quadratic polynomial interpolating $H(x)$ at the points $\{x_a, x_b, x_c\}$, then

$$P(x) = H[x_a] + H[x_a, x_b](x - x_a) + H[x_a, x_b, x_c](x - x_a)(x - x_b), \quad (5.74)$$

as may be verified by noting that $P(x_a) = H[x_a] = H(x_a)$,

$$\begin{aligned} P(x_b) = H[x_a] + H[x_a, x_b](x_b - x_a) &= H(x_a) + \frac{H(x_b) - H(x_a)}{x_b - x_a}(x_b - x_a) \\ &= H(x_b), \end{aligned}$$

and so on. Since $P(x)$ is constructed to interpolate H on the interval $[x_{j-1/2}, x_{j+1/2}]$, we choose $x_a = x_{j-1/2}$ and $x_b = x_{j+1/2}$. It remains to choose x_c. Note that x_c appears only in the factor $H[x_a, x_b, x_c]$ in (5.74), so the choice for the final point in the stencil influences the interpolating polynomial solely through this factor. The ENO strategy is to extend the interval in the direction that gives the smallest value for $H[x_a, x_b, x_c]$, or equivalently, the smallest coefficient for the function $(x - x_a)(x - x_b)$. Thus, using the identity $H[x_0, x_1, x_2] = H[x_2, x_0, x_1]$,

$$x_c = \begin{cases} x_{j-\frac{3}{2}} & \text{if } \left|H[x_{j-\frac{3}{2}}, x_{j-\frac{1}{2}}, x_{j+\frac{1}{2}}]\right| \le \left|H[x_{j-\frac{1}{2}}, x_{j+\frac{1}{2}}, x_{j+\frac{3}{2}}]\right|, \\ x_{j+\frac{3}{2}} & \text{otherwise.} \end{cases}$$

Recall that the numerical fluxes $F_{j+1/2}$ are actually computed from the derivative of (5.74), for example, if the interpolation stencil $\{x_{j-3/2}, x_{j-1/2}, x_{j+1/2}\}$ is selected,

$$\begin{aligned} F_{j+\frac{1}{2}} &= \frac{\partial P}{\partial x}(x_{j+\frac{1}{2}}) \\ &= H[x_{j-\frac{1}{2}}, x_{j+\frac{1}{2}}] + H[x_{j-\frac{3}{2}}, x_{j-\frac{1}{2}}, x_{j+\frac{1}{2}}]\left(x_{j+\frac{1}{2}} - x_{j-\frac{1}{2}}\right). \end{aligned} \quad (5.75)$$

No zero-order divided differences appear in the expressions for the fluxes. Taking $\phi_j = \psi(x_j)$ as the initial data, one may reconstruct the ENO fluxes beginning with the first-order divided difference $H[x_{j-1/2}, x_{j+1/2}]$. Using (5.69), we find that

$$H[x_{j-\frac{1}{2}}, x_{j+\frac{1}{2}}] = f(\phi_j),$$

and similarly that (5.75) evaluates to (5.72). Note that on a uniform mesh, it is not necessary to perform the division by $x - x_0$ when computing the divided differences, as this factor may be absorbed into the coefficients $c_{\ell,s}$ in (5.71). The $c_{\ell,s}$ for schemes of orders 2–5 are given in Table 5.2.

The numerical fluxes $F_{j+1/2}$ could conceivably be computed by building the ENO interpolation stencil outward from either the interval just to the right or the interval just to the left of the cell boundary at $x_{j+1/2}$. This starting interval is the only interval that is guaranteed to be in the final ENO stencil for $F_{j+1/2}$.

Table 5.2 Coefficients $c_{\ell,s}$ for the computation of $F_{j+1/2}$ using essentially nonoscillatory approximations of orders 2–5. The index ℓ is the left shift of the stencil with respect to grid point x_j

Order	ℓ	$s=0$	$s=1$	$s=2$	$s=3$	$s=4$
	−1	3/2	−1/2			
2	0	1/2	1/2			
	1	−1/2	3/2			
	−1	11/6	−7/6	1/3		
3	0	1/3	5/6	−1/6		
	1	−1/6	5/6	1/3		
	2	1/3	−7/6	11/6		
	−1	25/12	−23/12	13/12	−1/4	
	0	1/4	13/12	−5/12	1/12	
4	1	−1/12	7/12	7/12	−1/12	
	2	1/12	−5/12	13/12	1/4	
	3	−1/4	13/12	−23/12	25/12	
	−1	137/60	−163/60	137/60	−21/20	1/5
	0	1/5	77/60	−43/60	17/60	−1/20
5	1	−1/20	9/20	47/60	−13/60	1/30
	2	1/30	−13/60	47/60	9/20	−1/20
	3	−1/20	17/60	−43/60	77/60	1/5
	4	1/5	−21/20	137/60	−163/60	137/60

Therefore, to satisfy the Courant–Friedrichs–Lewy condition that the numerical domain of dependence include the domain of dependence of the true solution (see Sect. 3.2.3), one must choose the starting interval that is upstream with respect to the direction of signal propagation. If $\mathrm{d}f/\mathrm{d}\psi(x_{j+1/2}) \geq 0$, the ENO stencil should be built outward from $[x_{j-1/2}, x_{j+1/2}]$. On the other hand, if $\mathrm{d}f/\mathrm{d}\psi(x_{j+1/2}) < 0$, the stencil should start with the interval $[x_{j+1/2}, x_{j+3/2}]$; in this second case the grid point x_j may lie outside the final stencil, which corresponds to the $\ell = -1$ case listed in Table 5.2.

The performances of the ENO and flux-limiter methods are compared in Fig. 5.18 for the same two test problems considered in Fig. 5.12, one-dimensional scalar advection of a step and a well-resolved sine wave. The numerical details for these simulations are identical to those presented in connection with Fig. 5.12. Both ENO methods are discretized in time using the third-order strong-stability-preserving Runge–Kutta (SSPRK) scheme (2.48). The second-order ENO method (dashed lines) is more diffusive than the flux-limited solution obtained with the minmod limiter and would not be a good choice in practical applications. Much better results are obtained with the fourth-order ENO method. In the case of the step in Fig. 5.18a, the slope of the fourth-order ENO solution is steeper than that of the minmod flux-limited solution, but is gentler than that produced using the superbee limiter. The principle advantage of the fourth-order ENO method in comparison with the others is not in its treatment of the step, but rather is that, in contrast to flux-limiter and FCT methods (see also Figs. 5.12, 5.13) it does not degrade the solution at smooth extrema, and thereby produces a very accurate approximation to the sine wave in Fig. 5.18b.

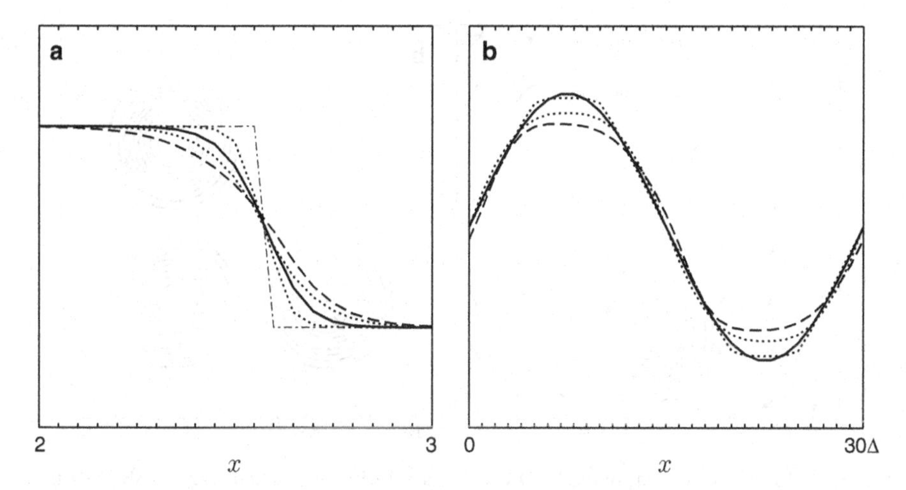

Fig. 5.18 Comparison of essentially nonoscillatory (ENO) and flux-limited approximations for **a** a propagating step, and **b** a well-resolved sine wave. Curves are for the second-order ENO approximation (*dashed line*), fourth-order ENO approximation (*solid line*), flux-limited solutions using the minmod and superbee limiters (both *dotted lines*, with the minmod limiter giving the more diffusive result). The exact solution is shown by the *thin dot-dashed line*, which is masked by the fourth-order ENO solution in **b**

5.7.3 WENO Methods

ENO methods rely on the identification of a single optimal stencil over which the flux is computed. Difficulties may arise, even in problems with smooth solutions, if the choice for the optimal stencil changes erratically from grid cell to grid cell owing to small variations in the solution. An example of such difficulties is shown in Fig. 5.19. The numerical domain used in these tests spans the square $0 \leq x, y \leq 1$, with spatial resolution $\Delta x = \Delta y = 0.01$. The initial tracer field, contoured in Fig. 5.19a, is

$$\psi(x, y, 0) = \tfrac{1}{2}[1 + \cos(\pi r/R)] + \gamma(x, y), \tag{5.76}$$

where

$$r(x, y) = \min\left(R, \sqrt{(x - x_c)^2 + (y - y_c)^2}\right), \tag{5.77}$$

$R = 0.15$, $x_c = 0.5$, $y_c = 0.3$, and γ is a random number in the interval $[0, 10^{-4}]$. The velocity field, shown by the vectors in Fig. 5.19a, produces solid-body rotation at an angular velocity of π radians per unit time. Time stepping is performed using the third-order SSPRK scheme (2.48) with a time step of 0.0025. The lateral boundary conditions are periodic.[12]

[12] Only the normal component of the velocity is actually periodic (in fact u is invariant in x, and v is invariant in y). Since the tracer concentration remains nearly zero on the boundaries, the solution is essentially identical to that for rotating flow in an unbounded spatial domain.

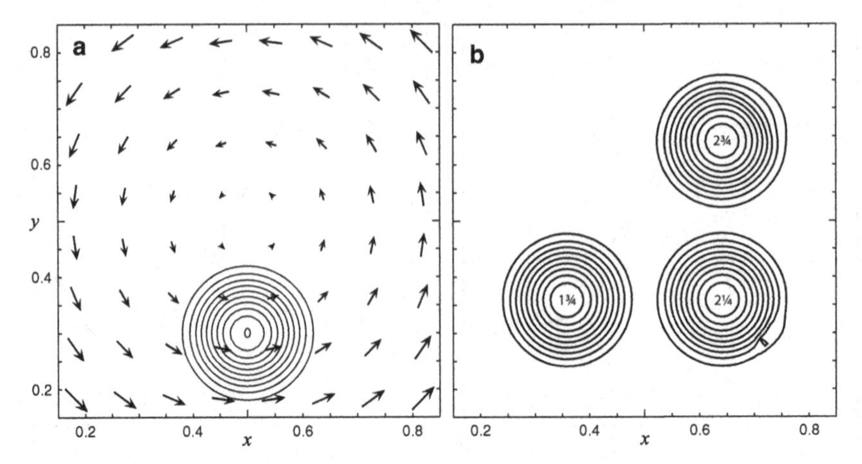

Fig. 5.19 Errors introduced by the ENO method: **a** velocity vectors and contours showing the initial position of the tracer, and **b** superimposed plots of the tracer at three later times. The contour interval is 0.1; the times at which the tracer distribution is plotted are indicated at its center. Only the central portion of the total domain is plotted

The tracer circles the center of the domain every two time units. Only minor errors are incurred during the first full rotation of the distribution, and the tracer field shown at $t = 1.75$ in Fig. 5.19b is almost perfect; however, as the second cycle begins, irregularities appear on the lower-right side of the tracer distribution at $t = 2.25$. The errors triggered around $t = 2.25$ seem to be a random occurrence; the error decreases considerably by $t = 2.75$ and the solution continues to recover toward circular symmetry until at least $t = 8$.

Problems such as that shown in Fig. 5.19 can be avoided using WENO methods, in which the actual flux at each cell interface is set to a weighted average of the fluxes computed over each individual ENO stencil. Recall that if an ENO method is kth-order accurate, the numerical flux $F_{j+1/2}$ is set to $p_\ell(x_{j+1/2})$, where ℓ is the particular left shift among k possible stencils that gives the smoothest approximation over the grid cell upstream of $x_{j+1/2}$. Assuming that $\mathrm{d}f/\mathrm{d}\psi(x_{j+1/2}) \geq 0$, the WENO approximation built from these same interpolation stencils has the form

$$F_{i+\frac{1}{2}} = \sum_{\ell=0}^{k-1} \omega_\ell p_\ell(x_{j+\frac{1}{2}}),$$

where the ω_ℓ are suitably defined weights, and using the same notation as in (5.71),

$$p_\ell(x_{j+\frac{1}{2}}) = \sum_{s=0}^{k-1} c_{\ell,s} f(\phi_{j-\ell+s}).$$

The weights are chosen to both minimize the influence of any p_ℓ computed over a stencil where the solution is discontinuous and to yield the highest possible order of accuracy where the solution is smooth.

First, consider the weights required to achieve the highest order of accuracy, and as an example suppose the underlying ENO method is second order. Then $p_1(x_{j+1/2})$ is the approximation to $F_{j+1/2}$ given in (5.72) and the leading-order term in the truncation error in the numerical flux divergence (5.65) is $-[(\Delta x)^3/3]\mathrm{d}^3 f/\mathrm{d}x^3$. Similarly, $p_0(x_{j+1/2})$ is the approximation in (5.73) and is associated with a leading-order error in the flux divergence of $[(\Delta x)^3/6]\mathrm{d}^3 f/\mathrm{d}x^3$. The linear combination of these two fluxes obtained by setting $\omega_1 = 1/3$ and $\omega_0 = 2/3$ will therefore be third order. More generally, a flux of order $2k-1$ can be obtained using an optimal linear combination of k, kth-order-accurate ENO fluxes.

Of course the actual WENO weights are not constants, but rather functions of the ϕ_j that reflect the presence of discontinuities. For each left shift ℓ in a method of order $2k-1$, define raw weights such that

$$\alpha_\ell = \frac{C_\ell^k}{(\beta_\ell + \epsilon)^n}, \tag{5.78}$$

where C_ℓ^k is the optimal weight for the stencil with left shift ℓ in a set of kth-order ENO approximations,[13] $\beta_\ell(\phi)$ becomes large if the solution is discontinuous on the stencil spanned by $p_\ell(x)$, and ϵ is a small positive constant that ensures α_ℓ will be bounded.[14] The exponent n is typically set to 2. The actual weights are then normalized so that they sum to unity:

$$w_\ell = \frac{\alpha_\ell}{\alpha_0 + \ldots + \alpha_{k-1}}$$

In regions where the solution is smooth, a WENO method of order $2k-1$ can be obtained if $w_\ell = C_\ell^k + O\left[(\Delta x)^{k-1}\right]$, and w_ℓ will satisfy this condition if

$$\beta_\ell = \sum_{n=1}^{k-1} \int_{x_{j-\frac{1}{2}}}^{x_{j+\frac{1}{2}}} (\Delta x)^{2n-1} \left(\frac{\mathrm{d}^n p_\ell}{\mathrm{d}x^n}\right)^2 \mathrm{d}x. \tag{5.79}$$

(Jiang and Shu 1996). Note that for each $x_{j+1/2}$, the β_ℓ from every stencil approximates the same mathematical expression. Thus, if the solution is smooth, all the β_ℓ will be approximately equal, and since $\sum_\ell C_\ell^k = 1$, the weights will indeed approach the optimal C_ℓ^k.

Under the continued assumption that $\mathrm{d}f/\mathrm{d}\psi(x_{j+1/2}) > 0$, the β_ℓ for the popular fifth-order WENO method are

$$\beta_0 = \tfrac{13}{12}\left[f(\phi_j) - 2f(\phi_{j+1}) + f(\phi_{j+2})\right]^2 + \tfrac{1}{4}\left[3f(\phi_j) - 4f(\phi_{j+1}) + f(\phi_{j+2})\right]^2,$$
$$\beta_1 = \tfrac{13}{12}\left[f(\phi_{j-1}) - 2f(\phi_j) + f(\phi_{j+1})\right]^2 + \tfrac{1}{4}\left[f(\phi_{j-1}) - f(\phi_{j+1})\right]^2,$$
$$\beta_2 = \tfrac{13}{12}\left[f(\phi_{j-2}) - 2f(\phi_{j-1}) + f(\phi_j)\right]^2 + \tfrac{1}{4}\left[f(\phi_{j-2}) - 4f(\phi_{j-1}) + 3f(\phi_j)\right]^2,$$

[13] In the specific example just considered, $C_0^2 = 2/3$ and $C_1^2 = 1/3$.

[14] ϵ may be set to 10^{-6} times a characteristic scale for β_ℓ in regions where the solution is smooth.

and the optimal weights are

$$C_0^3 = \frac{3}{10}, \quad C_1^3 = \frac{3}{5}, \quad C_2^3 = \frac{1}{10}.$$

The fluxes for each of these stencils, which may be determined from Table 5.2, are

$$\begin{aligned} p_0(x_{j+\frac{1}{2}}) &= \tfrac{1}{3} f(\phi_j) + \tfrac{5}{6} f(\phi_{j+1}) - \tfrac{1}{6} f(\phi_{j+2}), \\ p_1(x_{j+\frac{1}{2}}) &= -\tfrac{1}{6} f(\phi_{j-1}) + \tfrac{5}{6} f(\phi_j) + \tfrac{1}{3} f(\phi_{j+1}), \\ p_2(x_{j+\frac{1}{2}}) &= \tfrac{1}{3} f(\phi_{j-2}) - \tfrac{7}{6} f(\phi_{j-1}) + \tfrac{11}{6} f(\phi_j). \end{aligned} \tag{5.80}$$

The WENO formulae for $\mathrm{d}f/\mathrm{d}\psi(x_{j+1/2}) < 0$ are symmetric about $x_{j+1/2}$.

Figure 5.20a shows a comparison of the fifth-order WENO method with the fourth-order ENO and MC flux-limiter methods for the same propagating-step problem considered in Fig. 5.12a; the performance of all three methods is clearly similar. In contrast, larger differences are apparent between the solutions shown in Fig. 5.20b, in which the initial condition is the sum of 7.5 and $10\Delta x$ waves (for details refer to the discussion of Fig. 3.6). Here the MC flux-limiter method is clearly inferior to the other two, but the fourth-order ENO scheme slightly outperforms the fifth-order WENO method. One weakness of the WENO method is that the weights w_ℓ can be slow to approach their optimal values as the numerical resolution improves. The WENO solution for the two-wave problem can be noticeably improved by reducing n from 2 to 1 in (5.78), which shifts the w_ℓ closer to their

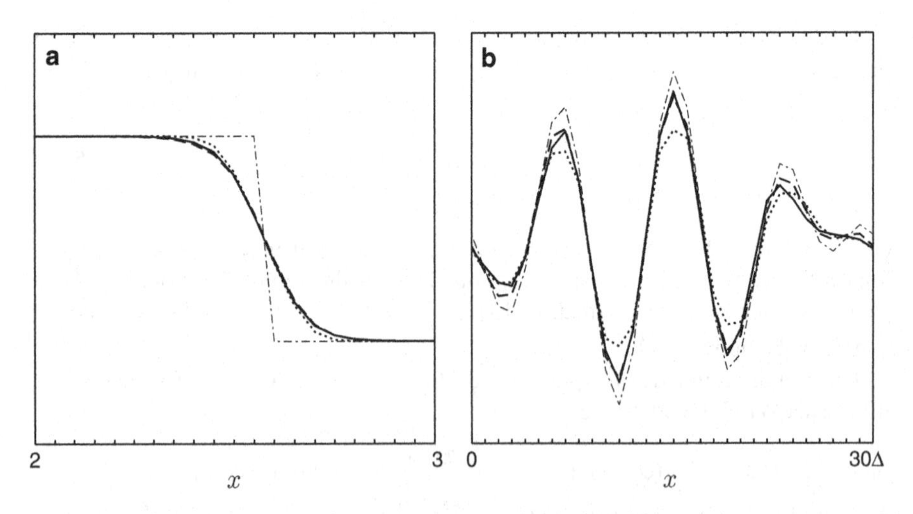

Fig. 5.20 Comparison of ENO, weighted essentially nonoscillatory (WENO), and flux-limited approximations for **a** a propagating step, and **b** the sum of equal -amplitude $7.5\Delta x$ and $10\Delta x$ waves. Curves are for the fourth-order ENO approximation (*dashed line*), fifth-order WENO approximation (*solid line*), and the flux-limited solution obtained using the MC limiter (*dotted line*). The exact solution is shown by the *thin dot-dashed line*

optimal weighting, although this can have a negative impact near strong discontinuities. Another way to improve the WENO solution to the two-wave problem is to use the optimal C_ℓ^k unless the variation among the β_ℓ is sufficient to push

$$\frac{\max_\ell \beta_\ell}{\min_\ell \beta_\ell + \epsilon}$$

above a threshold value of O(10) (Hill and Pullin 2004)[15].

5.8 Preserving Smooth Extrema

As the peak of a wave translates between two grid cells over an interval greater than a single time step, there should be time steps for which the maximum cell-averaged value increases (such as when the peak in the solution translates from a point on the boundary between two cells to exactly coincide with the center of one cell). Such increases in the amplitude of extrema are not allowed in FCT, TVD, or monotonicity-preserving methods, and as a consequence, smooth extrema are erroneously damped. Substantial errors may even develop near the crests and troughs of very well resolved waves (see Figs. 5.12b, 5.13b). One way to avoid this is through the use of WENO methods, which do not apply extra dissipation at smooth extrema, but WENO methods turn out to be considerably less efficient than FCT or flux-limiter schemes in many applications.

Several authors have therefore proposed criteria to identify smooth extrema and thereby avoid correcting or limiting the fluxes (or equivalently, limiting the polynomial reconstructions) necessary for their preservation. As before, let ϕ_{j+} be the value of the solution at the interface at $x_{j+1/2}$ as reconstructed from the cell-averaged values $\phi_j, \phi_{j\pm1}, \ldots$. Building on earlier criteria proposed by Sun et al. (1996) and by Nair et al. (1999), Zerroukat et al. (2005) suggested that ϕ_{j+} be judged as a spurious extremum (and therefore subject to limiting) if

$$(\phi_{j+} - \phi_j)(\phi_{j+1} - \phi_{j+}) < 0, \tag{5.81}$$

and at least one of the following inequalities is satisfied:

$$\begin{aligned} (\phi_j - \phi_{j-1})(\phi_{j+2} - \phi_{j+1}) &\geq 0, \\ (\phi_j - \phi_{j-1})(\phi_{j-1} - \phi_{j-2}) &\leq 0, \\ (\phi_{j+2} - \phi_{j+1})(\phi_{j+3} - \phi_{j+2}) &\leq 0, \\ (\phi_{j+} - \phi_j)(\phi_j - \phi_{j-1}) &\leq 0. \end{aligned} \tag{5.82}$$

[15] Once again ϵ is a small constant chosen to avoid division by zero.

Condition (5.81) identifies ϕ_{j+} as a local extremum. The following inequalities test the relation of this extremum to the solution in the surrounding grid cells, for example, flagging ϕ_{j+} as spurious if the slopes computed from cell average values to the right and left of $x_{j+1/2}$ both have the same sign via (5.82).

Colella and Sekora (2008) presented a different strategy for constructing a limiter for the piecewise-parabolic method that preserves accuracy at smooth extrema. They compared approximations to the second derivative at three nearby locations: x_j, x_{j+1}, and the cell interface at $x_{j+1/2}$. If the second derivative changes sign in the interval, ϕ_{j+} is set to $(\phi_j + \phi_{j+1})/2$, otherwise the second derivative with minimum magnitude is used to reconstruct the value at the cell interface in such a way that ϕ_{j+} matches the prediction from high-order interpolation when the solution is smooth and well resolved. A similar approach, again based on evaluating second derivatives at three adjacent points, is then used to adjust the structure of the parabola within each cell to eliminate any subcell extrema that are not part of the smooth solution.

As a third approach, Blossey and Durran (2008) suggested using a WENO-motivated smoothness metric to determine where limiters are applied. As discussed in the previous section, the key WENO smoothness indicator is the parameter β_ℓ, defined by (5.79). For a fifth-order WENO method, β_ℓ is the integral over the jth cell of the sum of the squares of the first and second derivatives of the ℓth interpolant. Being the sum of the norms of the first and second derivatives, β_ℓ is strongly sensitive to discontinuities in the solution. When the solution is smooth, the β_ℓ are nearly identical and the WENO method is high order. When discontinuities are present, there are large variations among the β_ℓ, and the WENO method adjusts the weight allotted to each stencil to reduce the influence of those spanning the discontinuity.

Motivated by (5.79), let γ_j be an approximation to

$$\frac{1}{2}\left((\Delta x)^2\frac{\partial^2\psi}{\partial x^2}(x_j)\right)^2 + \frac{1}{2}\left(2\Delta x\frac{\partial\psi}{\partial x}(x_j)\right)^2.$$

The factors of $1/2$ and 2 in the preceding expression are strictly for computational convenience since they allow γ_j to be efficiently calculated as

$$\begin{aligned}\gamma_j &= \left[\left(\phi_{j+1} - 2\phi_j + \phi_{j-1}\right)^2 + \left(\phi_{j+1} - \phi_{j-1}\right)^2\right]/2 \\ &= \left(\phi_{j+1} - \phi_j\right)^2 + \left(\phi_j - \phi_{j-1}\right)^2. \end{aligned} \tag{5.83}$$

Define a smoothness parameter $\lambda_{j+1/2}$ such that

$$\lambda_{j+\frac{1}{2}} = \frac{\max_{k\in K}\ \gamma_k}{\min_{k\in K}\ \gamma_k + \epsilon}, \tag{5.84}$$

where K is the upstream weighted set of indices

$$K = \begin{cases} [j-1, j, j+1], & \text{if } u_{j+\frac{1}{2}} \geq 0, \\ [j, j+1, j+2], & \text{otherwise,} \end{cases}$$

and ϵ is a small parameter that prevents division by zero. Flux correction (Sect. 5.4.2) or flux limiting (Sect. 5.5) is enforced only at those cell interfaces where $\lambda_{j+1/2}$ exceeds $\lambda_{\max}$; at all other interfaces the correction factor $C_{j+1/2}$ is set to unity. Similarly, polynomial modification (Sects. 5.6.2, 5.6.3), is enforced only in connection with flux computations at interfaces where $\lambda_{j+1/2}$ exceeds $\lambda_{\max}$. On the basis of empirical testing, Blossey and Durran (2008) suggested setting $\lambda_{\max}$ to 20. Examples of the performance of this approach, which will be referred to as "selective limiting," are given in Sect. 5.9.5.

5.9 Two Spatial Dimensions

The preceding discussion focused almost exclusively on problems in one spatial dimension. One effective way to extend one-dimensional methods to multiple dimensions is through fractional steps, and fractional steps have been used successfully in problems whose solutions contain discontinuities or poorly resolved gradients. A theoretical basis for the success of these methods was provided by Crandall and Majda (1980a), who showed that convergent approximations to the entropy-consistent solution of two-dimensional scalar conservation laws can be achieved using the method of fractional steps, provided that consistent conservation-form monotone schemes are used in each individual step.

Unsplit algorithms may, nevertheless, seem like the most natural way to simulate advection in a multidimensional flow, and stable, accurate, unsplit WENO approximations to the advection equation can easily be constructed. One simply includes a WENO-like approximation to the advective flux divergence along each coordinate axis and makes a suitable reduction in the time step to preserve stability (e.g., reducing it by $\sqrt{2}/2$ in the two-dimensional case – see Sect. 4.2.1). WENO methods are typically integrated using third-order Runge–Kutta time differencing, which, being a nonamplifying scheme, allows high-order methods to be constructed without modifying the stencils used along each coordinate in the one-dimensional problem. On the other hand, many forward-in-time methods use the Lax–Wendroff scheme to achieve second-order accuracy, and as discussed in Sect. 4.2.1, mixed spatial derivatives must then be computed as part of the approximation that cancels the $O(\Delta t)$ truncation error. Such mixed spatial derivatives are naturally generated in fractional-step methods and in the flux-limiter approach described in Sect. 5.9.2, but must be explicitly included in some other unsplit schemes.

We begin this section by considering two representative unsplit methods: FCT (Zalesak 1979) and a flux-limiter algorithm for two-dimensional nondivergent flow proposed by LeVeque (1996). Several other schemes with varying degrees of similarity have also appeared in the literature, including those by Smolarkiewicz (1984), Colella (1990), Saltzman (1994), Leonard et al. (1996), Lin and Rood (1996), Thuburn (1996), and Stevens and Bretherton (1996).

5.9.1 FCT in Two Dimensions

The extension of the flux-correction algorithm described in Sect. 5.4.2 to multidimensional problems is straightforward and was discussed in detail by Zalesak (1979). Only the two-dimensional case will be considered here, for which a monotone low-order solution could be computed using the two-dimensional upstream difference (4.31). As discussed in Sect. 4.2.1, however, the CTU method (4.36) is a better choice. Zalesak (1979) suggested using high-order fluxes computed parallel to the coordinate axis together with nonamplifying leapfrog–trapezoidal time differencing (2.71) and (2.72). A second-order, forward-in-time method can also be computed using an appropriate form of the upstream biased Lax–Wendroff method (4.38). The generalization of (4.38) to problems with spatially varying nondivergent winds will be discussed in Sect. 5.9.3.

Suppose that i and j are the grid-cell indices along the two spatial coordinates. In contrast to the one-dimensional case, there will now be four antidiffusive fluxes into each grid cell, and one must compute four coefficients $C_{i\pm 1/2,j}$ and $C_{i,j\pm 1/2}$ to limit these fluxes. The formulae for $C_{i\pm 1/2,j}$ and $C_{i,j\pm 1/2}$ are identical to those given for $C_{j\pm 1/2}$ in Sect. 5.4.2, except for the inclusion of the second dimension in the subscript notation and the computation of the total antidiffusive fluxes in and out of grid point i, j as

$$\begin{aligned}P_{i,j}^{+} &= \max\left(0, A_{i-\frac{1}{2},j}\right) - \min\left(0, A_{i+\frac{1}{2},j}\right)\\&\quad + \max\left(0, A_{i,j-\frac{1}{2}}\right) - \min\left(0, A_{i,j+\frac{1}{2}}\right),\end{aligned}$$

$$\begin{aligned}P_{i,j}^{-} &= \max\left(0, A_{i+\frac{1}{2},j}\right) - \min\left(0, A_{i-\frac{1}{2},j}\right)\\&\quad + \max\left(0, A_{i,j+\frac{1}{2}}\right) - \min\left(0, A_{i,j-\frac{1}{2}}\right).\end{aligned}$$

The formula for the permissible range of values for $\phi_{i,j}^{n+1}$ also needs to be generalized to two dimensions; the most natural choice is to define

$$\begin{aligned}\phi_{i,j}^{a} &= \max\left(\phi_{i,j}^{n}, \phi_{i,j}^{\mathrm{td}}\right),\\\phi_{i,j}^{b} &= \min\left(\phi_{i,j}^{n}, \phi_{i,j}^{\mathrm{td}}\right),\end{aligned}$$

and then to let

$$\begin{aligned}\phi_{i,j}^{\max} &= \max\left(\phi_{i,j}^{a}, \phi_{i,j-1}^{a}, \phi_{i,j+1}^{a}, \phi_{i-1,j}^{a}, \phi_{i+1,j}^{a}\right),\\\phi_{i,j}^{\min} &= \min\left(\phi_{i,j}^{b}, \phi_{i,j-1}^{b}, \phi_{i,j+1}^{b}, \phi_{i-1,j}^{b}, \phi_{i+1,j}^{b}\right).\end{aligned}$$

The preceding technique for determining $\phi^{\min}$ and $\phi^{\max}$ does not completely prevent the development of small undershoots and overshoots in situations where ϕ is being transported in a direction almost perpendicular to the gradient of ϕ. Nevertheless, the spurious oscillations are typically very small and can be completely eliminated using additional correction steps discussed by Zalesak.

5.9.2 Flux-Limiter Methods for Uniform Two-Dimensional Flow

As was the case for the one-dimensional flux-limiter methods discussed previously, the solution strategy is to use a monotone method to compute low-order fluxes near poorly resolved gradients and then to correct these low-order fluxes in regions where the solution is well resolved using fluxes obtained from a higher-order scheme. A finite-difference approximation to the equation governing the advection of a passive scalar in a two-dimensional flow (5.26) can be written in the conservation form

$$\phi_{i,j}^{n+1} = \phi_{i,j}^{n} - \frac{\Delta t}{\Delta s}\left[F_{i+\frac{1}{2},j}^{n} - F_{i-\frac{1}{2},j}^{n} + G_{i,j+\frac{1}{2}}^{n} - G_{i,j-\frac{1}{2}}^{n}\right]. \tag{5.85}$$

The terms $F_{i-1/2,j}$ and $G_{i,j-1/2}$ are approximations to the advective fluxes through the left and lower boundaries of the grid cell centered on $\phi_{i,j}$. The horizontal mesh spacing is Δs, and it is assumed to be equal along the x- and y-axes for notational simplicity. To present the method in its simplest form, we temporarily assume that both velocity components are positive and spatially uniform. The complete algorithm for an arbitrary nondivergent flow is presented in Sect. 5.9.3.

A simple monotone approximation to the advective flux is given by the upstream, or donor-cell, method, which for positive velocities yields

$$F_{i+\frac{1}{2},j}^{\mathrm{up}} = u\phi_{i,j}, \qquad G_{i,j+\frac{1}{2}}^{\mathrm{up}} = v\phi_{i,j}. \tag{5.86}$$

In the standard upwind method, these fluxes are transmitted parallel to the coordinate axes. Each flux induces a change in $\phi_{i,j}$ equal to the upstream value of ϕ times

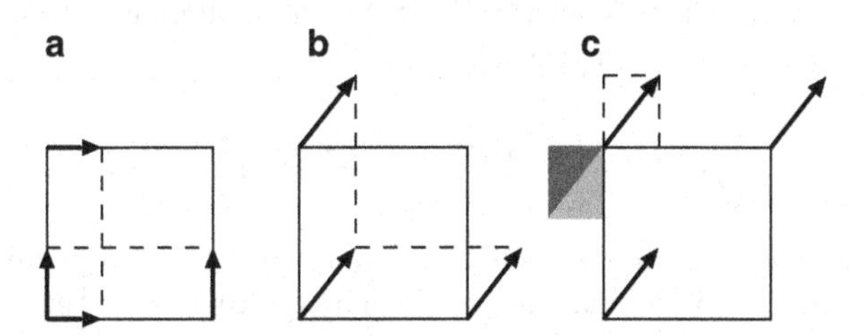

Fig. 5.21 **a** Transmission of the upstream fluxes parallel to the coordinate axes in grid cell (i, j); *heavy arrows* denote the vector displacements $u\Delta t$ and $v\Delta t$; the *square* indicates the grid-cell boundary. **b** Transmission of the fluxes parallel to the wind field; *heavy arrows* denote displacements over time Δt along the total wind vector. **c** Area in cell $(i-1, j)$ from which material is actually transmitted into cell $(i, j+1)$ through the receiving cell's lower face (*light shading*) and through its left face (*dark shading*)

the ratio of the area swept out by the incoming fluid divided by the total area of each grid cell. The situation is schematically illustrated in Fig. 5.21a.

In reality the fluxes are transmitted parallel to the velocity vector, as illustrated in Fig. 5.21b, and an improved monotone scheme can be obtained by accounting for the transmission of the fluxes at their correct angle to the coordinate axes. The transport of these off-axis fluxes through each grid cell can be accounted for by a modification of the basic upstream fluxes (5.86). As illustrated in Fig. 5.21c, the area swept out by the axis-parallel fluxes into cell $(i, j+1)$ incorrectly includes a rectangular region (dashed box) that is actually filled by material from a rectangular region of the same area originating in cell $(i-1, j)$. Half this material is transmitted through the lower face of cell $(i, j+1)$; the other half is transmitted through its left face. The ratio of the area of each triangular region to the area of the full grid cell is $0.5uv(\Delta t)^2/(\Delta s)^2$. Recalling that one factor of $\Delta t/\Delta s$ is already present in (5.85), one can modify the axis-parallel upstream flux through the lower boundary of cell $(i, j+1)$ to correct for this transverse propagation by setting

$$G^{\text{ctu}}_{i,j+\frac{1}{2}} = G^{\text{up}}_{i,j+\frac{1}{2}} - \frac{\Delta t}{2\Delta s} uv(\phi_{i,j} - \phi_{i-1,j}). \tag{5.87}$$

The upstream flux parallel to the x-axis into cell $(i, j+1)$ may be similarly modified to account for the remaining transverse flux from cell $(i-1, j)$ by setting

$$F^{\text{ctu}}_{i-\frac{1}{2},j+1} = F^{\text{up}}_{i-\frac{1}{2},j+1} - \frac{\Delta t}{2\Delta s} uv(\phi_{i-1,j+1} - \phi_{i-1,j}). \tag{5.88}$$

When the wind speed is constant, the method obtained by accounting for flux propagation along the wind vector is identical to the CTU method (4.36). In Sect. 4.2.1 the CTU method was derived for a governing equation in advective form using the method of characteristics to compute backward fluid-parcel trajectories. An alternative derivation can be performed using the finite-volume formalism by following trajectories backward from the corner of each grid cell and then computing the average value of ϕ within the rectangular volume occupied by that cell at the previous time step (Colella 1990). Using back trajectories to define the subareas A_1 through A_4 shown in Fig. 5.22, and assuming that the solution is piecewise constant within each grid cell,

$$\phi^{n+1}_{i,j} = \frac{1}{(\Delta s)^2}\left[A_1\phi^n_{i,j} + A_2\phi^n_{i,j-1} + A_3\phi^n_{i-1,j} + A_4\phi^n_{i-1,j-1}\right].$$

This finite-difference equation is not in conservation form, but it is equivalent to the conservation form (5.85), with the fluxes given by (5.86)–(5.88). The CTU method will be monotone whenever the subareas A_1 through A_4 are positive, which in the general case where u and v can have arbitrary sign requires that

$$\max(|u|, |v|)\frac{\Delta t}{\Delta s} \leq 1. \tag{5.89}$$

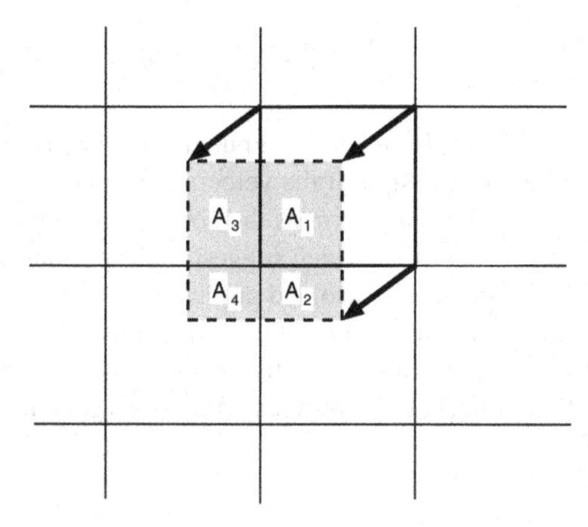

Fig. 5.22 Backward trajectories from cell (i, j) defining the departure volume in the corner transport upstream method

The CTU method must be first-order accurate because it is a linear monotone scheme (Godunov 1959). As proposed by LeVeque (1996), an essentially second-order scheme can be obtained using the same strategy employed for the one-dimensional flux-limiter methods: Corrective fluxes are added to F^{ctu} and G^{ctu} that make the scheme equivalent to the Lax–Wendroff method except in regions where the solution is poorly resolved and the corrective flux is reduced to minimize spurious oscillations. As discussed in Sect. 3.4.4, the Lax–Wendroff approximation to the constant-wind-speed two-dimensional advection equation has the form

$$\frac{\phi_{i,j}^{n+1} - \phi_{i,j}^{n}}{\Delta t} + \delta_{2x} u \phi_{i,j}^{n} + \delta_{2y} v \phi_{i,j}^{n} = \frac{\Delta t}{2} H(\phi^n),$$

where $H(\phi)$ is at least a first-order numerical approximation to

$$\psi_{tt} = u^2 \psi_{xx} + 2uv\psi_{xy} + v^2 \psi_{yy},$$

and the subscripts on ψ denote partial derivatives. The divergence of the off-axis fluxes in the CTU method generate a first-order approximation to the mixed partial derivative in the preceding expression. Thus, the only modifications that need to be made to convert the CTU scheme to the Lax–Wendroff method are to replace the one-sided approximations to $\partial\psi/\partial x$ and $\partial\psi/\partial y$ with centered second-order finite differences and to include an approximation to

$$\frac{\Delta t}{2} \left(u^2 \psi_{xx} + v^2 \psi_{yy} \right).$$

This can be accomplished by adding the following terms to the CTU fluxes:

$$F_{i+\frac{1}{2},j} = F_{i+\frac{1}{2},j}^{\text{ctu}} + \frac{|u|}{2} \left(1 - |u| \frac{\Delta t}{\Delta s} \right) (\phi_{i+1,j} - \phi_{i,j}),$$

$$G_{i,j+\frac{1}{2}} = G^{\mathrm{ctu}}_{i,j+\frac{1}{2}} + \frac{|v|}{2}\left(1 - |v|\frac{\Delta t}{\Delta s}\right)(\phi_{i,j+1} - \phi_{i,j}).$$

(Assuming that the CTU fluxes have been computed in the upstream direction, these formulae apply regardless of the sign of the velocity.) The preceding corrections to the CTU flux have exactly the same form as the corrections to the upstream flux in the one-dimensional problem (5.47), which suggests that spurious oscillations in the vicinity of discontinuities or poorly resolved gradients can be controlled if the corrections are limited using one of the flux-limiter functions discussed in Sect. 5.5.2. The resulting flux-limited approximation to the two-dimensional advection problem is neither TVD nor monotone, but the spurious oscillations generated by this scheme are extremely weak.

5.9.3 Nonuniform Nondivergent Flow

The generalization of this method to a nonuniform nondivergent velocity field is most easily presented as the algorithm in Table 5.3, in which the fluxes are initialized to zero at the beginning of each time step and then incrementally built up in the course of two passes through the numerical mesh. The velocities are assumed to be staggered such that $u_{i+1/2,j}$ and $v_{i,j+1/2}$ are displaced $(\Delta s/2, 0)$ and $(0, \Delta s/2)$ away from the grid point where $\phi_{i,j}$ is defined.[16]

As discussed in Sect. 4.4.2.2, approximations to the advective form of the transport equation (5.29) have the advantage of exactly preserving initial scalar fields that are horizontally uniform. If the flow is nondivergent, the algorithm in Table 5.3 can easily be recast in an algebraically equivalent advective form that retains the conservation properties of the original method. When u and v are positive, the upstream approximation to the spatial derivative operators in (5.29) is

$$\varDelta_{i,j} = u_{i-\frac{1}{2},j}\left(\frac{\phi_{i,j} - \phi_{i-1,j}}{\Delta x}\right) + v_{i,j-\frac{1}{2}}\left(\frac{\phi_{i,j} - \phi_{i,j-1}}{\Delta y}\right).$$

If the discretized velocity field approximates (5.28) such that

$$\frac{u_{i+\frac{1}{2},j} - u_{i-\frac{1}{2},j}}{\Delta x} + \frac{v_{i,j+\frac{1}{2}} - v_{i,j-\frac{1}{2}}}{\Delta y} = 0, \tag{5.90}$$

$\varDelta_{i,j}$ may be expressed in the equivalent form

$$\varDelta_{i,j} = \frac{F^{\mathrm{up}}_{i+\frac{1}{2},j} - F^{\mathrm{up}}_{i-\frac{1}{2},j}}{\Delta x} + \frac{G^{\mathrm{up}}_{i,j+\frac{1}{2}} - G^{\mathrm{up}}_{i,j-\frac{1}{2}}}{\Delta y},$$

[16] See Fig. 4.6 for an illustration of the same staggering scheme in a different context.

Table 5.3 Algorithm for executing one time step of LeVeque's two-dimensional flux-limited advection scheme

- *Initialize the fluxes to zero*
  for each i, j do
$$F^n_{i-\frac{1}{2},j} = 0, \quad G^n_{i,j-\frac{1}{2}} = 0 \qquad (\star)$$
- *Incrementally increase F and G due to fluxes through the left cell interface*
  for each i, j do
$$U = u^{n+\frac{1}{2}}_{i-\frac{1}{2},j}$$
$$R = \phi^n_{i,j} - \phi^n_{i-1,j}$$
if $U > 0$, then $I = i - 1$, else $I = i$
$$F^n_{i-\frac{1}{2},j} = F^n_{i-\frac{1}{2},j} + U\phi_{I,j} \qquad (\star\star)$$
if $U > 0$, then $I = i$, else $I = i - 1$
$$\text{if } v^{n+\frac{1}{2}}_{i,j+\frac{1}{2}} > 0, \text{ then } G^n_{I,j+\frac{1}{2}} = G^n_{I,j+\frac{1}{2}} - \frac{\Delta t}{2\Delta s} RUv^{n+\frac{1}{2}}_{i,j+\frac{1}{2}}$$
$$\text{if } v^{n+\frac{1}{2}}_{i,j-\frac{1}{2}} < 0, \text{ then } G^n_{I,j-\frac{1}{2}} = G^n_{I,j-\frac{1}{2}} - \frac{\Delta t}{2\Delta s} RUv^{n+\frac{1}{2}}_{i,j-\frac{1}{2}}$$
$$\text{if } \quad U > 0, \quad \text{then } \; r^n_{i-\frac{1}{2},j} = (\phi^n_{i-1,j} - \phi^n_{i-2,j})/R,$$
$$\text{else } \; r^n_{i-\frac{1}{2},j} = (\phi^n_{i+1,j} - \phi^n_{i,j})/R$$
$$F^n_{i-\frac{1}{2},j} = F^n_{i-\frac{1}{2},j} + \tfrac{|U|}{2}\left(1 - \tfrac{\Delta t}{\Delta s}|U|\right) C(r^n_{i-\frac{1}{2},j})R$$
- *Incrementally increase F and G due to fluxes through the bottom cell interface*
  (as above, switching the roles of i and j, u and v, and F and G)
- *Update ϕ*
  for each i, j do
$$\phi^{n+1}_{i,j} = \phi^n_{i,j} - \frac{\Delta t}{\Delta s}\left[F^n_{i+\frac{1}{2},j} - F^n_{i-\frac{1}{2},j} + G^n_{i,j+\frac{1}{2}} - G^n_{i,j-\frac{1}{2}}\right] \qquad (\star\star\star)$$

where F^{up} and G^{up} are the upstream fluxes defined in (5.86). The algorithm in Table 5.3 may therefore be modified to yield a conservative advective-form approximation by replacing the three lines marked by stars with

$$F^n_{i-\frac{1}{2},j} = 0, \quad G^n_{i,j-\frac{1}{2}} = 0, \quad \Delta^n_{i,j} = 0, \qquad (\star)$$

$$\Delta^n_{i,j} = \Delta^n_{i,j} + U(\phi_{I+1,j} - \phi_{I,j}), \qquad (\star\star)$$

$$\phi^{n+1}_{i,j} = \phi^n_{i,j} - \frac{\Delta t}{\Delta s}\left[\Delta^n_{i,j} + F^n_{i+\frac{1}{2},j} - F^n_{i-\frac{1}{2},j} + G^n_{i,j+\frac{1}{2}} - G^n_{i,j-\frac{1}{2}}\right]. \quad (\star\star\star)$$

No additional modifications of the second-order correction terms in F and G are required. The equivalence of the second-order corrections in the flux-and advective-form algorithms is a consequence of the nondivergence of the velocity field. Provided that the flow is steady, the advective form of the governing equation (5.29) implies that

$$\psi_{tt} = u(u\psi_x)_x + u(v\psi_y)_x + v(u\psi_x)_y + v(v\psi_y)_y,$$

whereas the flux form (5.26) implies that

$$\psi_{tt} = (u(u\psi)_x)_x + \left(u(v\psi)_y\right)_x + (v(u\psi)_x)_y + \left(v(v\psi)_y\right)_y.$$

If the flow is nondivergent, both of the preceding equations can be expressed as

$$\psi_{tt} = \left(u^2\psi_x\right)_x + \left(uv\psi_y\right)_x + (uv\psi_x)_y + \left(v^2\psi_y\right)_y.$$

This is the form of the second-order Lax–Wendroff correction that is actually approximated by the finite differences in both the advective and the flux-form algorithms.

5.9.4 Operator Splitting

Consider once again the problem of passive scalar transport in a nondivergent two-dimensional flow. Let $F^x_{i+1/2,j}(\phi)$ and $F^y_{i,j+1/2}(\phi)$ be one-dimensional fluxes through the "east" and "north" faces of the grid cell (i, j) divided by the normal velocity at the cell interface. For example, if the subgrid tracer distribution is reconstructed using the piecewise-parabolic method, $F^y_{i,j+1/2}(\phi)$ is obtained by dividing (5.63) by c and prepending i to the indices associated with ϕ. The simplest split approximation to (5.26) may then be written as

$$\phi^s_{i,j} = \phi^n_{i,j} - \Delta t\left[u_{i+\frac{1}{2},j}F^x_{i+\frac{1}{2},j}(\phi^n) - u_{i-\frac{1}{2},j}F^x_{i-\frac{1}{2},j}(\phi^n)\right]/\Delta x, \quad (5.91)$$

$$\phi^{n+1}_{i,j} = \phi^s_{i,j} - \Delta t\left[v_{i,j+\frac{1}{2}}F^y_{i,j+\frac{1}{2}}(\phi^s) - v_{i,j-\frac{1}{2}}F^y_{i,j-\frac{1}{2}}(\phi^s)\right]/\Delta y. \quad (5.92)$$

As discussed in Sect. 4.3.1, (5.91) and (5.92) are first-order accurate in time unless the operators F^x and F^y commute. If the operators do not commute, but the individual steps are at least second-order accurate in time, the full scheme can nevertheless be made second order using Strang splitting (i.e., by reversing the sequence of the x and y integrations every time step.)

A significant shortcoming of this simple splitting is that it will typically not preserve an initially uniform tracer field in a complex nondivergent flow. The problem is that all convergence induced by the velocity component parallel to the x coordinate must be canceled by divergence in the velocities parallel to the y coordinate, but the split scheme generally does not exactly reproduce this cancelation. In addition, any one-dimensional convergence or divergence in an individual split step makes it difficult to correctly impose one-dimensional flux or slope limiters. In regions where a single velocity component is convergent, positive extrema should amplify during the split step (5.91), but one-dimensional limiters may prevent such amplification.

Several other split schemes have been proposed that do preserve initially uniform tracer fields (Leonard et al. 1996; Clappier 1979). Here we will focus on the approach suggested by Easter (1993), in which the full mass-continuity equation is diagnostically integrated in the same time-split manner as the equations for the concentrations of all the individual tracers. To present Easter's "mass-consistent" algorithm it is helpful to switch to a slightly more general notation that allows for density variations within the fluid.

Let ρ be the density of a fluid mixture and denote the mass fraction[17] of a passive tracer by ψ. Then the equation for the local rate of a change in the tracer density may be written[18]

$$\frac{\partial \rho\psi}{\partial t} + \frac{\partial}{\partial x}(\rho u \psi) + \frac{\partial}{\partial y}(\rho v \psi) = 0. \tag{5.93}$$

If the fluid is assumed to be incompressible, as discussed in Sect. 8.1, the nondivergence condition (5.28) is used to derive an elliptic equation for the pressure that closes the set of governing equations. Nevertheless, the continuity equation for two-dimensional flow,

$$\frac{\partial \rho}{\partial t} + \frac{\partial}{\partial x}(\rho u) + \frac{\partial}{\partial y}(\rho v) = 0, \tag{5.94}$$

still holds and may be used to predict ρ in an incompressible flow where the initial density is not uniform throughout the fluid. Even when the density is uniform, (5.94) can still prove useful in time-split integrations.

A mass-consistent time-split approximation to (5.93) and (5.94) may be written as

$$\rho^s_{i,j} = \rho^n_{i,j} - \frac{\Delta t}{\Delta x}\left[(\rho u)^n_{i+\frac{1}{2},j} - (\rho u)^n_{i-\frac{1}{2},j}\right], \tag{5.95}$$

$$(\rho\phi)^s_{i,j} = (\rho\phi)^n_{i,j} - \frac{\Delta t}{\Delta x}\left[(\rho u)^n_{i+\frac{1}{2},j} F^x_{i+\frac{1}{2},j}(\phi^n) - (\rho u)^n_{i-\frac{1}{2},j} F^x_{i-\frac{1}{2},j}(\phi^n)\right] \tag{5.96}$$

$$\phi^s_{i,j} = (\rho\phi)^s_{i,j} / \rho^s_{i,j}, \tag{5.97}$$

$$\rho^{n+1}_{i,j} = \rho^s_{i,j} - \frac{\Delta t}{\Delta y}\left[(\rho v)^n_{i,j+\frac{1}{2}} - (\rho v)^n_{i,j-\frac{1}{2}}\right], \tag{5.98}$$

$$(\rho\phi)^{n+1}_{i,j} = (\rho\phi)^s_{i,j} - \frac{\Delta t}{\Delta y}\left[(\rho v)^n_{i,j+\frac{1}{2}} F^y_{i,j+\frac{1}{2}}(\phi^s) - (\rho v)^n_{i,j-\frac{1}{2}} F^y_{i,j-\frac{1}{2}}(\phi^s)\right], \tag{5.99}$$

$$\phi^{n+1}_{i,j} = (\rho\phi)^{n+1}_{i,j} / \rho^{n+1}_{i,j}. \tag{5.100}$$

[17] The mass fraction is the ratio of the mass of the tracer to the total mass of the fluid.

[18] In contrast to the usage throughout most of this chapter, in this section $\rho\psi$ is the tracer density. Elsewhere the same quantity is typically denoted by ψ.

Assuming the flow is nondivergent, the interface velocities in the preceding expressions should be identical to those satisfying (5.90) at time t^n. If the density is uniform, the ρ in the preceding expressions may be interpreted as pseudodensities and ρ^n (but not ρ^s) may be set to unity.[19] Strang splitting should be used to preserve second-order accuracy in time by reversing the order of the x and y integrations every time step.

Changes in tracer density due solely to one-dimensional convergence and divergence are removed in steps (5.97) and (5.100), so horizontally uniform initial fields of ϕ are preserved. As a consequence of (5.97), limiters can be imposed when reconstructing piecewise-polynomial subcell distributions of ϕ^s without worrying about any temporary increases in extrema that might be produced by one-dimensional convergence or divergence in step (5.96). Alternatively, in FCT implementations of (5.95)–(5.100), fluxes such as $(\rho v)^n_{i,j+1/2} F^y_{i,j+1/2}(\phi^s)$ may be limited without preventing the physically correct amplification of extrema by one-dimensional convergence or divergence using the method described in Skamarock (2006) or Blossey and Durran (2008). Examples of the performance of a piecewise-parabolic-method-based implementation of this method are given in the next section.

The preceding operator-splitting strategy can be easily generalized to more complicated problems. If more than one tracer is being simulated, additional expressions of the form (5.96), (5.97), (5.99), and (5.100) must be evaluated for each tracer, but the equations predicting ρ^s and ρ^{n+1} need only be solved once per time step. The influence of advective fluxes along a third spatial dimension can be included by adding another three steps analogous to (5.95)–(5.97), and reversing the sequence of the x–y–z integrations every time step following (4.59). This method can also be used in a slightly modified form with Courant numbers $|u\Delta t/\Delta x|$ and $|v\Delta t/\Delta y|$ larger than unity (see Sect. 7.2).

5.9.5 A Numerical Example

In the following, the performance of various time-split implementations of the piecewise-parabolic method will be compared in a test problem in which a passive tracer is advected in a nondivergent deformational flow. The spatial domain is the square $0 \le x \le 1$, $0 \le y \le 1$, and the initial concentration of the tracer is a spatially uniform value of ψ_0, the unit-amplitude cube

$$\psi(x,y,0) = \begin{cases} 1 & \text{if } \max(|x-0.3|, |y-0.5|) \le 0.15 \\ 0 & \text{otherwise,} \end{cases} \qquad (5.101)$$

or the cosine bell

$$\psi(x,y,0) = \frac{1}{2}[1+\cos(\pi r)],$$

[19] When $\rho^n = 1$, (5.98) and (5.100) are unnecessary, but they are retained here to emphasize how an identical set of operations can be performed in each split step.

where

$$r(x, y) = \min\left(1,\ 4\sqrt{\left(x - \frac{1}{4}\right)^2 + \left(y - \frac{1}{4}\right)^2}\right).$$

In all cases, the velocity field is a swirling shear flow defined such that

$$u(x, y) = \sin^2(\pi x)\sin(2\pi y)\cos(\pi t/5),$$
$$v(x, y) = -\sin^2(\pi y)\sin(2\pi x)\cos(\pi t/5).$$

The initial cosine-bell tracer distribution and the structure of the flow field are both plotted in Fig. 5.23a. The tracer distribution is most highly deformed at $t = 2.5$, at which time it appears as the long thin arc shown in Fig. 5.23b, which was obtained from a numerical simulation on a high-resolution mesh. Since the velocity periodically reverses direction, every fluid parcel returns to its original position after five time units, and the correct tracer distribution at $t = 5$ is identical to the initial field. The accuracy of the numerical solutions obtained at $t = 5$ can therefore be evaluated by comparing them with the initial tracer distribution. This same problem was considered by LeVeque (1996).

Pairs of numerical solutions were obtained using horizontal grid intervals of 0.02 and 0.01. At the time of maximum deformation, the width of the arc is reduced to approximately 0.05 in Fig. 5.23b, so the tracer distribution is very poorly resolved on the 0.02 grid. The time step was chosen to that the maximum Courant number during the split steps, $\max(|u\Delta t/\Delta x|, |v\Delta t/\Delta y|)$, was unity.

The importance of mass consistency in split approximations to (5.26) is revealed by the simulations initialized with $\psi(x, y, 0) = \psi_0 = 1$, shown in Fig. 5.24. These

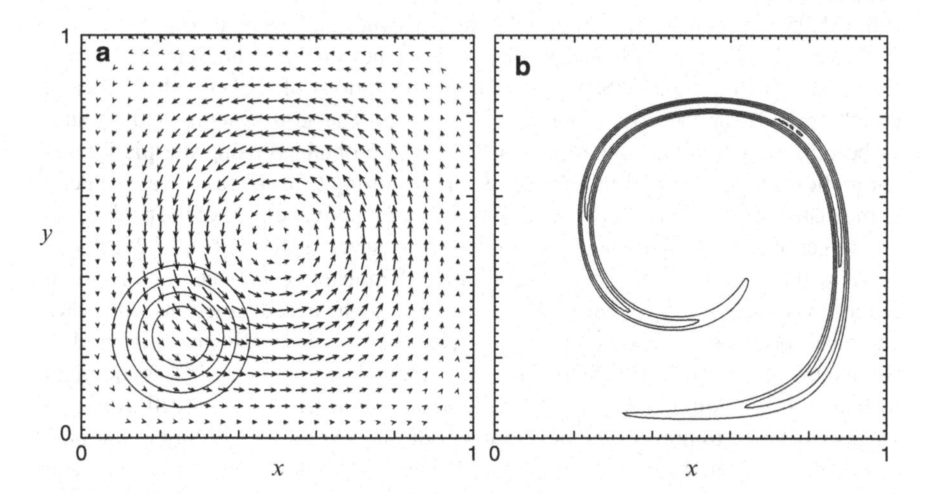

Fig. 5.23 **a** Velocity vectors and tracer concentration field at $t = 0$. **b** Tracer concentration field at $t = 2.5$. The tracer is contoured at intervals of 0.2 beginning with the 0.2 contour line. The length of each vector is proportional to the speed of the flow

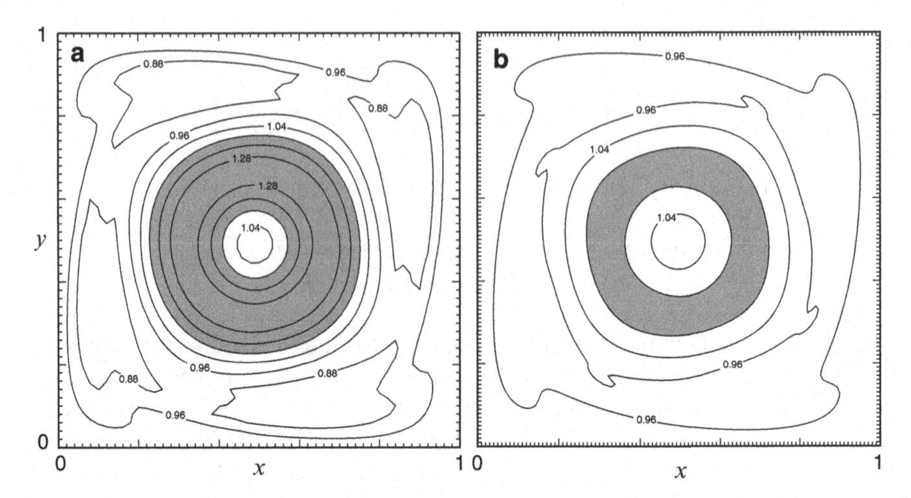

Fig. 5.24 Contours of ϕ at $t = 5$ showing the development of spurious deviations from its initial value of unity using non-mass-corrected Strang splitting and a horizontal grid interval of **a** 0.02 and **b** 0.01. The contour interval is 0.08; regions in which $\phi > 1.12$ are shaded *gray*

simulations were conducted using the one-dimensional piecewise-parabolic method in a Strang-split implementation of the simple formulae (5.91) and (5.92). The fluxes were computed from (5.63) without using limiters to modify the shape of the reconstructed parabolas. Despite the flow being well resolved and reversing symmetrically in time, large perturbations in the numerical solution ϕ appear by the end of the simulation. In particular, a ring of fluid develops in which ϕ exceeds ψ_0 by roughly 15 and 30% in the 0.01- and 0.02-resolution cases, respectively. When the same unlimited piecewise-parabolic method is applied in the mass-consistent algorithm (5.95)–(5.100), ϕ remains equal to its constant initial value of ψ_0.

A second look at the influence of mass consistency on the solution is provided in Fig. 5.25, which compares Strang-spilt implementations of (5.91) and (5.92) and (5.95)–(5.100) for the cosine-bell test at $t = 5$. The grid spacing is 0.02 and, as before, the one-dimensional reconstructions are obtained using the piecewise-parabolic method without limiting. The simple splitting (Fig. 5.25a) leaves the peak in the distribution too far "north," and also produces more damping of the peak value and larger negative undershoots than the mass-consistent splitting (Fig. 5.25b).

As will be discussed in Sect. 5.9.6, negative chemical tracer concentrations, such as those visible in Fig. 5.25, can produce unstable, unphysical interactions with other chemical species in a reacting flow. Two approaches for eliminating negative undershoots are compared in Fig. 5.26. The simulation shown in Fig. 5.26a is identical to that in Fig. 5.25b, except that every parabola reconstructed from cell-averaged values is limited to prevent the development of new extrema according to the procedure described at the end of Sect. 5.6.3. This globally applied limiter does indeed prevent the formation of spurious negative concentrations, but it also reduces the peak amplitude of the cosine bell, relative to the unlimited case, by 35%.

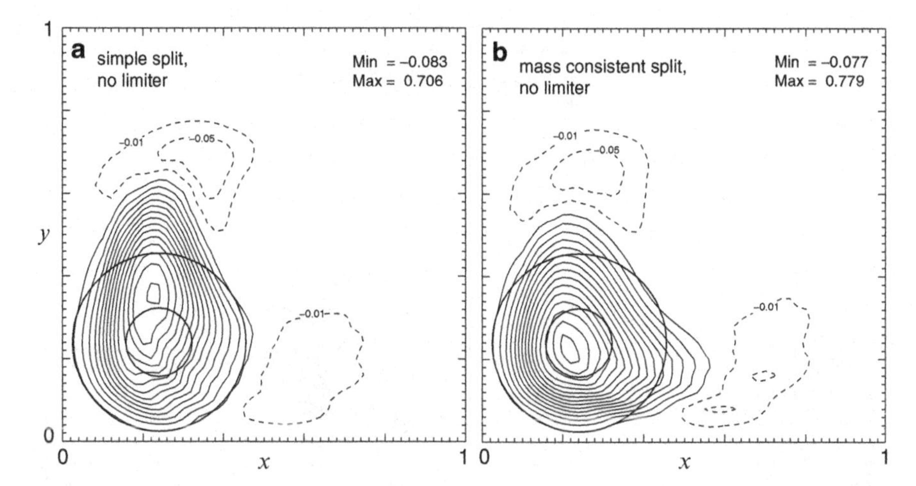

Fig. 5.25 Comparison of the exact solution at $t = 5$ with numerical solutions obtained using a Strang-split piecewise-parabolic-method scheme that is not mass-consistent (**a**) or is mass-consistent (**b**). The *heavy circles* are the 0.05 and 0.75 contour lines of the exact solution. The *thin solid lines* are contours of the numerical solution at intervals of 0.05, beginning with the 0.05 contour. The *dashed lines* are the $-\mathbf{0.01}$ and $-\mathbf{0.05}$ contours. The maximum and minimum values of ϕ at $t = 5$ are noted in the *upper right* of **a** and **b**

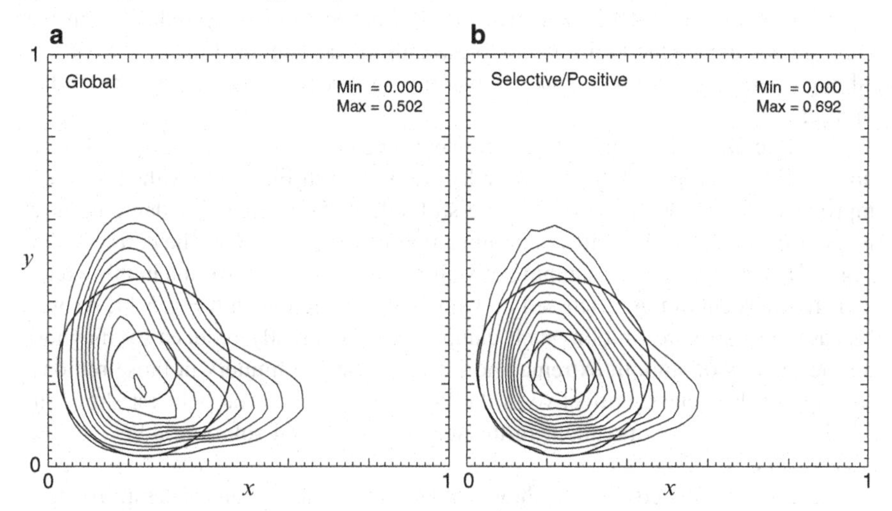

Fig. 5.26 As in Fig. 5.25b except the piecewise parabolas are **a** limited throughout the entire domain, or **b** selectively limited together with FCT-based positivity preservation

The strong damping generated by the global limiter can be greatly reduced by restricting the locations at which the reconstructed piecewise parabolas are limited using the approaches presented in Sect. 5.8. The simulation shown in Fig. 5.26b is identical to that shown in Fig. 5.26a except limiting was only applied to the piecewise-parabolic-method fluxes at interfaces where $\lambda_{i+1/2}$, computed from (5.84) exceeded 20. In addition, to avoid very small negative undershoots,

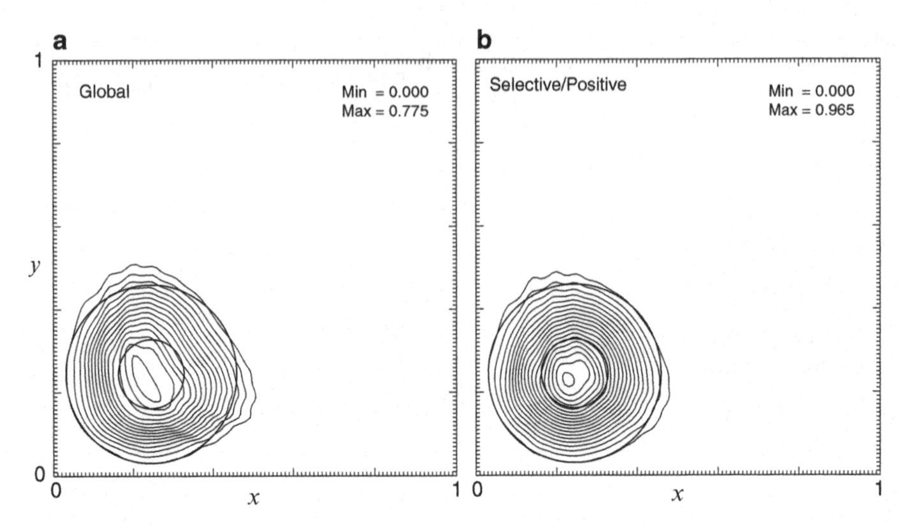

Fig. 5.27 As in Fig. 5.26 except the grid cell intervals are reduced by a factor of 0.5

all differences between this selectively limited piecewise-parabolic-method flux and a separately evaluated upstream flux were corrected to guarantee positivity as discussed in Sect. 5.10.1. Selective limiting produces an 11% reduction in the amplitude of the cosine bell relative to the unlimited solution (Fig. 5.25b), which, although nontrivial, is much smaller than the 35% reduction generated by global limiting.

The simulations leading to Fig. 5.26 were repeated with twice the spatial resolution (grid intervals of 0.01); the results are plotted in Fig. 5.27. Although much improved, the case with global limiting still suffers from excessive damping near the peak of the cosine bell; its maximum amplitude is just 77% of the correct value (Fig. 5.27a). The excessive damping of well-resolved smooth extrema in this example is reminiscent of that discussed previously in connection with Fig. 5.11. As was the case at coarser resolution, such damping can be greatly reduced by diagnosing the regions of smooth extrema and avoiding the use limiters in those regions. This approach is clearly effective in the high-resolution cosine-bell test shown in Fig. 5.27b, where the maximum in the selectively limited solution is 96% of the correct value.

The tests with the cosine bell show that selective limiting can successfully identify regions where limiters *should not* be applied, but they do not demonstrate the algorithm identifies locations where limiters *should* be applied, because any negative concentrations are eliminated by the positivity-preserving FCT scheme (the same strategy that should be followed in actual simulations of reacting flow). Therefore, the ability of selective limiting to prevent overshoots was assessed using the unit cube (5.101) to define the initial tracer concentration in the swirling flow test. Except for the change in the initial condition, the simulations shown in Fig. 5.28 are identical to those plotted in Fig. 5.27. There are no overshoots in the globally limited solution shown in Fig. 5.28a; rather the maximum ϕ is reduced relative to the true

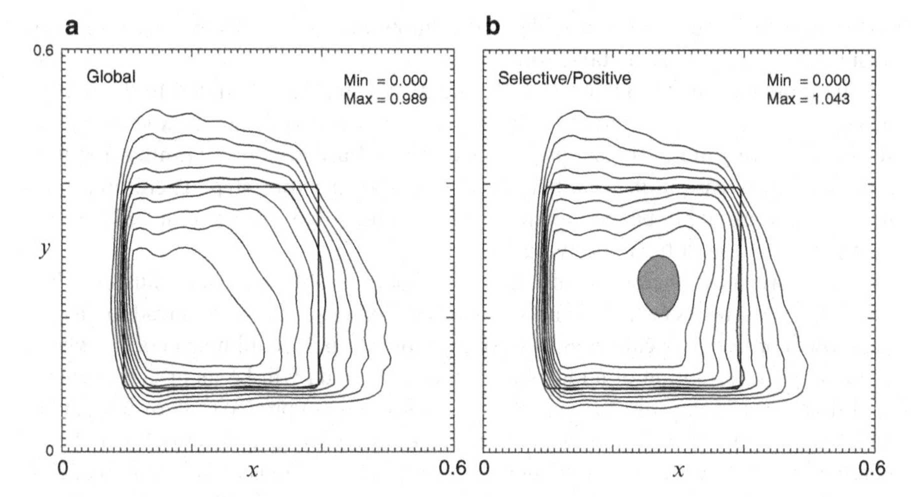

Fig. 5.28 Comparison of the exact solution at $t = 5$ with numerical solutions for the unit-amplitude cube obtained using **a** global limiting, or **b** selective limiting with positivity preservation. The *thin lines* are contours at intervals of 0.1; *shading* indicates regions where ϕ exceeds 1.0. The *heavy lines* show the initial and final locations of the 0.5 contour. Only the subdomain $0 \leq x, y \leq 0.6$ is plotted. The maximum and minimum values of ϕ at $t = 5$ are noted in the *upper right* of **a** and **b**

solution by 0.011. Selective limiting (Fig. 5.28b) allows a small overshoot of magnitude 0.047, but in most respects produces a better solution since it avoids the excessive damping in the "northeast" corner of the cube produced by the global limiter.

The preceding examples illustrate how selective limiting can greatly reduce the damping generated by global limiters in the vicinity of smooth extrema while still restricting overshoots to very small values. As in situations where even small undershoots are unacceptable (because, for example, they represent negative chemical concentrations), the small negative undershoots allowed by selective limiting were completely eliminated using the positivity-preserving FCT approach discussed in Sect. 5.10.1.

5.9.6 When Is a Limiter Necessary?

The use of flux limiters or limiters in polynomial reconstructions can be essential to ensure the convergence of numerical approximations to problems with shocks or discontinuous solutions. On the other hand, in an advection problem such as that considered in the preceding section, the initially smooth concentration field never develops a discontinuity in a finite time; there are no spurious weak solutions, and limiters are not required to guarantee convergence. The use of a limiter in numerical simulations of advective transport is optional and can be considered as a device for converting one type of error, namely, undershoots and overshoots, into a less easily quantifiable but more acceptable form.

The most obvious disadvantages of limiters are that, as illustrated in the previous section, they can introduce significant errors in regions of smooth extrema. Even selective limiting modestly damps smooth extrema. Furthermore, all limiters require a nontrivial increase in the number of computations per time step. Therefore, limiters might reasonably be avoided in problems where they are not required for the correct simulation of the underlying physics.

There are nevertheless many fluid transport problems where numerically generated overshoots or undershoots couple with other processes and lead to unstable growth or nonphysical behaviors. An example of such coupling can occur when simulating the evolution of atmospheric clouds. A cloud will be incorrectly generated if an error in the simulation of water-vapor transport produces an overshoot in which the water-vapor mixing ratio exceeds the saturation mixing ratio. Latent heat is released as the water vapor condenses to form the spurious cloud, and this heat generates buoyancy perturbations that feed back on the flow field, thereby altering the subsequent evolution of the system.

A more catastrophic coupling between numerically generated undershoots and other physical processes can occur if negative chemical concentrations are generated in simulations of chemically reacting flows. The mixing ratio of a chemical species should never drop below zero, but undershoots in regions of steep poorly resolved gradients may produce negative chemical concentrations that destabilize the integration by triggering impossible chemical reactions. As a mathematically simple example, consider the following pair of equations describing the advection and interaction of two chemical species:

$$\frac{\partial \psi_1}{\partial t} + c\frac{\partial \psi_1}{\partial x} = -r\psi_1\psi_2, \tag{5.102}$$

$$\frac{\partial \psi_2}{\partial t} + c\frac{\partial \psi_2}{\partial x} = r\psi_1\psi_2. \tag{5.103}$$

Here ψ_1 and ψ_2 represent the concentration of each chemical species, and r is the rate at which they react, transforming ψ_1 into ψ_2. Suppose there is a nonzero background concentration of ψ_1 throughout the domain and that ψ_2 drops very rapidly to zero outside some localized "plume." If leapfrog-time centered-space differencing is used to simulate the downwind transport of the plume, small dispersive ripples will appear at the edge of the plume, and regions will develop where $\psi_2 < 0$, $\psi_1 > 0$. In the absence of the chemical reactions, these negative regions would remain small and relatively insignificant. However, at any point where $\psi_2 < 0$ and $\psi_1 > 0$, the chemical reaction terms in (5.102) and (5.103) drive ψ_2 more negative while simultaneously increasing ψ_1, thereby amplifying the undershoot and ultimately destabilizing the numerical integration.

Global limiters may be used to avoid the generation of spurious negative concentrations, but as previously noted, they tend to erroneously damp smooth extrema. As an alternative, several *positive-definite* advection schemes have been specifically formulated that do not significantly modify positive extrema (including at locations where overshoots may be a problem) but do prevent the generation of false negatives. These methods will be discussed in the next section.

5.10 Schemes for Positive-Definite Advection

A *positive-definite* advection scheme is a method that never generates a negative value from nonnegative initial data.[20] Any monotone scheme is positive definite, but there are no other simple relationships between the sets of methods that are positive definite and those that are monotonicity preserving or TVD. TVD schemes need not be positive definite, and positive-definite schemes need not be TVD.

Early attempts to construct positive-definite advection schemes involved "filling algorithms," in which the solution obtained after each integration step was corrected by filling in any negative values. To conserve the total mass of the advected species, negatives cannot simply be set to zero; compensating mass must removed from positive regions. There are a variety of filling algorithms designed for this purpose. Some filling algorithms attempt to fill local negative regions from adjacent positive areas (Mahlman and Sinclair 1977). This may be a physically satisfying way to remove dispersive undershoots, but it requires a great deal of logical testing that cannot be performed efficiently on vector computers. In other approaches the compensating mass is removed from the entire field by multiplying the value at every grid point by the ratio of the total original mass to the total nonnegative mass. Multiplicative compensation is computationally efficient, but it preferentially damps the regions of highest tracer concentration. Other filling algorithms are reviewed by Rood (1987). Although empirical testing has shown the value of filling algorithms, the theoretical basis for these schemes is largely undeveloped.

5.10.1 An FCT Approach

The FCT algorithms presented in Sects. 5.4.2 and 5.9.1 do not allow new minima to be smaller than those predicted by the low-order monotone scheme, so, like a monotone method, these FCT schemes are positive definite. The general FCT algorithm can, however, be greatly simplified if all that is required is a positive-definite result. As noted by Smolarkiewicz (1989), any numerical conservation law of the form

$$\frac{\phi_j^{n+1} - \phi_j^n}{\Delta t} + \left(\frac{F_{j+\frac{1}{2}} - F_{j-\frac{1}{2}}}{\Delta x} \right) = 0 \tag{5.104}$$

[20] *Negative-definite* schemes may be similarly defined as any method that never generates positive values from nonpositive initial data. Any positive-definite scheme can be trivially converted to a negative-definite method.

can be converted to a positive-definite method. To illustrate the approach in its simplest form, temporarily suppose that the fluxes are always positive (as would be the case if (5.104) were used to approximate an advection problem involving nonnegative flow velocities and tracer concentrations). Then (5.104) will be positive definite if the actual fluxes are replaced by corrected fluxes $C_{j+1/2}F_{j+1/2}$, in which the correction factor is defined by

$$C_{j+\frac{1}{2}} = \min\left[1, \left(\phi_j^n \frac{\Delta x}{\Delta t}\right) / F_{j+\frac{1}{2}}\right].$$

This correction ensures that the outgoing flux is not large enough to drive ϕ_j^{n+1} negative.

Now consider the general case, in which the fluxes may have arbitrary sign. The flux-correction coefficient can be determined by omitting steps 1–5 of the Zalesak correction algorithm presented in Sect. 5.4.2 and modifying steps 6 and 7 as follows. Let P_j be the total flux out of grid volume j,

$$P_j = \max\left(0, F_{j+\frac{1}{2}}\right) - \min\left(0, F_{j-\frac{1}{2}}\right) + \epsilon, \tag{5.105}$$

where ϵ is a small number (such as 10^{-6}) added to ensure that P_j is not zero. Also let Q_j be the maximum outward flux that can be supported without forcing ϕ_j^{n+1} negative,

$$Q_j = \phi_j^n \frac{\Delta x}{\Delta t}. \tag{5.106}$$

Determine the ratio by which the fluxes $F_{j+1/2}$ and $F_{j-1/2}$ would need to be reduced to ensure that a negative is not be created at grid volume j,

$$R_j = \min\left(1, Q_j / P_j\right) \tag{5.107}$$

Finally, choose the actual limiter for the corrected flux $C_{j+1/2}F_{j+1/2}$ such that negatives are not created in the volume from which that flux is actually removing material:

$$C_{j+\frac{1}{2}} = \begin{cases} R_j & \text{if } F_{j+\frac{1}{2}} \geq 0, \\ R_{j+1} & \text{if } F_{j+\frac{1}{2}} < 0. \end{cases} \tag{5.108}$$

In contrast to the general FCT procedure, there is no initial step involving a low-order monotone scheme because it is not necessary to use a low-order solution to estimate the permissible range of values for ϕ_j^{n+1}. One simply sets $\phi_j^{\min} = 0$, imposes no constraint on $\phi_j^{\max}$, and corrects the fluxes to avoid generating values less than $\phi_j^{\min}$. Clearly, it is possible to further generalize this procedure by setting both $\phi_j^{\min}$ and $\phi_j^{\max}$ to any pair of arbitrarily specified constants.

Although it is conceptually simpler not to bother evaluating a low-order monotone solution, very little computational effort is required to do so when simulating advection, and that effort is often worthwhile. Since the monotone solution is positive definite, it can be used to reduce the impact of the preceding algorithm on the high-order fluxes as follows. Let ϕ^{td} denote the solution predicted using upstream differencing to update the current values of ϕ^n (as in the notation used with standard FCT and flux-limiter methods), and define a correction flux as the difference between the high-order and upstream fluxes,

$$F^{\text{cor}}_{j+\frac{1}{2}} = F^{\text{h}}_{j+\frac{1}{2}} - F^{\text{up}}_{j+\frac{1}{2}}.$$

The fluxes used in the conservation law (5.104) become

$$F_{j+\frac{1}{2}} = F^{\text{up}}_{j+\frac{1}{2}} + C_{j+\frac{1}{2}} F^{\text{cor}}_{j+\frac{1}{2}},$$

where $C_{j+1/2}$ deviates from unity only where necessary to prevent the development of negative ϕ, and may be evaluated using (5.105)–(5.108) with ϕ_j^n replaced by ϕ_j^{td} in (5.106) and all the fluxes F replaced by F^{cor}.

One peculiarity of this method is that the shape of the solution will change if a spatially uniform background field is added to the initial tracer concentration. This behavior differs from that of the true solution, in which the time tendency of the tracer concentration is determined only by the velocity field and the derivatives of the tracer-concentration field. Most of the other previously discussed methods for representing discontinuities and steep gradients avoid this dependence on the mean background concentration by using a different formulation of the nonlinear flux corrector. For example, the nonlinear correction used in the flux-limited scheme described in Sect. 5.5 is computed as a function of the ratio of the slopes of the numerical solution on each side of an individual grid point, and this ratio is independent of the magnitude of any horizontally uniform background concentration.

The sensitivity of the solution to changes in the background concentration depends on the magnitude of the undershoots that develop when the background concentration is zero. WENO methods and the selectively limited piecewise-parabolic method produce very small undershoots and, therefore, the preceding approach can be used to eliminate all negative values generated by these methods without making them very sensitive to the background concentration.

5.10.2 Antidiffusion via Upstream Differencing

One unique way to obtain a positive-definite advection scheme is to use upstream differencing to apply an antidiffusive correction to a previously computed monotone solution (Smolarkiewicz 1983). The first step of the Smolarkiewicz algorithm is a standard upstream difference in conservation form,

$$\phi_j^* = \phi_j^n - \left[F(\phi_j^n, \phi_{j+1}^n, c_{j+\frac{1}{2}}) - F(\phi_{j-1}^n, \phi_j^n, c_{j-\frac{1}{2}}) \right], \tag{5.109}$$

where

$$F(\phi_j, \phi_{j+1}, c) = \left[\left(\frac{c+|c|}{2}\right)\phi_j + \left(\frac{c-|c|}{2}\right)\phi_{j+1}\right]\frac{\Delta t}{\Delta x}.$$

The novel aspect of the Smolarkiewicz scheme is that the antidiffusion step is performed using a second upstream difference. Defining an "antidiffusion velocity"

$$\tilde{c}_{j+\frac{1}{2}} = \frac{(|c_{j+\frac{1}{2}}|\Delta x - c^2_{j+\frac{1}{2}}\Delta t)}{(\phi^*_j + \phi^*_{j+1} + \epsilon)}\left(\frac{\phi^*_{j+1} - \phi^*_j}{\Delta x}\right)$$

(where ϵ is a small positive number whose presence guarantees that the denominator will be nonzero), the antidiffusion step is

$$\phi^{n+1}_j = \phi^*_j - \left[F(\phi^*_j, \phi^*_{j+1}, \tilde{c}_{j+\frac{1}{2}}) - F(\phi^*_{j-1}, \phi^*_j, \tilde{c}_{j-\frac{1}{2}})\right]. \quad (5.110)$$

Since the first step (5.109) is the standard upstream method, it is monotone, positive definite, and highly diffusive. If c is constant, the first step provides a second-order approximation to the modified equation

$$\frac{\partial \psi}{\partial t} + \frac{\partial c\psi}{\partial x} = \frac{\partial}{\partial x}\left(K\frac{\partial \psi}{\partial x}\right),$$

in which K is the numerical diffusivity:

$$K = \frac{|c|\Delta x - c^2 \Delta t}{2}.$$

The second step compensates for this diffusion by subtracting a finite-difference approximation to the leading-order truncation error associated with the upstream method. In particular, the second step (5.110) is a numerical approximation to

$$\frac{\partial \psi}{\partial t} = -\frac{\partial \tilde{c}\psi}{\partial x},$$

in which

$$\tilde{c} = \begin{cases} \dfrac{K}{\psi}\dfrac{\partial \psi}{\partial x}, & \text{if } \psi > 0; \\ 0, & \text{if } \psi = 0. \end{cases}$$

Although the second step utilizes upstream differencing, the ϕ^{n+1} are highly nonlinear functions of the ϕ^*, and the second step is not monotone. The second step will, nevertheless, be positive definite provided that

$$\left| \frac{\tilde{c}_{j+\frac{1}{2}} \Delta t}{\Delta x} \right| \leq \frac{1}{2} \quad \text{for all } j,$$

which guarantees that even when both antidiffusive velocities are directed out of a particular grid cell, the antidiffusive fluxes will be too weak to generate a negative value. The preceding condition is satisfied whenever the initial ϕ are nonnegative and the maximum Courant number associated with the physical velocity field satisfies

$$\left| \frac{c_{j+\frac{1}{2}} \Delta t}{\Delta x} \right| \leq 1 \quad \text{for all } j.$$

Then

$$\left| \frac{\tilde{c}_{j+\frac{1}{2}} \Delta t}{\Delta x} \right| \leq \frac{|c_{j+\frac{1}{2}}| \Delta t}{\Delta x} \left(1 - \frac{|c_{j+\frac{1}{2}}| \Delta t}{\Delta x} \right) \left(\frac{|\phi^*_{j+1} - \phi^*_j|}{|\phi^*_{j+1} + \phi^*_j + \epsilon|} \right)$$

$$\leq \frac{1}{4} \left(\frac{|\phi^*_{j+1} - \phi^*_j|}{|\phi^*_{j+1} + \phi^*_j + \epsilon|} \right).$$

Since the first step is monotone, all the ϕ^* are nonnegative and

$$\left| \frac{\tilde{c}_{j+\frac{1}{2}} \Delta t}{\Delta x} \right| \leq \frac{1}{4}.$$

The Smolarkiewicz scheme can easily be extended to multidimensional problems (Smolarkiewicz 1984) and can be made monotonicity preserving by applying limiters in the antidiffusion step (Smolarkiewicz and Grabowski 1990). Note that like the FCT approach discussed in the preceding section, the Smolarkiewicz scheme will change the shape of the solution if a spatially uniform background field is added to the initial tracer concentration.

5.11 Curvilinear Coordinates

If the physical boundary constraining a fluid is nonrectangular, it can be advantageous to solve the governing equations in a curvilinear coordinate system that follows the boundary. In other circumstances, it is possible to simplify the problem by using cylindrical or spherical coordinates to exploit certain symmetries in the fluid system. When the governing equations are expressed in non-Cartesian coordinates, additional "metric" terms arise. These terms should be approximated in a way that preserves the conservation properties of the numerical scheme and the ability of the scheme to represent discontinuities and poorly resolved gradients. One elegant way to treat the metric terms is to begin with the equation formulated for an arbitrary curvilinear coordinate system and to apply one of the preceding methods directly to the transformed system (e.g., Smolarkiewicz and Margolin 1993).

For example, suppose that $(x_1, \ldots, x_n)$ is a position vector in Cartesian coordinates, that $(\tilde{x}_1, \ldots, \tilde{x}_n)$ is the corresponding vector in curvilinear coordinates, and that there is a smooth mapping between the two systems for which the Jacobian of the transformation $J = \text{Det}(\partial x_i / \partial \tilde{x}_j)$ is nonsingular. Then the velocities in the curvilinear coordinates are related to the Cartesian velocities such that

$$\tilde{v}_i = \frac{\partial \tilde{x}_i}{\partial x_k} v_k,$$

where repeated subscripts are summed. The divergence of the velocity vector transforms as

$$\frac{\partial v_i}{\partial x_i} = \frac{1}{J} \frac{\partial}{\partial \tilde{x}_k} (J \tilde{v}_k). \tag{5.111}$$

(See Gal-Chen and Somerville 1975.)

The equations governing the transport of a passive tracer in two-dimensional nondivergent flow may be expressed in curvilinear coordinates in either advective or flux form. Let $(x_1, x_2) = (x, y)$ and $(u_1, u_2) = (u, v)$. The advective form

$$\frac{d\psi}{dt} \equiv \frac{\partial \psi}{\partial t} + \tilde{u} \frac{\partial \psi}{\partial \tilde{x}} + \tilde{v} \frac{\partial \psi}{\partial \tilde{y}} = 0$$

can be derived from first principles using the definition of the total derivative in the transformed coordinates. The flux form

$$\frac{\partial \psi}{\partial t} + \frac{1}{J} \frac{\partial}{\partial \tilde{x}} (J \tilde{u} \psi) + \frac{1}{J} \frac{\partial}{\partial \tilde{y}} (J \tilde{v} \psi) = 0,$$

where

$$J = \frac{\partial x}{\partial \tilde{x}} \frac{\partial y}{\partial \tilde{y}} - \frac{\partial x}{\partial \tilde{y}} \frac{\partial y}{\partial \tilde{x}},$$

can also be derived from first principles using the expression for the divergence in transformed coordinates (5.111). The flux form implies conservation of ψ (provided that coordinate transformation is time independent) and is ready for direct approximation by a numerical conservation law. The numerical fluxes can be limited or corrected as discussed previously to preserve monotonicity.

The proper formulation of a numerical approximation of the advective form is more subtle. As discussed previously, it is important to create a finite-difference approximation to the advective form that is algebraically equivalent to the flux form. On a staggered mesh, one typically requires the velocities to satisfy the incompressible continuity equation,

$$\frac{1}{J_{i,j}} \left[\left(\delta_x (J_{i,j} \tilde{u}_{i,j}) + \delta_y (J_{i,j} \tilde{v}_{i,j}) \right) \right] = 0,$$

where it is assumed that $\tilde{u}_{i+1/2,j}$ is located $\Delta x/2$ to the "east" and $\tilde{v}_{i,j+1/2}$ is $\Delta y/2$ to the "north" of the grid point where $\phi_{i,j}$ is defined. The numerical representation the advective form should then approximate

$$\frac{\partial \psi}{\partial t} + \frac{1}{J}\left(\tilde{u}J\frac{\partial \psi}{\partial \tilde{x}} + \tilde{v}J\frac{\partial \psi}{\partial \tilde{y}}\right) = 0,$$

where the common factor of J is not canceled out of the numerator and denominator because it is evaluated at different locations on the staggered mesh. The evaluation of J at these slightly different grid points is required to make the finite-difference method in advective form algebraically equivalent to a scheme in flux form.

Problems

1. Consider two sets of equations that might be supposed to govern one-dimensional shallow-water flow:

$$\frac{\partial u}{\partial t} + \frac{\partial}{\partial x}\left(\frac{u^2}{2} + gh\right) = 0, \qquad \frac{\partial h}{\partial t} + \frac{\partial hu}{\partial x} = 0,$$

and

$$\frac{\partial hu}{\partial t} + \frac{\partial}{\partial x}\left(hu^2 + g\frac{h^2}{2}\right) = 0, \qquad \frac{\partial h}{\partial t} + \frac{\partial hu}{\partial x} = 0.$$

(a) Under what conditions do these systems have identical solutions?

(b) Give an example, including initial conditions and expressions for the time-dependent solutions, for which these systems have different solutions.

(c) In those situations where these systems have different solutions, which one serves as the correct mathematical model for shallow-water flow? (*Hint:* The correct choice must be determined from fundamental physical principles.)

2. Compute the speed at which the unit-amplitude jump (5.3) must propagate to be a weak solution to the conservation law

$$\frac{\partial \psi^2}{\partial t} + \frac{\partial}{\partial x}\left(\frac{2\psi^3}{3}\right) = 0.$$

How does this speed compare with that at which the same jump is propagated by the inviscid Burgers equation? Explain whether the difference in the speed of these jumps is consistent with the sign of the inequality in the entropy condition for solutions to Burgers's equation (5.13)?

3. Show that if $\psi(x,0) \geq 0$, the solution to

$$\frac{\partial \psi}{\partial t} + \frac{\partial}{\partial x}[c(x)\psi] = 0$$

remains nonnegative for all $t \geq 0$. Assume that c and ψ have continuous derivatives in order to simplify the argument. (*Hint:* To develop negative ψ, there must be a first time t_0 and some point x_0 for which $\psi(x_0, t_0) = 0$ and $\psi_t(x_0, t_0) < 0$. Show that this is impossible.) Does this result generalize to problems in two and three spatial dimensions?

4. Use the results of Problem 3 to show that if

$$\frac{\partial \psi}{\partial t} + \frac{\partial}{\partial x}[c(x)\psi] = 0, \qquad \frac{\partial \varphi}{\partial t} + \frac{\partial}{\partial x}[c(x)\varphi] = 0,$$

and $\psi(x,0) \geq \varphi(x,0)$, then $\psi(x,t) \geq \varphi(x,t)$ for all x and $t > 0$.

5. Suppose the constant-wind-speed advection equation (5.20) is approximated using the Lax–Friedrichs scheme

$$\phi_j^{n+1} = \frac{1}{2}(\phi_{j+1}^n + \phi_{j-1}^n) - \frac{\mu}{2}(\phi_{j+1}^n - \phi_{j-1}^n), \tag{5.112}$$

where $\mu = c\Delta t/\Delta x$. Compare the implicit numerical diffusion generated by this scheme with that produced by upstream differencing. Show that the ratio of the leading-order numerical diffusion in the upstream scheme to that in the Lax–Friedrichs method is $\mu/(1+\mu)$.

6. Suppose that the constant-wind-speed advection equation (5.20) is approximated using the scheme

$$\phi_j^{n+1} = \left(1 - c\Delta t\, \delta_{2x} + \gamma(\Delta x)^2 \delta_x^2\right) \phi_j^n,$$

where γ is a user-specified parameter determining the amount of numerical smoothing.

(a) What is the largest value of γ for which the scheme can be monotone?

(b) Suppose that γ is specified as some value γ_0 for which the scheme can be monotone. For what values of $\mu = c\Delta t/\Delta x$ will the scheme actually be monotone?

(c) For what value of γ is this scheme equivalent to the Lax–Friedrichs scheme (5.112)?

7. Explain why a scheme that is monotonicity preserving need not be TVD (or more precisely, total variation nonincreasing). Explain why being TVD does not imply that a scheme is monotone.

8. Show that no new maxima or minima can develop in smooth solutions to the conservation law (5.7).

9. Show that Harten's criterion (5.38) ensuring that schemes of the form (5.37) are TVD is not sufficient to guarantee that they are monotone.

10. Show that the superbee flux limiter (5.44) is equivalent to the following slope limiter:

$$\sigma_j = \begin{cases} 0 & \text{if } ab \le 0, \\ \operatorname{sgn}(a) \max\left[0, \min(|a|, 2|b|), \min(2|a|, |b|)\right] & \text{otherwise,} \end{cases}$$

where

$$a = \frac{\phi_{j+1} - \phi_j}{\Delta x}, \qquad b = \begin{cases} \dfrac{\phi_j - \phi_{j-1}}{\Delta x} & \text{if } c \ge 0, \\ \dfrac{\phi_{j+2} - \phi_{j+1}}{\Delta x} & \text{if } c < 0. \end{cases}$$

Compare the geometric limits placed on the slope by the superbee limiter with those imposed by the minmod and MC limiters.

11. Suppose that Lax–Wendroff solutions are sought to a one-dimensional advection equation (5.20) and that the velocity $c(t)$ depends on time but not on x.

(a) Derive an expression for $\partial^2\psi/\partial t^2$ in terms of the spatial derivatives of ψ and functions of the velocity field.

(b) Show that a fully second order Lax–Wendroff approximation to this problem can be obtained using (5.47) with c replaced by $(c^{n+1} + c^n)/2$.

12. Show that if the solution is infinitely differentiable, weights w_ℓ can indeed be determined to create an order $2k - 1$ approximation to the flux $F_{j+1/2}$ from the kth-order-accurate ENO approximations $p_\ell(x_{j+1/2})$, where $\ell = 0, 1, \ldots, k - 1$.

13. Show that, considered in isolation, the antidiffusion step (5.110) of the Smolarkiewicz positive-definite advection scheme is not monotone.

14. Suppose that $f(s)$ is a continuously differentiable function of s and that $\psi(x,t)$ is a solution to the scalar conservation law (5.7). Show that the characteristic curves for this hyperbolic partial differential equation are straight lines.

15. *Compute solutions to the advection equation (5.20) on the periodic domain $0 \le x \le 1$ subject to the initial condition $\psi(x,0) = \sin^6(2\pi x)$. Let $c = 0.1$.

(a) Compare the exact solution with numerical solutions obtained using forward, Lax–Wendroff, and flux-limited methods. In the flux-limited methods compute the low-order flux using the upstream scheme and the high-order flux using the Lax–Wendroff method, but try three different flux limiters: the MC, the minmod, and the superbee. Perform the simulations using a Courant number $c\Delta t/\Delta x = 0.5$ and $\Delta x = 1/40$. As part of your discussion submit two plots of the solution at time $t = 20$, one comparing the exact solution with that obtained using the three different flux limiters, and one comparing the exact, upstream, Lax–Wendroff, and MC flux-limited solutions. Scale the vertical axis so that $-0.4 \leq \psi(x) \leq 1.4$.

(b) Repeat the preceding simulations for the initial condition

$$\psi(x,0) = \begin{cases} 1 & \text{if } |x - \frac{1}{2}| \leq \frac{1}{4}; \\ 0 & \text{otherwise.} \end{cases}$$

Again submit two plots of the solution at time $t = 20$, one comparing the exact solution with that obtained using the three different flux limiters, and one comparing the exact, upstream, Lax–Wendroff, and MC flux-limited solutions. Discuss your results.

16. *Determine the effective order of accuracy of the minmod, MC, and superbee flux-limited approximations to the advection equation considered in Problem 15 except use the very smooth initial data $\psi(x,0) = \sin(2\pi x)$. In addition, compute results for the Zalesak FCT method using upstream differencing for the low-order solution and the Lax–Wendroff scheme for the higher-order solution. Also try the iterative FCT scheme discussed at the end of Sect. 5.4.2 using the preceding noniterated FCT solution for the low-order scheme during the second iteration. Keep the Courant number fixed at 0.5, and use $\Delta x = 1/20$, 1/40, 1/80, 1/160, and 1/320. Compute the L_2 norm of the difference between the exact and approximate solutions and plot the log of the error versus the log of Δx to estimate the power of Δx that is proportional to the error as $\Delta x \to 0$. Compare these results with the theoretical order of accuracy for the standard upstream and Lax–Wendroff methods.

17. *Compare simulations of the geostrophic adjustment problem described in Problem 12 in Chap. 4 obtained using a flux-limiter scheme with the MC limiter and FCT. Consider both the initial conditions: the discontinuous step and the slightly smoothed step. To use the constant-wind-speed advection algorithms presented in this chapter, transform the governing equations to an equivalent system for the unknown functions $u+g\eta/c$, $u-g\eta/c$, and v (where $c^2 = gH$). Use upstream differencing and the Lax–Wendroff scheme for the monotone and second-order methods. Treat the Coriolis terms in the transformed system via operator splitting.

Chapter 6
Series-Expansion Methods

Series-expansion methods that are potentially useful in geophysical fluid dynamics include the spectral, pseudospectral, finite-element, and discontinuous Galerkin methods. The spectral method plays a particularly important role in global atmospheric models, in which the horizontal structure of the numerical solution is often represented as a truncated series of spherical harmonics. Finite-element methods, on the other hand, are not commonly used in multidimensional wave propagation problems because they generally require the solution of implicit algebraic systems and are therefore not as efficient as competing explicit methods. The discontinuous Galerkin method is a combination of the spectral and finite-element methods and also has many similarities with finite-volume methods. All of these series-expansion methods share a common foundation that will be discussed in the next section.

6.1 Strategies for Minimizing the Residual

Suppose F is an operator involving spatial derivatives of ψ, and that solutions are sought to the partial differential equation

$$\frac{\partial \psi}{\partial t} + F(\psi) = 0, \tag{6.1}$$

subject to the initial condition $\psi(x, t_0) = f(x)$ and to boundary conditions at the edges of some spatial domain S. The basic idea in all series-expansion methods is to approximate the spatial dependence of ψ as a linear combination of a finite number of predetermined expansion functions. Let the general form of the series expansion be written as

$$\phi(x,t) = \sum_{k=1}^{N} a_k(t)\varphi_k(x), \tag{6.2}$$

where $\varphi_1, \ldots, \varphi_N$ are predetermined expansion functions satisfying the required boundary conditions. Then the task of solving (6.1) is transformed into a problem

D.R. Durran, *Numerical Methods for Fluid Dynamics: With Applications to Geophysics*, Texts in Applied Mathematics 32, DOI 10.1007/978-1-4419-6412-0_6,

of calculating the unknown coefficients $a_1(t), \ldots, a_N(t)$ in a way that minimizes the error in the approximate solution. One might hope to obtain solvable expressions for the $a_k(t)$ by substituting (6.2) into the governing equation (6.1). For example, if F is a linear function of $\partial^n \psi/\partial x^n$ with constant coefficients and (6.2) is a truncated Fourier series, direct substitution will yield a system of ordinary differential equations for the evolution of the $a_k(t)$. Unfortunately, direct substitution yields a solvable system of equations for the expansion coefficients only when the φ_k are eigenfunctions of the differential operator F, i.e., direct substitution works in precisely those special cases for which analytic solutions are available. This, of course, is a highly restrictive limitation.

In the general case where the φ_k are not eigenfunctions of F, it is impossible to specify $a_1(t), \ldots, a_N(t)$ such that an expression of the form (6.2) exactly satisfies (6.1). As an example, suppose $F(\psi) = \psi \partial\psi/\partial x$ and the expansion functions are the Fourier components $\varphi_k = e^{ikx}$, $-N \le k \le N$. If this Fourier series is substituted into (6.1), the nonlinear product in $F(\psi)$ introduces spatial variations at wave numbers that were not present in the initial truncated series, e.g., $F(e^{iNx}) = iNe^{i2Nx}$. A total of $4N + 1$ equations are obtained after substituting the expansion (6.2) into (6.1) and requiring that the coefficients of each Fourier mode sum to zero. It is not possible to choose the $2N + 1$ Fourier coefficients in the original expansion to satisfy these $4N + 1$ equations simultaneously. The best one can do is to select the expansion coefficients to minimize the error.

Since the actual error in the approximate solution $\|\psi - \phi\|$ cannot be determined, the most practical way to try to minimize the error is to minimize the *residual*,

$$R(\phi) = \frac{\partial \phi}{\partial t} + F(\phi), \tag{6.3}$$

which is the amount by which the approximate solution fails to satisfy the governing equation. Three different strategies are available for constraining the size of the residual. Each strategy leads to a system of N coupled *ordinary* differential equations for the time-dependent coefficients $a_1(t), \ldots, a_N(t)$. This transformation of the partial differential equation into a system of ordinary differential equations is similar to that which occurs in grid-point methods when the spatial derivatives are replaced with finite differences.

One strategy for constraining the size of the residual is to pick the $a_k(t)$ to minimize the square of the L_2 norm of the residual:

$$(\|R(\phi)\|_2)^2 = \int_S [R(\phi(x))]^2 \, dx.$$

A second approach, referred to as collocation, is to require the residual to be zero at a discrete set of grid points:

$$R(\phi(j\Delta x)) = 0 \quad \text{for all } j = 1, \ldots, N.$$

The third strategy, known as the Galerkin approximation, requires the residual to be *orthogonal* to each of the expansion functions, i.e.,

$$\int_S R(\phi(x))\varphi_k(x)\,\mathrm{d}x = 0 \quad \text{for all } k = 1, \ldots, N. \tag{6.4}$$

Different series-expansion methods rely on one or more of the preceding approaches. The collocation strategy is used in the pseudospectral method and in some finite-element formulations, but not in the spectral method. The L_2-minimization and Galerkin criteria are equivalent when applied to a problem of the form (6.1), and are the basis of the spectral method. The Galerkin approximation is also used extensively in finite-element schemes and, not surprisingly, in the discontinuous Galerkin method.

The equivalence of the L_2-minimization criterion and the Galerkin approximation can be demonstrated as follows. According to (6.3), the residual depends on both the instantaneous values of the expansion coefficients and their time tendencies. The expansion coefficients are determined at the outset from the initial conditions and are known at the beginning of any subsequent integration step. The criteria for constraining the residual are not used to obtain the instantaneous values of the expansion coefficients, but rather to determine their time evolution. If the rate of change of the kth expansion function is calculated to minimize $(\|R(\phi)\|_2)^2$, a necessary criterion for a minimum may be obtained by differentiation with respect to the quantity $\mathrm{d}a_k/\mathrm{d}t \equiv \dot{a}_k$

$$\begin{aligned}
0 &= \frac{d}{d(\dot{a}_k)}\left\{\int_S (R(\phi))^2\,\mathrm{d}x\right\} \\
&= \frac{d}{d(\dot{a}_k)}\left\{\int_S \left[\sum_{n=1}^{N} \dot{a}_n\varphi_n + F\left(\sum_{n=1}^{N} a_n\varphi_n\right)\right]^2 \mathrm{d}x\right\} \\
&= 2\int_S \left[\sum_{n=1}^{N} \dot{a}_n\varphi_n + F\left(\sum_{n=1}^{N} a_n\varphi_n\right)\right]\varphi_k\,\mathrm{d}x \qquad (6.5) \\
&= 2\int_S R(\phi)\varphi_k\,\mathrm{d}x. \qquad (6.6)
\end{aligned}$$

The second derivative of $(\|R(\phi)\|_2)^2$ with respect to $\dot{a}_k$ is $2(\|\varphi_k\|_2)^2$, which is positive. Thus, the extremum condition (6.6) is associated with a true minimum of $(\|R(\phi)\|_2)^2$, and the Galerkin requirement is identical to the condition obtained by minimizing the L_2 norm of the residual.

As derived in (6.5), the Galerkin approximation and the L_2-minimization of the residual both lead to a system of ordinary differential equations for the expansion coefficients of the form

$$\sum_{n=1}^{N} M_{n,k}\frac{da_n}{dt} = -\int_S \left[F\left(\sum_{n=1}^{N} a_n\varphi_n\right)\varphi_k\right]\mathrm{d}x \quad \text{for all } k = 1, \ldots, N, \tag{6.7}$$

where, following terminology from continuum mechanics,

$$M_{n,k} = \int_S \varphi_n \varphi_k \,\mathrm{d}x$$

are the components of the *mass matrix* $\mathbf{M}$. The initial conditions for the preceding system of differential equations are obtained by choosing $a_1(t_0), \ldots, a_N(t_0)$ such that $\phi(x, t_0)$ provides the "best" approximation to $f(x)$. The possible strategies for constraining the initial error are identical to those used to ensure that the residual is small. As before, the choice that minimizes the L_2 norm of the initial error also satisfies the Galerkin requirement that the initial error be orthogonal to each of the expansion functions,

$$\int_S \left(\sum_{n=1}^{N} a_n(t_0)\varphi_n(x) - f(x) \right) \varphi_k(x) \,\mathrm{d}x = 0 \quad \text{for all } k = 1, \ldots, N,$$

or equivalently,

$$\sum_{n=1}^{N} M_{n,k} a_n = \int_S f(x)\varphi_k(x) \,\mathrm{d}x \quad \text{for all } k = 1, \ldots, N. \tag{6.8}$$

6.2 The Spectral Method

The characteristic that distinguishes the spectral method from the finite-element method is that the expansion functions form an orthogonal set. Since the expansion functions are orthogonal, $M_{n,k}$ is zero unless $n = k$, and the system of differential equations for the coefficients (6.7) reduces to

$$\frac{da_k}{dt} = -\frac{1}{M_{k,k}} \int_S \left[F\left(\sum_{n=1}^{N} a_n \varphi_n \right) \varphi_k \right] \mathrm{d}x \quad \text{for all } k = 1, \ldots, N. \tag{6.9}$$

This is a particularly useful simplification, since explicit algebraic equations for each $a_k(t + \Delta t)$ may be easily obtained when the time derivatives in (6.9) are replaced with finite differences. In contrast, the finite-difference approximation of the time derivatives in the more general form (6.7) introduces a coupling between all the expansion coefficients at the new time level, and the solution of the resulting implicit system of algebraic equations may require considerable computation. The orthogonality of the expansion functions also reduces the expression for the initial value of each expansion coefficient (6.8) to

$$a_k(t_0) = \frac{1}{M_{k,k}} \int_S f(x)\varphi_k(x) \,\mathrm{d}x. \tag{6.10}$$

The choice of some particular family of orthogonal expansion functions is largely dictated by the geometry of the problem and by the boundary conditions. Fourier series are well suited to rectangular domains with periodic boundary conditions. Chebyshev polynomials are a possibility for nonperiodic domains. Associated Legendre functions are useful for representing the latitudinal dependence of a function on the spherical Earth. Since Fourier series lead to the simplest formulae, they will be used to illustrate the elementary properties of the spectral method. The special problems associated with spherical geometry will be discussed in Sect. 6.4.

6.2.1 Comparison with Finite-Difference Methods

In Chap. 2, a variety of finite-difference methods were tested on the one-dimensional advection equation

$$\frac{\partial \psi}{\partial t} + c\frac{\partial \psi}{\partial x} = 0. \tag{6.11}$$

Particular emphasis was placed on the simplest case, in which c was constant. When c is constant, it is easy to find expansion functions that are eigenfunctions of the spatial derivative term in (6.11). As a consequence, the problem of advection by a constant wind is almost too simple for the spectral method. Nevertheless, the constant-wind case reveals some of the fundamental strengths and weaknesses of the spectral method and allows a close comparison between the spectral method and finite-difference schemes.

Suppose, therefore, that c is constant and solutions are sought to (6.11) on the periodic domain $-\pi \leq x \leq \pi$, subject to the initial condition $\psi(x,0) = f(x)$. A Fourier series expansion

$$\phi(x,t) = \sum_{k=-N}^{N} a_k(t)\mathrm{e}^{\mathrm{i}kx} \tag{6.12}$$

is the natural choice for this problem. Since individual Fourier modes are eigenfunctions of the differential operator in (6.11), evolution equations for the Fourier coefficients of the form

$$\frac{da_k}{dt} + \mathrm{i}cka_k = 0 \tag{6.13}$$

may be obtained by directly substituting (6.12) into the advection equation. In this atypically simple case, the residual is zero, and it is not necessary to adopt any particular procedure to minimize its norm. Nevertheless, (6.13) can also be obtained through the Galerkin requirement that the residual be orthogonal to each of the expansion functions. To apply the Galerkin formulation it is necessary to generalize the definition of orthogonality to include complex-valued functions. Two complex-valued functions $g(x)$ and $h(x)$ are *orthogonal* over the domain S if

$$\int_S g(x)h^*(x)\,\mathrm{d}x = 0,$$

where $h^*(x)$ denotes the complex conjugate of $h(x)$.[1] As an example, note that for integer values of n and m,

$$\int_{-\pi}^{\pi} e^{inx} e^{-imx}\, dx = \begin{cases} \left.\dfrac{e^{i(n-m)x}}{i(n-m)}\right|_{-\pi}^{\pi} = 0, & \text{if } m \neq n; \\ 2\pi, & \text{if } m = n, \end{cases}$$

which is just the well-known orthogonality condition for two Fourier modes. Using this orthogonality relation and setting $f(\psi) = c\partial\psi/\partial x$, with c constant, reduces (6.9) to (6.13).

If the ordinary differential equation (6.13) is solved analytically (in practical applications it must be computed numerically), solutions have the form $a_k(t) = \exp(-ickt)$. Thus, in the absence of time-differencing errors, the frequency of the kth Fourier mode is identical to the correct value for the continuous problem $\omega = ck$. The spectral approximation does not introduce phase-speed or amplitude errors – even in the shortest wavelengths! The ability of the spectral method to correctly capture the amplitude and phase speed of the shortest resolvable waves is a significant advantage over conventional grid-point methods, in which the spatial derivative is approximated by finite differences, yet surprisingly the spectral method is not necessarily a good technique for modeling short-wavelength disturbances. The problem lies in the fact that it is only those waves retained in the truncated series expansion that are correctly represented in the spectral solution. If the true solution has a great deal of spatial structure on the scale of the shortest wavelength in the truncated series expansion, the spectral representation will not accurately approximate the true solution.

The problems with the representation of short-wavelength features in the spectral method are illustrated in Fig. 6.1, which shows ten grid-point values forming a $2\Delta x$-wide spike against a zero background on a periodic domain with a uniformly spaced grid. Also shown is the curve defined by the truncated Fourier series passing through those ten grid-point values. The Fourier series approximation to the $2\Delta x$ spike exhibits large oscillations about the zero background state on both sides of the spike. Now suppose that the data in Fig. 6.1 represent the initial condition for a constant-wind-speed advection problem. The overshoots and undershoots associated with the Fourier approximation will not be apparent at the initial time if the data are sampled only at the points on the discrete mesh. If time-differencing errors are neglected, the grid-point values will also be exact at those subsequent times at which the initial distribution has translated an integral number of grid intervals. The grid-point values will, however, reveal the oscillatory error at intermediate times. The worst errors in the grid-point values will occur at times when the solution has traveled $n + 1/2$ grid intervals, where n is any integer. In this particular example, the error on the discrete grid does not accumulate with time; it oscillates instead, achieving

[1] Multiplication by the complex conjugate ensures that if $g(x) = a(x) + ib(x)$ with a and b real, then

$$\int_S g(x)g^*(x)\, dx = \int_S (a^2 + b^2)\, dx.$$

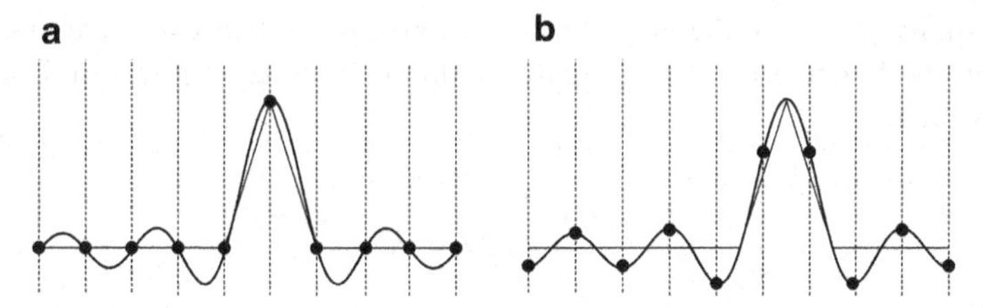

Fig. 6.1 **a** Ten periodic grid-point values exhibiting a piecewise-linear $2\Delta x$ spike, and the truncated Fourier series approximation passing through those ten points. **b** Values of the truncated Fourier series sampled at the same grid-point location after translating the curve one half grid point to the right

a minimum when the solution has translated an integral number of grid intervals. The maximum error is limited by the error generated when the initial condition is projected onto the truncated Fourier series.

6.2.1.1 The Finite Fourier Transform

There is a simple relationship between the nine[2] independent grid-point values in Fig. 6.1 and the nine coefficients determining the truncated Fourier series passing through those points. If the Fourier series expansion of a real-valued function is truncated at wave number N, the set of Fourier coefficients contains $2N + 1$ pieces of data. Assuming that the Fourier expansion functions are periodic on the domain $0 \le x \le 2\pi$, an equivalent amount of information is contained by the $2N + 1$ function values $\phi(x_j, t)$, where

$$x_j = j\left(\frac{2\pi}{2N+1}\right) \quad \text{and} \quad j = 1, \ldots, 2N+1.$$

It is obvious that the set of Fourier coefficients $a_{-N}(t), \ldots, a_N(t)$ defines the grid-point values $\phi(x_j, t)$ through the relation

$$\phi(x_j, t) = \sum_{k=-N}^{N} a_k(t) \mathrm{e}^{\mathrm{i}kx_j}. \tag{6.14}$$

Although it is not as self-evident, the $2N + 1$ Fourier coefficients may also be determined from the $2N + 1$ grid-point values. An exact algebraic expression for $a_n(t)$ can be obtained by noting that

$$\begin{aligned}\sum_{j=1}^{2N+1} \phi(x_j, t)\mathrm{e}^{-\mathrm{i}nx_j} &= \sum_{j=1}^{2N+1}\left(\sum_{m=-N}^{N} a_m(t)\mathrm{e}^{\mathrm{i}mx_j}\right)\mathrm{e}^{-\mathrm{i}nx_j} \\ &= \sum_{m=-N}^{N} a_m(t)\left(\sum_{j=1}^{2N+1} \mathrm{e}^{\mathrm{i}mx_j}\mathrm{e}^{-\mathrm{i}nx_j}\right).\end{aligned} \tag{6.15}$$

[2] The tenth grid-point value is redundant information because the solution is periodic.

Further simplification of the preceding equation is possible because the final summation in (6.15) obeys an orthogonality condition on the discrete mesh. Using the definition of x_j,

$$\sum_{j=1}^{2N+1} \mathrm{e}^{\mathrm{i}mx_j}\mathrm{e}^{-\mathrm{i}nx_j} = \sum_{j=1}^{2N+1} \left(\mathrm{e}^{\frac{\mathrm{i}2\pi(m-n)}{2N+1}}\right)^j. \tag{6.16}$$

If $m = n$, then (6.16) sums to $2N + 1$; for $m \neq n$ the formula for the sum of a finite geometric series,

$$1 + r + r^2 + \cdots + r^n = \frac{1 - r^{n+1}}{1 - r}, \tag{6.17}$$

may be used to reduce (6.16) to

$$\sum_{j=1}^{2N+1} \mathrm{e}^{\mathrm{i}mx_j}\mathrm{e}^{-\mathrm{i}nx_j} = \frac{\mathrm{e}^{\frac{\mathrm{i}2\pi(m-n)}{2N+1}}\left(1 - \mathrm{e}^{\mathrm{i}2\pi(m-n)}\right)}{1 - \mathrm{e}^{\frac{\mathrm{i}2\pi(m-n)}{2N+1}}} = 0. \tag{6.18}$$

Using these orthogonality properties, (6.15) becomes

$$a_n(t) = \frac{1}{2N+1} \sum_{j=1}^{2N+1} \phi(x_j, t)\mathrm{e}^{-\mathrm{i}nx_j}. \tag{6.19}$$

The relations (6.19) and (6.14), known as *finite Fourier transforms*, are discretized analogues to the standard Fourier transform and its inverse. The integrals in the continuous transforms are replaced by finite sums in the discrete expressions. These formulae, or more specifically the mathematically equivalent fast Fourier transform (FFT) algorithms, are essential for obtaining efficient spectral solutions in many practical applications where it is advantageous to transform the solution back and forth between wave-number space and physical space once during the execution of every time step.

6.2.1.2 The Ordering of the Coefficients in Typical FFTs

Although finite Fourier transforms of vectors with an odd number of elements yield formulae most directly analogous to those for continuous Fourier series, all efficient FFTs use an even number of grid points. The ordering of the individual Fourier modes in typical FFTs is also very different from the ordering in more conventional complex Fourier series. In the following we obtain relations equivalent to (6.14) and (6.19) in the form in which they are employed in typical FFTs.

Let $\phi_j = \phi(x_j)$ be the vector of M grid-point values on the periodic mesh

$$x_j = \frac{2\pi(j-1)}{M}, \qquad j = 1, 2, \ldots, M.$$

Assuming M is even, define $N = M/2$ and $\omega_M = e^{2\pi\mathrm{i}/M}$. Note $(\omega_M)^M = 1$, implying that ω_M is an Mth root of unity. One may express the discrete Fourier series as

$$\phi_j = \sum_{k=-N+1}^{N} a_k e^{ikx_j} = \sum_{k=-N+1}^{N} a_k e^{ik2\pi(j-1)/M} = \sum_{k=-N+1}^{N} a_k \omega_M^{k(j-1)},$$

or splitting up the summation, as

$$\phi_j = \sum_{k=-N+1}^{-1} a_k \omega_M^{k(j-1)} + \sum_{k=0}^{N} a_k \omega_M^{k(j-1)}. \tag{6.20}$$

Using
$$\omega_M^{(k-M)(j-1)} = \omega_M^{k(j-1)} 1^{1-j} = \omega_M^{k(j-1)},$$

we can write (6.20) as

$$\phi_j = \sum_{k=N+1}^{2N-1} a_{k-M} \omega_M^{k(j-1)} + \sum_{k=0}^{N} a_k \omega_M^{k(j-1)}. \tag{6.21}$$

Defining a new vector of coefficients b_k such that

$$b_1 = a_0, b_2 = a_1, \ldots, b_{N+1} = a_N, b_{N+2} = a_{-N+1}, b_{N+3} \\ = a_{-N+2}, \ldots, b_{2N} = a_{-1},$$

(6.21) takes the compact form

$$\phi_j = \sum_{k=1}^{M} b_k \omega_M^{(k-1)(j-1)}. \tag{6.22}$$

Also, using the discrete orthogonality condition, one can show that

$$b_k = \frac{1}{M} \sum_{j=1}^{M} \phi_j \omega_M^{-(k-1)(j-1)}. \tag{6.23}$$

As in (6.23), a typical FFT routine yields an array in which the Fourier coefficients a_k are ordered by increasing positive wave number up through element $M/2+1$ (for M even) and then ordered by increasing negative wave number through element M. Note that the equations corresponding to (6.22) and (6.23) may vary among different FFT routines because neither the placement of the $1/M$ normalization factor nor the choice of the equation containing the negative exponent is standardized.

6.2.1.3 The Equivalent Grid-Point Method

If c is constant, the spectral solution to the advection equation (6.11) can be recast in the form of an equivalent finite-difference method. Observe that

$$\begin{aligned} \frac{d\phi(x_j,t)}{dt} &= \sum_{n=-N}^{N} \frac{da_n}{dt} e^{inx_j} \\ &= -\sum_{n=-N}^{N} inca_n(t) e^{inx_j} \end{aligned}$$

$$= -c \sum_{n=-N}^{N} in \left(\frac{1}{2N+1} \sum_{k=1}^{2N+1} \phi(x_k, t) e^{-inx_k} \right) e^{inx_j}$$

$$= -c \sum_{k=1}^{2N+1} C_{j,k} \, \phi(x_k, t),$$

where

$$C_{j,k} = \frac{1}{2N+1} \sum_{n=-N}^{N} in e^{in(x_j - x_k)}.$$

The finite-difference coefficient $C_{j,k}$ depends only on the difference between j and k, and is zero if $j = k$. If $j \neq k$, a simpler expression for $C_{j,j+\ell}$ can be obtained by defining

$$s = x_j - x_{j+\ell} = -\ell \left(\frac{2\pi}{2N+1} \right),$$

in which case

$$C_{j,j+\ell} = \frac{1}{2N+1} \frac{d}{ds} \left(\sum_{n=-N}^{N} e^{ins} \right) = \frac{1}{2N+1} \frac{d}{ds} \left(e^{-iNs} \sum_{n=0}^{2N} (e^{is})^n \right).$$

Using (6.17) to sum the finite geometric series, differentiating, and noting that $e^{i(2N+1)s} = 1$, the preceding expression becomes

$$C_{j,j+\ell} = \frac{e^{i(N+\frac{1}{2})s}}{2 \sin\left(\frac{s}{2}\right)} = \frac{(-1)^{\ell+1}}{2 \sin\left(\frac{\ell\pi}{2N+1}\right)},$$

which implies that since $C_{j,j+\ell} = -C_{j,j-\ell}$, the equivalent finite-difference formula is centered in space.

Two grid-point values are used in the centered second-order finite-difference approximation to $\partial\psi/\partial x$. A centered fourth-order difference utilizes four points; the sixth-order difference requires six grid points. Every grid point on the numerical mesh (except the central point) is involved in the spectral approximation of $\partial\psi/\partial x$. As will be shown in the next section, the use of all these grid points allows the spectral method to compute derivatives of smooth functions with very high accuracy. Merilees and Orszag (1979) have compared the weighting coefficients for the spectral method on a 17-point periodic grid with the weighting coefficients for centered second-order through 16th-order finite differences. Their calculations appear in Table 6.1, which shows that the influence of remote grid points on the spectral calculation is much greater than the remote influence in any of the finite-difference formulae. The large degree of remote influence in the spectral method has been a source of concern, since the true domain of dependence for the constant-wind-speed advection equation is a straight line. Practical evidence suggests that this remote influence is not a problem provided that enough terms are retained in the truncated Fourier series to adequately resolve the spatial variations in the solution.

Table 6.1 Comparison of weight accorded to each grid point as a function of its distance to the central grid point in centered finite differences and in a spectral method employing 17 expansion coefficients

Δx away from central point	2nd order	4th order	6th order	16th order	Spectral
1	0.500	0.667	0.750	0.889	1.006
2		−0.083	−0.150	−0.311	−0.512
3			0.017	0.113	0.351
4				−0.035	−0.274
5				0.009	0.232
6				−0.001	−0.207
7				0.000	0.192
8				−0.000	−0.186

6.2.1.4 Order of Accuracy

The accuracy of a finite difference is characterized by the truncation error, which is computed by estimating a smooth function's values at a series of grid points through the use of Taylor series, and by substituting those Taylor series expansions into the finite-difference formula. The discrepancy between the finite-difference calculation and the true derivative is the truncation error and is usually proportional to some power of the grid interval. A conceptually similar characterization of accuracy is possible for the computation of spatial derivatives via the spectral method.

The basic idea is to examine the difference between the actual derivative of a smooth function and the approximate derivative computed from the spectral representation of the same function. Suppose that a function $\psi(x)$ is periodic on the domain $-\pi \leq x \leq \pi$ and that the first few derivatives of ψ are continuous. Then ψ and its first derivative can be represented by the convergent Fourier series

$$\psi(x) = \sum_{k=-\infty}^{\infty} a_k e^{ikx}, \tag{6.24}$$

and

$$\frac{\partial \psi}{\partial x} = \sum_{k=-\infty}^{\infty} ika_k e^{ikx}.$$

If ψ is represented by a spectral approximation, the series will be truncated at some wave number N, but the Fourier coefficients for all $|k| \leq N$ will be identical to those in the infinite series (6.24).[3] Thus, the error in the spectral representation of $\partial\psi/\partial x$ is

[3] To ensure that the a_k are identical in both the infinite and the truncated Fourier series, it is necessary to compute the integral in the Fourier transform,

$$a_k = \frac{1}{2\pi}\int_{-\pi}^{\pi} \psi(x) e^{-ikx}\, dx,$$

with sufficient accuracy to avoid aliasing error.

$$E = \sum_{|k|>N} \mathrm{i}ka_k \mathrm{e}^{\mathrm{i}kx}.$$

If the pth derivative of ψ is piecewise continuous, and all lower-order derivatives are continuous, the Fourier coefficients satisfy the inequality

$$|a_k| \leq \frac{C}{|k|^p}, \tag{6.25}$$

where C is a positive constant (see Problem 10). Thus,

$$|E| \leq 2 \sum_{k=N+1}^{\infty} \frac{C}{|k|^{p-1}} \leq 2C \int_N^{\infty} \frac{\mathrm{d}s}{s^{p-1}} = \frac{2C}{p-2}\left(\frac{1}{N^{p-2}}\right).$$

As demonstrated in the preceding section, a $2N + 1$ mode spectral representation of the derivative is equivalent to a finite-difference formula involving $2N + 1$ grid points equally distributed throughout the domain. The spectral computation is therefore equivalent to a finite-difference computation with grid spacing $\Delta x_{\mathrm{e}} = 2\pi/(2N + 1)$. Thus, $\Delta x_{\mathrm{e}} \propto N^{-1}$ and

$$|E| \leq \tilde{C}(\Delta x_{\mathrm{e}})^{p-2}, \tag{6.26}$$

where $\tilde{C}$ is another constant. It follows that the effective order of accuracy of the spectral method is determined by the smoothness of ψ. If ψ is infinitely differentiable, the truncation error in the spectral approximation goes to zero faster than any finite power of Δx_{e}. In this sense, spatial derivatives are represented with infinite-order accuracy by the spectral method.

The preceding error analysis suggests that if a Fourier series approximation to $\psi(x)$ (as opposed to $\mathrm{d}\psi/\mathrm{d}x$) is truncated at wave number N, the error will be $O(1/N^{p-1})$. This error estimate is actually too pessimistic. As noted by Gottlieb and Orszag (1977, p. 26), (6.25) can be tightened, because if the pth derivative of ψ is piecewise continuous and all lower-order derivatives are continuous,

$$|a_k| \ll \frac{1}{|k|^p} \qquad \text{as} \qquad k \to \pm\infty. \tag{6.27}$$

Away from the points where $\mathrm{d}^p\psi/\mathrm{d}x^p$ is discontinuous, the maximum-norm error in the truncated Fourier series decays at a rate similar to the magnitude of the first few neglected Fourier coefficients, and according to (6.27) this rate is faster than $O(1/N^p)$. In practice, the error is $O(1/N^{p+1})$ away from the points where $\mathrm{d}^p\psi/\mathrm{d}x^p$ is discontinuous and $O(1/N^p)$ near the discontinuities (Fornberg 1996, p. 13).

6.2.1.5 Time Differencing

The spectral representation of the spatial derivatives reduces the original partial differential equation to the system of ordinary differential equations (6.9). In most practical applications, this system must be solved numerically. The time-differencing schemes discussed in Chap. 2 provide a number of possibilities, among which the leapfrog and Adams–Bashforth methods are the most common choices. Integrations performed using the spectral method typically require smaller time steps than those used when spatial derivatives are computed with low-order finite differences. This decrease in the maximum allowable time step is a natural consequence of the spectral method's ability to correctly resolve the spatial gradient in a $2\Delta x$ wave.

To better examine the source of this time-step restriction, consider the case of advection by a constant wind speed c. When c is constant, the time dependence of the kth Fourier component is governed by (6.13), which is just the oscillation equation (2.19) with $\kappa = -ck$. The maximum value of $|k|$ is $N = \pi/\Delta x_\mathrm{e} - 1/2 \approx \pi/\Delta x_\mathrm{e}$, where Δx_e is the equivalent mesh size introduced in connection with (6.26) and $\pi/\Delta x_\mathrm{e} \gg 1/2$ if the total domain $[-\pi, \pi]$ is divided into at least ten grid intervals. If the oscillation equation is integrated using the leapfrog scheme, the stability requirement is $|\kappa \Delta t| \leq 1$. Thus, the time step in a stable leapfrog spectral solution must satisfy $|c\Delta t/\Delta x_\mathrm{e}| \leq 1/\pi$.

Now suppose the advection equation (6.11) is approximated using a centered second-order spatial difference. The time evolution of the approximate solution at the jth grid point is, once again, governed by the oscillation equation. In this case, however, $\kappa = -c \sin(k\Delta x)/\Delta x$. The misrepresentation of the shorter wavelengths by the finite difference reduces the maximum value of $|\kappa|$ to $c/\Delta x$, and the leapfrog stability criterion relaxes to $|c\Delta t/\Delta x| \leq 1$. If higher-order finite differences are used, the error in the shorter wavelengths is reduced, and the maximum allowable value of $|c\Delta t/\Delta x|$ decreases to 0.73 for a fourth-order difference, and to 0.63 for a sixth-order difference. Machenhauer (1979) noted that as the order of a centered finite-difference approximation approaches infinity, the maximum value of $|c\Delta t/\Delta x|$ for which the scheme is stable approaches $1/\pi$. The maximum stable time step for the leapfrog spectral method is therefore consistent with the interpretation of the spectral method as an infinite-order finite-difference scheme.

6.2.2 Improving Efficiency Using the Transform Method

The computational effort required to obtain spectral solutions to the advection equation ceases to be trivial if there are spatial variations in the wind speed. In such circumstances, the Galerkin requirement (6.9) becomes

$$\frac{da_k}{dt} = -\frac{\mathrm{i}}{2\pi} \sum_{n=-N}^{N} n a_n \int_{-\pi}^{\pi} c(x,t) \mathrm{e}^{\mathrm{i}(n-k)x}\, dx. \tag{6.28}$$

Although it may be possible to evaluate the integrals in (6.28) exactly for certain special flows, in most instances the computation must be done by numerical quadrature. In a typical practical application, $c(x,t)$ would be available at the same spatial resolution as $\psi(x,t)$; indeed, many models might include equations that simultaneously predict c and ψ. Suppose, therefore, that c is given by the Fourier series

$$c(x,t) = \sum_{m=-N}^{N} c_m(t)\mathrm{e}^{\mathrm{i}mx}. \tag{6.29}$$

Substitution of this series expansion into (6.28) gives

$$\frac{da_k}{dt} = -\frac{\mathrm{i}}{2\pi} \sum_{n=-N}^{N} \sum_{m=-N}^{N} n c_m a_n \int_{-\pi}^{\pi} \mathrm{e}^{\mathrm{i}(n+m-k)x}\,\mathrm{d}x,$$

which reduces, by the orthogonality of the Fourier modes, to

$$\frac{da_k}{dt} = -\sum_{\substack{m+n=k \\ |m|,\,|n|\le N}} \mathrm{i} n c_m a_n. \tag{6.30}$$

The notation below the summation indicates that the sum should be performed for all indices n and m such that $|n| \le N$, $|m| \le N$, and $n + m = k$.

Although the expression (6.30) is relatively simple, it is not suitable for implementation in large, high-resolution numerical models. The number of arithmetic operations required to evaluate the time derivative of one Fourier coefficient via (6.30) is proportional to the total number of Fourier coefficients, $M \equiv 2N + 1$. The total number of operations required to advance the solution one time step is therefore $O(M^2)$. On the other hand, the number of calculations required to evaluate a finite-difference formula at an individual grid point is independent of the total number of grid points. Thus, assuming there are M points on the numerical grid, a finite-difference solution may be advanced one time step with just $O(M)$ arithmetic operations. Spectral models are therefore less efficient than finite-difference models when the approximate solution is represented by a large number of grid points – or equivalently, a large number of Fourier modes. Moreover, the relative difference in computational effort increases rapidly with increases in M. As a consequence, spectral models were limited to just a few Fourier modes until the development of the transform method by Orszag (1970) and Eliasen et al. (1970).

The key to the transform method is the efficiency with which FFTs can be used to swap the solution between wave-number space and physical space. Only $O(M \log M)$ operations are needed to convert a set of M Fourier coefficients, representing the Fourier transform of $\phi(x)$, into the M grid-point values $\phi(x_j)$.[4] The basic idea behind the transform method is to compute product terms like $c\partial\psi/\partial x$ by transforming c and $\partial\psi/\partial x$ from wave-number space to physical space (which takes $O(M \log M)$ operations), then multiplying c and $\partial\psi/\partial x$ at each grid point (requiring $O(M)$ operations), and finally transforming the product back to wave-number space (which again uses $O(M \log M)$ operations). The total number of operations required to evaluate $c\partial\psi/\partial x$ via the transform technique is therefore $O(M \log M)$, and when the number of Fourier components is large, it is far more efficient to

[4] To be specific, if M is a power of 2, a transform can be computed in $2M \log_2 M$ operations using the FFT algorithm.

perform these $O(M \log M)$ operations than the $O(M^2)$ operations necessary for the direct computation of (6.30) in wave-number space. To appreciate the degree to which the transform method can improve efficiency, suppose the spectral method is used in a two-dimensional problem in which the spatial dependence along each coordinate is represented by 128 Fourier modes; then an order-of-magnitude estimate of the increase in speed allowed by the transform method is

$$O\left(\frac{128 \times 128}{\log_2(128 \times 128)}\right) = O(1000).$$

The transform method is implemented as follows. Suppose that one wishes to determine the Fourier coefficients of the product of $\phi(x)$ and $\chi(x)$ such that

$$\phi(x)\chi(x) = \sum_{k=-K}^{K} p_k \mathrm{e}^{\mathrm{i}kx},$$

where ϕ and χ are periodic on the interval $0 \leq x \leq 2\pi$ and

$$\phi(x) = \sum_{m=-K}^{K} a_m \mathrm{e}^{\mathrm{i}mx}, \quad \chi(x) = \sum_{n=-K}^{K} b_n \mathrm{e}^{\mathrm{i}nx} \tag{6.31}$$

As just discussed, it is more efficient to transform ϕ and χ to physical space, compute their product in physical space, and transform the result back to wave-number space than to compute p_k from the "convolution sum"

$$p_k = \sum_{\substack{m+n=k \\ |m|,|n|\leq K}} a_m b_n.$$

The values of p_k obtained with the transform technique will be identical to those computed by the preceding summation formula, provided that there is sufficient spatial resolution to avoid aliasing error[5] during the computation of the product terms on the physical-space mesh. Suppose that the physical-space mesh is defined such that

$$x_j = \frac{2\pi j}{2N+1}, \quad \text{where} \quad j = 1, \ldots, 2N+1. \tag{6.32}$$

It might appear natural to chose $N = K$, thereby equating the number of grid points on the physical mesh with the number of Fourier modes. It is, however, necessary to choose $N > K$ to avoid aliasing error.

The amount by which N must exceed K may be most easily determined using the graphical diagram shown in Fig. 6.2, which is similar to Fig. 4.8 in Sect. 4.4.1. The wave number is plotted along the horizontal axis; without loss of generality, only positive wave numbers will be considered. The cutoff wave number in the original

[5] Aliasing error occurs when a short-wavelength fluctuation is sampled at discrete intervals and misinterpreted as a longer-wavelength oscillation. As discussed in Sect. 4.4.1, aliasing error can be generated in attempting to evaluate the product of two poorly resolved waves on a numerical mesh.

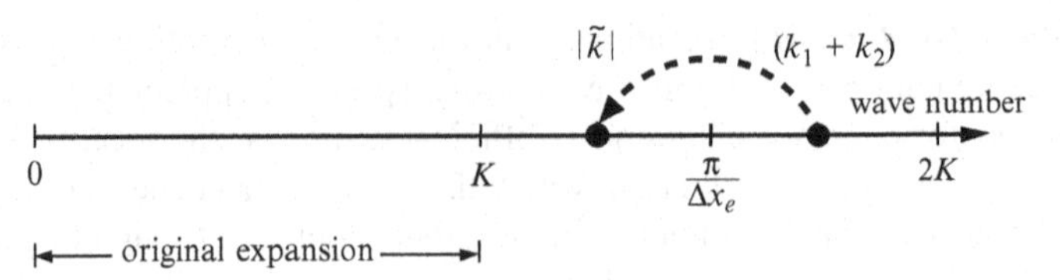

Fig. 6.2 Aliasing of $k_1 + k_2$ into $\tilde{k}$ such that $|\tilde{k}|$ appears as the symmetric reflection of $k_1 + k_2$ about the cutoff wave number on the high-resolution physical mesh

expansion is K, and the cutoff wave number on the high-resolution physical mesh is $\pi/\Delta x_\mathrm{e}$. Any aliasing error that results from the multiplication of waves with wave numbers k_1 and k_2 will appear at wave number $\tilde{k} = k_1 + k_2 - 2\pi/\Delta x_\mathrm{e}$. The goal is to choose a sufficiently large value for $\pi/\Delta x_\mathrm{e}$ to guarantee that no finite-amplitude signal is aliased into those wave numbers retained in the original Fourier expansion, which lie in the interval $-K \leq k \leq K$. The highest wave number that will have nonzero amplitude after the binary product has been computed on the physical mesh is $2K$. Thus, there will be no aliasing error if

$$K < \left| 2K - \frac{2\pi}{\Delta x_\mathrm{e}} \right| = \frac{2\pi}{\Delta x_\mathrm{e}} - 2K.$$

Using the definition of Δx_e implied by (6.32), the criteria for the elimination of aliasing error reduces to $N > (3K - 1)/2$.

The preceding result may be verified algebraically by considering the formula for $\tilde{p}_k$, the kth component of the finite Fourier transform computed from the grid-point values of $\phi\chi$ on the physical mesh,

$$\tilde{p}_k \equiv \frac{1}{M} \sum_{j=1}^{M} \phi(x_j)\chi(x_j)\mathrm{e}^{-\mathrm{i}kx_j}. \tag{6.33}$$

Here $M = 2N + 1$ is the total number of grid points on the physical mesh. Let the values of $\phi(x_j)$ and $\chi(x_j)$ appearing in the preceding formula be expressed in the form

$$\phi(x_j) = \sum_{m=-N}^{N} a_m \mathrm{e}^{\mathrm{i}mx_j}, \qquad \chi(x_j) = \sum_{n=-N}^{N} b_n \mathrm{e}^{\mathrm{i}nx_j},$$

where those values of a_n and b_m that were not included in the original series expansions (6.31) are zero, i.e.,

$$a_\ell = b_\ell = 0 \quad \text{for} \quad K < |\ell| \leq N. \tag{6.34}$$

Substituting these expressions for $\phi(x_j)$ and $\chi(x_j)$ into (6.33), one obtains

$$\tilde{p}_k = \sum_{m=-N}^{N} \sum_{n=-N}^{N} a_m b_n \left(\frac{1}{M} \sum_{j=1}^{M} \mathrm{e}^{\mathrm{i}(m+n-k)x_j} \right). \tag{6.35}$$

Since $x_j = 2\pi j/M$, each term in the inner summation in (6.35) is unity when $m + n - k$ is 0, M, or $-M$. The inner summation is zero for all other values of m and n by the discrete orthogonality condition (6.18). Thus, (6.35) may be written

$$\tilde{p}_k = \sum_{\substack{m+n=k \\ |m|,|n|\le N}} a_m b_n + \sum_{\substack{m+n=k+M \\ |m|,|n|\le N}} a_m b_n + \sum_{\substack{m+n=k-M \\ |m|,|n|\le N}} a_m b_n, \qquad (6.36)$$

where the last two terms represent aliasing error, only one of which can be nonzero for a given value of k. The goal is to determine the minimum resolution required on the physical mesh (or equivalently, the smallest M) that will prevent aliasing errors from influencing the value of p_k^* associated with any wave number retained in the original Fourier expansion. Any aliasing into a negative wave number will arise through the summation

$$\sum_{\substack{m+n=k+M \\ |m|,|n|\le N}} a_m b_n.$$

If follows from (6.34) that $a_m b_n = 0$ if $m + n > 2K$, so for a given wave number k all the terms in the preceding summation will be zero if $m+n = k+M > 2K$. Thus, there will be no aliasing error in p_k^* for those wave numbers retained in the original expansion if M satisfies $-K + M > 2K$. An equivalent condition expressed in terms of the wave number N is $N > (3K - 1)/2$, which is the same result obtained using Fig. 6.2. A similar argument may be used to show that this same condition also prevents the third term in (6.36) from generating aliasing error in the retained wave numbers. The choice $N = 3K/2$ is therefore sufficient to ensure that $p_k = \tilde{p}_k$ for all $|k| \le K$ and guarantee that the transform method yields the same algebraic result as the convolution sum in wave-number space. To maximize the efficiency of the FFTs used in practical applications, N is often chosen to be the smallest integer exceeding $(3k - 1)/2$ that contains no prime factor larger than 5.

The procedure used to implement the transform method may be summarized as follows. To be concrete, suppose that a solution to the variable-wind-speed advection equation is sought on the periodic domain $0 \le x \le 2\pi$ and that $c(x,t)$ is being simultaneously predicted by integrating a second unspecified equation. Let both ϕ and c be approximated by Fourier series expansions of the form (6.14) and (6.29) with cutoff wave numbers $N = K$.

1. Pad the coefficients in the Fourier expansions of c and ϕ with zeros by defining $a_k = c_k = 0$, for $K < |k| \le 3K/2$.
2. Multiply each a_k by ik to compute the derivative of ϕ in wave-number space.
3. Perform two inverse FFTs to obtain $c(x_j)$ and $\partial\phi(x_j)/\partial x$ on the physical-space grid, whose nodal points are located at $x_j = 2\pi j/(3K + 1)$.
4. Compute the product $c(x_j)\partial\phi(x_j)/\partial x$ on the physical-space grid.
5. (If terms representing additional forcing are present in the governing equation, and if those terms are more easily evaluated on the physical mesh than in wave-number space, evaluate those terms now and add the result to $c(x_j)\partial\phi(x_j)/\partial x$.)

6. FFT $c(x_j)\partial\phi(x_j)/\partial x$ to obtain the total forcing at each wave number, i.e., to get the right-hand side of (6.30). Discard the forcing at wave numbers for which $|k| > K$.
7. Step the Fourier coefficients forward to the next time level using an appropriate time-differencing scheme.

Note that the transform method allows processes that are difficult to describe mathematically in wave-number space to be conveniently evaluated during the portion of the integration cycle when the solution is available on the physical mesh. For example, if ϕ represents the concentration of water vapor, any change in ϕ produced by the condensation or evaporation of water depends on the degree to which the vapor pressure at a given grid point exceeds the saturation vapor pressure. The degree of supersaturation is easy to determine in physical space but very difficult to assess in wave-number space.

6.2.3 Conservation and the Galerkin Approximation

The mathematical equations describing nondissipative physical systems often conserve domain averages of quantities such as energy and momentum. When spectral methods are used to approximate such systems, the numerical solution replicates some of the important conservation properties of the true solution. To examine the conservation properties of the spectral method for a relatively general class of problems, consider those partial differential equations of the form (6.1) for which the forcing has the property that $\overline{vF(v)} = 0$, where the overbar denotes the integral over the spatial domain and v is any sufficiently smooth function that satisfies the boundary conditions.

An example of this type of problem is the simulation of passive tracer transport by nondivergent flow in a periodic spatial domain, which is governed by the equation

$$\frac{\partial \psi}{\partial t} + \mathbf{v} \cdot \nabla \psi = 0.$$

In this case $F(v) = \mathbf{v} \cdot \nabla v$. One can verify that $\overline{vF(v)} = 0$ if v is any periodic function with continuous first derivatives by noting that

$$\int_D v(\mathbf{v} \cdot \nabla v)\, \mathrm{d}V = \frac{1}{2} \int_D \nabla \cdot (v^2 \mathbf{v}) - v^2 (\nabla \cdot \mathbf{v})\, \mathrm{d}V = 0,$$

where the second equality follows from periodicity and the nondivergence of the velocity field.

If ϕ is an approximate spectral solution to (6.1) in which the time dependence is not discretized, then

$$\frac{\partial \phi}{\partial t} + F(\phi) = R(\phi), \tag{6.37}$$

where $R(\phi)$ denotes the residual. Suppose that the partial differential equation being approximated is a conservative system for which $\overline{\phi F(\phi)} = 0$, then multiplying (6.37) by ϕ and integrating over the spatial domain yields

$$\frac{1}{2}\overline{\frac{\partial \phi^2}{\partial t}} = \overline{\phi R(\phi)}.$$

The right side of the preceding equation is zero because ϕ is a linear combination of the expansion functions and $R(\phi)$ is orthogonal to each individual expansion function. As a consequence,

$$\frac{d}{dt}\|\phi\|_2 = 0, \tag{6.38}$$

implying that spectral approximations to conservative systems are not subject to nonlinear instability because (6.38) holds independent of the linear or nonlinear structure of $F(\psi)$. The only potential source of numerical instability is in the discretization of the time derivative.

Neglecting time-differencing errors, spectral methods will also conserve $\overline{\phi}$ provided that $\overline{F(\upsilon)} = 0$, where once again υ is any sufficiently smooth function that satisfies the boundary conditions. The conservation of $\overline{\phi}$ can be demonstrated by integrating (6.37) over the domain to obtain

$$\frac{\partial \overline{\phi}}{\partial t} = \overline{R(\phi)} \propto \overline{R(\phi)\varphi_0} = 0,$$

where φ_0 is the lowest-wave-number orthogonal expansion function, which is a constant.

6.3 The Pseudospectral Method

The spectral method uses orthogonal expansion functions to represent the numerical solution and constrains the residual error to be orthogonal to each of the expansion functions. As discussed in Sect. 6.1, there are alternative strategies for constraining the size of the residual. The pseudospectral method utilizes one of these alternative strategies: the collocation approximation, which requires the residual to be zero at every point on some fixed mesh. Spectral and pseudospectral methods might both represent the solution with the same orthogonal expansion functions; however, as a consequence of the collocation approximation, the pseudospectral method is basically a grid-point scheme – series expansion functions are used only to compute derivatives.

To illustrate the pseudospectral procedure, suppose that solutions are sought to the advection equation (6.11) on the periodic domain $0 \leq x \leq 2\pi$ and that the approximate solution ϕ and the spatially varying wind speed $c(x)$ are represented by Fourier series truncated at wave number K:

$$\phi(x,t) = \sum_{n=-K}^{K} a_n \mathrm{e}^{\mathrm{i}nx}, \quad c(x,t) = \sum_{m=-K}^{K} c_m \mathrm{e}^{\mathrm{i}mx}.$$

The collocation requirement at grid point j is

$$\sum_{n=-K}^{K} \frac{da_n}{dt} \mathrm{e}^{\mathrm{i}nx_j} + \sum_{m=-K}^{K} c_m \mathrm{e}^{\mathrm{i}mx_j} \sum_{n=-K}^{K} \mathrm{i}na_n \mathrm{e}^{\mathrm{i}nx_j} = 0. \tag{6.39}$$

Enforcing $R(\phi(x_j)) = 0$ at $2K + 1$ points on the physical-space grid leads to a solvable linear system for the time derivatives of the $2K + 1$ Fourier coefficients. In the case of the Fourier spectral method, the most efficient choice for the location of these points is the equally spaced mesh

$$x_j = j\left(\frac{2\pi}{2K+1}\right), \quad j = 1, 2, \ldots, 2K+1. \tag{6.40}$$

There is no need actually to solve the linear system for the $\mathrm{d}a_k/\mathrm{d}t$. It is more efficient to write (6.39) in the equivalent form

$$\frac{d\phi}{dt}(x_j) + c(x_j)\frac{\partial \phi}{\partial x}(x_j) = 0, \tag{6.41}$$

where

$$\frac{\partial \phi}{\partial x}(x_j) = \sum_{n=-K}^{K} \mathrm{i}na_n \mathrm{e}^{\mathrm{i}nx_j}. \tag{6.42}$$

The grid-point nature of the pseudospectral method is apparent in (6.41), which is similar to the time-tendency equations[6] that arise in differential–difference approximations to the advection equation, except that the derivative is computed in a special way. Instead of using finite differences, one calculates the spatial derivative at each time step by first computing the Fourier coefficients through the discrete Fourier transform

$$a_k(t) = \frac{1}{2K+1} \sum_{j=1}^{2K+1} \phi(x_j, t)\mathrm{e}^{-\mathrm{i}kx_j},$$

then differentiating each Fourier mode analytically and inverse transforming according to (6.42). This procedure requires two FFTs per time step.

The advantage of the pseudospectral method relative to conventional finite-difference schemes is that provided the solution is smooth, the pseudospectral method is more accurate. As discussed in Sect. 3.2.1, the error in the Fourier

[6] As in conventional spectral and finite-difference techniques, the time derivative would be discretized using the leapfrog scheme, the Adams–Bashforth scheme, or some other appropriate method.

approximation to the derivative of an infinitely differentiable function will decrease more rapidly than any finite power of Δx as the grid resolution is increased. Thus, like the spectral method, the pseudospectral method is essentially an infinite-order finite-difference scheme. The disadvantage of the pseudospectral method is that it requires more computation than conventional finite-difference schemes when both methods are used with the same spatial resolution. If M is the total number of grid points, the FFTs in the pseudospectral computation require $O(M \log(M))$ operations per time step, whereas conventional finite-difference methods need only $O(M)$ operations. The extra work per time step may, however, be easily offset if the increased accuracy of the pseudospectral representation allows the computations to be performed on a coarser mesh.

The advantage of the pseudospectral method relative to the spectral method is that the pseudospectral method requires less computation. The increase in efficiency of the pseudospectral method is achieved by allowing aliasing error in the computation of the products of spatially varying functions. As a consequence of this aliasing error, the residual need not be orthogonal to the individual expansion functions, and the pseudospectral method does not possess the conservation properties discussed in Sect. 6.2.3. In particular, the pseudospectral method is subject to nonlinear instability.

The difference in aliasing between the pseudospectral and spectral methods can be evaluated by multiplying (6.39) by e^{-ikx_j} and summing over all j to obtain

$$\sum_{n=-K}^{K} \frac{da_n}{dt} \sum_{j=1}^{2K+1} e^{i(n-k)x_j} + \sum_{n=-K}^{K} \sum_{m=-K}^{K} inc_m a_n \sum_{j=1}^{2K+1} e^{i(n+m-k)x_j} = 0.$$

Using the definition of x_j (6.40) and the discrete-mesh orthogonality condition (6.18), the preceding equation reduces to

$$\frac{da_k}{dt} + \sum_{\substack{m+n=k \\ |m|,|n|\le K}} inc_m a_n + \sum_{\substack{m+n=k+M \\ |m|,|n|\le K}} inc_m a_n + \sum_{\substack{m+n=k-M \\ |m|,|n|\le K}} inc_m a_n = 0,$$

where $M = 2K + 1$. As when spectral computations are performed using the transform technique, the last two terms represent aliasing error, only one of which can be nonzero for a given value of k. In contrast to the spectral method, however, these aliasing terms do not disappear, because in the pseudospectral method the number of grid points on the physical mesh is identical to the number of Fourier wave numbers, and therefore none of the c_m and a_n need be zero.

One might suppose that aliasing error always decreases the accuracy of the solution, but the impact of aliasing error on accuracy depends on the problem. As an example, suppose that spectral and pseudospectral approximations are computed to the solution of the viscous Burgers equation

$$\frac{\partial \psi}{\partial t} + \psi \frac{\partial \psi}{\partial x} = \nu \frac{\partial^2 \psi}{\partial x^2} \tag{6.43}$$

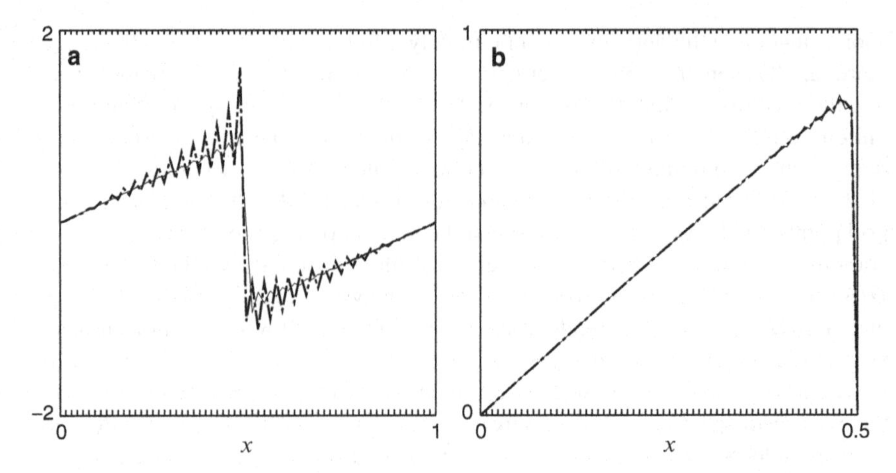

Fig. 6.3 Spectral (*solid line*) and pseudospectral (*dot-dashed line*) solutions to the viscous Burgers equation at $t = 0.4$: **a** truncation at wave number 64π; **b** truncation at wave number 128π. Only the left half of the domain is shown in **b**

on the periodic domain $0 \leq x \leq 1$ subject to the initial condition $\psi(x,0) = \sin(2\pi x)$. Spectral and pseudospectral solutions to this problem are shown in Fig. 6.3 at $t = 0.4$ for the case $\nu = 0.002$. In these computations the time differencing for the nonlinear advection term was leapfrog, and the diffusion term was integrated using a forward difference over an interval of $2\Delta t$. The time step was selected such that

$$\frac{\Delta t}{\Delta x} \max_x [\psi(x,0)] = 0.1 .$$

The true solution to the inviscid problem gradually steepens around the point $x = 1/2$ and becomes discontinuous at $t = (2\pi)^{-1}$ (see Sect. 4.5.1), but the viscous dissipation in (6.43) prevents the gradient at $x = 1/2$ from collapsing to a true discontinuity and gradually erodes the amplitude of the solution so that $\psi(x,t) \to 0$ as $t \to \infty$. Figure 6.3a shows the approximate solutions to (6.43) obtained using a spectral truncation at wave number 64π, which is equivalent to a grid spacing of $\Delta x = 1/64$. Both the spectral and the pseudospectral solutions have trouble resolving the steep gradient at $x = 1/2$ and develop significant $2\Delta x$ noise. The amplitude of the $2\Delta x$ ripples remains bounded in the spectral solution, but aliasing error generates a rapidly growing instability in the $2\Delta x$ component of the pseudospectral solution.

Rather different results are, however, obtained if the same problem is repeated with twice the spatial resolution. When the cutoff wave number is 128π, the pseudospectral method remains stable and actually generates a more accurate solution than that obtained with the spectral method. A close-up comparison of the two solutions over the subdomain $0 \leq x \leq 1/2$ appears in Fig. 6.3b. The $2\Delta x$ ripples in the spectral solution are of distinctly larger amplitude than those appearing in the pseudospectral solution. The pseudospectral solution remains stable because the rate at which viscous damping erodes a $2\Delta x$ wave increases by a factor of 4 as the spatial resolution is doubled from $\Delta x = 1/64$ to $1/128$, and when $\Delta x = 1/128$, the rate of energy removal from the $2\Delta x$ wave by viscous damping exceeds the rate at which $2\Delta x$ waves are amplified by aliasing error. The superiority of the pseudospectral solution over the spectral solution in Fig. 6.3b highlights the fact that although conservation of $\|\phi\|_2$ implies stability, it does not imply better accuracy.

The influence of aliasing error on accuracy is largely a matter of chance. Although it is certainly an error when the pseudospectral method misrepresents interactions between $2\Delta x$ and $3\Delta x$ waves as an aliased contribution to a $6\Delta x$ wave, it is also an error when the spectral method simply neglects the interactions between these same short waves, since the product of $2\Delta x$ and $3\Delta x$ disturbances should properly appear in a $6\Delta x/5$ wave. In Burgers's equation, and in many other fluid-flow problems, there is a cascade of energy to smaller scales. An accurate conservative scheme, such as the spectral method, replicates this down-scale energy transfer except that the cascade is terminated at the shortest scales resolved in the numerical simulation. In the absence of viscous dissipation, the spectral approximation to Burgers's equation conserves energy, and the energy that cascades down scale simply accumulates in the shortest resolvable modes. To simulate the continued cascade of energy into the unresolvable scales of motion, it is necessary to remove energy from the shortest resolvable waves. The energy-removal algorithm constitutes a parameterization of the influence of unresolved short wavelengths on the resolved modes and should be designed to represent the true behavior of the physical system as closely as possible.

Whatever the exact details of the energy-removal scheme, if it prevents the unphysical accumulation of energy at the short wavelengths in the spectral solution, the same energy-removal scheme will often stabilize a pseudospectral solution to the same problem. In the case shown in Fig. 6.3b, for example, the amount of viscous dissipation required to stabilize the pseudospectral solution is less than that required to remove the spurious ripples from the spectral solution. Although it is not generally necessary to filter the solution this heavily, aliasing error can be completely eliminated by removing all energy at wavelengths shorter than or equal to $3\Delta x$ after each time step, or equivalently by removing the highest one third of the resolved wave numbers. If such a filter is used in combination with a pseudospectral method truncated at wave number M, the resulting algorithm is identical to that for a Galerkin spectral method truncated at wave number $2M/3$ in which the nonlinear terms are computed via the transform method.

6.4 Spherical Harmonics

The two-dimensional distribution of a scalar variable on the surface of a sphere can be efficiently approximated by a truncated series of spherical harmonic functions. Spherical harmonics can also be used to represent three-dimensional fields defined within a volume bounded by two concentric spheres if grid points or finite elements are used to approximate the spatial structure along the radial coordinate and thereby divide the computational domain into a series of nested spheres. Let λ be the longitude, θ the latitude, and define $\mu = \sin\theta$. If ψ is a smooth function of λ and μ, it can be represented by a convergent expansion of spherical harmonic functions of the form

$$\psi(\lambda,\mu) = \sum_{m=-\infty}^{\infty} \sum_{n=|m|}^{\infty} a_{m,n} Y_{m,n}(\lambda,\mu), \tag{6.44}$$

where each spherical harmonic function $Y_{m,n}(\lambda,\mu) = P_{m,n}(\mu)\mathrm{e}^{\mathrm{i}m\lambda}$ is the product of a Fourier mode in λ and an associated Legendre function in μ.

The associated Legendre functions are generated from the Legendre polynomials using the relation

$$P_{m,n}(\mu) = \left[\frac{(2n+1)}{2}\frac{(n-m)!}{(n+m)!}\right]^{1/2} (1-\mu^2)^{m/2} \frac{d^m}{d\mu^m} P_n(\mu), \tag{6.45}$$

where

$$P_n(\mu) = \frac{1}{2^n n!}\frac{d^n}{d\mu^n}\left[(\mu^2-1)^n\right] \tag{6.46}$$

is the Legendre polynomial of degree n, and the formula that results after substituting (6.46) into (6.45) is valid for $|m| \le n$. Note that when m is odd,[7] the associated Legendre functions are not polynomials in μ.

The leading factor in (6.45) normalizes $P_{m,n}$ so that

$$\int_{-1}^{1} P_{m,n}(\mu) P_{m,s}(\mu)\,\mathrm{d}\mu = \delta_{ns}, \tag{6.47}$$

where $\delta_{ns} = 1$ if $n = s$ and is zero otherwise. As a consequence, the orthogonality relation for the spherical harmonics becomes

$$\frac{1}{2\pi}\int_{-1}^{1}\int_{-\pi}^{\pi} Y_{m,n}(\lambda,\mu) Y^*_{r,s}(\lambda,\mu)\,\mathrm{d}\lambda\,\mathrm{d}\mu = \delta_{mr}\delta_{ns}, \tag{6.48}$$

where $Y^*_{r,s}$ is the complex conjugate of $Y_{r,s}$. The associated Legendre functions have the property that $P_{-m,n}(\mu) = (-1)^m P_{m,n}(\mu)$, which implies that $Y_{-m,n} = (-1)^m Y^*_{m,n}$, and thus the expansion coefficients for any approximation to a real-valued function satisfy

$$a_{-m,n} = (-1)^m a^*_{m,n}. \tag{6.49}$$

Two recurrence relations satisfied by the associated Legendre functions that will be used in the subsequent analysis are

$$\mu P_{m,n} = \varepsilon_{m,n+1} P_{m,n+1} + \varepsilon_{m,n} P_{m,n-1} \tag{6.50}$$

[7] The m index of the associated Legendre function $P_{m,n}$ indicates the "order," whereas the n index indicates the "degree," which is the same as the degree of the embedded Legendre polynomial.

and

$$(1-\mu^2)\frac{dP_{m,n}}{d\mu} = -n\varepsilon_{m,n+1}P_{m,n+1} + (n+1)\varepsilon_{m,n}P_{m,n-1}, \tag{6.51}$$

where

$$\varepsilon_{m,n} = \left(\frac{n^2-m^2}{4n^2-1}\right)^{1/2}.$$

The spherical harmonics are eigenfunctions of the Laplacian operator on the sphere such that

$$\nabla^2 Y_{m,n} = \frac{-n(n+1)}{a^2} Y_{m,n}, \tag{6.52}$$

where a is the radius of the sphere and the horizontal Laplacian operator in spherical coordinates is

$$\begin{aligned}\nabla^2 &= \frac{1}{a^2\cos^2\theta}\left[\frac{\partial^2}{\partial\lambda^2} + \cos\theta\frac{\partial}{\partial\theta}\left(\cos\theta\frac{\partial}{\partial\theta}\right)\right]\\ &= \frac{1}{a^2(1-\mu^2)}\frac{\partial^2}{\partial\lambda^2} + \frac{1}{a^2}\frac{\partial}{\partial\mu}\left[(1-\mu^2)\frac{\partial}{\partial\mu}\right].\end{aligned}$$

The eigenvalue associated with each $Y_{m,n}$ can be used to define a total wave number by analogy to the situation on a flat plane, where

$$\nabla^2 e^{i(mx+ny)} = -(m^2+n^2)e^{i(mx+ny)}$$

and the total wave number is $(m^2+n^2)^{1/2}$. The total wave number associated with $Y_{m,n}$ is $(n^2+n)^{1/2}/a$, which is independent of the zonal wave number m. The nonintuitive absence of m in the formula for the total wave number arises, in part, because the effective meridional wave number of a spherical harmonic depends on both m and n.

6.4.1 Truncating the Expansion

The gross structure of $Y_{m,n}(\lambda,\mu)$ may be appreciated by examining the distribution of its nodal lines on the surface of the sphere. The nodal lines for the those modes for which $0 \le m \le n \le 3$ are schematically diagrammed in Fig. 6.4. The zonal structure of $Y_{m,n}$ is that of a simple Fourier mode $e^{im\lambda}$, so there are $2m$ nodal lines intersecting a circle of constant latitude. The distribution of the nodal lines along a line of constant longitude is more complex. The formula for the meridional structure of each of the modes shown in Fig. 6.4 is given in Table 6.2. $P_{m,n}(\mu)$ is proportional to

$$(1-\mu^2)^{m/2}\frac{d^{n+m}}{d\mu^{n+m}}(\mu^2-1)^n. \tag{6.53}$$

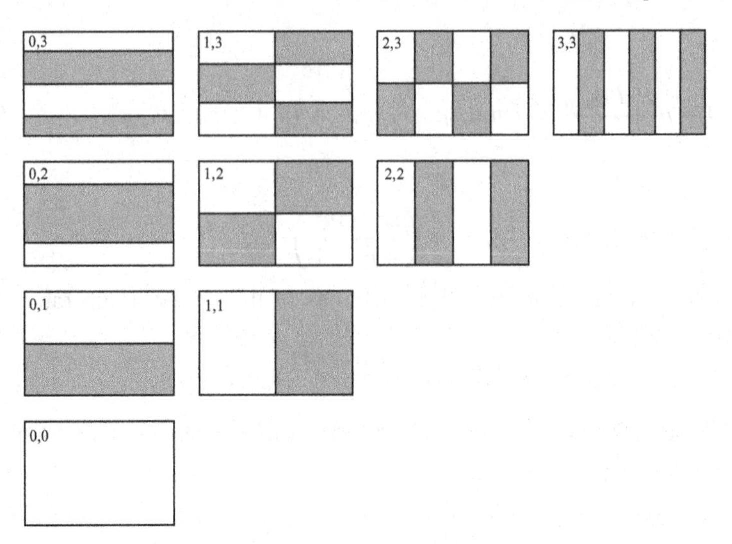

Fig. 6.4 The distribution of the nodal lines for the spherical harmonics $Y_{m,n}$ in the set $0 \leq m \leq n \leq 3$. The horizontal axis in each map is linear in λ and includes the domain $-\pi \leq \lambda \leq \pi$. The vertical axis is linear in $\sin\theta$ and includes the domain $-\pi/2 \leq \theta \leq \pi/2$. The m, n index of each mode is indicated in the *upper-left corner* of each map

Table 6.2 Meridional structure of the spherical harmonics appearing in Fig. 6.4

n	$P_{0,n}$	$P_{1,n}$	$P_{2,n}$	$P_{3,n}$
3	$\sqrt{\frac{7}{8}}(5\mu^3 - 3\mu)$	$\sqrt{\frac{21}{32}}(5\mu^2 - 1)\sqrt{1-\mu^2}$	$\frac{\sqrt{105}}{4}(\mu - \mu^3)$	$\frac{\sqrt{70}}{8}(1-\mu^2)^{\frac{3}{2}}$
2	$\sqrt{\frac{5}{8}}(3\mu^2 - 1)$	$\frac{\sqrt{15}}{2}\mu\sqrt{1-\mu^2}$	$\frac{\sqrt{15}}{4}(1-\mu^2)$	–
1	$\sqrt{\frac{3}{2}}\mu$	$\frac{\sqrt{3}}{2}\sqrt{1-\mu^2}$	–	–
0	$1/\sqrt{2}$	–	–	–

The first factor has no zeros between the north and south poles. The second factor is a polynomial of degree $n - m$, and all of its $n - m$ zeros lie between the poles. Thus, the modes with zero meridional wave number are $Y_{s,s}$, whereas the modes with zero zonal wave number are $Y_{0,s}$. Figure 6.4 also provides a graphical illustration of the reason why expansions in spherical harmonics are constructed without needing to define and include modes with $m > n$.

In all practical applications the infinite series (6.44) must be truncated to create a numerical approximation of the form

$$\psi(\lambda, \mu) = \sum_{m=-M}^{M} \sum_{n=|m|}^{N(m)} a_{m,n} Y_{m,n}(\lambda, \mu). \tag{6.54}$$

The *triangular* truncation, in which $N(m) = M$, is unique among the various possible truncations because it is the only one that provides uniform spatial resolution over the entire surface of the sphere. The approximation to $\psi(\lambda, \mu)$ obtained using a triangular truncation is invariant to an arbitrary rotation of the latitude and longitude coordinates about the center of the sphere. This invariance follows from the fact that any spherical harmonic of degree less than or equal to M (i.e., for which $n \leq M$) can be exactly expressed as a linear combination of the spherical harmonics in an Mth-order triangular truncation defined with respect to the arbitrarily rotated coordinates. To be specific, if $Y_{m,n}$ is a spherical harmonic with $n \leq M$, and λ', μ', and $Y'_{r,s}$ are coordinates and spherical harmonics defined with respect to an arbitrarily rotated polar axis, then there exist a set of expansion coefficients $b_{r,s}$ such that

$$Y_{m,n}(\lambda, \mu) = \sum_{r=-M}^{M} \sum_{s=|r|}^{M} b_{r,s} Y'_{r,s}(\lambda', \mu')$$

(Courant and Hilbert 1953, p. 535).

In spite of its elegance, the triangular truncation may not be optimal in situations where the characteristic scale of the approximated field exhibits a systematic variation over the surface of the sphere. In the Earth's atmosphere, for example, the perturbations in the geopotential height field in the tropics are much weaker than those in the middle latitudes. A variety of alternative truncations have therefore been used in low-resolution ($M < 30$) global atmospheric models. The most common alternative is the *rhomboidal* truncation, in which $N(m) = |m| + M$ in (6.54). The set of indices (m, n) retained in triangular and rhomboidal truncations with approximately the same number of degrees of freedom are compared in Fig. 6.5. Only the right half-plane is shown in Fig. 6.5, since whenever $Y_{m,n}$ is included in the truncation, $Y_{-m,n}$ is also retained. In comparison with the triangular truncation, the rhomboidal truncation neglects two families of modes with large n, those for which $n - m \approx 0$ and those for which $n - m \approx n$. The first of these families is composed of high-zonal-wave-number modes that are equatorially trapped, since the first factor in (6.53) has an mth-order zero at each pole. The second family of neglected modes have small

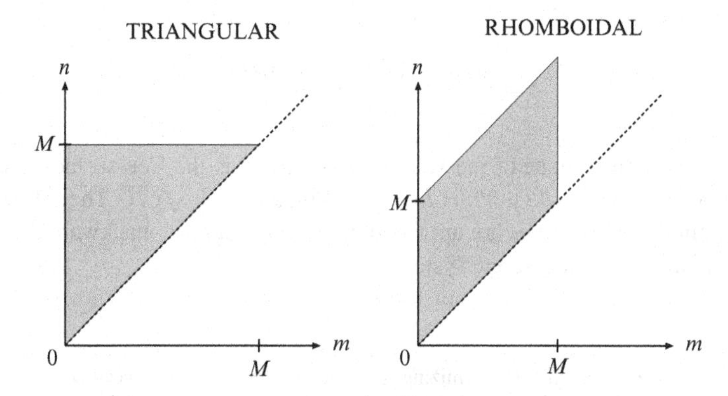

Fig. 6.5 *Shading* indicates the portion of the $m \geq 0$ half-plane in which the m and n indices are retained in Mth-order triangular and rhomboidal truncations

zonal wave number but fine meridional structure near the poles. As a consequence, the spatial resolution in a rhomboidal truncation is somewhat concentrated in the middle latitudes, which may be suitable for low-resolution models of the Earth's atmosphere but less appropriate for more general applications. Other truncations in which $N(m)$ is a more complex function of m have also been proposed to improve the efficiency of low-resolution climate models (Kiehl et al. 1996). At present there does not seem to be a clear consensus about which truncation is most suitable for use in low-resolution atmospheric models. The triangular truncation is, however, the universal choice in high-resolution global weather forecasting.

6.4.2 Elimination of the Pole Problem

Explicit finite-difference approximations to the equations governing fluid motion on a sphere can require very small time steps to maintain stability if the grid points are distributed over the sphere on a uniform latitude–longitude mesh. This time-step restriction arises because the convergence of the meridians near the poles greatly reduces the physical distance between adjacent nodes on the same latitude circle, and as a consequence, the Courant–Friedrichs–Lewy (CFL) condition is far more restrictive near the poles than in the tropics. Several approaches have been used to circumvent this problem (Williamson 1979), but they all have at least one cumbersome aspect. One of the most elegant solutions to the pole problem is obtained using spherical harmonic expansion functions in a spectral or pseudospectral approximation.[8]

A simple context in which to compare the stability criteria obtained using spherical harmonics and finite differences is provided by the shallow-water equations linearized about a resting basic state of depth H on a nonrotating sphere:

$$\frac{\partial \delta}{\partial t} + g\nabla^2 h = 0,$$

$$\frac{\partial h}{\partial t} + H\delta = 0,$$

where

$$\delta = \frac{1}{a\cos\theta}\left[\frac{\partial u}{\partial \lambda} + \frac{\partial v\cos\theta}{\partial \theta}\right]$$

is the horizontal divergence of the velocity field and h is the free-surface displacement. Let $\delta^n = \delta(\lambda, \theta, n\Delta t)$, $h^n = h(\lambda, \theta, n\Delta t)$, and $c = \sqrt{gH}$. Then if the time derivatives in these equations are approximated using forward–backward differencing, one obtains the semidiscrete system

[8] Another attractive approach for minimizing the pole problem in global weather forecasting is provided by the semi-Lagrangian semi-implicit scheme discussed in Sect. 7.4.2.

$$\frac{\delta^{n+1} - \delta^n}{\Delta t} + g\nabla^2 h^n = 0, \tag{6.55}$$

$$\frac{h^{n+1} - h^n}{\Delta t} + H\delta^{n+1} = 0, \tag{6.56}$$

or, after eliminating δ^{n+1} and δ^n,

$$\frac{h^{n+1} - 2h^n + h^{n-1}}{(\Delta t)^2} - c^2\nabla^2 h^n = 0. \tag{6.57}$$

Suppose that the spatial structure of h^n is represented by spherical harmonics in the triangular truncation

$$h(\lambda, \mu, n\Delta t) = \sum_{r=-M}^{M} \sum_{s=|r|}^{M} b_{r,s}^n Y_{r,s}(\lambda, \mu),$$

where as before $\mu = \sin\theta$. Since the spherical harmonics are eigenfunctions of the Laplacian operator on the sphere, a solvable system of equations for the expansion coefficients $b_{r,s}^n$ can be obtained by substituting this expansion into (6.57) and using (6.52) to arrive at

$$\frac{b_{r,s}^{n+1} - 2b_{r,s}^n + b_{r,s}^{n-1}}{(\Delta t)^2} = -\frac{c^2 s(s+1)}{a^2} b_{r,s}^n.$$

Assuming that the expansion coefficients have a time dependence proportional to $\mathrm{e}^{-\mathrm{i}\omega n\Delta t}$, the dispersion relation for each mode is

$$\sin^2\left(\frac{\omega\Delta t}{2}\right) = s(s+1)\left(\frac{c\Delta t}{2a}\right)^2.$$

The scheme will be stable when the frequency associated with the highest total wavenumber is real, or

$$\frac{c\Delta t\sqrt{M(M+1)}}{2a} < 1. \tag{6.58}$$

Now suppose that the Laplacian operator in (6.57) is evaluated using finite differences on a latitude–longitude grid in which $\Delta\theta$ and $\Delta\lambda$ are uniform over the globe. Then near the poles, the highest-frequency components in the numerical solution will be forced by short-wavelength spatial variations around a latitude circle. Approximating the second derivative with respect to longitude in (6.57) as $(a\cos\theta)^{-2}\delta_\lambda^2$ and substituting a Fourier mode in time and longitude of the form

$$\hat{h}(\theta)\mathrm{e}^{\mathrm{i}(rm\Delta\lambda - \omega n\Delta t)}$$

into the resulting semidiscrete equation yields

$$\frac{4\hat{h}}{(\Delta t)^2}\sin^2\left(\frac{\omega\Delta t}{2}\right) = \frac{4c^2\hat{h}}{(a\Delta\lambda\cos\theta)^2}\sin^2\left(\frac{r\Delta\lambda}{2}\right) - \frac{c^2}{a^2\cos\theta}\frac{\partial}{\partial\theta}\left(\cos\theta\frac{\partial\hat{h}}{\partial\theta}\right).$$

For those modes with zero meridional wave number, a necessary condition for the reality of ω and the stability of this semidiscrete approximation is that

$$\frac{c\Delta t}{a\Delta\lambda\cos\theta} \leq 1.$$

Let M be the highest zonal wave number (in radians) resolved on the numerical mesh; then $M\Delta\lambda = \pi$, and the preceding stability condition may be expressed as

$$\frac{cM\Delta t}{a\pi\cos\theta} \leq 1.$$

A comparison of this condition with (6.58) shows that the maximum stable time step that can be used with the finite-difference method on the portion of the mesh where $\theta \to \pm\pi/2$ is far smaller than that which can be used in a spectral model employing spherical harmonic expansion functions with the same cutoff wave number.

The restrictions on the maximum stable time step can be removed altogether by using trapezoidal time differencing instead of the forward–backward scheme in (6.55) and (6.56). This is not a particularly efficient approach when the Laplacian is approximated using finite differences, since the trapezoidal approximation generates a large system of implicit algebraic equations that must be solved at every time step. Trapezoidal time differencing can, however, be implemented very efficiently in spectral approximations that use spherical harmonic expansion functions, because the spherical harmonics are eigenfunctions of the horizontal Laplacian operator on the sphere. As a consequence, the expansion coefficient for each $Y_{r,s}$ can be computed independently of the other modes, and the implicit coupling introduced by trapezoidal time differencing only generates a trivial two-variable system involving the amplitudes of the divergence and the free-surface elevation of each mode. The ease with which trapezoidal approximations to (6.55) and (6.56) can be integrated using spherical harmonics can be used to great advantage in formulating semi-implicit time-differencing approximations to the nonlinear equations governing fluid flow on a sphere (see Sects. 8.2.3, 8.6.5).

6.4.3 Gaussian Quadrature and the Transform Method

In most practical applications some of the forcing terms in the governing equations contain products of two or more spatially varying functions. Unless the total number of modes retained in the series expansion is very small, a variant of the transform method described in Sect. 6.2.2 must be used to efficiently apply spectral methods to such problems. The transform between grid-point values and the spectral coefficients of the spherical harmonic functions is, however, more cumbersome and computationally less efficient than the FFT. The lack of highly efficient transforms is one drawback associated with the use of spherical harmonic expansion functions in global spectral models. Even so, it is far more efficient to use the transform method than to use the alternative "interaction coefficient" method, in which the forcing is computed from a summation of products of pairs of the spectral coefficients (Orszag 1970; Eliasen et al. 1970).

If $\psi(\lambda, \mu)$ is approximated by a truncated series of spherical harmonics of the form (6.54), the transformation from the set of spectral coefficients to points on a latitude–longitude grid can be computed using the relation

$$\psi(\lambda, \mu) = \sum_{m=-M}^{M} \hat{a}_m(\mu) \mathrm{e}^{\mathrm{i}m\lambda}, \tag{6.59}$$

where

$$\hat{a}_m(\mu) = \sum_{n=|m|}^{N(m)} a_{m,n} P_{m,n}(\mu). \tag{6.60}$$

The first summation (6.59) is a discrete Fourier transform with respect to the longitudinal coordinate λ that can be efficiently evaluated using FFTs to obtain $2M+1$ grid-point values around each latitude circle in $O[M \log M]$ operations. The second summation (6.60) is essentially an inner product requiring $O\left[N^2\right]$ operations to evaluate $\hat{a}_m$ at N different latitudes. The lack of a fast transform for the latitude coordinate makes the spherical harmonic spectral model less efficient than spectral models that use two-dimensional Fourier series.

The inverse transform, from physical space to spectral coordinates, is accomplished as follows. The orthogonality properties of the spherical harmonics (6.48) imply that

$$a_{m,n} = \frac{1}{2\pi} \int_{-1}^{1} \int_{-\pi}^{\pi} \psi(\lambda, \mu) Y_{m,n}^{*}(\lambda, \mu) \, \mathrm{d}\lambda \, \mathrm{d}\mu,$$

or equivalently,

$$a_{m,n} = \int_{-1}^{1} \hat{a}_m(\mu) P_{m,n}(\mu) \, \mathrm{d}\mu, \tag{6.61}$$

where

$$\hat{a}_m(\mu) = \frac{1}{2\pi} \int_{-\pi}^{\pi} \psi(\lambda, \mu) \mathrm{e}^{-\mathrm{i}m\lambda} \, \mathrm{d}\lambda. \tag{6.62}$$

The last of these integrals is a Fourier transform that can be evaluated from data on a discrete mesh using FFTs. After the Fourier transform of ψ has been computed, the integral (6.61) can be evaluated using Gaussian quadrature. Provided that one avoids aliasing error, it is possible to numerically evaluate both (6.61) and (6.62) without introducing errors beyond those associated with the original truncation of the spherical harmonic expansion at some finite wave number. Before discussing how to avoid aliasing error, it may be helpful to review Gaussian quadrature.

6.4.3.1 Gaussian Quadrature

As will soon be demonstrated, when evaluated in connection with the transform method, the integrand in (6.61) is a polynomial in μ, which is fortuitous, because efficient formulae exist for exactly computing the definite integral of a polynomial. For example, suppose that $f(x)$ is a polynomial in x of degree less than or equal to $m-1$ and that m arbitrarily spaced grid points x_j are distributed over the domain $a \leq x \leq b$. Then $f(x)$ can be expressed in the form of a Lagrange interpolating polynomial

$$f(x) = \sum_{j=1}^{m} f(x_j)p_j(x), \tag{6.63}$$

where

$$p_j(x) = \prod_{\substack{k=1 \\ k\neq j}}^{m} \frac{(x-x_k)}{(x_j-x_k)}. \tag{6.64}$$

If weights w_j are defined such that

$$w_j = \int_a^b p_j(x)\,\mathrm{d}x, \tag{6.65}$$

it follows from (6.63) that

$$\int_a^b f(x)\,\mathrm{d}x = w_1 f(x_1) + w_2 f(x_2) + \cdots + w_m f(x_m), \tag{6.66}$$

and that (6.65) and (6.66) give the exact integral of all polynomials $f(x)$ of degree less than or equal to $m-1$.

The preceding formula achieves exact results for polynomials up to degree $m-1$ without imposing any constraint on the location of the x_j within the interval $[a,b]$. Gaussian quadrature, on the other hand, obtains exact results for polynomials up to degree $2m-1$ without adding more terms to the quadrature formula by choosing the x_j to be the zeros of the Legendre polynomial of degree m. The role played by Legendre polynomials in Gaussian quadrature is essentially independent of their relation to the associated Legendre functions and spherical harmonics. The property of the Legendre polynomials that is important for Gaussian quadrature is that these polynomials satisfy the orthogonality condition[9]

$$\int_{-1}^{1} P_m(x)P_n(x)\,\mathrm{d}x = \frac{2\delta_{mn}}{2n+1}. \tag{6.67}$$

---

[9] Other commonly used sets of orthogonal polynomials are orthogonal with respect to a nonconstant weight function. In the case of Chebyshev polynomials, for example,

$$\int_{-1}^{1} T_m(x)T_n(x)(1-x^2)^{-1/2}\,\mathrm{d}x = 0,$$

unless $m = n$.

To appreciate how this judicious choice for the x_j increases the accuracy of (6.65) and (6.66), suppose that $f(x)$ is a polynomial of degree $2m - 1$ defined on the domain $[-1, 1]$. (More general domains $[a, b]$ can easily be transformed to the interval $[-1, 1]$.) Let $P_m(x)$ be the Legendre polynomial on $[-1, 1]$ of degree m. If $q(x)$ and $r(x)$ are, respectively, the quotient and the remainder obtained when dividing f by P_m, then $f = qP_m + r$, where both q and r are polynomials of degree less than or equal to $m - 1$. Since the polynomial q can be expressed as a linear combination of the Legendre polynomials of degree less than or equal to $m - 1$, all of which are orthogonal to P_m,

$$\int_{-1}^{1} f(x)\,\mathrm{d}x = \int_{-1}^{1} q(x)P_m(x)\,\mathrm{d}x + \int_{-1}^{1} r(x)\,\mathrm{d}x = \int_{-1}^{1} r(x)\,\mathrm{d}x. \tag{6.68}$$

Furthermore, if the x_j are chosen as the zeros of P_m,

$$\begin{aligned}\sum_{j=1}^{m} w_j f(x_j) &= \sum_{j=1}^{m} w_j q(x_j)P_m(x_j) + \sum_{j=1}^{m} w_j r(x_j)\\ &= \sum_{j=1}^{m} w_j r(x_j)\\ &= \int_{-1}^{1} r(x)\,\mathrm{d}x,\end{aligned} \tag{6.69}$$

where the third equality is obtained because r is a polynomial of degree less than or equal to $m - 1$. It follows from (6.68) and (6.69) that

$$\int_{-1}^{1} f(x)\,\mathrm{d}x = \sum_{j=1}^{m} w_j f(x_j),$$

and that m-point Gaussian quadrature is exact for polynomials up to degree $2m - 1$. The weights for m-point Gaussian quadrature over the domain $[-1, 1]$ may be expressed in terms of the corresponding nodal points as

$$w_j = \frac{2}{(1 - x_j^2)\left[P_m'(x_j)\right]^2} = \frac{2(1 - x_j^2)}{\left[mP_{m-1}(x_j)\right]^2}. \tag{6.70}$$

Formulae for the x_j are not known in closed form and must be computed numerically. This can be done using Newton's method (Dahlquist and Björck 1974) with first guesses for the m zeros of $P_m(x)$ given by the roots of the mth-degree Chebyshev polynomial

$$\tilde{x}_j = -\cos\left(\frac{2j - 1}{2m}\pi\right) \quad \text{for} \quad 1 \le j \le m. \tag{6.71}$$

6.4.3.2 Avoiding Aliasing Error

The product of two or more truncated spectral harmonic expansions contains high-order Fourier modes in λ and high-order functions in μ that are not present in the original truncation. When using the transform method, it is important to retain enough zonal wave numbers in the Fourier transforms and enough meridional grid points in the Gaussian quadrature to ensure that the transform procedure does not generate errors in any of the modes retained in the original truncated expansions.

Suppose that one wishes to compute the spectral coefficients of the binary product $\psi\chi$, where ψ and χ are given by the truncated spherical harmonic expansions

$$\psi(\lambda,\mu) = \sum_{p=-M}^{M}\sum_{q=|p|}^{N(p)} a_{p,q}Y_{p,q}(\lambda,\mu), \tag{6.72}$$

$$\chi(\lambda,\mu) = \sum_{r=-M}^{M}\sum_{s=|r|}^{N(r)} b_{r,s}Y_{r,s}(\lambda,\mu). \tag{6.73}$$

Let $c_{m,n}$ be the coefficient of $Y_{m,n}$ in the spherical harmonic expansion of $\psi\chi$. Without loss of generality consider the case $m \geq 0$, since the coefficients for which $m < 0$ can be obtained using (6.49). Then

$$c_{m,n} = \int_{-1}^{1} \hat{c}_m(\mu)P_{m,n}(\mu)\,\mathrm{d}\mu, \tag{6.74}$$

where

$$\hat{c}_m(\mu) = \frac{1}{2\pi}\int_{-\pi}^{\pi} \psi(\lambda,\mu)\chi(\lambda,\mu)\mathrm{e}^{-\mathrm{i}m\lambda}\,\mathrm{d}\lambda. \tag{6.75}$$

The last of the preceding integrals is a Fourier transform, and as discussed in Sect. 6.2.2, the discrete Fourier transform of binary products of Fourier series truncated at wave number M can be evaluated without aliasing error if the transforms are computed using a minimum of $(3M-1)/2$ wave numbers. To maximize the efficiency of the FFTs in practical applications, the actual cutoff wave number may be chosen as the smallest product of prime factors no larger than 5 that exceeds $(3M-1)/2$. This criterion for the cutoff wave number can be alternatively expressed as a requirement that the physical mesh include a minimum of $3M+1$ grid points around each latitude circle.

Now consider the evaluation of (6.74). The associated Legendre functions have the form

$$P_{m,n}(\mu) = (1-\mu^2)^{m/2}Q_{m,n}(\mu),$$

where $Q_{m,n}$ is a polynomial in μ of degree $n-m$. Since $P_{m,n}$ is not a polynomial when m is odd, it is not obvious that Gaussian quadrature can be used to integrate (6.74) without error. Nevertheless, it turns out that the complete integrand $\hat{c}_m(\mu)P_{m,n}(\mu)$ is a polynomial in μ whose maximum degree can be determined as follows. Substituting the finite series expansions for ψ and χ into (6.75) and using the orthogonality of the Fourier modes,

$$\hat{c}_m(\mu) = \sum_{\substack{p+r=m \\ |p|,|r|\le M}} \left(\sum_{q=|p|}^{N(p)} a_{p,q} P_{p,q}(\mu) \right) \left(\sum_{s=|r|}^{N(r)} b_{r,s} P_{r,s}(\mu) \right),$$

where the notation below the first summation indicates that the sum should be performed for all indices p and r such that $|p| \le M$, $|r| \le M$, and $p + r = m$. Each term in $\hat{c}_m(\mu) P_{m,n}(\mu)$ is therefore a function of the form

$$(1 - \mu^2)^{(p+r+m)/2} Q_{p,q}(\mu) Q_{r,s}(\mu) Q_{m,n}(\mu).$$

Since the indices in the preceding expression satisfy $p + r + m = 2m$, each term is a polynomial in μ of degree

$$2m + (q - p) + (s - r) + (n - m) = q + s + n.$$

The highest degree of the polynomials in the integrand of (6.74) is the maximum value of $q + s + n$, which is dependent on the type of truncation used in the expansions (6.72) and (6.73). In the case of a triangular truncation, this maximum is simply $3M$, and the exact evaluation of (6.74) by Gaussian quadrature requires a minimum of $(3M + 1)/2$ meridional grid points. In the case of a rhomboidal truncation, the maximum value of $q+s+n$ is $3M+p+r+m = 5M$, and $(5M+1)/2$ meridional grid points are required for an exact quadrature.

6.4.4 Nonlinear Shallow-Water Equations

Two additional considerations that arise in using spherical harmonic expansion functions in practical applications are the evaluation of derivatives with respect to the meridional coordinate and the representation of the vector velocity field. The treatment of these matters can be illustrated by considering the algorithm proposed by Bourke (1972) for integrating the nonlinear shallow-water equations on a rotating sphere,

$$\frac{\partial \mathbf{u}}{\partial t} + \mathbf{u} \cdot \nabla \mathbf{u} + f \mathbf{k} \times \mathbf{u} + \nabla \Phi = 0, \tag{6.76}$$

$$\frac{\partial \Phi}{\partial t} + \nabla \cdot \Phi \mathbf{u} = 0. \tag{6.77}$$

Here $\mathbf{u} = u\mathbf{i} + v\mathbf{j}$, where u and v are the eastward and northward velocity components, $f = 2\Omega \sin\theta$ is the Coriolis parameter, $\mathbf{k}$ is the vertical unit vector, Φ is the gravitational constant times the free-surface displacement, and ∇ is the horizontal gradient operator.

6.4.4.1 Prognostic Equations for Vorticity and Divergence

The velocity components u and v are not conveniently approximated by a series of spherical harmonic functions because artificial discontinuities in u and v are present at the poles unless the wind speed at the pole is zero. This problem arises because the direction defined as "east" switches by 180° as an observer traveling northward along a meridian steps across the pole. The same vector velocity that is recorded as "westerly" at a point on the Greenwich meridian is recorded as "easterly" at a point on the international dateline. In a similar way, a southerly velocity becomes a northerly velocity as the observer crosses the pole. This difficulty can be commonly avoided by replacing the prognostic equations for u and v by equations for the vorticity and divergence and by rewriting all remaining expressions involving $\mathbf{v}$ in terms of the transformed velocities

$$U = u \cos\theta, \qquad V = v \cos\theta.$$

Both U and V are zero at the poles and are free of discontinuities. It is also convenient to separate Φ into a constant mean $\overline{\Phi}$ and a perturbation $\Phi'(\lambda, \mu, t)$.

Equations for the divergence δ and the vertical component of vorticity ζ can be derived by substituting for $\mathbf{u} \cdot \nabla \mathbf{u}$ in (6.76) using the identity

$$\mathbf{u} \cdot \nabla \mathbf{u} = (\nabla \times \mathbf{u}) \times \mathbf{u} + \frac{1}{2} \nabla (\mathbf{u} \cdot \mathbf{u}) \tag{6.78}$$

to yield

$$\frac{\partial \mathbf{u}}{\partial t} + (\nabla \times \mathbf{u}) \times \mathbf{u} + \frac{1}{2} \nabla (\mathbf{u} \cdot \mathbf{u}) + f \mathbf{k} \times \mathbf{u} + \nabla \Phi = \mathbf{0}. \tag{6.79}$$

Taking the divergence of the preceding expression gives

$$\frac{\partial \delta}{\partial t} = \mathbf{k} \cdot \nabla \times (\zeta + f)\mathbf{u} - \nabla^2 \left(\Phi' + \frac{\mathbf{u} \cdot \mathbf{u}}{2} \right), \tag{6.80}$$

and taking the vertical component of the curl of (6.79) yields

$$\frac{\partial \zeta}{\partial t} = -\nabla \cdot (\zeta + f)\mathbf{u}. \tag{6.81}$$

The horizontal velocity may be expressed in terms of a stream function ψ and a velocity potential χ as

$$\mathbf{u} = \mathbf{k} \times \nabla \psi + \nabla \chi,$$

in which case the vertical component of the vorticity is

$$\zeta = \mathbf{k} \cdot \nabla \times \mathbf{u} = \nabla^2 \psi, \tag{6.82}$$

and the divergence is

$$\delta = \nabla \cdot \mathbf{u} = \nabla^2 \chi. \tag{6.83}$$

The governing equations (6.77), (6.80), and (6.81) can be concisely expressed in spherical coordinates by defining the operator

$$\mathscr{H}(A,B) = \frac{1}{a}\left(\frac{1}{1-\mu^2}\frac{\partial A}{\partial \lambda} + \frac{\partial B}{\partial \mu}\right).$$

Using the relations

$$\nabla \cdot \alpha \mathbf{u} = \mathscr{H}(\alpha U, \alpha V) \tag{6.84}$$

and

$$\mathbf{k} \cdot \nabla \times \alpha \mathbf{u} = \mathscr{H}(\alpha V, -\alpha U), \tag{6.85}$$

(6.77)–(6.81) become

$$\frac{\partial \nabla^2 \chi}{\partial t} = \mathscr{H}(V\nabla^2\psi, -U\nabla^2\psi) - 2\Omega\left(\frac{U}{a} - \mu\nabla^2\psi\right) - \nabla^2\left(\Phi' + \frac{U^2+V^2}{2(1-\mu^2)}\right), \tag{6.86}$$

$$\frac{\partial \nabla^2 \psi}{\partial t} = -\mathscr{H}(U\nabla^2\psi, V\nabla^2\psi) - 2\Omega\left(\frac{V}{a} + \mu\nabla^2\chi\right), \tag{6.87}$$

$$\frac{\partial \Phi'}{\partial t} = -\mathscr{H}(U\Phi', V\Phi') - \overline{\Phi}\nabla^2\chi. \tag{6.88}$$

The preceding system of equations for ψ, χ, and Φ' can be closed using the diagnostic relations

$$U = (1-\mu^2)\mathscr{H}(\chi, -\psi) \quad \text{and} \quad V = (1-\mu^2)\mathscr{H}(\psi, \chi). \tag{6.89}$$

6.4.4.2 Implementation of the Transform Method

The basic strategy used to implement the transform method for spherical harmonic expansion functions is the same as that used with simpler Fourier series, which is to compute binary products in physical space and then transform the result back to spectral space. During this procedure, those terms involving derivatives are evaluated to within the truncation error of the spectral approximation using the known properties of the expansion functions. The zonal derivative of each spherical harmonic is simply $imY_{m,n}$. The horizontal Laplacian is easily evaluated using (6.52), and the meridional derivative can be determined using the recurrence relation (6.51).

Prognostic equations for the spectral coefficients associated with the vorticity, the divergence, and the free-surface displacement can be derived as follows. Let

$$\psi(\lambda,\mu) = a^2 \sum_{m=-M}^{M} \sum_{n=|m|}^{N(m)} \psi_{m,n} Y_{m,n}(\lambda,\mu), \tag{6.90}$$

$$\chi(\lambda,\mu) = a^2 \sum_{m=-M}^{M} \sum_{n=|m|}^{N(m)} \chi_{m,n} Y_{m,n}(\lambda,\mu), \tag{6.91}$$

$$\Phi'(\lambda,\mu) = \sum_{m=-M}^{M} \sum_{n=|m|}^{N(m)} \Phi_{m,n} Y_{m,n}(\lambda,\mu).$$

Since the nonlinear products in the governing equations (6.86)–(6.88) are $U\nabla^2\psi$, $V\nabla^2\psi$, $U\Phi'$, $V\Phi'$, and U^2+V^2, it is also convenient to define expansion coefficients for U and V. These coefficients can be diagnostically computed from the expansion coefficients for ψ and χ as follows. Let $U_{m,n}$ and $V_{m,n}$ be the spectral expansion coefficients for U/a and V/a; then

$$\begin{aligned} U_{m,n} &= \frac{1}{2\pi}\int_{-1}^{1}\int_{-\pi}^{\pi} \frac{1}{a^2}\left(\frac{\partial\chi}{\partial\lambda} - (1-\mu^2)\frac{\partial\psi}{\partial\mu}\right) Y^*_{m,n}\,\mathrm{d}\lambda\,\mathrm{d}\mu \\ &= \mathrm{i}m\chi_{m,n} + (n-1)\varepsilon_{m,n}\psi_{m,n-1} - (n+2)\varepsilon_{m,n+1}\psi_{m,n+1}, \end{aligned}$$

where the second equality follows from (6.51) and the orthogonality of the spherical harmonics. Similarly,

$$V_{m,n} = \mathrm{i}m\psi_{m,n} - (n-1)\varepsilon_{m,n}\chi_{m,n-1} + (n+2)\varepsilon_{m,n+1}\chi_{m,n+1}.$$

Note that nonzero values of $\psi_{m,n}$ and $\chi_{m,n}$ imply nonzero values of $U_{m,n+1}$ and $V_{m,n+1}$, so the expansions for U and V must be truncated at one higher degree than those for ψ and χ, i.e.,

$$U(\lambda,\mu) = a \sum_{m=-M}^{M} \sum_{n=|m|}^{N(m)+1} U_{m,n} Y_{m,n}(\lambda,\mu),$$

$$V(\lambda,\mu) = a \sum_{m=-M}^{M} \sum_{n=|m|}^{N(m)+1} V_{m,n} Y_{m,n}(\lambda,\mu).$$

After products such as $U\nabla^2\psi$ have been computed on the physical mesh, the right sides of (6.86)–(6.88) are transformed back to wave-number space. The first step of this transformation is performed using FFTs. Suppose, for notational convenience, that $\hat{A}_m, \hat{B}_m, \ldots, \hat{E}_m$ are the Fourier transforms of the preceding binary products such that

$$U\nabla^2\psi = \sum_{m=-M}^{M} \hat{A}_m e^{\mathrm{i}m\lambda}, \quad V\nabla^2\psi = \sum_{m=-M}^{M} \hat{B}_m e^{\mathrm{i}m\lambda}$$

$$U\Phi' = \sum_{m=-M}^{M} \hat{C}_m e^{\mathrm{i}m\lambda}, \quad V\Phi' = \sum_{m=-M}^{M} \hat{D}_m e^{\mathrm{i}m\lambda}$$

$$\frac{1}{2}(U^2+V^2) = \sum_{m=-M}^{M} \hat{E}_m e^{\mathrm{i}m\lambda}.$$

The remaining step in the transformation back to wave-number space is computed by Gaussian quadrature. The only nontrivial quadratures are those related to the transform of $(U^2 + V^2)/(1 - \mu^2)$ and of functions of the form μR and $\mathscr{H}(R, S)$, where R and S are functions of μ and λ. Functions of the form μR can be transformed analytically using the recurrence relation (6.50). As an example, consider the term in (6.87) proportional to $\mu\nabla^2\chi$, whose (m, n)th spectral coefficient is

$$\frac{1}{2\pi}\int_{-1}^{1}\int_{-\pi}^{\pi} \mu\nabla^2\chi Y_{m,n}^{*}\, d\lambda\, d\mu = \int_{-1}^{1}\left[\sum_{s=|m|}^{N(m)} s(s+1)\chi_{m,s}\mu P_{m,s}\right] P_{m,n}\, d\mu$$
$$= (n-1)n\varepsilon_{m,n}\chi_{m,n-1} + (n+1)(n+2)\varepsilon_{m,n+1}\chi_{m,n+1}.$$

Now consider the transform of $\mathscr{H}(R, S)$, where R and S are binary products of $\nabla^2\psi$ or ϕ' and U or V. Let the Fourier transforms of $R(\lambda, \mu)$ and $S(\lambda, \mu)$ be denoted by $\hat{R}_m(\mu)$ and $\hat{S}_m(\mu)$, and define the coefficient of the (m, n)th component of the spherical harmonic expansion for $\mathscr{H}(R, S)$ to be $\mathscr{G}_{m,n}(\hat{R}_m, \hat{S}_m)$. Then

$$\mathscr{G}_{m,n}(\hat{R}_m, \hat{S}_m) = \frac{1}{a}\int_{-1}^{1}\left(\frac{im}{1-\mu^2}\hat{R}_m + \frac{\partial \hat{S}_m}{\partial\mu}\right) P_{m,n}\, d\mu. \tag{6.92}$$

S contains a factor of either U or V, and since U and V are zero at $\mu = \pm 1$, (6.92) may be integrated by parts to obtain

$$\mathscr{G}_{m,n}(\hat{R}_m, \hat{S}_m) = \frac{1}{a}\int_{-1}^{1}\left(\frac{im}{1-\mu^2}\hat{R}_m P_{m,n} - \hat{S}_m\frac{\partial P_{m,n}}{\partial\mu}\right) d\mu.$$

The derivative in the preceding equation can be evaluated exactly using (6.51), and the result can be integrated exactly wherever it appears in (6.86)–(6.88) using Gaussian quadrature over the same number of nodes as required for the transformation of simple binary products. The exactness of this integral follows from the same type of argument used in connection with the ordinary binary product (6.74); the integrand will consist of a sum of terms of the form

$$(1-\mu^2)^{m-1} Q_{p,q}(\mu) Q_{m-p,s}(\mu) Q_{m,n}(\mu), \tag{6.93}$$

where $Q_{m,n}$ is a polynomial in μ of degree $n - m$. Except for the case $m = 0$, (6.93) is a polynomial of sufficiently low degree that it can be computed exactly. When $m = 0$, it is easier to consider the equivalent integral (6.92). Because $m = 0$, the first term in the integrand is zero, and the second is the sum of terms of the form

$$Q_{p,q}(\mu) Q_{-p,s}(\mu) \frac{d}{d\mu} P_{0,n}(\mu).$$

The last factor is the derivative of the nth-degree Legendre polynomial, and the entire expression is once again a polynomial of sufficiently low degree to be integrated exactly by Gaussian quadrature.

Finally, consider the transform of $1/2(U^2+V^2)/(1-\mu^2)$, whose (m,n)th component will be denoted by $E_{m,n}$. From the definition of $\hat{E}_m$,

$$E_{m,n} = \int_{-1}^{1} \frac{\hat{E}_m}{1-\mu^2} P_{m,n}\, \mathrm{d}\mu. \tag{6.94}$$

The numerator in the preceding integrand is an ordinary binary product, and as argued in connection with (6.74), it must be a polynomial of sufficiently low degree that it can be integrated exactly by Gaussian quadrature over the same nodes used for the other transforms. Since both U and V have zeros at $\mu = \pm 1$, the polynomial in the numerator has roots at $\mu = \pm 1$ and must be exactly divisible by $(1-\mu^2)$. As a consequence, the entire integrand in (6.94) is a polynomial that can be integrated without error using Gaussian quadrature.

Using the preceding relations, the time tendencies of the spectral coefficients of the divergence, vorticity, and free-surface displacement become

$$\begin{aligned}-n(n+1)\frac{d\chi_{m,n}}{dt} &= \mathcal{G}_{m,n}\left(\hat{B}_m, -\hat{A}_m\right) + \frac{n(n+1)}{a^2}\left(\Phi_{m,n} + E_{m,n}\right)\\ &\quad - 2\Omega\big[U_{m,n} + (n-1)n\varepsilon_{m,n}\psi_{m,n-1} + (n+1)(n+2)\varepsilon_{m,n+1}\psi_{m,n+1}\big],\end{aligned}$$

$$\begin{aligned}-n(n+1)\frac{d\psi_{m,n}}{dt} &= -\mathcal{G}_{m,n}\left(\hat{A}_m, \hat{B}_m\right) - 2\Omega V_{m,n}\\ &\quad + 2\Omega\big[(n-1)n\varepsilon_{m,n}\chi_{m,n-1} + (n+1)(n+2)\varepsilon_{m,n+1}\chi_{m,n+1}\big],\end{aligned} \tag{6.95}$$

and

$$\frac{d\Phi_{m,n}}{dt} = -\mathcal{G}_{m,n}\left(\hat{C}_m, \hat{D}_m\right) + n(n+1)\overline{\Phi}\chi_{m,n}.$$

The extension of this algorithm to three-dimensional models for the simulation of global atmospheric flow is discussed in Sect. 8.6.2 and in Machenhauer (1979).

6.5 The Finite-Element Method

The finite-element method has not been widely used to obtain numerical solutions to hyperbolic partial differential equations because it generates implicit equations for the unknown variables at each new time level. The most efficient methods for the solution of wave-propagation problems are generally schemes that update the unknowns at each subsequent time level through the solution of explicit algebraic equations. Nevertheless, in some atmospheric applications computational efficiency can be improved by using semi-implicit differencing to integrate a subset of the complete equations via the implicit trapezoidal method while the remaining terms in the governing equations are integrated explicitly (see Sect. 8.2), and the finite-element method can be used to efficiently approximate the vertical structure of the flow in such models (Staniforth and Daley 1977). In addition, the finite-element method is easily adapted to problems in irregularly shaped domains, and as a consequence, it has been used in several oceanic applications to model tides and currents in bays and coastal regions (Foreman and Thomson 1997).

In contrast to the situation with hyperbolic partial differential equations, the finite-element method is very widely used to solve time-independent problems. The tendency of finite-element approximations to produce implicit algebraic equations is not a disadvantage in steady-state problems since the finite-difference approximations to such problems also generate implicit algebraic equations. Moreover, in most steady-state systems the fundamental physical problem can be stated in a variational form naturally suited for solution via the finite-element technique (Strang and Fix 1973). Our interest lies in the application of the finite-element method to time-dependent wavelike flows for which variational forms do not naturally arise. The most useful variational criteria for the equations governing most wavelike flows are simply obtained by minimizing the residual as defined by (6.3). The possible strategies for minimizing the residual are those discussed in Sect. 6.1. The collocation strategy will not be examined here since, at least for piecewise-linear expansion functions, it leads to methods that are identical to simple finite differences. More interesting algorithms with better conservation properties can be achieved using the Galerkin requirement that the residual be orthogonal to each expansion function or equivalently, by minimizing $(\|R(\phi)\|_2)^2$.

As discussed in Sect. 6.1, enforcement of the Galerkin requirement leads to the system of ordinary differential equations

$$\sum_{n=1}^{N} M_{n,k}\frac{da_n}{dt} = -\int_S \left[F\left(\sum_{n=1}^{N} a_n\varphi_n\right)\varphi_k\right] \mathrm{d}x \quad \text{for } k = 1,\ldots,N, \tag{6.96}$$

where

$$M_{n,k} = \int_S \varphi_n\varphi_k \, \mathrm{d}x.$$

The difference between the spectral method and the Galerkin form of the finite-element method lies in the choice of expansion functions. In the spectral method, the expansion functions form an orthogonal set, and each φ_k is nonzero over most of the spatial domain. The orthogonality of the spectral expansion functions ensures that $M_{n,k}$ is zero unless $n = k$, greatly simplifying the left side of (6.96). However, since the spectral expansion functions are nonzero over most of the domain, the evaluation of the right side of (6.96) involves considerable computation.

In the finite-element method, the expansion functions are not usually orthogonal, but each φ_n is nonzero only over a small, localized portion of the total domain. An example of a finite-element expansion function is given by the *chapeau* (or "hat") function shown in Fig. 6.6. In the case of the chapeau function, the total domain is partitioned into N nodes, and φ_k is defined as a piecewise-linear function equal to unity at the kth node and zero at every other node. If the series expansion (6.2) utilizes chapeau functions, the resulting sum will be a piecewise-linear approximation to the true function $\psi(x)$. Because the finite-element expansion functions are not

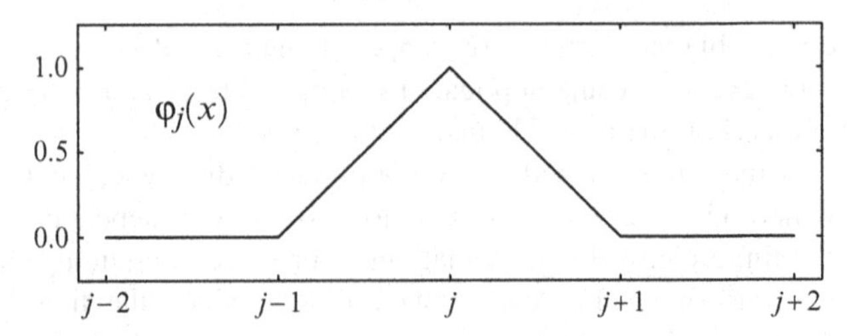

Fig. 6.6 The chapeau expansion function ϕ_j. The x-axis is labeled in units of Δx

orthogonal, the left side of (6.96) constitutes an implicit relationship between the $\mathrm{d}a_k/\mathrm{d}t$ at a small number of adjacent nodes, and a sparse linear system must be solved every time step.

Because it is necessary to solve a system of linear algebraic equations on every time step, the computational effort required by the Galerkin finite-element method typically exceeds that associated with finite-difference and spectral methods. Nevertheless, in comparison with the spectral method, the finite-element approach does reduce the computation required to evaluate the right side of (6.96). Since the finite-element expansion functions are nonzero only over a small portion of the total domain, the number of arithmetic operations required to evaluate the right side of (6.96) is $O(N)$, which is comparable to that involved in the calculation of conventional finite differences and can be considerably less than the $O(N \log N)$ operations required to evaluate the same expression using the spectral transform method.

6.5.1 Galerkin Approximation with Chapeau Functions

If the wind speed is constant, the Galerkin approximation to the one-dimensional advection equation (6.11) requires that

$$\sum_{n=1}^{N} M_{n,k} \frac{da_n}{dt} + c \sum_{n=1}^{N} a_n \int_S \frac{d\varphi_n}{dx} \varphi_k \,\mathrm{d}x = 0 \quad \text{for} \quad k = 1, \ldots, N. \tag{6.97}$$

Assuming that the φ_n are chapeau functions and shifting the x-origin to coincide with the left edge of each interval of integration, the integrals involving products of the expansion functions become

$$M_{j-1,j} = M_{j+1,j} = \int_S \varphi_{j+1}\varphi_j \,\mathrm{d}x = \int_0^{\Delta x} \left(\frac{x}{\Delta x}\right)\left(\frac{\Delta x - x}{\Delta x}\right) \mathrm{d}x = \frac{\Delta x}{6},$$

$$M_{j,j} = 2\int_0^{\Delta x} \left(\frac{\Delta x - x}{\Delta x}\right)^2 \mathrm{d}x = \frac{2\Delta x}{3},$$

$$-\int_S \frac{d\varphi_{j-1}}{dx}\varphi_j \,\mathrm{d}x = \int_S \frac{d\varphi_{j+1}}{dx}\varphi_j \,\mathrm{d}x = \int_0^{\Delta x} \left(\frac{1}{\Delta x}\right)\left(\frac{\Delta x - x}{\Delta x}\right) \mathrm{d}x = \frac{1}{2}.$$

Since all other integrals involving products of the expansion functions or their derivatives are zero, (6.97) reduces to

$$\frac{d}{dt}\left(\frac{a_{j+1}+4a_j+a_{j-1}}{6}\right)+c\left(\frac{a_{j+1}-a_{j-1}}{2\Delta x}\right)=0. \tag{6.98}$$

This scheme may be analyzed as if it were a standard differential–difference equation because a_j, the coefficient of the jth chapeau function, is also the nodal value of the approximate solution $\phi(x_j)$. In fact, (6.98) is identical to the fourth-order compact differential–difference approximation to the advection equation (3.51), whose properties were previously discussed in Sect. 3.3.4. In particular, the scheme's spatial truncation error is $O\left[(\Delta x)^4\right]$ and its phase speed in the limit of good resolution is

$$c^* \approx c\left(1-(k\Delta x)^4/180\right). \tag{6.99}$$

This scheme also performs very well at moderately poor spatial resolution. As was shown in Fig. 3.10, (6.98) generates less phase-speed error in moderately short waves than does explicit fourth- or sixth-order spatial differencing.

Now suppose that the time derivatives in (6.98) are approximated using leapfrog time differencing. The discrete-dispersion relation becomes

$$\sin(\omega\Delta t)=\frac{3\mu\sin(k\Delta x)}{\cos(k\Delta x)+2}.$$

When $|\mu|<1/\sqrt{3}$, the right side of the preceding equation is bounded by unity and the scheme is stable. As was the case with the spectral method, the finite-element method better approximates the spatial derivative of coarsely resolved waves (such as the $3\Delta x$ wave) and, thus, the finite-element approximation to the advection equation captures higher-frequency oscillations and the maximum stable time step is reduced relative to that allowed by a centered second-order finite-difference approximation to the spatial derivative.

One way to circumvent this time-step restriction is to use trapezoidal time differencing. The trapezoidal method is unconditionally stable, is more accurate than leapfrog differencing, and does not support a computational mode. Despite these advantages, the trapezoidal scheme is not used in most finite-difference approximations to wave-propagation problems because it leads to implicit equations. The implicit nature of trapezoidal differencing is not, however, a problem in this application, because (6.98) is already a linear system of implicit equations for the $\mathrm{d}a_j/\mathrm{d}t$ and trapezoidal differencing does not increase the bandwidth of the coefficient matrix. The trapezoidal method is, however, less attractive in more general applications where systems of equations must be solved. For example, the implicit coupling among the various nodal values remains tridiagonal when the linearized shallow-water system (4.1) and (4.2) is approximated using explicit time differencing and the chapeau-function finite-element method, but if the leapfrog differencing is replaced by the trapezoidal method, the u^{n+1} and h^{n+1} become implicit functions of each other, and the resulting linear system has a larger bandwidth.

6.5.2 Petrov–Galerkin and Taylor–Galerkin Methods

Another way to increase the maximum stable time step of finite-element approximations to wave-propagation problems is to generalize the orthogonality condition satisfied by the residual. As an alternative to the standard Galerkin requirement that the residual be orthogonal to each of the expansion functions, one may define a different set of "test" functions and require the residual to be orthogonal to each of these test functions. This approach, known as the *Petrov–Galerkin* method, can yield schemes that are stable for Courant numbers as large as unity, and can greatly increase the accuracy of computations performed at Courant numbers near the stability limit. The Petrov–Galerkin method does not, however, share all of the desirable conservation properties of the standard Galerkin method.

Let ϑ_k be an arbitrary member of the set of test functions. If the time derivative is approximated by a forward difference, the Petrov–Galerkin formula for the differential–difference approximation to the general partial differential equation (6.1) is

$$\int_S \left[\sum_{j=1}^{N} \left(\frac{a_j^{n+1} - a_j^n}{\Delta t} \right) \varphi_j + F\left(\sum_{j=1}^{N} a_j^n \varphi_j \right) \right] \vartheta_k \, \mathrm{d}x = 0 \quad \text{for all } k.$$

As a specific example, suppose that a Petrov–Galerkin approximation is sought to the advection equation (6.11) with c constant and nonnegative. Let the expansion functions $\varphi_j(x)$ be the chapeau functions defined previously and, as suggested by Morton and Parrott (1980), define a family of test functions of the form $\vartheta_k = (1-\nu)\varphi_k + \nu\chi_k$, where ν is a tunable parameter and χ_j is the localized sawtooth function

$$\chi_j(x) = \begin{cases} 6(x - x_{j-1})/\Delta x - 2, & \text{if } x \in [x_{j-1}, x_j]; \\ 0, & \text{otherwise.} \end{cases}$$

(χ_j is normalized so that its integral over the domain is unity.) Using these test functions, one may express the Petrov–Galerkin approximation to the constant-wind-speed advection equation in terms of the nodal values as

$$\left[1 + \tfrac{1}{6}(1-\nu)\tilde{\delta}_x^2\right]\left(a^{n+1} - a^n\right) + \mu\tilde{\delta}_{2x}a^n = \tfrac{1}{2}\mu\nu\tilde{\delta}_x^2 a^n, \tag{6.100}$$

where $\tilde{\delta}_x = \Delta x\, \delta_x$ is a nondimensional finite-difference operator and $\mu = c\Delta t/\Delta x$ is the Courant number. Morton and Parrott (1980) used the energy method to show this scheme is stable for $0 \le \mu \le \nu \le 1$. The amplification factor for this scheme is

$$A = 1 - \frac{\mathrm{i}\mu \sin(k\Delta x) - \mu\nu[\cos(k\Delta x) - 1]}{1 + (1-\nu)[\cos(k\Delta x) - 1]/3}.$$

In the limit $k\Delta x \to 0$

$$A = 1 - \mathrm{i}\mu k\Delta x - \frac{1}{2}\mu\nu(k\Delta x)^2 + \frac{1}{6}\mu\nu(k\Delta x)^3 + O\left[(k\Delta x)^4\right],$$

which matches the correct amplification factor, $e^{-i\omega\Delta t} = e^{-i\mu k\Delta x}$, through first order except that the scheme is second-order accurate when $\nu = \mu$. Clearly one should choose $\nu = \mu$ since this allows the largest stable time step and gives the best accuracy. When $\nu = \mu$, (6.100) is closely related to the standard Lax–Wendroff approximation (3.72); the only difference appears in the linear operator (i.e., the mass matrix) acting on the forward-time difference. As noted by Morton and Parrott (1980), if ν is set equal to μ, the truncation error in (6.100) is always less than that for the standard Lax–Wendroff scheme; the improvement is particularly pronounced for small values of μ. On the other hand, the standard Lax–Wendroff method is stable for $|\mu| \leq 1$, whereas the Galerkin–Petrov method (6.100) is a upstream method that requires $\mu \geq 0$ for stability. A formula analogous to (6.100) can be derived for negative flow velocities using test functions in which the sawtooth component has a negative slope.

A better scheme than that just derived via the Petrov–Galerkin approach can be obtained using the *Taylor–Galerkin* method. The Taylor–Galerkin method does not require the specification of a second set of test functions and yields centered-in-space methods. In the Taylor–Galerkin approach, the time derivative is discretized before invoking the finite-element formalism to approximate the spatial derivatives. Donea et al. (1987) presented Taylor–Galerkin approximations to several hyperbolic problems in which the time discretization is Lax–Wendroff, leapfrog, or trapezoidal. In the following we will focus on Lax–Wendroff-type approximations to the constant-wind-speed advection equation.

If the spatial dependence of the solution is not discretized, an $O\left[(\Delta t)^2\right]$-accurate Lax–Wendroff approximation to the constant-wind-speed advection equation has the form

$$\frac{\phi^{n+1} - \phi^n}{\Delta t} + c\frac{\partial \phi^n}{\partial x} = \frac{c^2 \Delta t}{2}\frac{\partial^2 \phi^n}{\partial x^2}$$

(see Sect. 3.4.4). Suppose that the spatial structure in the preceding differential–difference equation is approximated using the Galerkin finite-element method with chapeau expansion functions, then the function value at each node satisfies

$$\left[1 + \frac{1}{6}\tilde{\delta}_x^2\right]\left(a^{n+1} - a^n\right) + \mu\tilde{\delta}_{2x}a^n = \frac{1}{2}\mu^2\tilde{\delta}_x^2 a^n, \tag{6.101}$$

where once again $\tilde{\delta}_x = \Delta x\, \delta_x$ and $\mu = c\Delta t/\Delta x$. The stability condition for this scheme is $|\mu| \leq 1/\sqrt{3}$, which is identical to that for the leapfrog approximation to (6.98). The leading-order errors in the modified equations[10] for this method and the standard finite-difference Lax–Wendroff method (3.72) are shown in Table 6.3. The leading-order errors in both schemes are second order and generate numerical dispersion. The dispersion, or phase-speed error, in each method is the net result of accelerative time-differencing error and decelerative space-differencing error. These errors partially cancel in the standard finite-difference Lax–Wendroff method, and are eliminated entirely when $|\mu| = 1$. On the other

[10] The "modified equation" is discussed in Sect. 3.4.2.

Table 6.3 Modified equations for Lax–Wendrof-type finite-difference and finite-element approximations and the Taylor–Galerkin method

Finite difference (3.72)
$\psi_t + c\psi_x = -(c/6)(\Delta x)^2(1-\mu^2)\psi_{xxx} - (c/8)(\Delta x)^3\mu(1-\mu^2)\psi_{xxxx} + \ldots$
Finite element (6.101)
$\psi_t + c\psi_x = (c/6)(\Delta x)^2\mu^2\psi_{xxx} - (c/24)(\Delta x)^3\mu(1-3\mu^2)\psi_{xxxx} + \ldots$
Taylor–Galerkin finite-element (6.102)
$\psi_t + c\psi_x = -(c/24)(\Delta x)^3\mu(1-\mu^2)\psi_{xxxx} + \ldots$

After Donea et al. (1987)
Subscripts denote partial derivatives.

hand, the leading-order error in the Lax–Wendroff finite-element method is due entirely to accelerative time-differencing error. The decelerative phase error generated by the finite-element approximation to the spatial derivatives is $O\left[(\Delta x)^4\right]$ (cf. (6.99)), and as a consequence, there is no beneficial cancelation between time-differencing error and space-differencing error in the leading-order error for the Lax–Wendroff finite-element method.

Donea (1984) observed that much better results can be obtained using a third-order Lax–Wendroff approximation. Expanding the true solution to the constant-wind-speed advection equation at time $(n+1)\Delta t$ in a Taylor series about its value at time $n\Delta t$, and using the governing equation to replace the first- and second-order time derivatives by expressions involving derivatives with respect to x, gives

$$\psi^{n+1} - \psi^n = -c\Delta t\frac{\partial \psi^n}{\partial x} + \frac{(c\Delta t)^2}{2}\frac{\partial^2 \psi^n}{\partial x^2} + \frac{c^2(\Delta t)^3}{6}\left(\frac{\partial^3 \psi}{\partial t\,\partial x^2}\right)^n + O\left[(\Delta t)^4\right],$$

where ψ^n is the value of the true solution at $t = n\Delta t$. The mixed third-order derivative in the preceding equation is not replaced by an expression proportional to $\partial^3\psi/\partial x^3$ because the finite-element approximation to such a term would require smoother expansion functions than the piecewise-linear chapeau functions. Instead, the derivative with respect to time in $\partial^3\psi/(\partial t\,\partial x^2)$ can be conveniently approximated by a forward difference to obtain the following $O\left[(\Delta t)^3\right]$-accurate approximation to the advection equation:

$$\left(1 - \frac{(c\Delta t)^2}{6}\frac{\partial^2}{\partial x^2}\right)\left(\phi^{n+1} - \phi^n\right) + c\Delta t\frac{\partial \phi^n}{\partial x} = \frac{(c\Delta t)^2}{2}\frac{\partial^2 \phi^n}{\partial x^2},$$

in which $\phi^n(x)$ is a semidiscrete approximation to $\psi(n\Delta t, x)$ Using chapeau functions to approximate the spatial dependence of ϕ^n and demanding that the residual be orthogonal to each expansion function, one obtains the Taylor–Galerkin formula for the function value at each node

$$\left[1 + \frac{1}{6}(1-\mu^2)\tilde{\delta}_x^2\right]\left(a^{n+1} - a^n\right) + \mu\tilde{\delta}_{2x}a^n = \frac{1}{2}\mu^2\tilde{\delta}_x^2 a^n. \qquad (6.102)$$

This scheme is stable for $|\mu| \leq 1$. Examination of the modified equation for (6.102), which appears in Table 6.3, shows that, in contrast to the Lax–Wendroff finite-difference and finite-element methods, the Taylor–Galerkin method is free from second-order dispersive errors. The leading-order error in the Taylor–Galerkin method is third-order and weakly dissipative. The difference between the Taylor–Galerkin scheme (6.102), the Petrov–Galerkin method (6.100), and the Lax–Wendroff finite-element method (6.101) involves only minor perturbations to the coefficients in the tridiagonal mass matrix. All three schemes require essentially the same computation per time step, but the Taylor–Galerkin method is the most accurate and is stable over the widest range of Courant numbers.

6.5.3 Quadratic Expansion Functions

Higher-degree expansion functions are widely used in finite-element approximations to elliptic partial differential equations. Higher-degree expansion functions are not, however, commonly used in finite-element simulations of wavelike flow. One serious disadvantage of higher-degree expansion functions is that they ordinarily increase the implicit coupling in the equations for the time evolution of the expansion coefficients. Another disadvantage is that the accuracy obtained using higher-degree expansion functions in hyperbolic problems is generally lower than that which can be achieved using the same expansion functions in finite-element approximations to elliptic and parabolic partial differential equations (Strang and Fix 1973). In fact, the order of accuracy of the function values at the nodes given by quadratic and Hermite-cubic finite-element approximations to hyperbolic equations is lower than that obtained with piecewise-linear chapeau functions. High orders of accuracy can be obtained using cubic splines (Thomée and Wendroff 1974), but splines introduce a nonlocal coupling between the coefficients of the finite-element expansion functions that makes them too inefficient for most applications involving wavelike flows. One final difficulty associated with the use of higher-degree expansion functions in hyperbolic problems is that their behavior can be complicated and difficult to rigorously analyze. With these concerns in mind, let us examine the behavior of quadratic finite-element approximations to the constant-wind-speed advection equation.

Suppose the piecewise-linear approximation generated by the superposition of chapeau expansion functions is replaced by piecewise-quadratic functions of the form

$$q(x) = C_1 + C_2 x + C_3 x^2. \tag{6.103}$$

In contrast to linear interpolation, the three coefficients C_1, C_2, and C_3 cannot be uniquely determined by the two nodal values at the ends of each element. The most straightforward way to proceed is to extend the piecewise-quadratic function across an interval of $2\Delta x$ and to choose C_1, C_2, and C_3 so that (6.103) matches the function values at the "midpoint" node and at both "endpoint" nodes. Suppose that a function assumes the values a_1 and a_3 at the endpoint nodes x_1 and x_3, and assumes the value b_2 at the midpoint node x_2. The quadratic Lagrange interpolating polynomial that assumes these values at the nodes is

$$\frac{(x-x_2)(x-x_3)}{(x_1-x_2)(x_1-x_3)}a_1+\frac{(x-x_1)(x-x_3)}{(x_2-x_1)(x_2-x_3)}b_2+\frac{(x-x_1)(x-x_2)}{(x_3-x_1)(x_3-x_2)}a_3.$$

On the interval $x_1 \le x \le x_3$, the preceding expression is algebraically identical to

$$a_1\varphi_1^{\mathrm{e}}(x)+b_2\varphi_2^{\mathrm{m}}(x)+a_3\varphi_3^{\mathrm{e}}(x),$$

where φ_j^{e} is the endpoint quadratic expansion function

$$\varphi_j^{\mathrm{e}}(x)=\begin{cases}1-3\dfrac{|x-x_j|}{2\Delta x}+\dfrac{1}{2}\left(\dfrac{x-x_j}{\Delta x}\right)^2, & \text{if } |x-x_j|\le 2\Delta x,\\ 0, & \text{otherwise,}\end{cases}\tag{6.104}$$

and φ_j^{m} is the midpoint quadratic expansion function

$$\varphi_j^{\mathrm{m}}(x)=\begin{cases}\left(\dfrac{x-x_{j-1}}{\Delta x}\right)\left(2-\dfrac{x-x_{j-1}}{\Delta x}\right), & \text{if } |x-x_j|\le \Delta x,\\ 0, & \text{otherwise.}\end{cases}\tag{6.105}$$

These expansion functions are plotted in Fig. 6.7. The jth endpoint expansion function is zero outside an interval of length $4\Delta x$ centered at the node x_j; it is unity at x_j and is zero at every other node. The jth midpoint expansion function is zero outside an interval of length $2\Delta x$ centered at x_j; it is equal to unity at x_j and is zero at the other nodes. As was the case with chapeau expansion functions, the coefficient of the jth quadratic expansion function is also the value of the approximate solution at the jth node.

Let a finite-element approximation to the solution to the constant-wind-speed advection equation (6.11) be constructed from the preceding quadratic expansion functions such that

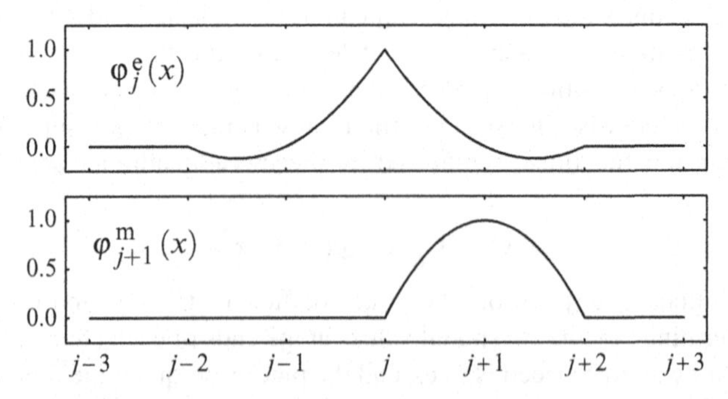

Fig. 6.7 Quadratic expansion functions for φ_j^{e}, an endpoint node centered at grid point j, and $\varphi_{j+1}^{\mathrm{m}}$, a midpoint node centered at $j+1$. The x-axis is labeled in units of Δx

$$\phi(x,t) = \sum_{j \text{ odd}} a_j(t)\varphi_j^{\text{e}}(x) + \sum_{\ell \text{ even}} b_\ell(t)\varphi_\ell^{\text{m}}(x). \tag{6.106}$$

Enforcing the Galerkin requirement that the residual be orthogonal to each expansion function yields two families of equations for the evolution of the expansion coefficients in the constant-wind-speed advection problem. The equations centered at the endpoint nodes are

$$\frac{d}{dt}\left(\frac{-a_{j-2} + 2b_{j-1} + 8a_j + 2b_{j+1} - a_{j+2}}{10}\right) + c\left(\frac{b_{j+1} - b_{j-1}}{\Delta x} - \frac{a_{j+2} - a_{j-2}}{4\Delta x}\right) = 0, \tag{6.107}$$

whereas those centered on the midpoints are

$$\frac{d}{dt}\left(\frac{a_{\ell-1} + 8b_\ell + a_{\ell+1}}{10}\right) + c\left(\frac{a_{\ell+1} - a_{\ell-1}}{2\Delta x}\right) = 0. \tag{6.108}$$

As formulated above, the quadratic finite-element method (QFEM) requires more work per time step per element than the linear finite-element scheme because a pentadiagonal matrix must be inverted to evaluate the time derivatives in (6.107) and (6.108), whereas the mass matrix associated with (6.98) is only tridiagonal. Steppler (1987) suggested a more efficient alternative in which the expansion functions are the set of piecewise-linear chapeau functions centered at every node together with an additional set of functions representing the quadratic corrections at every second node.

Although (6.107) and (6.108) are expressions for the coefficients of the QFEM expansion functions, they can be alternatively interpreted as finite-difference approximations for the function values at each node. The truncation error in the function values at the nodes can therefore be assessed by a conventional Taylor series analysis which shows that both (6.107) and (6.108) are $O\left[(\Delta x)^2\right]$-accurate finite-difference approximations to the one-dimensional advection equation. The truncation error at the nodes is considerably worse than the $O\left[(\Delta x)^4\right]$ error obtained using chapeau expansion functions!

In fact, the error in QFEM solutions to some constant-wind-speed advection problems can be significantly smaller than that obtained using chapeau functions over the same number of nodes. There are two reasons why the preceding comparison of truncation errors can be misleading. The first reason is that, unlike finite-difference approximations, finite-element methods involve an explicit assumption about the functional dependence of the solution between the nodal points, and the error at all nonnodal points is $O\left[(\Delta x)^2\right]$ for both the linear finite-element method and the QFEM. In the case of chapeau expansion functions, the function values between the nodal points are obtained by linear interpolation and are therefore only $O\left[(\Delta x)^2\right]$ accurate. In general, a smooth function can be interpolated to $O\left[(\Delta x)^{n+1}\right]$ by piecewise polynomials of degree n. Thus, if the nodal values could be specified with negligible error, quadratic expansion functions could provide $O\left[(\Delta x)^3\right]$ accuracy between the nodes. But since the QFEM method only predicts the nodal values to $O\left[(\Delta x)^2\right]$, the accuracy between the nodes is also limited to $O\left[(\Delta x)^2\right]$. As discussed in detail by Cullen and Morton (1980), it is necessary to specify how the error will be measured (e.g., pointwise errors at the nodes or the square integral of the error over the entire spatial domain) before attempting to determine the truncation error in finite-element approximations.

The second, and perhaps more important, reason why the preceding comparison of truncation error can be misleading is that it does not provide reliable information about the errors in the poorly resolved waves, and in many fluid-dynamical applications, the total error can be dominated by the errors in the shortest waves. The error in QFEM solutions to the constant-wind-speed advection equation can be evaluated as a function of the spatial resolution by examining the phase-speed and amplitude errors in semidiscrete wave solutions to (6.107) and (6.108). There is, however, no single wave of the form $\phi_j(t) = \mathrm{e}^{\mathrm{i}(kj\Delta x - \omega t)}$ that will simultaneously satisfy (6.107) and (6.108). If the initial disturbance consists of a single wave, at subsequent times the approximate numerical solution will split into two traveling disturbances such that the wave amplitude at the endpoint nodes is systematically different from that at the midpoint nodes. The nodal values in each traveling disturbance may be expressed in the form

$$a_j(t) = \mathrm{e}^{\mathrm{i}k(j\Delta x - c^* t)}, \qquad b_\ell(t) = r_\mathrm{a}\mathrm{e}^{\mathrm{i}k(\ell\Delta x - c^* t)}, \tag{6.109}$$

where r_a is the ratio of the amplitude at a midpoint node to the amplitude at an endpoint node, and c^* is the phase speed. Substitution of (6.109) into (6.107) and (6.108) yields

$$\frac{1}{5}\left(8 - 2\cos 2\theta + 4r_\mathrm{a}\cos\theta\right)c^* - \frac{c}{\theta}\left(4r_\mathrm{a}\sin\theta - \sin 2\theta\right) = 0$$

and

$$\frac{1}{5}\left(\cos\theta + 4r_\mathrm{a}\right)c^* - \frac{c}{\theta}\sin\theta = 0,$$

where $\theta = k\Delta x$. Solutions to this system of two equations in the two unknowns c^* and r_a have the form

$$\frac{c^*}{c} = \frac{\sin\theta}{\theta\left(1 + \sin^2\theta\right)}\left(-2\cos\theta \pm \left(9 + \sin^2\theta\right)^{1/2}\right) \tag{6.110}$$

and

$$r_\mathrm{a} = \frac{1}{4}\left(5\frac{c}{c^*}\frac{\sin\theta}{\theta} - \cos\theta\right). \tag{6.111}$$

Two difficulties arise in connection with the interpretation of these results. The first is the apparent presence of two distinct sets of solutions: "physical" and "nonphysical" modes associated, respectively, with the positive and negative roots in (6.110). Yet, as recognized by Cullen (1982), none of the solutions are extraneous because the set of solutions given by the positive root in (6.110) is identical to the set of solutions given by the negative root. The relation between these two sets of solutions may be derived as follows.

Let those quantities related to the positive root in (6.110) be denoted with a subscript "p" and those related to the negative root be denoted by a subscript "n." Also let $\omega^* = kc^*$ be the frequency of the QFEM solution. Elementary trigonometric identities imply that for $k \in [0, \pi/\Delta x]$, i.e., for the positive wave numbers resolvable on the numerical grid,

$$\omega_{\mathrm{p}}^*(k) = \omega_{\mathrm{n}}^*(k - \pi/\Delta x) \quad \text{and} \quad r_{\mathrm{a_p}}(k) = -r_{\mathrm{a_n}}(k - \pi/\Delta x), \tag{6.112}$$

and for $k \in [-\pi/\Delta x, 0]$

$$\omega_{\mathrm{p}}^*(k) = \omega_{\mathrm{n}}^*(k + \pi/\Delta x) \quad \text{and} \quad r_{\mathrm{a_p}}(k) = -r_{\mathrm{a_n}}(k + \pi/\Delta x).$$

Now consider a negative-root mode with endpoint values a_j and midpoint values b_ℓ. Consistent with (6.106), the indices j are even and ℓ are odd. Using (6.112)

$$a_j = \mathrm{e}^{\mathrm{i}\left[(k-\pi/\Delta x)j\Delta x - \omega_{\mathrm{n}}^*(k-\pi/\Delta x)t\right]} = (-1)^j \mathrm{e}^{\mathrm{i}kj\Delta x}\mathrm{e}^{-\mathrm{i}\omega_{\mathrm{p}}^*(k)t} = \mathrm{e}^{\mathrm{i}\left[kj\Delta x - \omega_{\mathrm{p}}^*(k)t\right]},$$

and

$$\begin{aligned} b_\ell &= r_{\mathrm{a_n}}(k - \pi/\Delta x)\mathrm{e}^{\mathrm{i}\left[(k-\pi/\Delta x)\ell\Delta x - \omega_{\mathrm{n}}^*(k-\pi/\Delta x)t\right]} \\ &= -r_{\mathrm{a_p}}(k)(-1)^\ell \mathrm{e}^{\mathrm{i}k\ell\Delta x}\mathrm{e}^{-\mathrm{i}\omega^*{}_{\mathrm{p}}(k)t} \\ &= r_{\mathrm{a_p}}(k)\mathrm{e}^{\mathrm{i}\left[k\ell\Delta x - \omega^*{}_{\mathrm{p}}(k)t\right]}, \end{aligned}$$

which demonstrates that for $k \in [0, \pi/\Delta x]$ the negative-root mode with wave number $k - \pi/\Delta x$ is identical to the positive-root mode with wave number k. Similarly, for $k \in [-\pi/\Delta x, 0]$, the negative-root mode with wave number $k + \pi/\Delta x$ is identical to the positive-root mode with wave number k.

We now arrive at the second difficulty in interpreting our results: How can both of the phase speeds given by the two roots of (6.110) simultaneously describe the behavior of individual solutions to (6.107) and (6.108)? The answer lies in the fact that solutions of the form (6.109) are not conventional semidiscrete Fourier modes that propagate without changing shape between each pair of adjacent nodes. Instead, (6.109) is a semidiscrete approximation to a function of the form $g(x)\mathrm{e}^{\mathrm{i}(kx-\omega t)}$, where the factor $g(x)$ is introduced to account for the extra spatial dependence associated with the midpoint-node coefficient r_{a}.

The nature of the QFEM eigenmodes can be more easily understood by expressing them as the sum of two *conventional* semidiscrete Fourier modes traveling at different speeds and moving in opposite directions. As a first step, (6.109) is written in the alternative form

$$\hat{q}_n(t) = \frac{1}{2}\left[(1 + r_{\mathrm{a}}) + (-1)^n(1 - r_{\mathrm{a}})\right]\mathrm{e}^{\mathrm{i}[kn\Delta x - \omega^*(k)t]}, \qquad n = 0, 1, \ldots 2N.$$

Even values of n give the nodal values a_j; odd values yield the b_ℓ. Then for a positive-root eigenmode,

$$\begin{aligned}\hat{q}_n(t) &= \left(\frac{1+r_{\mathrm{a_p}}}{2}\right)\mathrm{e}^{\mathrm{i}\left[kn\Delta x-\omega_{\mathrm{p}}^*(k)t\right]} + \left(\frac{1-r_{\mathrm{a_p}}}{2}\right)\mathrm{e}^{\mathrm{i}\left[(k-\pi/\Delta x)n\Delta x-\omega_{\mathrm{p}}^*(k)t\right]}\\ &= \left(\frac{1+r_{\mathrm{a_p}}}{2}\right)\left[\mathrm{e}^{\mathrm{i}\left[kn\Delta x-\omega_{\mathrm{p}}^*(k)t\right]} + \left(\frac{1-r_{\mathrm{a_p}}}{1+r_{\mathrm{a_p}}}\right)\mathrm{e}^{\mathrm{i}\left[(k-\pi/\Delta x)n\Delta x-\omega_{\mathrm{n}}^*(k-\pi/\Delta x)t\right]}\right],\end{aligned}$$

where (6.112) is used to obtain the last equality and without loss of generality it is assumed that $k \in [0, \pi/\Delta x]$. Defining $q = 2\hat{q}/(1+r_{\mathrm{a_p}})$ and $\beta = (1-r_{\mathrm{a_p}})/(1+r_{\mathrm{a_p}})$, one may write the preceding expression in the form

$$q_n(t) = \mathrm{e}^{\mathrm{i}k\left[n\Delta x-c_{\mathrm{p}}^*(k)t\right]} + \beta\mathrm{e}^{\mathrm{i}(k-\pi/\Delta x)\left[n\Delta x-c_{\mathrm{n}}^*(k-\pi/\Delta x)t\right]}. \tag{6.113}$$

Here $c_{\mathrm{p}}^*(k)$ and $c_{\mathrm{n}}^*(k-\pi/\Delta x)$ are given, respectively, by the positive and negative roots in (6.110), and the value of $r_{\mathrm{a_p}}$ used to evaluate β is obtained by substituting $c_{\mathrm{p}}^*(k)/c$ into (6.111). If $-\pi/\Delta x \le k \le 0$, the preceding expression for q_n is modified by replacing $k-\pi/\Delta x$ with $k+\pi/\Delta x$.

According to (6.113), *the eigenmodes of the QFEM approximation to the one-dimensional advection equation* (6.11) *are the superposition of a wave moving downstream at speed* c_{p}^* *and a second nonphysical wave moving upstream at speed* c_{n}^*. Neither the upstream-moving wave nor the downstream-moving wave can satisfy (6.107) and (6.108) without the simultaneous presence of the second wave whose amplitude, relative to the first wave, is determined by β. Values of $c_{\mathrm{p}}^*(k)/c$, $c_{\mathrm{n}}^*(k-\pi/\Delta x)/c$, and $\beta(k)$ are plotted as a function of k in Fig. 6.8.

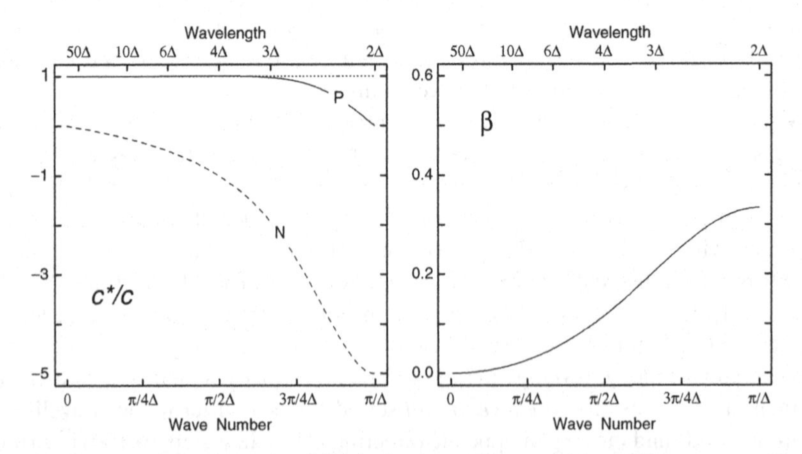

Fig. 6.8 Normalized phase speeds $c_{\mathrm{p}}^*(k)/c$ for the positive-root (physical) component (*P*) and $c_{\mathrm{n}}^*(k-\pi/\Delta x)/c$ for the negative-root (nonphysical) component (*N*) of a quadratic finite-element method (QFEM) eigenmode for the constant-wind-speed advection equation as a function of wave number k. Also shown is β, the ratio of the amplitude of the negative-root component to the amplitude of the positive-root component

Each of the two waves that compose the QFEM eigenmode is a conventional semidiscrete Fourier mode with a well-defined phase speed and group velocity. The phase speeds given by the two roots of (6.110) are the phase speeds of these individual Fourier modes. No single phase speed precisely describes the motion of a QFEM eigenmode, although since β is small for eigenmodes longer than about $4\Delta x$, well-resolved eigenmodes appear to translate with almost no change in form at speed c_p^*.

The group velocity, $c_g^* = \partial\omega/\partial k$, for each individual Fourier mode in (6.113) satisfies

$$\frac{c_g^*}{c} = \frac{\pm\cos\theta\left(2 + 7\cos^2\theta\right) + 2\left(2 - 3\cos^2\theta\right)\sqrt{10 - \cos^2\theta}}{\left(2 - \cos^2\theta\right)^2\sqrt{10 - \cos^2\theta}},$$

where the positive root is associated with $\partial\omega_\mathrm{p}/\partial k$ and the negative root is associated with $\partial\omega_\mathrm{n}/\partial k$. Note that $c_{g_\mathrm{p}}(k) = c_{g_\mathrm{n}}(k \pm \pi/\Delta x)$, since $\cos\theta = -\cos(\theta \pm \pi)$, implying that both of the individual modes in (6.113) propagate at the same group velocity. Thus, although there is some ambiguity in the precise determination of the phase speed of a single QFEM eigenmode, each eigenmode does have a well-defined group velocity. This group velocity is plotted as a function of wavenumber in Fig. 6.9.

The data in Figs. 6.8 and 6.9 suggest that, in many respects, a QFEM approximation to (6.11) yields qualitatively similar results to those obtained with linear finite elements or finite differences: long, well-resolved waves will be treated accurately, whereas disturbances with wavelengths near $2\Delta x$ will be subject to substantial error. As the numerical resolution improves, the QFEM can, nevertheless, yield much better solutions to (6.11) than those obtained using linear elements (Gresho and Sani

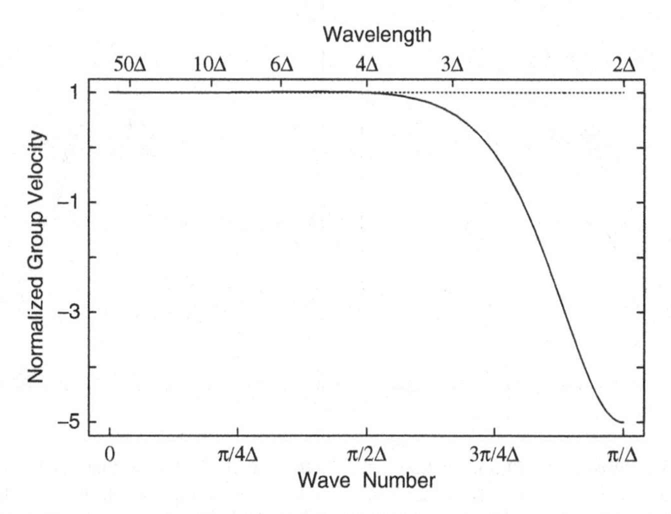

Fig. 6.9 Normalized group velocity c_g^*/c for the QFEM eigenmodes as a function of k

1998). The advantage of the QFEM method seems to be attributable to its low phase-speed error and relative freedom from numerical dispersion. The phase-speed error in the downstream-moving (physical) component of the QFEM solution is very small, both for relatively short waves, such as a $3\Delta x$ wave, and in the limit of good spatial resolution, for which

$$\frac{c_{\mathrm{p}}^*}{c} \approx 1 + \frac{(k\Delta x)^4}{270}.$$

The most serious inaccuracies in QFEM approximations to (6.11) are associated with the nonphysical upstream-propagating mode, whose amplitude is at least 10% that of the physical mode for all wavelengths shorter than $4\Delta x$. Moreover, the amplitude of the nonphysical mode decays rather slowly with increasing numerical resolution. In the limit $k\Delta x \to 0$,

$$\beta \approx \frac{(k\Delta x)^2}{24}.$$

How is the difference in the eigenmode structure to linear and quadratic finite-element approximations to (6.11) made manifest in practical applications? One example is provided by the comparison of finite-difference, linear finite-element, and quadratic finite-element solutions to (6.11) shown in Fig. 6.10. These solutions were obtained using trapezoidal time differencing with a very small Courant number ($c\Delta t/\Delta x = 1/16$), so essentially all the error is produced by the spatial discretization. Solutions were computed on the periodic domain $0 \leq x \leq 3$ subject to the initial condition

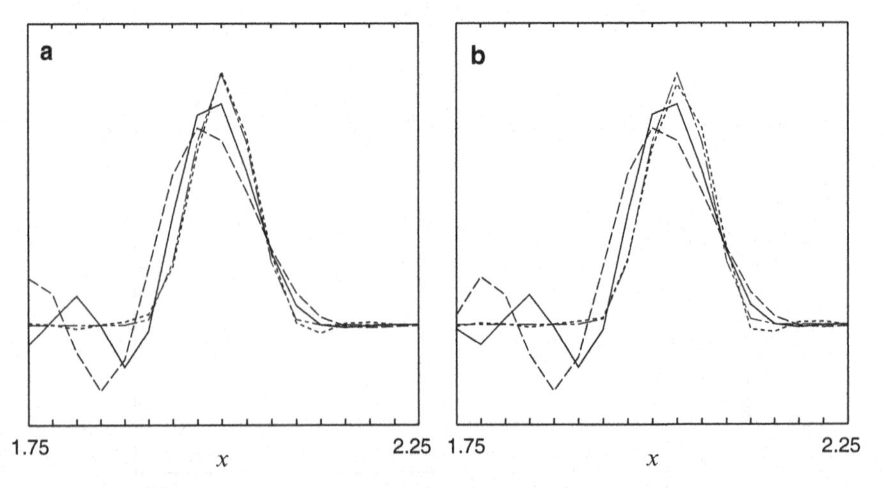

Fig. 6.10 Comparison of solutions to the constant-wind-speed advection equation at **a** $t = 10$ and **b** $t = 10\frac{5}{16}$: quadratic finite-element solution (*short-dashed line*), linear finite-element solution (*solid line*), fourth-order explicit finite-difference solution (*long-dashed* line), and exact solution (*thin dot-dashed line*)

$$\psi(x,0) = \begin{cases} \frac{1}{4}(\cos(8\pi(x-1)) + 1)^2, & \text{if } |x-1| \le \frac{1}{8}, \\ 0, & \text{otherwise}. \end{cases}$$

To facilitate the comparison with the finite-difference method, the nodal values were initialized by collocation, i.e., $a_j(0) = \psi(j\Delta x, 0)$. The horizontal mesh spacing is $\Delta x = 1/32$, implying that the total width of the initial spike is $8\Delta x$, which is sufficiently narrow to reveal short-wavelength errors without allowing the solution to be completely dominated by $2\Delta x$ disturbances. The wind speed is $c = 0.1$. The solution at $t = 10$ is shown in Fig. 6.10a, at which time the peak in the true solution is centered at $x = 2$. Only the central portion of the total domain is shown in Fig. 6.10. For simplicity, the QFEM solution is plotted as a piecewise-linear function between the nodes. The superiority of the QFEM solution over the linear finite-element solution is clearly evident. The linear finite-element solution is, nevertheless, substantially better than the solution obtained with explicit fourth-order finite differences.

The nature of the amplitude error in the QFEM solution can be seen by comparing the plots in Fig. 6.10. The exact solution propagates exactly one grid interval between the times shown in the plots in Fig. 6.10. There are essentially no changes in the shapes of the linear finite-element and the finite-difference solutions over this short period of time, but the QFEM solution is damped noticeably. This damping is followed by reamplification as the solution translates another Δx, and the amplitude of the peak in the QFEM solution continues to oscillate as it moves alternatively over the midpoint and endpoint nodes. Nevertheless, even when the QFEM solution looks its worst, it is still much better than the solutions generated by the other schemes. Although the amplitude of the nonphysical component of each QFEM eigenmode remains small in this linear constant-coefficient test problem, there is no guarantee that it will not be amplified by wave–wave interactions and contribute to aliasing error in nonlinear problems. For example, Cullen (1982) reported that, in those regions where the solution is smooth, quadratic elements give worse results than linear elements in finite-element-method approximations to the inviscid Burgers equation.

The results shown in Fig. 6.10 are consistent with the comparison of the phase speeds for each scheme plotted in Fig. 6.11. The phase speeds of the $3\Delta x$ or $4\Delta x$ waves are captured much better by the QFEM than by the linear finite-element method or fourth-order finite differencing. Yet the QFEM is probably not an optimal choice for this problem. Also plotted in Fig. 6.11 are the phase speeds produced when the spatial derivative in the constant-wind-speed advection equation is approximated using Lele's tridiagonal compact finite-difference formula (3.53). The compact scheme exhibits essentially the same accuracy as the QFEM for the poorly resolved waves, but in many respects the compact scheme is superior because it has no nonphysical mode, its truncation error is $O\left[(\Delta x)^4\right]$, and it more easily implemented as an efficient tridiagonal implicit system. On the other hand, the QFEM does benefit from the excellent conservation and stability properties of all Galerkin methods (see Sect. 6.2.3).

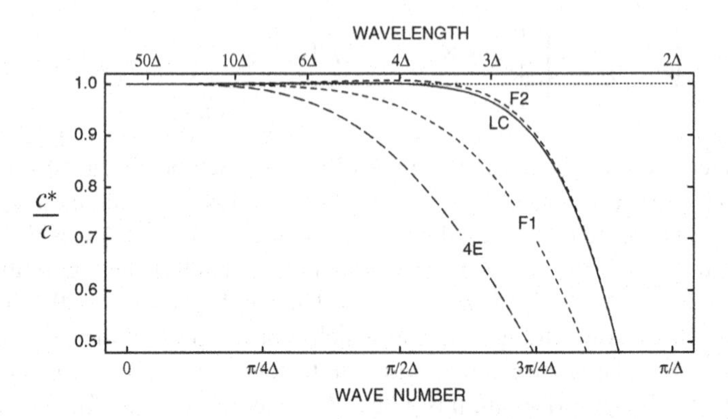

Fig. 6.11 Phase-speed error as a function of spatial resolution for linear finite-elements (*F1*) and quadratic finite-elements: physical mode (*F2*), explicit centered fourth-order differences (*4E*), and Lele's fourth-order tridiagonal compact scheme (*LC*)

6.5.4 Two-Dimensional Expansion Functions

The construction of finite-element approximations to problems in two or more spatial dimensions is straightforward. In the following we will briefly consider the two-dimensional case. The simplest two-dimensional expansion functions are nonzero only within some rectangular region. One of the simplest types of interpolation that can be performed on a rectangular mesh is bilinear interpolation in which the function is estimated as

$$C_1 + C_2 x + C_3 y + C_4 xy.$$

The four coefficients $C_1, \ldots, C_4$ can be uniquely determined within each rectangle by the function values at the four vertices. Bilinear interpolation reduces to linear interpolation along lines parallel to the x or y coordinate axes. Individual expansion functions for bilinear interpolation, sometimes known as "pagoda" functions, may be expressed as the product of a chapeau function with respect to x times a second chapeau function with respect to y. Each pagoda function is unity at a central node and drops to zero at the eight surrounding nodes.

If the two-dimensional constant-wind-speed advection equation

$$\frac{\partial \psi}{\partial t} + U\frac{\partial \psi}{\partial x} + V\frac{\partial \psi}{\partial y} = 0$$

is approximated using the pagoda-function finite-element method, evaluation of (6.96) yields the following system of ordinary differential equations:

$$\mathscr{A}^x \mathscr{A}^y \frac{da_{i,j}}{dt} + U\mathscr{A}^y \delta_{2x} a_{i,j} + V\mathscr{A}^x \delta_{2y} a_{i,j} = 0, \tag{6.114}$$

where $a_{i,j}$ is the expansion coefficient of the (i,j)th pagoda function or equivalently, the approximate solution at the (i,j)th node, and $\mathscr{A}^x$ and $\mathscr{A}^y$ are the averaging operators

$$\mathscr{A}^x a_{i,j} = \frac{1}{6}(a_{i+1,j} + 4a_{i,j} + a_{i-1,j}),$$

$$\mathscr{A}^y a_{i,j} = \frac{1}{6}(a_{i,j+1} + 4a_{i,j} + a_{i,j-1}).$$

The mass matrix generated by the product of the averaging operators $\mathscr{A}^x\mathscr{A}^y$ couples the time tendencies at nine different nodes and is a band matrix with a very wide bandwidth. One technique, known as "mass lumping," that has been occasionally advocated to eliminate this implicit coupling diagonalizes the mass matrix via some essentially arbitrary procedure (such as summing the coefficients in each row of the mass matrix and assigning the result to the diagonal). As discussed by Gresho et al. (1978) and Donea et al. (1987), mass lumping degrades the accuracy of finite-element approximations to hyperbolic problems.

An efficient solution to (6.114) can, nevertheless, be obtained by organizing the computations as follows. First, evaluate the spatial derivatives at every nodal point by solving the family of tridiagonal systems

$$\mathscr{A}^x \left(\frac{\partial \phi}{\partial x}\right)_{i,j} = \delta_{2x} a_{i,j} \quad \text{and} \quad \mathscr{A}^y \left(\frac{\partial \phi}{\partial y}\right)_{i,j} = \delta_{2y} a_{i,j} \tag{6.115}$$

for $(\partial\phi/\partial x)_{i,j}$ and $(\partial\phi/\partial y)_{i,j}$. Then the time tendency at each nodal point is given by the uncoupled equations

$$\frac{da_{i,j}}{dt} + U\left(\frac{\partial \phi}{\partial x}\right)_{i,j} + V\left(\frac{\partial \phi}{\partial y}\right)_{i,j} = 0. \tag{6.116}$$

Note that the time derivative in (6.116) must be approximated using explicit time differencing to avoid implicit algebraic equations in the fully discretized approximation. The solution algorithm given by equations (6.115) and (6.116) is exactly that which would be most naturally used to solve the two-dimensional advection equation using the compact finite-difference operator (3.49).

A variety of other two-dimensional expansion functions can also be defined, including higher-degree piecewise polynomials on a rectangular grid and piecewise-linear functions on a triangular grid. If the computational domain itself is rectangular, it appears that the most efficient schemes are obtained using rectangular elements (Staniforth 1987). On the other hand, if the computational domain is highly irregular, it can be advantageous to approximate the solution using a network of triangular elements. This approach has been used when modeling tidal currents in bays (Lynch and Gray 1979).

When finite-element expansion functions are defined on triangular grids, several polynomial expressions in x and y must be integrated over triangular domains to evaluate the coefficients in the Galerkin approximation (6.96). The calculation of these coefficients is facilitated if each expansion function is defined with respect to the local coordinate system (ξ, η) illustrated in Fig. 6.12. Within each triangular element, the linearly interpolated expansion functions have the general form

$$\alpha\eta + \beta\xi + \gamma = 0,$$

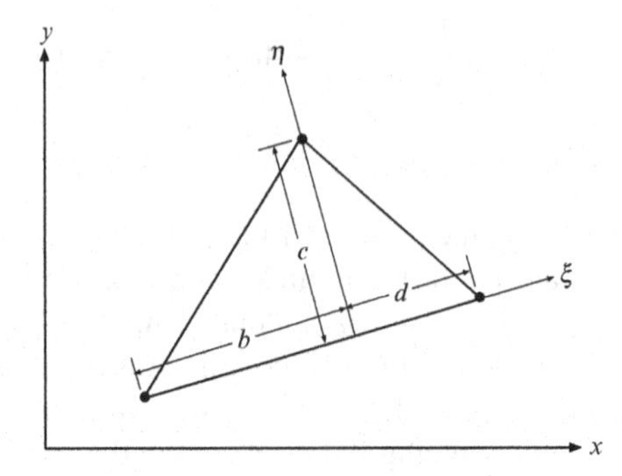

Fig. 6.12 Local-coordinate system for integrating polynomial expressions over a triangle

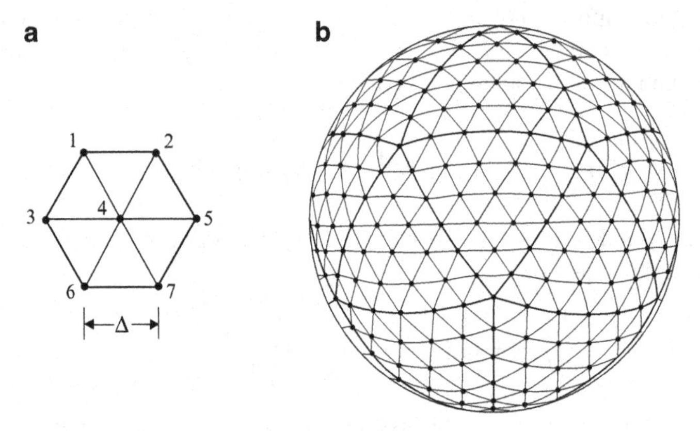

Fig. 6.13 **a** Hexagonal element formed from equilateral triangles. **b** Subdivision of a spherical icosahedron into an almost uniform triangular grid

where α, β, and γ are determined by the values at the vertices of the triangle. Polynomial expressions in ξ and η can be integrated over the triangular domain T using the helpful formula

$$\int\!\!\int_T \xi^r \eta^s \, \mathrm{d}\xi \, \mathrm{d}\eta = c^{s+1} \left(d^{\,r+1} - (-b)^{r+1} \right) \frac{r!s!}{(r+s+2)!},$$

where b, c, and d are the positive dimensions indicated in Fig. 6.12.

Suppose that solutions to the one-dimensional advection equation (6.11) are sought in a domain that has been divided into a uniform grid of equilateral triangles. Then every node not lying along the boundary is surrounded by six triangular elements whose union is a hexagon. These hexagons may be used to define finite-element expansion functions that are unity at the center of the hexagon and zero at each of the surrounding nodes. Let the nodes be numbered as shown in Fig. 6.13. Assume that the x-axis is parallel to the line segment connecting nodes 3, 4, and 5, and let Δ denote the distance between any pair of nodes.

After considerable algebra, one can show that (6.96) reduces to

$$\frac{1}{2}\left(\frac{da_1}{dt}+\frac{da_2}{dt}+\frac{da_3}{dt}+6\frac{da_4}{dt}+\frac{da_5}{dt}+\frac{da_6}{dt}+\frac{da_7}{dt}\right) \\ +\frac{c}{6\Delta}[(a_2-a_1)+2(a_5-a_3)+(a_7-a_6)]=0.$$

One simple nonrectangular domain that can be covered by a quasi-homogeneous lattice of equilateral triangles is the surface of a sphere. A perfectly uniform covering can be obtained using the 12 nodes that are the vertices of a regular icosahedron inscribed within the sphere. If there are more than 12 nodes, the coverage will not be perfectly uniform, but an approximately homogeneous distribution of triangles can be achieved as follows. Beginning with a regular icosahedron inscribed within the sphere, the edges of the icosahedron are projected along great-circle arcs to the surface of the sphere. The resulting spherical triangles are further subdivided into a large number of smaller, almost uniform, triangular elements as illustrated in Fig. 6.13b. All nodes on this mesh, except for the original 12 vertices of the icosahedron, are surrounded by six triangular elements whose union is a hexagon. The original 12 vertices of the icosahedron are surrounded by only five triangular elements, and at these special nodes the elements are pentagonal. The distance between adjacent nodes may vary by as much as 25% over the surface of the sphere, and is smallest in the vicinity of the vertices of the inscribed icosahedron. Williamson (1968) and Sadourny et al. (1968) provided additional details about the properties of geodesic spherical grids. Further discussion of triangular grids in global finite-element models is presented in Cullen (1974), Cullen and Hall (1979), and Priestley (1992).

6.6 The Discontinuous Galerkin Method

QFEM or higher-degree finite-element methods lead to relatively complex mass matrices which make the solution of (6.7) too inefficient in problems that are well suited for integration by explicit time differences. In such cases, a better way to obtain high-order accuracy is to partition the domain into a set of h relatively coarse-grained elements and to represent the structure within each element using a polynomial basis of degree p. Two closely related h–p methods are the spectral element method and the discontinuous Galerkin method.

The primary difference between these two approaches is that the Galerkin criterion is enforced globally in the spectral element method, which requires the integral in (6.7) to be computed over the entire domain, whereas the computations in the discontinuous Galerkin method are localized such that information only needs to be shared between neighboring elements. The discontinuous Galerkin method is, therefore, well adapted to massively parallel computer architectures. Another advantage of the discontinuous Galerkin approach is that it yields approximations to conservation laws that are both locally (over each element) and globally conservative, whereas spectral element methods, which typically provide approximations to advective form operators, are only globally conservative. On the other hand, an advantage of the spectral element method is that one can easily require continuity of the solution across element boundaries, which allows it to be more naturally generalized to treat diffusion and other higher-than-first-order differential operators. In the following we will focus on the discontinuous Galerkin method, because of its suitability for the simulation of low-viscosity flow on highly parallel computer architectures. Karniadakis and Sherwin (2005) provided an excellent and thorough discussion of the spectral element method.

Consider the scalar conservation law

$$\frac{\partial \psi}{\partial t} + \frac{\partial}{\partial x} F(\psi) = 0 \tag{6.117}$$

subject to the initial condition $\psi(x, t_0) = g(x)$. Suppose the global domain is divided into nonoverlapping elements such that the jth element has width Δx_j and covers the subdomain $S_j = [x_j - \Delta x_j/2, x_j + \Delta x_j/2]$. Let the approximate solution over the jth element be denoted as $\tilde{\phi}_j(x)$ and expressed as the sum of a finite set of expansion functions $\varphi_k(x)$, such that

$$\tilde{\phi}_j(x,t) = \sum_{n=0}^{N} a_n(t)\varphi_n(x), \quad x \in S_j. \tag{6.118}$$

Enforcing the Galerkin criteria (6.4) separately in each element requires that, for all k,

$$\int_{S_j} \frac{\partial \tilde{\phi}_j}{\partial t} \varphi_k \,\mathrm{d}x = -\int_{S_j} \frac{\partial F(\tilde{\phi}_j)}{\partial x} \varphi_k \,\mathrm{d}x. \tag{6.119}$$

Integrating the right side by parts, the preceding expression becomes

$$\int_{S_j} \frac{\partial \tilde{\phi}_j}{\partial t} \varphi_k \,\mathrm{d}x = -\int_{S_j} \frac{\partial}{\partial x} \left[F(\tilde{\phi}_j)\varphi_k \right] \mathrm{d}x + \int_{S_j} F(\tilde{\phi}_j) \frac{d\varphi_k}{dx} \,\mathrm{d}x. \tag{6.120}$$

It is convenient to map the subdomain S_j to the interval $[-1, 1]$ using the transformation

$$\xi = \frac{2(x - x_j)}{\Delta x_j},$$

in which case

$$\mathrm{d}x = \frac{\Delta x_j}{2} \mathrm{d}\xi, \quad \text{and} \quad \frac{\partial}{\partial x} = \frac{2}{\Delta x_j} \frac{\partial}{\partial \xi}.$$

Using (6.118) and this transformation, (6.120) becomes

$$\sum_{n=0}^{N} M_{n,k} \frac{da_n}{dt} = \int_{-1}^{1} F(\tilde{\phi}_j) \frac{d\varphi_k}{d\xi} \,\mathrm{d}\xi - F(\tilde{\phi}_j(1))\varphi_k(1) + F(\tilde{\phi}_j(-1))\varphi_k(-1), \tag{6.121}$$

where the components of the mass matrix are

$$M_{n,k} = \frac{\Delta x_j}{2} \int_{-1}^{1} \varphi_n \varphi_k \, \mathrm{d}\xi, \quad 0 \le n, k \le N. \tag{6.122}$$

One advantage of (6.121) over (6.119) is that it only requires derivatives of the expansion functions, which can generally be evaluated analytically, whereas the derivative of $F(\tilde{\phi}_j)$ is needed for (6.119). In addition, (6.121) provides a convenient way to couple adjacent elements through the specification of fluxes across the element boundaries.

The exact fluxes in (6.121) are replaced by approximate fluxes $\hat{F}$ that depend only on the numerical solution just to the right and left of each element boundary, so the approximation to $F[\tilde{\phi}_j(1)]$ takes the form $\hat{F}[\tilde{\phi}_j(1), \tilde{\phi}_{j+1}(-1)]$. As in finite-volume methods, the numerical flux must be consistent in the sense that for any constant ψ_0, $F(\psi_0) = \hat{F}(\psi_0, \psi_0)$. One possible choice for $\hat{F}$ is the Godunov flux (5.53), which in the current context may be expressed as

$$\hat{F}(r,s) = \begin{cases} \min\limits_{r \le \chi \le s} F(\chi) & \text{if } r \le s, \\ \max\limits_{s \le \chi \le r} F(\chi) & \text{otherwise}. \end{cases} \tag{6.123}$$

Several alternative formulations for $\hat{F}$ are given in Cockburn and Shu (2001). For problems with continuous solutions, the sensitivity to the specification of the fluxes at the element boundaries decreases significantly as the degree of the polynomial approximation within each element increases, because as p increases any jumps at the element boundaries become very small.

6.6.1 Modal Implementation

Discontinuous Galerkin methods may be implemented using either a *modal* or a *nodal* representation of the polynomial expansion functions within each element. The modal approach, reviewed in more detail by Cockburn and Shu (2001), is considered in this section, followed by a discussion of the nodal approach in Sect. 6.6.2. The most common family of polynomials to choose for the φ_k in the modal discontinuous Galerkin method is the set of Legendre polynomials of degree zero through N, which, as noted in connection with (6.67), are orthogonal over the interval $[-1, 1]$. Substituting (6.67) into (6.122) yields

$$M_{n,k} = \frac{\Delta x_j}{2n+1} \delta_{nk}.$$

The first five Legendre polynomials,

$$1, \quad \xi, \quad (3\xi^2 - 1)/2, \quad (5\xi^3 - 3\xi)/2, \quad (35\xi^4 - 30\xi^2 + 3)/8,$$

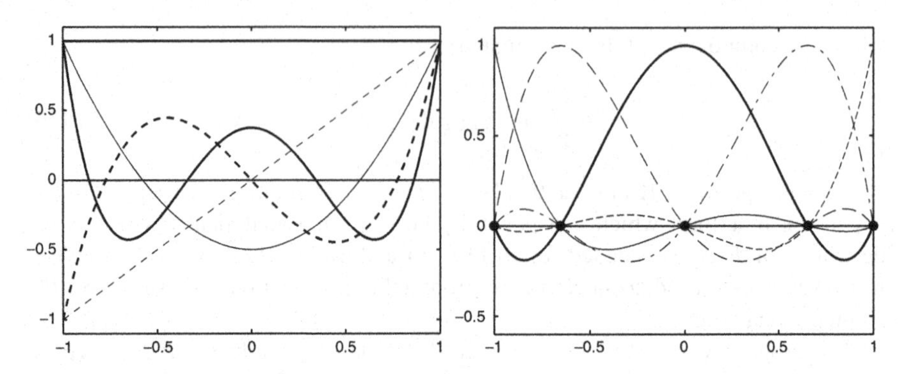

Fig. 6.14 **a** Modal basis consisting of the Legendre polynomials P_0 through P_4. **b** Lagrange interpolating polynomials of degree 4 and the Gauss–Legendre–Lobatto nodes for a five-point nodal expansion

are plotted in Fig. 6.14a. Higher-degree Legendre polynomials can be obtained from the recursion relation

$$(k+1)P_{k+1}(\xi) = (2k+1)\xi P_k(\xi) - kP_{k-1}(\xi).$$

Even-degree P_k are symmetric; odd degrees are asymmetric, i.e.,

$$P_k(-\xi) = (-1)^k P_k(\xi). \tag{6.124}$$

At the right endpoint, $P_k(1) = 1$ and $P_k'(1) = k(k+1)/2$; these may be used in conjunction with (6.124) to obtain the corresponding values at the left endpoint. The derivative in the interior of the interval may be computed using

$$\frac{\xi^2 - 1}{k} P_k'(\xi) = \xi P_k(\xi) - P_{k-1}(\xi).$$

When the φ_k are Legendre polynomials, the evolution equation for the kth expansion coefficient (6.121) reduces to

$$\frac{da_k}{dt} = \frac{2k+1}{\Delta x_j} \left[\int_{-1}^{1} F(\tilde{\phi}_j) \frac{dP_k}{d\xi} \, d\xi - \hat{F}(\tilde{\phi}_j, \tilde{\phi}_{j+1}) + (-1)^k \hat{F}(\tilde{\phi}_{j-1}, \tilde{\phi}_j) \right]. \tag{6.125}$$

In practical applications, the integral in (6.125) is evaluated by quadrature. As discussed in Sect. 6.4.3.1, if the integrand is a polynomial and L nodal points are used to approximate the integral, one can obtain a result that is exact for polynomials through degree $2L-1$ by choosing the nodes to be the L zeros of P_L. (When the nodes are zeros of a Legendre polynomial, the numerical integration is referred to as Gaussian quadrature or, more specifically, Gauss–Legendre quadrature.) If F is a linear function of $\tilde{\phi}_j$ and the polynomial expansion is truncated at degree N, the integrand in (6.125) will consist of polynomials of degree less than or equal to $2N-1$, so the integral can be evaluated exactly using Gauss–Legendre quadrature with N nodes. The weight at each node is given by (6.70), where the location of the jth zero may be determined numerically using (6.71) as a first guess. On the other hand, if F is a function of $\tilde{\phi}_j^2$ (or in more general systems, contains binary products of the unknown variables), the integrand contains polynomials up to degree $3N - 1$, and $3N/2$ nodes are required to ensure the integral is exact.

The initial conditions are projected onto the expansion functions in each interval using the standard Galerkin requirement that any initial error must be orthogonal to each of the expansion functions. If $g_j(\xi)$ is the initial condition on S_j mapped to the interval $[-1, 1]$, the Galerkin requirement is

$$\int_{-1}^{1} \left[\sum_{n=0}^{N} a_n(0) P_n - g_j \right] P_k \frac{\Delta x_j}{2} \, \mathrm{d}\xi = 0, \quad 0 \le k \le N,$$

or

$$a_k(0) = \frac{2k+1}{2} \int_{-1}^{1} g_j P_k \, \mathrm{d}\xi, \quad 0 \le k \le N. \tag{6.126}$$

The preceding integral may be approximated using Gauss–Legendre quadrature, although the number of nodes in the quadrature should exceed N.[11]

6.6.2 Nodal Implementation

Suppose $N+1$ nodes $\xi_0, \xi_1, \dots \xi_N$ are spread over the interval $[-1, 1]$. The ends of the interval are included in the set of nodes to facilitate the evaluation of the fluxes; therefore, $\xi_0 = -1$ and $\xi_N = 1$. The remaining nodes are chosen to maximize the accuracy with which the integrals in (6.121) and (6.122) can be approximated by quadrature. The set of nodes yielding exact results for polynomial integrands of the highest possible degree consists of the zeros of the Jacobi polynomial $P_{N-1}^{1,1}(\xi)$. Jacobi polynomials of degree m and n satisfy the orthogonality relation

$$\int_{-1}^{1} (1-\xi)^{\alpha} (1+\xi)^{\beta} P_m^{\alpha,\beta}(\xi) P_n^{\alpha,\beta}(\xi) \, \mathrm{d}\xi = 0, \quad \text{if} \quad m \ne n. \tag{6.127}$$

After setting $\alpha = \beta = 1$ in (6.127), one can use a derivation directly analogous to that presented for Gauss–Legendre quadrature in Sect. 6.4.3 to demonstrate that quadratures based on the zeros of $P_{N-1}^{1,1}(\xi)$ yield exact results for polynomials through degree $2N - 1$ (see Problem 18). A numerical integration in which the function is evaluated at the endpoints ± 1 and at the zeros of a Jacobi polynomial is known as Gauss–Legendre–Lobatto (GLL) quadrature.

[11] If L nodes are used in a Gauss–Legendre-quadrature approximation to (6.126) and $L \le N$, $a_L(0)$ will evaluate to zero regardless of the actual functional form of $g_j(\xi)$.

In the nodal discontinuous Galerkin method, the expansion functions within each element are the Nth-degree Lagrange polynomials interpolating the $N+1$ GLL nodes,

$$\varphi_k(\xi) = \prod_{\substack{n=0 \\ n \neq k}}^{N} \frac{(\xi - \xi_n)}{(\xi_k - \xi_n)}. \tag{6.128}$$

Note that φ_k has a value of zero at every node except at ξ_k, where its value is unity. If the integral in (6.122) is evaluated by GLL quadrature at the nodal points $\xi_0, \xi_1, \ldots \xi_N$, the approximate mass matrix $\widetilde{\mathbf{M}}$ will be diagonal, since

$$\widetilde{\mathbf{M}}_{n,k} = \sum_{i=0}^{N} w_i \varphi_n(\xi_i)\varphi_k(\xi_i) = \sum_{i=0}^{N} w_i \delta_{in}\delta_{ik} = w_n \delta_{nk}, \tag{6.129}$$

where the w_n are weights satisfying

$$w_n = \int_{-1}^{1} \varphi_n(\xi)\, \mathrm{d}\xi.$$

For expansion functions of the form (6.128), the integrand in (6.122) is a polynomial of degree $2N$. Thus, although the quadrature (6.129) is indeed approximate, an exact result could be obtained using GLL quadrature with just one more nodal point. One might therefore suppose that the exact mass matrix is well approximated by $\widetilde{\mathbf{M}}$, but this is not so. As an example, consider the case $N = 4$, for which

$$\mathbf{M} = \begin{pmatrix} 0.0889 & 0.0259 & -0.0296 & 0.0259 & -0.0111 \\ 0.0259 & 0.4840 & 0.0691 & -0.0605 & 0.0259 \\ -0.0296 & 0.0691 & 0.6321 & 0.0691 & -0.0296 \\ 0.0259 & -0.0605 & 0.0691 & 0.4840 & 0.0259 \\ -0.0111 & 0.0259 & -0.0296 & 0.0259 & 0.0889 \end{pmatrix}$$

and

$$\widetilde{\mathbf{M}} = \begin{pmatrix} 0.1000 & 0 & 0 & 0 & 0 \\ 0 & 0.5444 & 0 & 0 & 0 \\ 0 & 0 & 0.7111 & 0 & 0 \\ 0 & 0 & 0 & 0.5444 & 0 \\ 0 & 0 & 0 & 0 & 0.1000 \end{pmatrix}.$$

Computational efficiency is sometimes increased in finite-element methods through *mass lumping*, in which the actual mass matrix is replaced with a diagonal matrix by summing entries across each row of the original matrix and placing that sum on the diagonal of the lumped matrix. Using the preceding example, one may verify that $\widetilde{\mathbf{M}}$ is obtained if the exact mass matrix $\mathbf{M}$ is "lumped." The same result holds for arbitrary N (Karniadakis and Sherwin 2005, p. 57). In contrast to the modal discontinuous Galerkin method, in which the mass matrix is diagonal because the expansion functions within each element are truly orthogonal, in the nodal discontinuous Galerkin approach the mass matrix is effectively diagonalized by mass lumping.

The derivatives of the Legendre polynomials are related to the Jacobi polynomials such that

$$P_N'(\xi) = \frac{1}{2}(N+1)P_{N-1}^{1,1}(\xi), \tag{6.130}$$

so the GLL nodes are also the zeros of $(1-\xi)(1+\xi)P_N'(\xi)$. The zeros of $P_N'(\xi)$ are not known analytically, but may be found numerically using the zeros of the Chebyshev polynomial of degree $N-1$ as a first guess, i.e., by using (6.71) with $m = N-1$. The individual expansion functions may be expressed in the form

$$\varphi_k(\xi) = \begin{cases} 1, & \text{if } \xi = \xi_k, \\ \dfrac{(\xi-1)(\xi+1)P_N'(\xi)}{N(N+1)P_N(\xi_k)(\xi_k-\xi)}, & \text{otherwise} \end{cases} \tag{6.131}$$

(see Problem 19). The five GLL nodes and expansion functions for the case $N = 4$ are plotted in Fig. 6.14b.

Evaluating the integral in (6.121) using GLL quadrature requires the values of $\mathrm{d}\varphi_k/\mathrm{d}\xi$ at the nodal points, which are

$$D_{k,n} \equiv \frac{d\varphi_k}{d\xi}(\xi_n) = \begin{cases} -N(N+1)/4, & \text{if } k = n = 0, \\ 0, & \text{if } k = n \neq 0, N, \\ N(N+1)/4, & \text{if } k = n = N, \\ \dfrac{P_N(\xi_n)}{P_N(\xi_k)(\xi_n-\xi_k)}, & \text{if } k \neq n\,. \end{cases}$$

Recalling that $\varphi_m(\xi_n) = \delta_{mn}$, the nodal implementation of (6.121) reduces to

$$w_k \frac{\Delta x_j}{2}\frac{da_k}{dt} = \sum_{n=0}^{N} F[\tilde{\phi}_j(\xi_n)]D_{k,n}w_n - \hat{F}(\tilde{\phi}_j, \tilde{\phi}_{j+1})a_N\delta_{kN} + \hat{F}(\tilde{\phi}_{j-1}, \tilde{\phi}_j)a_0\delta_{0k},$$

where the exact fluxes F have been replaced by the previously discussed numerical fluxes $\hat{F}$, and the GLL quadrature weights are

$$w_k = \frac{2}{N(N+1)[P_N(\xi_k)]^2}, \qquad k = 0, 1, \ldots, N.$$

Both $D_{k,n}$ and w_k are independent of the solution and may be evaluated once and stored for subsequent use during the integration.

As in the modal approach, the initial values $a_k(0)$ are determined by requiring the initial error to be orthogonal to every expansion function. Again letting $g_j(\xi)$ be the initial condition on S_j mapped to the interval $[-1, 1]$, this requires

$$\int_{-1}^{1}\left[\sum_{n=0}^{N} a_n(0)\varphi_n - g_j\right]\varphi_k \frac{\Delta x_j}{2}\,\mathrm{d}\xi = 0, \quad 0 \le k \le N.$$

Approximating this integral by quadrature on the $N+1$ GLL nodes and again recalling that $\varphi_m(\xi_n) = \delta_{mn}$, one finds $a_k(0) = g_j(\xi_k)$.

6.6.3 An Example: Advection

Suppose $F(\psi) = c\psi$, so (6.117) reduces to the one-dimensional scalar advection equation, and that the wind speed c is constant and nonnegative. Then the Godunov flux (6.123) reduces to the upstream flux $\hat{F}[\tilde{\phi}_j(1), \tilde{\phi}_{j+1}(-1)] = c\tilde{\phi}_j(1)$. As a test case, consider the two-wave problem first introduced in connection with Fig. 3.7. The domain $0 \le x \le 1$ is periodic, $c = 1$, and

$$\psi(x,0) = \sin(6\pi x) + \sin(8\pi x).$$

In contrast to the earlier example, where the solution was only integrated until it translated across 40% of the domain, we will examine solutions at time 10, when the waves have propagated around the domain ten times. The third-order strong-stability-preserving Runge–Kutta (SSPRK) method (2.48) is used to integrate the solution forward in time. In all cases the time step was chosen so that $c\Delta t/\Delta\tilde{x} = 0.1$, where for the modal discontinuous Galerkin simulations $\Delta\tilde{x} = \Delta x/N$ is the element width divided by the highest degree retained in the polynomial expansion, and for the nodal discontinuous Galerkin simulations $\Delta\tilde{x} = \Delta x_{\min}$ is the smallest internode spacing. This relatively small Courant number was chosen in an attempt to reduce the size of the time-differencing error in comparison with those errors introduced by the spatial discretization.

Solutions obtained using the nodal discontinuous Galerkin method are shown in Fig. 6.15.

As apparent in Fig. 6.15a, the solution generated using eight elements with four nodes per element is a reasonable, but not highly accurate approximation to the correct solution. If the number of elements is doubled, while reducing the number of nodes per element to three (which increases the total number of nodes from 32 to 48), the solution, shown in Fig. 6.15b, becomes slightly less accurate and also takes longer to compute (L_2 errors and execution times for these simulations are given in Fig. 6.16). In contrast, if the number of elements remains fixed at eight and the number of nodes is increased to five (giving a total of 40 nodes), one obtains the much more accurate result shown in Fig. 6.15c. This is a basic property of discontinuous Galerkin methods: as the number of nodes per element is increased, the rate of convergence to a smooth solution is faster than any fixed power of the effective grid spacing. These discontinuous Galerkin results may be compared with a basic centered fourth-order solution integrated using the three-stage third-order SSPRK method. The accuracy of the finite-difference solution is less than that of the eight-element, five-node discontinuous Galerkin solution, but is better than those in Fig. 6.15a and b.

Further illustration of the efficiency of discontinuous Galerkin methods with a relatively low number of elements and many nodes is shown in Fig. 6.16, in which the L_2 error in the two-wave problem is plotted as a function of the execution time required to obtain solutions at $t = 10$. Clearly the most efficient way to reduce the error is to use relatively few elements and more nodes. In all cases plotted in Fig. 6.16, the time stepping is computed using the three-stage SSPRK scheme with $c\Delta t/\Delta x_{\min} = 0.1$, a strategy similar to the actual practice in many geophysical

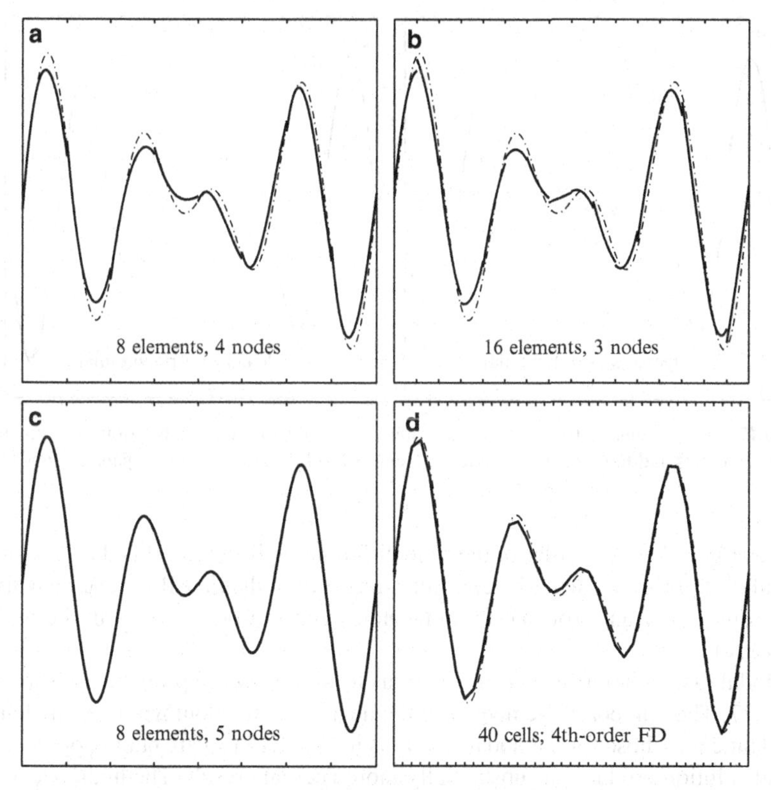

Fig. 6.15 Discontinuous Galerkin nodal solutions to the two-wave advection problem using **a** eight elements and four nodes, **b** 16 elements and three nodes, and **c** eight elements and five nodes. A 40-grid-cell solution obtained using centered fourth-order finite differences is shown in **d**

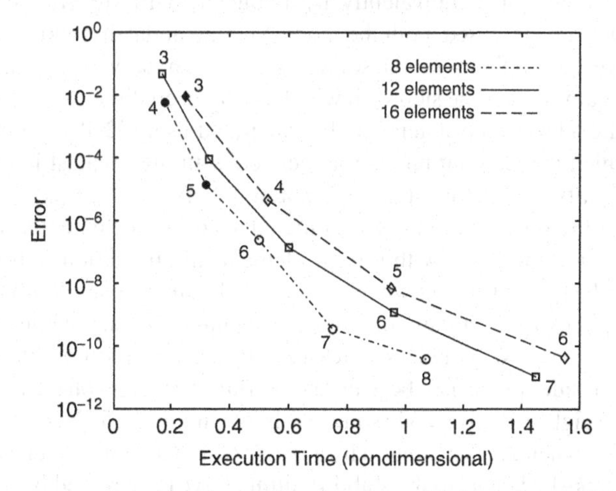

Fig. 6.16 L_2 error as a function of execution time for nodal discontinuous Galerkin methods using eight (*dashed-dotted line*), 12 (*solid line*), and 16 (*dashed line*) elements. The number of nodes is indicated by points along each line. The *three solid points* indicate the cases shown in Fig. 6.15a–c

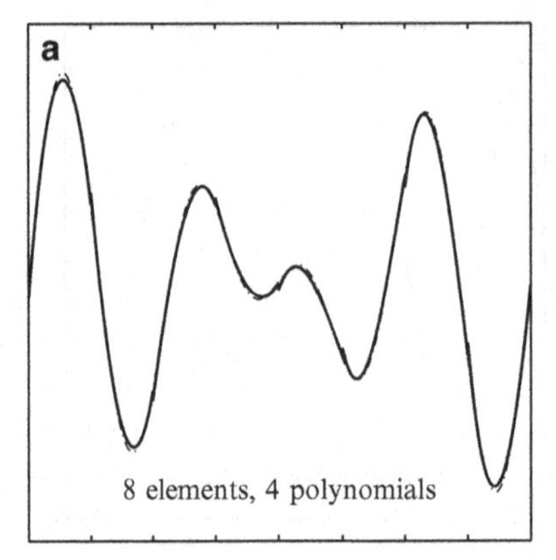

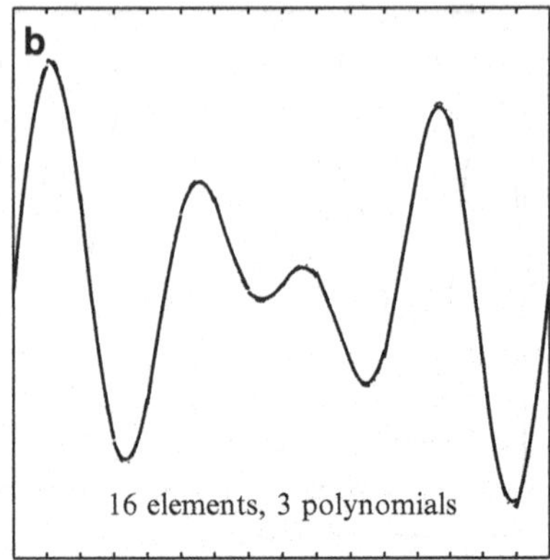

Fig. 6.17 Discontinuous Galerkin modal solutions to the two-wave advection problem using **a** eight elements and four expansion functions, and **b** 16 elements and three expansion functions

applications where the order of the time differencing is not as refined as that of the spatial discretization. Nevertheless, since the order of the time differencing is fixed, it does introduce an error that becomes nontrivial in those cases with the highest number of nodes.

Modal discontinuous Galerkin solutions to the two-wave problem are shown in Fig. 6.17. The number of elements and the degrees of freedom within each element are identical to those for the nodal solutions in Fig. 6.15a and b, but as apparent, the modal solution produces a substantially more accurate result. The higher accuracy of the modal approach likely arises because its mass matrix (6.122) is evaluated exactly, whereas the mass matrix for the nodal approach is approximated as diagonal through mass lumping (or equivalently, by a nonexact GLL quadrature). Nevertheless, as h and p are increased, both the modal and the nodal discontinuous Galerkin methods converge rapidly to smooth solutions with essentially exponential accuracy.

The maximum stable time step with which nodal discontinuous Galerkin approximations to the advection problem can be integrated using (2.48) can be estimated as that for which the Courant number associated with the smallest internode spacing becomes unity, i.e., Δt must approximately satisfy $c\Delta t/\Delta x_{\mathrm{min}} \leq 1$. Since the distribution of the nodes clusters toward the edges of each element as the order of the approximation increases within each element, the limitation imposed by this condition can be quite stringent when N is large. Hesthaven and Warburton (2008) provided an extensive discussion of ways to maintain accuracy while relaxing this stability requirement. Nevertheless, at least in this particular test problem, accuracy, not stability considerations may be a more important constraint on the time step. The L_2 error in the eight-cell, five-node solution shown in Fig. 6.15c increases by a modest factor of 2.9 when Δt is increased by a factor of 4, but jumps by a factor of 135 if $c\Delta t/\Delta x_{\mathrm{min}}$ is pushed close to the stability limit as Δt is increased by a factor of 9.

Returning to the influence of node clustering on the maximum stable time step, note that in Fig. 6.14b, the distribution of the nodes is only modestly skewed toward the edges of the element. If Δx_{avg} is the average internode spacing, the stability condition for $N = 5$ becomes $c\Delta t/\Delta x_{\text{avg}} < 0.67$, which may be compared with the condition $c\Delta t/\Delta x < 0.73$ for a leapfrog-time, centered-fourth-order-space finite difference approximation to the advection equation.

A simple estimate of the maximum stable time step for the modal discontinuous Galerkin method is available when both the order and the number of stages of the SSPRK scheme are $N + 1$ and the degree of the highest polynomial expansion function is N. In this case Δt should be chosen so that

$$C_{\max} \equiv \max_j \left(\frac{c\Delta t}{\Delta x_j} \right) \leq \frac{1}{2N + 1}$$

(Cockburn and Shu 2001). Since the region of stability expands as more stages are included in the SSPRK method, the preceding expression will overestimate the region of stability if the maximum degree of the polynomials in the expansion is increased while the time differencing is left unchanged. Yet in practice, one might integrate using the third-order SSPRK method regardless of the degree of the Legendre polynomial expansion. Table 6.4 gives the maximum Courant number for which stable integrations of modal discontinuous Galerkin approximations to the advection equation may be obtained using polynomials of degree N and the third-order SSPRK method (2.48). These empirically determined values of $C_{\max}$ agree with those given by Cockburn and Shu (2001).

If one were to construct a finite-difference scheme with the same number of degrees of freedom as in a modal discontinuous Galerkin method with polynomials up to degree N, the average grid spacing used for the finite differences would be $\Delta x_{\text{avg}} = \Delta x/N$. Courant numbers computed with respect to Δx_{avg} (instead of the entire width of the element) may be used to more easily compare the time-step limitations of modal discontinuous Galerkin methods with those for finite-difference schemes. Maximum values of $c\Delta t/\Delta x_{\text{avg}}$ are therefore listed in the last row of Table 6.4.

The maximum stable time step with which nodal discontinuous Galerkin approximations to the advection equation can be integrated is larger than that for the modal approach. For example, using the third-order SSPRK method (2.48) to integrate discontinuous Galerkin approximations to the advection equation with five degrees of freedom per element, empirically determined stability criteria require $NC_{\max}$ to be bounded by 0.35 and 0.67 for the modal and nodal methods, respectively. For comparison, the maximum stable Courant number with which the centered-fourth-order finite difference can be integrated with the same third-order SSPRK scheme is 1.26.

Table 6.4 Maximum Courant numbers for which third-order three-stage strong-stability-preserving Runge–Kutta integrations of the advection equation are stable for modal discontinuous Galerkin polynomial expansions of degree N

	2	3	4	5	6	7	8
$C_{\max}$	0.209	0.130	0.089	0.066	0.051	0.040	0.033
$NC_{\max}$	0.418	0.390	0.356	0.330	0.306	0.280	0.264

The discontinuous Galerkin method has the flexibility to be applied in several contexts beyond the simple example considered here. The fluxes at the cell boundaries and the polynomial degree within each element in modal discontinuous Galerkin methods may be limited in a manner similar to that in finite-volume methods to obtain schemes that are total variation diminishing or otherwise well behaved in the presence of steep gradients (Cockburn and Shu 2001). Indeed, the discontinuous Galerkin method can be considered a generalization of finite-volume methods to include piecewise-polynomial representations of arbitrarily high degree within each cell. If the element widths are adaptively narrowed in the vicinity of steep gradients, the discontinuous Galerkin method can provide high-order approximations in smooth portions of the flow, together with sharp nonovershooting approximations in regions of steep gradients. Another powerful property of discontinuous Galerkin methods is that, like the finite-element method, they are well suited for problems in complex geometries. As a geophysical example, the shallow-water equations on the sphere were solved using a nodal discontinuous Galerkin method by Giraldo et al. (2002) and using a modal discontinuous Galerkin approach by Nair et al. (2005).

Problems

1. Show that for $n > 1$, a wavelength of $(n+1)\Delta x/n$ aliases into a wavelength of $-(n+1)\Delta x$ if it is sampled on a uniform mesh with a grid spacing of Δx. Sketch an example for $n = 2$ and explain how the change in sign of the wave number influences the relative phases of the correct and aliased waves near any mesh point where the wave amplitude is zero.

2. Suppose that the spectral method is used to integrate a system including the equation

$$\frac{\partial \phi}{\partial t} + \cdots + \phi\chi\psi = 0,$$

and that the term $\phi(x,t)\chi(x,t)\psi(x,t)$ is to be evaluated using the transform technique. If $2K+1$ modes are retained in the Fourier expansions for ϕ, χ, and ψ, derive an expression determining the minimum number of grid points that must be present on the physical-space grid to avoid aliasing error in the product $\phi\chi\psi$.

3. In all practical applications, finite Fourier transforms are computed using the FFT algorithm Cooley and Tukey (1965). Suppose that the periodic spatial domain $0 \le x \le 2\pi$ is discretized so that

$$x_j = \frac{2\pi}{M} j, \quad \text{where} \quad j = 1, \ldots, M.$$

To be efficient, the FFT algorithm requires that M be the product of small prime numbers. Maximum efficiency is obtained when M is a power of 2; thus, most FFT codes assume that M is an even number. When the total number of grid points on the physical mesh is even, the finite Fourier transform and inverse transform are given by the relations

$$a_n(t) = \frac{1}{2N} \sum_{j=1}^{2N} \phi(x_j, t) \mathrm{e}^{-inx_j}$$

and

$$\phi(x_j, t) = \sum_{k=-N+1}^{N} a_k(t) \mathrm{e}^{ikx_j}.$$

The $2N$ data points in physical space uniquely define $2N$ Fourier coefficients. However, in contrast to (6.14), the wave number $k = -N$ does not appear in the expansion. Explain why the $-N$ wave number is retained in finite Fourier transforms when there is an odd number of points on the physical mesh and is dropped when the total number of points is even.

4. Solutions to the two-dimensional advection equation

$$\frac{\partial \psi}{\partial t} + u \frac{\partial \psi}{\partial x} + v \frac{\partial \psi}{\partial y} = 0$$

are sought in a domain that is periodic in both x and y. The velocity field is nondivergent.

(a) Show that the domain integral of ψ^3 is conserved by the exact solution to the unapproximated governing equations.

(b) If the effects of time-differencing errors are neglected, is the Galerkin spectral method guaranteed to yield an approximate solution to this problem that conserves the domain integral of ψ^3? Explain your answer.

5. Solutions are sought to the equation

$$\frac{\partial \psi}{\partial t} = \frac{\partial}{\partial x} \left(\nu(x) \frac{\partial \psi}{\partial x} \right)$$

on the periodic domain $0 \leq x \leq 2\pi$ using a series-expansion method in which

$$\psi(x, t) = \sum_{|m| \leq N} r_m(t) \mathrm{e}^{imx}, \qquad \nu(x) = \sum_{|n| \leq N} s_n \mathrm{e}^{inx}.$$

(a) If the solution is to be obtained using a Galerkin spectral method, derive the ordinary differential equation for $\mathrm{d}r_m/\mathrm{d}t$.

(b) Write down a trapezoidal approximation to the system of ordinary differential equations derived in (a). How would the efficiency with which this trapezoidal approximation can be computed change if ν did not depend on x?

6. Present an algorithm for the solution of the equation described in Problem 5 using a pseudospectral method and second-order Adams–Bashforth time differencing. Do not assume that ν is independent of x.

7. Pseudospectral solutions to the constant-wind-speed advection equation are to be obtained using leapfrog time differencing such that

$$\frac{\phi_j^{n+1} - \phi_j^{n-1}}{2\Delta t} + c_0 \left(\frac{\partial \phi^n}{\partial x}\right)_j = 0.$$

Suppose that the usual formula for calculating the derivative,

$$\left(\frac{\partial \phi^n}{\partial x}\right)_j = \sum_{|k|\leq N} \mathrm{i}k a_k \mathrm{e}^{\mathrm{i}kx_j},$$

is replaced by the modified expression

$$\left(\frac{\partial \phi^n}{\partial x}\right)_j = \sum_{|k|\leq N} \mathrm{i}\left[\frac{\sin(kc_0\Delta t)}{c_0\Delta t}\right] a_k \mathrm{e}^{\mathrm{i}kx_j}.$$

(a) Determine the phase-speed error and the maximum stable time step for the modified scheme.

(b) What limits the practical utility of this otherwise attractive scheme?

8. Using (6.47), verify the orthogonality relation for the spherical harmonics (6.48). Also use the relation $P_{-m,n}(\mu) = (-1)^m P_{m,n}(\mu)$ to show that $Y^*_{-m,n} = (-1)^m Y_{m,n}$ and that the expansion coefficients for any approximation to a real-valued function satisfy $a_{-m,n} = (-1)^m a^*_{m,n}$.

9. Express the associated Legendre function $P_{4,4}(\mu)$ as an algebraic function of μ (thereby producing an expression similar to the expressions in Table 6.2).

10. Derive (6.25) by repeatedly integrating

$$a_k = \frac{1}{2\pi}\int_{-\pi}^{\pi} \psi(x)\mathrm{e}^{-ikx}\,\mathrm{d}x$$

by parts.

11. *Consider the family of functions periodic on the interval [0, 1]

$$\psi(x) = \begin{cases} \cos^n\left[2\pi\left(x - \frac{1}{2}\right)\right] & \text{if } \left|x - \frac{1}{2}\right| < \frac{1}{4}, \\ 0 & \text{otherwise.} \end{cases}$$

Evaluate the rates of convergence of the Fourier series expansions to this family of functions for the cases $n = 0$, 1, and 2. Compute Fourier series expansions truncated at progressively higher wavenumbers K, such that $K = 2^m\pi$, for $m = 3, 4, \ldots, 8$. Compute the error in the expansion using both the maximum norm over the entire interval and the maximum norm in the region $|x - 1/2| \le 1/8$. Evaluate these maximum norms using grid-point values on a mesh with $\Delta x = 1/1024$, and plot the logarithm of the error versus the logarithm of Δx. How do the rates of convergence of the Fourier approximation to these functions compare with the rates suggested in Sect. 6.2.1.4?

12. Show that the L_2 norm of the solution to the viscous Burgers equation (6.43) on the periodic domain $0 \le x \le 1$,

$$\|\psi\|_2 = \left[\int_0^1 \psi^2 \, dx\right]^{1/2},$$

is bounded by its value at the initial time.

13. *Use the spectral and pseudospectral methods to compute numerical solutions to the viscous Burgers equation (6.43) subject to the initial condition $\psi(x, 0) = \sin(2\pi x)$. Set $\nu = 0.002$. Approximate the time derivative using leapfrog time differencing for the advection term and forward differencing for the diffusion. Initialize the leapfrog scheme with a single forward time step. Use a time step such that

$$\frac{\Delta t}{\Delta x} \max_x(\psi(x, 0)) = 0.16.$$

(a) Use $\Delta x = 1/64$ and 64 Fourier modes (which yields a cutoff wave number of 64π on this spatial domain). Show the solutions at $t = 0.40$ on a scale $-4 \le \psi \le 4$. Which scheme performs better? How seriously does aliasing error affect the stability and accuracy of the pseudospectral solution?

(b) Repeat the preceding simulations using $\Delta x = 1/128$ and 128 Fourier modes. Show the solutions at $t = 0.40$ in the subdomain $0 \le x \le 1/2$, $0 \le \psi \le 1$. How seriously does aliasing error degrading the stability and accuracy of the pseudospectral solution?

(c) Why is there an improvement in the pseudospectral solution between the simulations in (a) and (b)?

(d) If the spatial resolution is increased to 256 Fourier modes, both the spectral and the pseudospectral solutions become unstable. Why? Devise a way around this instability and obtain an approximation to the solution at $t = 0.40$. Again plot this solution on the subdomain $0 \le x \le 1/2, 0 \le \psi \le 1$.

14. Solutions to the coupled advection/chemical reaction equations

$$\frac{\partial \phi}{\partial t} + c\frac{\partial \phi}{\partial x} = \phi\psi, \qquad \frac{\partial \psi}{\partial t} + c\frac{\partial \psi}{\partial x} = -\phi\psi$$

are to be obtained using the Galerkin finite-element approach. Assume that the expansion functions are chapeau functions and that the approximate expressions for ψ and ϕ are

$$\phi \approx \sum_m a_m \varphi_m, \quad \psi \approx \sum_n b_n \varphi_n.$$

Using (6.98), we know that the first equation will have the form

$$\frac{\Delta x}{6}\left(\frac{da_{j+1}}{dt}+4\frac{da_j}{dt}+\frac{da_{j-1}}{dt}\right)+c\left(\frac{a_{j+1}-a_{j-1}}{2}\right)=X,$$

where X represents the Galerkin approximation to $\phi\psi$ Evaluate X in terms of the expansion coefficients a_k and b_k.

15. Determine the Galerkin finite-element approximation to $\psi\partial\psi/\partial x$ using chapeau expansion functions. Show that the result is identical to the conservative finite-difference operator appearing in (4.109).

16. *Compute solutions to Problem 13 in Chap. 4 using the spectral and pseudospectral methods. Use the same numerical parameters specified in that problem except choose Δt so that the Courant number based on the maximum wind speed for the shortest wavelength retained in the spectral truncation is 0.3. Do not use any type of smoother. Use leapfrog time differencing, taking a single forward step to obtain the solution at the first time level.

(a) Obtain solutions using 64 Fourier modes to approximate ψ and $c(x)$. Show your results at $t = 1.5$ and 3.0 as directed in Problem 13 in Chap. 4. Also show the two solutions at some time when the pseudospectral method is clearly showing some aliasing error. (*Hint:* This only happens for a limited period of time during the integration.)

(b) Now retry the solution with 128 Fourier modes and compare your results with those obtained in (a) and, if available, with the finite-difference solutions computed for Problem 13 in Chap. 4.

17. *Compute chapeau-function finite-element-method solutions to Problem 16 using the previously specified numerical parameters. Try spatial resolutions of $\Delta x = 1/32$ and $1/64$. If a_j is the amplitude at the jth node, the Galerkin chapeau-function approximation to the variable-wind-speed advection equation is

$$\frac{d}{dt}\left(a_{j-1}+4a_j+a_{j+1}\right)+\left(c_{j-1}+2c_j\right)\left(\frac{a_j-a_{j-1}}{\Delta x}\right)$$
$$+\left(c_{j+1}+2c_j\right)\left(\frac{a_{j+1}-a_j}{\Delta x}\right)=0.$$

(a) Initialize the problem by setting a_j and c_j equal to the exact function values at the nodes.

(b) Initialize the problem by projecting the exact data onto the nodes using the Galerkin (or least-squares) formula (6.8).

18. Suppose $g(x)$ is a polynomial that is to be integrated over the interval $[-1, 1]$ using Gaussian quadrature at $N + 1$ nodes, $x_0, x_1, \ldots, x_N$. That is,

$$\int_{-1}^{1} g(x)\,\mathrm{d}x \approx \sum_{j=0}^{N} w_j g(x_j),$$

where

$$w_j = \int_{-1}^{1} p_j(x)\,\mathrm{d}x.$$

Here $p_j(x)$ is given by (6.64) and is the Lagrangian interpolating polynomial with unit amplitude at node x_j and zero amplitude at all other nodes. Suppose that the set of nodes includes the endpoints ± 1. Show that choosing the $N-1$ interior nodes to be the zeros of the Jacobi polynomial $P_{N-1}^{1,1}(x)$ will ensure that the quadrature is exact for all polynomials of degree less than or equal to $2N-1$.

19. Derive the expression for φ_k in (6.131).

(a) As a first step show that Lagrange interpolating polynomials may be alternatively expressed as

$$\prod_{\substack{n=0\\n\neq k}}^{N} \frac{(\xi - \xi_n)}{(\xi_k - \xi_n)} = \frac{g(\xi)}{g'(\xi_k)(\xi - \xi_k)},$$

where $g(\xi) = \prod_{n=0}^{N}(\xi - \xi_n)$.

(b) Now complete the derivation of (6.131) using the fact that Legendre polynomials are solutions to the differential equation

$$\frac{d}{dx}\left[(1-x^2)\frac{d}{dx}P_n(x)\right] + n(n+1)P_n(x) = 0.$$

Chapter 7
Semi-Lagrangian Methods

Most of the fundamental equations in fluid dynamics can be derived from first principles in either a *Lagrangian* form or an *Eulerian* form. Lagrangian equations describe the evolution of the flow that would be observed following the motion of an individual parcel of fluid. Eulerian equations describe the evolution that would be observed at a fixed point in space (or at least at a fixed point in a coordinate system such as the rotating Earth whose motion is independent of the fluid). If $S(x,t)$ represents the sources and sinks of a chemical tracer $\psi(x,t)$, the evolution of the tracer in a one-dimensional flow field may be alternatively expressed in Lagrangian form as

$$\frac{d\psi}{dt} = S, \tag{7.1}$$

or in Eulerian form as

$$\frac{\partial \psi}{\partial t} + u\frac{\partial \psi}{\partial x} = S.$$

The mathematical equivalence of these two equations follows from the definition of the total derivative,

$$\frac{d}{dt} = \frac{\partial}{\partial t} + \frac{dx}{dt}\frac{\partial}{\partial x},$$

and the definition of the velocity,

$$\frac{dx}{dt} = u. \tag{7.2}$$

One strategy for the solution of (7.1) as an initial-value problem would be to choose a regularly spaced distribution of fluid parcels at the initial time, assign a value for ψ to each of these fluid parcels from the initial condition, and then integrate the ordinary differential equations (7.1) and (7.2) to determine the location and the tracer concentration of each parcel as a function of time. The difficulty with this strategy is that in most practical applications the distribution of the fluid parcels eventually becomes highly nonuniform, and the numerical approximation of $\psi(x,t)$ becomes inaccurate in regions where the fluid parcels are widely separated. In theory, this situation can be improved by adding new parcels to those regions

D.R. Durran, *Numerical Methods for Fluid Dynamics: With Applications to Geophysics*, Texts in Applied Mathematics 32, DOI 10.1007/978-1-4419-6412-0_7,

where the initial parcels have become widely separated and removing parcels from regions where the parcels have become too concentrated. It is, however, difficult to create a simple algorithm for adding and removing fluid parcels in response to their evolving distribution within the fluid.

A much better scheme for regulating the number and distribution of the fluid parcels can be obtained by choosing a completely new set of parcels at every time step. The parcels making up this set are those arriving at each node on a regularly spaced grid at the end of each step. As noted by Wiin-Nielsen (1959), this approach, known as the *semi-Lagrangian method*, keeps the fluid parcels evenly distributed throughout the fluid and facilitates the computation of spatial derivatives via finite differences. As an illustration of this approach, let $t^n = n\Delta t$ and $x_j = j\Delta x$; then a semi-Lagrangian approximation to (7.1) can be written using the trapezoidal scheme

$$\frac{\phi\left(x_j, t^{n+1}\right) - \phi\left(\tilde{x}_j^n, t^n\right)}{\Delta t} = \frac{1}{2}\left[S\left(x_j, t^{n+1}\right) + S\left(\tilde{x}_j^n, t^n\right)\right], \tag{7.3}$$

where ϕ is the numerical approximation to ψ, and $\tilde{x}_j^n$ is the estimated x coordinate of the departure point of the trajectory originating at time t^n and arriving at (x_j, t^{n+1}). The value of $\tilde{x}_j^n$ is computed by numerically integrating (7.2) backward over a time interval of Δt starting from the initial condition $x(t^{n+1}) = x_j$. Then, since the endpoint of the backward trajectory is unlikely to coincide with a grid point, $\phi(\tilde{x}_j^n, t^n)$ and $S(\tilde{x}_j^n, t^n)$ must be obtained by interpolation.

Semi-Lagrangian methods are of considerable practical interest because in some applications they are more efficient than competing Eulerian schemes. Another advantage of the semi-Lagrangian approach is that it is easy to use in problems with nonuniform grids. In addition, semi-Lagrangian schemes avoid the primary source of nonlinear instability in most geophysical wave-propagation problems because the nonlinear advection terms appearing in the Eulerian form of the momentum equations are eliminated when those equations are expressed in a Lagrangian frame of reference.

As a result of the pioneering work by Robert (1981, 1982), semi-Lagrangian semi-implicit methods have become one of the most popular architectures used in global weather forecast models. An extensive review of the application of semi-Lagrangian methods to atmospheric problems was provided by Staniforth and Côté (1991). Semi-Lagrangian methods are also used in a variety of other fluid-dynamical applications, where they are sometimes referred to as *Lagrangian–Eulerian* or *Eulerian–Lagrangian* methods (e.g., Hirt et al. 1974; Oliveira and Baptista 1995).

7.1 The Scalar Advection Equation

The stability and accuracy of Eulerian finite-difference methods were first examined in Chap. 2 by studying the constant-wind-speed advection equation. We will begin the analysis of semi-Lagrangian schemes by considering the same problem, and then investigate the additional considerations that arise when variations in the velocity field make the backward trajectory calculation nontrivial.

7.1.1 Constant Velocity

A semi-Lagrangian approximation to the advection equation for a passive tracer can be written in the form

$$\frac{\phi(x_j, t^{n+1}) - \phi(\tilde{x}_j^n, t^n)}{\Delta t} = 0, \tag{7.4}$$

where $\tilde{x}_j^n$ again denotes the departure point of a trajectory originating at time t^n and arriving at (x_j, t^{n+1}). If the velocity is constant, the backward trajectory computation is trivial, and letting U denote the wind speed,

$$\tilde{x}_j^n = x_j - U\Delta t.$$

Let p be the integer part of $U\Delta t/\Delta x$ and without loss of generality suppose that $U \geq 0$; then $\tilde{x}_j^n$ lies in the interval $x_{j-p-1} \leq x < x_{j-p}$, as shown in Fig. 7.1. Defining

$$\alpha = \frac{x_{j-p} - \tilde{x}_j^n}{\Delta x}$$

and approximating $\phi(\tilde{x}_j^n, t^n)$ by linear interpolation, (7.4) becomes

$$\phi_j^{n+1} = (1-\alpha)\phi_{j-p}^n + \alpha\phi_{j-p-1}^n, \tag{7.5}$$

where $\phi_j^n = \phi(x_j, t^n)$. Note that if Δt is small enough that

$$0 \leq U\frac{\Delta t}{\Delta x} \leq 1,$$

(7.5) reduces to the formula for Eulerian upstream differencing.

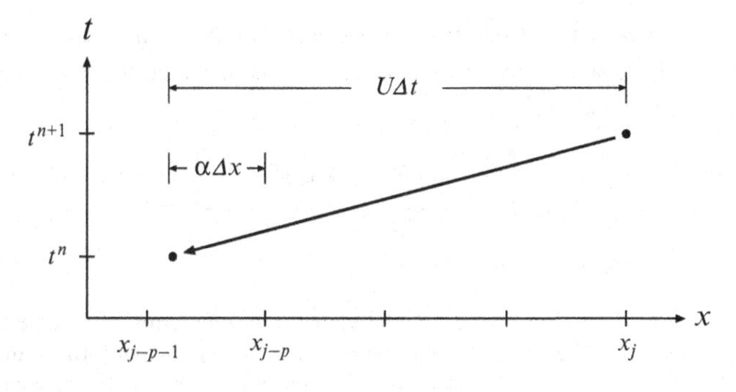

Fig. 7.1 Backward trajectory from (x_j, t^{n+1}) to $(\tilde{x}_j^n, t^n)$

7.1.1.1 Stability

Following Bates and McDonald (1982), the stability of the preceding semi-Lagrangian approximation can be analyzed using von Neumann's method. Substituting a solution of the form $\phi_j^n = A_k^n e^{i(kj\Delta x)}$ into (7.5), one obtains

$$A_k = \left[1 - \alpha(1 - e^{-ik\Delta x})\right] e^{-ikp\Delta x},$$

from which it follows that

$$|A_k|^2 = 1 - 2\alpha(1-\alpha)(1 - \cos k\Delta x).$$

This expression is the same as that obtained for the amplification factor associated with upstream differencing, except that α has replaced the Courant number in (3.23). On the basis of the analysis given in Sect. 3.2.2, the amplification factor for all waves resolved on the numerical mesh will be less than or equal to unity, provided that

$$0 \le \alpha \le 1,$$

which is always satisfied because the estimated departure point always lies between grid points x_{j-p} and x_{j-p-1}. It is possible to take arbitrarily large time steps without violating the Courant–Friedrichs–Lewy (CFL) condition because the backward trajectory calculation ensures that the numerical domain of dependence includes the domain of dependence of the true solution.

Errors will be made in the backward trajectory calculation in practical applications where the wind speed is not constant. These errors will affect the accuracy of the solution, and if they do not go to zero as $\Delta x \to 0$ and $\Delta t \to 0$, they may prevent the numerical solution from converging to the correct solution. Nevertheless, as long as the interpolation is performed using data from the two grid points surrounding the estimated departure point, the maximum norm of the solution will not grow with time.

7.1.1.2 Accuracy

The truncation error of the preceding semi-Lagrangian scheme can be determined by substituting appropriate Taylor series expansions of the continuous solution into the finite-difference scheme[1]

$$\frac{\phi_j^{n+1} - \left[(1-\alpha)\phi_{j-p}^n + \alpha\phi_{j-p-1}^n\right]}{\Delta t} = 0. \tag{7.6}$$

[1] According to the discussion in Sect. 2.2.3, the global truncation error is of same the order as the leading-order errors in the numerical approximation to the differential form of the governing equation (7.6) and is one power of Δt lower than the truncation error in the integrated form (7.5).

In the case of the unforced scalar advection equation, it is helpful to perform the Taylor series expansions about the point $(\tilde{x}_j^n, t^n)$ because this isolates the errors in the trajectory calculations from those generated by the interpolation of the tracer field. Let $\psi_d = \psi(\tilde{x}_j^n, t^n)$; then the error produced by linear interpolation is

$$(1-\alpha)\psi_{j-p}^n + \alpha\psi_{j-p-1}^n = \psi_d + \alpha(1-\alpha)\frac{(\Delta x)^2}{2}\frac{\partial^2\psi}{\partial x^2}\bigg|_d + O\left[(\Delta x)^3\right]. \quad (7.7)$$

Since the wind speed is constant, the backward trajectory is exact and $\psi_j^{n+1} = \psi_d$. This can be verified by expanding ψ_j^{n+1} in a Taylor series. Defining $s = x_j - \tilde{x}_j^n$,

$$\begin{aligned}\psi_j^{n+1} &= \psi_d + \Delta t \frac{\partial\psi}{\partial t}\bigg|_d + s\frac{\partial\psi}{\partial x}\bigg|_d \\ &\quad + \frac{(\Delta t)^2}{2}\frac{\partial^2\psi}{\partial t^2}\bigg|_d + s\Delta t\frac{\partial^2\psi}{\partial t\,\partial x}\bigg|_d + \frac{s^2}{2}\frac{\partial^2\psi}{\partial x^2}\bigg|_d + \cdots, \quad (7.8)\end{aligned}$$

and since $s = U\Delta t$, the preceding equation reduces to

$$\begin{aligned}\psi_j^{n+1} &= \psi_d + \Delta t\left(\frac{\partial}{\partial t} + U\frac{\partial}{\partial x}\right)\psi\bigg|_d + \frac{(\Delta t)^2}{2}\left(\frac{\partial}{\partial t} + U\frac{\partial}{\partial x}\right)^2\psi\bigg|_d + \cdots \\ &= \psi_d. \quad (7.9)\end{aligned}$$

Substituting (7.7) and (7.9) into (7.6) yields

$$\frac{\psi_j^{n+1} - \left[(1-\alpha)\psi_{j-p}^n + \alpha\psi_{j-p-1}^n\right]}{\Delta t} \approx -\frac{\alpha(1-\alpha)}{2}\frac{(\Delta x)^2}{\Delta t}\frac{\partial^2\psi}{\partial x^2}\bigg|_d. \quad (7.10)$$

If the Courant number is held constant as $\Delta t \to 0$ and $\Delta x \to 0$, the truncation error is clearly $O(\Delta x)$. If $\Delta x/\Delta t \to 0$ as $\Delta x \to 0$ and $\Delta t \to 0$, the error is no larger than $O(\Delta x)$. It may appear that the semi-Lagrangian scheme could be inconsistent in the limit $\Delta t/\Delta x \to 0$, but once the Courant number drops below unity, $\alpha = U\Delta t/\Delta x$, and using

$$\frac{\partial^2\psi}{\partial t^2} = U^2\frac{\partial^2\psi}{\partial x^2},$$

the leading-order truncation error reduces to

$$\frac{\Delta t}{2}\frac{\partial^2\psi}{\partial t^2} - U\frac{\Delta x}{2}\frac{\partial^2\psi}{\partial x^2}.$$

This is identical to the leading-order truncation error in Eulerian upstream differencing (3.9). The global truncation error of the semi-Lagrangian scheme (7.5) is therefore of first order in space and time.

Consistent with (7.10), the preceding semi-Lagrangian scheme is exact whenever the Courant number is an integer, i.e., whenever the departure point exactly coincides with a grid point. This may be compared with Eulerian upstream differencing, which is exact for a Courant number of unity (see Sects. 3.4.1, 3.4.2). In practical applications the wind speed and therefore the Courant number are functions of space and time. Stability constraints require the Eulerian upstream method to be integrated using a time step that ensures that the maximum Courant number will be less than 1 at every point within the computational domain. As a consequence of this restriction on Δt, the domain-averaged Courant number is often substantially less than the optimal value of unity. In contrast, the unconditional stability of the semi-Lagrangian scheme allows the time step to be chosen such that the average value of the Courant number is unity – or any integer – thereby reducing the average truncation error throughout the computational domain.

7.1.1.3 Higher-Order Interpolation

Upstream differencing generates too much numerical diffusion to be useful in practical computations involving Eulerian problems with smooth solutions. A similar situation holds in the Lagrangian framework, where linearly interpolating the tracer field also generates too much diffusion. Higher-order interpolation is therefore used in most semi-Lagrangian approximations to equations with smooth solutions. If x_{j-p} is the nearest grid point to the estimated departure point and a quadratic polynomial is fit to the three closest grid-point values of the tracer field,

$$\phi(\tilde{x}_j^n, t^n) = \frac{\alpha}{2}(1+\alpha)\phi_{j-p-1}^n + (1-\alpha^2)\phi_{j-p}^n - \frac{\alpha}{2}(1-\alpha)\phi_{j-p+1}^n, \tag{7.11}$$

where, as before, $(p+\alpha)\Delta x = U\Delta t$, except that p is now chosen such that $|\alpha| \leq 1/2$. Substituting the preceding equation into (7.4) yields a semi-Lagrangian scheme that approximates the constant wind-speed advection equation to $O\left[(\Delta x)^3/\Delta t\right]$, which gives second-order accuracy. In the limit $\Delta t/\Delta x \to 0$ this scheme is identical to the Lax–Wendroff method (3.72).

Cubic interpolation is widely used in practical applications. If p is the integer part of $U\Delta t/\Delta x$ with $U > 0$ and the cubic is defined to match ϕ at the four closest grid-point values to the departure point, then

$$\begin{aligned}\phi(\tilde{x}_j^n, t^n) = &-\frac{\alpha(1-\alpha^2)}{6}\phi_{j-p-2}^n + \frac{\alpha(1+\alpha)(2-\alpha)}{2}\phi_{j-p-1}^n \\ &+ \frac{(1-\alpha^2)(2-\alpha)}{2}\phi_{j-p}^n - \frac{\alpha(1-\alpha)(2-\alpha)}{6}\phi_{j-p+1}^n.\end{aligned} \tag{7.12}$$

The preceding equation is expressed in the form of a Lagrange interpolating polynomial and is an efficient choice if several fields are to be interpolated to the same departure point, since the coefficients of the ϕ_i need not be recalculated for each field. If only one field is being interpolated, (7.12) can be evaluated more efficiently by writing it as a Newton polynomial (see Dahlquist and Björck 1974 and Problem 2).

The leading-order truncation error in a cubic semi-Lagrangian approximation to the constant-wind-speed advection equation is third order in the perturbations (specifically, $O\left[(\Delta x)^4/\Delta t\right]$). More generally, the local error produced by pth-order polynomial interpolation is $O\left[(\Delta x)^{p+1}\right]$, and the global truncation error in the corresponding semi-Lagrangian approximation to the constant-wind-speed problem is $O\left[(\Delta x)^{p+1}/\Delta t\right]$ (McDonald 1984). In most applications, the motivation for using high-order polynomials to interpolate the tracer field is not to accelerate the convergence of the numerical solution to the correct solution as $\Delta x \to 0$ and $\Delta t \to 0$, but rather to improve the accuracy of the marginally resolved waves. This is the same reason that high-order finite differences are often used to approximate the spatial derivative in Eulerian schemes for the solution of the advection equation.

7.1.1.4 Two Spatial Dimensions

A semi-Lagrangian approximation to the advection equation in a two-dimensional flow may be expressed as

$$\frac{\phi(x_{m,n}, y_{m,n}, t^{s+1}) - \phi(\tilde{x}^s_{m,n}, \tilde{y}^s_{m,n}, t^s)}{\Delta t} = 0, \tag{7.13}$$

where $(\tilde{x}^s_{m,n}, \tilde{y}^s_{m,n})$ denotes the departure point of a trajectory originating at time t^s and arriving at $(x_{m,n}, y_{m,n}, t^{s+1})$. Let U and V denote the x and y velocity components and, as before, suppose they are constant and nonnegative. Denote the integer parts of $U\Delta t/\Delta x$ and $V\Delta t/\Delta y$ by p and q, respectively, and define $\alpha = U\Delta t/\Delta x - p$ and $\beta = V\Delta t/\Delta y - q$. Then a first-order approximation to $\phi(\tilde{x}^s_{m,n}, \tilde{y}^s_{m,n}, t^s)$ can be obtained using bilinear interpolation, which yields the semi-Lagrangian scheme

$$\begin{aligned}\phi^{s+1}_{m,n} &= (1-\alpha)\left[(1-\beta)\phi^s_{m-p,n-q} + \beta\phi^s_{m-p,n-q-1}\right] \\ &\quad + \alpha\left[(1-\beta)\phi^s_{m-p-1,n-q} + \beta\phi^s_{m-p-1,n-q-1}\right].\end{aligned}$$

When the Courant numbers along the x- and y-axes are less than unity, $p = q = 0$, and the preceding expression reduces to the corner transport upstream method (4.35).

A second-order approximation can be obtained using biquadratic interpolation. To abbreviate the notation, let $\phi_C = \phi_{m-p,n-q}$ and denote the surrounding points using the compass directions (north, northeast, east, etc.) such that $\phi_N = \phi_{m-p,n-q+1}$, $\phi_{NE} = \phi_{m-p+1,n-q+1}$, etc. If p and q are now chosen such that $|\alpha| \le 1/2$ and $|\beta| \le 1/2$, the resulting semi-Lagrangian scheme has the form

$$\begin{aligned}\phi^{s+1}_{n,m} &= \frac{\alpha}{2}(1+\alpha)\left[\frac{\beta}{2}(1+\beta)\phi^s_{SW} + (1-\beta^2)\phi^s_W - \frac{\beta}{2}(1-\beta)\phi^s_{NW}\right] \\ &\quad + (1-\alpha^2)\left[\frac{\beta}{2}(1+\beta)\phi^s_S + (1-\beta^2)\phi^s_C - \frac{\beta}{2}(1-\beta)\phi^s_N\right] \\ &\quad - \frac{\alpha}{2}(1-\alpha)\left[\frac{\beta}{2}(1+\beta)\phi^s_{SE} + (1-\beta^2)\phi^s_E - \frac{\beta}{2}(1-\beta)\phi^s_{NE}\right].\end{aligned}$$

If the Courant numbers along the x- and y-axes are less than unity, this scheme reduces to the upstream biased Lax–Wendroff method (4.38). Von Neumann stability analysis can be used to show that the preceding bilinear and biquadratic semi-Lagrangian schemes are unconditionally stable (Bates and McDonald 1982).

The preceding interpolation formula generalizes to higher-order polynomials and three dimensions in a straightforward way. Since the evaluation of a three-dimensional high-order interpolating polynomial requires considerable computation, the exact formulae are sometimes approximated. Ritchie et al. (1995), for example, simplified the full expression for three-dimensional cubic interpolation by neglecting the "corner" points.

A more efficient way to use high-order polynomials in multidimensional problems is to use "cascade interpolation" (Leslie and Purser 1995; Nair et al. 1999). In the two-dimensional case, the values at the departure points are obtained by performing two one-dimensional interpolations: once using values parallel to one of the coordinate axes, and once along curves representing the upstream Lagrangian displacement of the grid lines parallel to the other coordinate axis. An important advantage of cascade interpolation is that it facilitates the use of standard limiters in connection with each one-dimensional interpolation to preserve positivity or prevent the development of overshoots and undershoots (Nair et al. 1999; Zerroukat et al. 2005).

The basic interpolation sequence is illustrated in Fig. 7.2, which shows a portion of the rectilinear computational grid and the upstream position of that same grid one time step earlier (the deformed mesh). The situation in Fig. 7.2 is one where

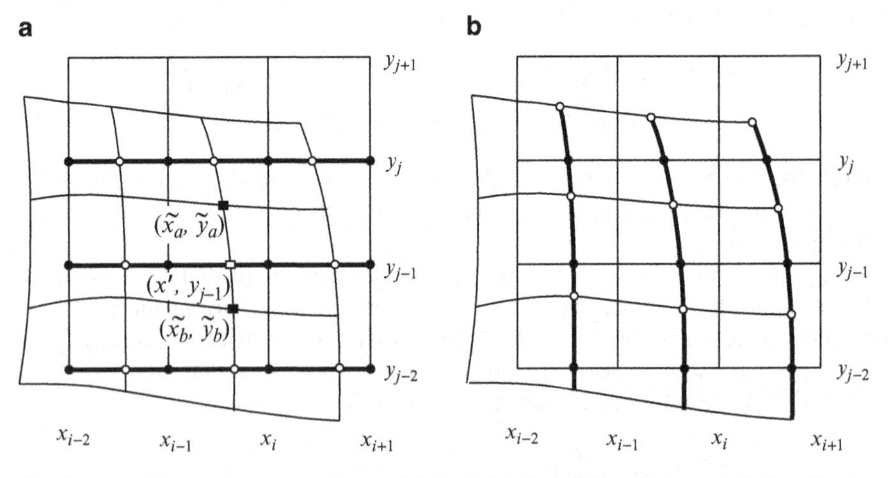

Fig. 7.2 Semi-Lagrangian cascade interpolation. The lattice of departure points defines the deformed mesh; the data are available each time step on the rectilinear grid. **a** First set of interpolations using data at the points indicated by the *solid circles* along the *heavy horizontal lines* to estimate ϕ at the locations of the *open circles*. (and *open square*). **b** Second set of interpolations using data at the points denoted by *solid circles* along the *heavy quasi-vertical curves* to estimate ϕ at the locations of the *open circles*

a nonuniform flow is generally toward the "northeast" and the Courant numbers parallel to each coordinate are less than unity, but identical formulae apply if the lattice of departure points actually translates across several grid cells in a single time step. The values of ϕ at the vertices forming the lattice of departure points (the open circles in Fig. 7.2b) are estimated from the values of ϕ at the vertices on the rectilinear grid (the solid circles in Fig. 7.2a) as follows.

The first set of interpolations runs along lines parallel to the x-axis and provides estimates of ϕ at the open circles (and the open square) in Fig. 7.2a. These open circles are at the intersections of the fixed grid lines parallel to the x-axis and curves whose downstream translation over time Δt will bring them precisely over the fixed grid lines paralleling the y-axis. As a preliminary step, the x coordinate of each of these intersections must be determined. Let $(\tilde{x}_a, \tilde{y}_a)$ and $(\tilde{x}_b, \tilde{y}_b)$ be the coordinates for the pair of departure points indicated by the solid squares in Fig. 7.2a, as determined by back trajectory calculations, and let (x', y_{j-1}) be the intersection of the curve connecting these two departure points with the line $y = y_{j-1}$ (indicated by the open square). In most applications, sufficient accuracy can be obtained using linear interpolation to evaluate x', in which case

$$x' = \tilde{x}_b + \frac{y_j - \tilde{y}_b}{\tilde{y}_a - \tilde{y}_b} (\tilde{x}_a - \tilde{x}_b). \tag{7.14}$$

Once the x coordinates of the points indicated by the open circles have been computed, the values of ϕ at each of these points can be estimated by one-dimensional interpolation using data from the original rectilinear grid along each heavy black line in Fig. 7.2a. After this first set of interpolations, estimates of ϕ are available at the black circles along each of the heavy curves in Fig. 7.2b, and the ϕ values at the departure points (open circles) are be obtained by a second set of interpolations along those curves. This second set of interpolations can be conducted using the distance along each black curve as the independent variable. The distances along each curve are evaluated using linear interpolation via formulae similar to (7.14); see Nair et al. (1999) for details.

7.1.2 Variable Velocity

Now consider the case where the velocity is a function of space and time, as would be necessary to produce the deformation of the mesh shown in Fig. 7.2. The backward trajectory of each fluid parcel must then be estimated by a nontrivial numerical integration. The truncation error in the variable-velocity case can again be determined by expanding $\psi(x, t)$ in a Taylor series about the estimated departure point and evaluating an expression of the form

$$\frac{1}{\Delta t}\left(\psi_j^{n+1} - \psi_d\right) + \frac{1}{\Delta t}\left(\psi_d - \sum_{k=-r}^{s} \beta_k \psi_{j-p+k}^n\right), \tag{7.15}$$

where $\psi_d = \psi(\tilde{x}_j^n, t^n)$ and the summation represents an $(r+s)$-order polynomial interpolation of ψ^n to the departure point. The first term in the preceding expression is determined by the error in the trajectory calculation, and the second term is determined by the error in the interpolation of ψ^n to the estimated departure point. The error generated in interpolating ψ to the estimated departure point is the same as that for the constant-velocity case, but the estimated departure point will not generally coincide with the true departure point, and as a consequence, the ψ_j^{n+1} will no longer be identical to ψ_d. To determine the difference between ψ_j^{n+1} and ψ_d, let x^n denote the position of a fluid parcel at time t^n and suppose that backward trajectories are computed subject to the initial condition $x^{n+1} = x_j$.

First suppose the trajectory is computed using Euler's method

$$\tilde{x}_j^n = x^{n+1} - u(x^{n+1}, t^n)\Delta t.$$

As before, define $s = x_j - \tilde{x}_j^n = x^{n+1} - \tilde{x}_j^n$. Then

$$s = u(x^{n+1}, t^n)\Delta t \tag{7.16}$$

$$= \Delta t \left[u(\tilde{x}_j^n, t^n) + s\frac{\partial u}{\partial x}(\tilde{x}_j^n, t^n) + O\left(s^2\right) \right] \tag{7.17}$$

$$= u_d \Delta t + O\left[(\Delta t)^2\right], \tag{7.18}$$

where $u_d = u(\tilde{x}_j^n, t^n)$ and the last equality is obtained by substituting (7.16) into (7.17). The difference between ψ_j^{n+1} and ψ_d is then determined by substituting (7.18) into (7.8) to obtain

$$\psi_j^{n+1} = \psi_d + \Delta t \left(\frac{\partial}{\partial t} + u\frac{\partial}{\partial x} \right) \psi \bigg|_d + O\left[(\Delta t)^2\right]$$
$$= \psi_d + O\left[(\Delta t)^2\right]. \tag{7.19}$$

According to (7.15), the contribution from the trajectory calculation to the global truncation error in the semi-Lagrangian scheme is $\left(\psi_j^{n+1} - \psi_d\right)/\Delta t$, so the error generated by the Euler method is $O(\Delta t)$.

A second-order-accurate result can be obtained using the two-stage trajectory calculation

$$x_* = x^{n+1} - u(x^{n+1}, t^n)\Delta t/2, \tag{7.20}$$

$$\tilde{x}_j^n = x^{n+1} - u(x_*, t^{n+\frac{1}{2}})\Delta t. \tag{7.21}$$

As before, the initial condition is $x^{n+1} = x_j$. This differs slightly from the classical Runge–Kutta midpoint method discussed in Sect. 2.3 in that the first stage uses the velocity $u(x^{n+1}, t^n)$ rather than $u(x^{n+1}, t^{n+1})$; the latter is more convenient if u^{n+1} is being predicted at the same time as ϕ^{n+1}. The second-order accuracy of this calculation can be verified as follows. Substitute (7.20) into (7.21) and let the superscript $n+1$ denote evaluation at $(x(t^{n+1}), t^{n+1})$. Then

$$\tilde{x}_j^n = x^{n+1} - \Delta t\, u\left(x^{n+1} - u(x^{n+1}, t^n)\Delta t/2,\ t^{n+\frac{1}{2}}\right) \tag{7.22}$$

$$= x^{n+1} - \Delta t\left[u^{n+1} - \frac{\Delta t}{2}\left(u\frac{\partial u}{\partial x}\right)^{n+1} - \frac{\Delta t}{2}\left(\frac{\partial u}{\partial t}\right)^{n+1} + O\left[(\Delta t)^2\right]\right]$$

$$= x^{n+1} - \Delta t\, u^{n+1} + \frac{(\Delta t)^2}{2}\left(\frac{du}{dt}\right)^{n+1} + O\left[(\Delta t)^3\right]$$

$$= x^{n+1} - \Delta t\left(\frac{dx}{dt}\right)^{n+1} + \frac{(\Delta t)^2}{2}\left(\frac{d^2x}{dt^2}\right)^{n+1} + O\left[(\Delta t)^3\right]. \tag{7.23}$$

Since the right side of (7.23) matches the Taylor series expansion of $\tilde{x}_j^n$ about $x(t^{n+1})$ to within an error of $O\left[(\Delta t)^3\right]$, the global truncation error in the back-trajectory calculation is $O\left[(\Delta t)^2\right]$ (Iserles 1996, p. 7).

Now consider the error generated when the Runge–Kutta scheme is used to compute the back trajectory in a semi-Lagrangian scheme. In many practical applications the velocity data are available only at discrete points on a space–time grid, and to evaluate (7.21), $u(x_*, t^{n+1/2})$ must be estimated by interpolation or extrapolation. Before examining the errors introduced by such interpolation and extrapolation, consider those cases where the velocity can be evaluated exactly, so the only errors arising in the trajectory calculations are those generated by the Runge–Kutta scheme itself. Using the definition $s = x_j - \tilde{x}_j^n$, (7.22) becomes

$$s = \Delta t\, u\Big(\tilde{x}_j + s - u(\tilde{x}_j + s, t^n)\Delta t/2,\ t^n + \Delta t/2\Big).$$

Thus, $s = u_d\Delta t + O\left[(\Delta t)^2\right]$, which may be substituted into the right side of the preceding equation to yield

$$s = u_d\Delta t + \frac{(\Delta t)^2}{2}\left(\frac{\partial}{\partial t} + u\frac{\partial}{\partial x}\right)u\bigg|_d + O\left[(\Delta t)^3\right]. \tag{7.24}$$

Substituting (7.24) into (7.8) gives

$$\psi_j^{n+1} = \psi_d + \Delta t\left(\frac{\partial}{\partial t} + u\frac{\partial}{\partial x}\right)\psi\bigg|_d + \frac{(\Delta t)^2}{2}\left(\frac{\partial}{\partial t} + u\frac{\partial}{\partial x}\right)^2\psi\bigg|_d + O\left[(\Delta t)^3\right],$$

which implies that the Runge–Kutta scheme (7.20) and (7.21) generates an $O\left[(\Delta t)^2\right]$ contribution to the total error in the semi-Lagrangian approximation.

Now suppose that the velocity data are available only at discrete locations on the space–time mesh. Ideally, the velocity at time $t^{n+1/2}$ would be computed by interpolation between times t^n and t^{n+1}. Such interpolation cannot, however, be performed when semi-Lagrangian methods are used to solve prognostic equations for the velocity itself, because the velocity at t^{n+1} will be needed for the trajectory calculations before it has been computed. This problem is generally avoided by extrapolating the velocity field forward in time using data from the two previous time levels such that

$$u(t^{n+\frac{1}{2}}) = \frac{3}{2}u(t^n) - \frac{1}{2}u(t^{n-1}). \tag{7.25}$$

Suppose that the extrapolated velocity field at $u(t^{n+1/2})$ is then linearly interpolated to x_* using data at the nearest spatial nodes, and let u_* denote this interpolated and extrapolated velocity. Since linear interpolation and extrapolation are second-order accurate,

$$u(x_*, t^{n+\frac{1}{2}}) = u_* + O\left[(\Delta x)^2\right] + O\left[(\Delta t)^2\right].$$

Substituting the preceding expression into (7.22) shows that the use of u_* instead of the exact velocity adds an $O\left[\Delta t(\Delta x)^2)\right] + O\left[(\Delta t)^3\right]$ error to the back-trajectory calculation, and thereby contributes a term of $O\left[(\Delta x)^2\right] + O\left[(\Delta t)^2\right]$ to the global truncation error in the semi-Lagrangian solution. A fully second order semi-Lagrangian scheme can therefore be obtained by (a) using (7.20) to estimate the midpoint of the back trajectory, (b) computing u_* by linearly interpolating and extrapolating the velocity field, (c) determining the departure point from

$$\tilde{x}_j^n = x^{n+1} - u_* \Delta t,$$

(d) evaluating $\phi(\tilde{x}_j^n, t^n)$ using quadratic interpolation, and (e) setting ϕ_j^{n+1} to this value.

A variety of other schemes have been proposed to compute back trajectories. One popular scheme, the second-order implicit midpoint method,

$$\tilde{x}_j^n = x_j - u\left((x_j + \tilde{x}_j^n)/2, t^{n+\frac{1}{2}}\right) \Delta t, \tag{7.26}$$

is typically solved by iteration. The Runge–Kutta scheme (7.20) and (7.21) can be considered a two-step iterative approximation to (7.26), but even if the implicit midpoint method is iterated to convergence, its formal order of accuracy is no greater than that obtained using (7.20) and (7.21). A second alternative scheme can be used if the velocities are being calculated as prognostic variables during the integration. Then the value of $\mathrm{d}u/\mathrm{d}t$ can be saved after the evaluation of the forcing terms in the momentum equation and employed in subsequent trajectory calculations using the second-order scheme

$$x_* = x^{n+1} - u(x^{n+1}, t^n)\Delta t,$$
$$\tilde{x}_j^n = x^{n+1} - u(x_*, t^n)\Delta t + \frac{(\Delta t)^2}{2}\frac{du}{dt}(x_*, t^n)$$

(Krishnamurti et al. 1990; Smolarkiewicz and Pudykiewicz 1992).

Higher-order schemes for the computation of back trajectories have also been devised (Temperton and Staniforth 1987). Although a third-order trajectory scheme must be employed as part of any fully third order semi-Lagrangian method, higher-order trajectory computations are not widely used. In many of the applications where semi-Lagrangian methods are most advantageous, it is easier to accurately compute the back trajectory than it is to accurately interpolate all the resolved scales in the tracer field. As a consequence, second-order schemes are often used for the back trajectory calculation even when the tracer field is interpolated using cubic or higher-order polynomials. Moreover, in those problems where there is nonzero forcing in the Lagrangian reference frame, the time integral of the forcing is seldom approximated to more than second-order accuracy, and in such circumstances the use of a higher-order scheme to compute the back trajectory will not reduce the overall time-truncation error of the semi-Lagrangian scheme below $O\left[(\Delta t)^2\right]$.

7.2 Finite-Volume Integrations with Large Time Steps

As is the case for many Eulerian advective-form approximations, numerical solutions to the semi-Lagrangian approximation to the advection equation (7.13) do not typically conserve the domain integral of a passive tracer. Finite-volume methods provide a natural way to develop mass-conservative approximations, and they have been generalized to allow simulations of advective transport using large time steps via two different approaches. The first approach is genuinely semi-Lagrangian: the departure points for the vertices surrounding an "arrival" volume of fluid are computed by backward trajectories and the total mass within the arrival volume at time t^{n+1} is set equal to the total mass in the volume defined by the departure points at t^n. The second approach is Eulerian and simply involves computing the time-averaged fluxes through the sides of the volume over intervals Δt for which the Courant number exceeds unity.

In both approaches it is necessary to reconstruct the subcell distribution of $\tilde{\phi}(x,t^n)$ from the cell-averaged values ϕ^n. (In this section ϕ^n will refer to cell-averaged quantities, in contrast to its use throughout the rest of this chapter.) That reconstruction is often performed using piecewise-parabolic or piecewise-cubic polynomials, and the resulting reconstructions may be limited to avoid undershoots and overshoots (see Sect. 5.6.3). Both the true semi-Lagrangian and large-time-step Eulerian approaches are easily implemented in one dimension, but challenges arise when either method is extended to two- or three-dimensional problems.

Beginning with Rančić (1992), several mass-conservative semi-Lagrangian schemes have appeared in the atmospheric science literature. Accurate and efficient mass-conservative semi-Lagrangian implementations have been recently created using generalizations of cascade interpolation (Sect. 7.1.1.4) to sequentially build up approximations to the integral of $\tilde{\phi}(\mathbf{x},t^n)$ over the departure volume (e.g., Nair et al. 2002; Zerroukat et al. 2002; Norman and Nair 2008). In contrast to true semi-Lagrangian methods, Lin and Rood (1996) and Leonard et al. (1996) proposed mass-conservative hybrid schemes in which intermediate estimates from semi-Lagrangian advective-form integrations parallel to each coordinate axis are used in a final Eulerian flux-form integration. As noted by Lauritzen (2007), care should be taken to ensure consistency between the advective and flux differencing operators in these schemes to avoid excessive damping when using Courant numbers much larger than unity and low-order operators.

One relatively simple way to implement a mass-conservative scheme capable of stable integrations using large time steps is through the time-split Eulerian finite-volume method proposed by Skamarock (2006). After making provision for the larger time steps, this is essentially the method defined by (5.95)–(5.100). The modifications required to use large time steps are as follows. Suppose the velocities are positive and consider the y-direction fluxes. As a first step, the integer shift s and fractional Courant number $\tilde{\mu}$ for each cell interface are chosen to satisfy

$$\tilde{\mu}\rho^n_{i,j-s} + \sum_{k=1}^{s} \rho^n_{i,j-k+1} = \frac{\Delta t}{\Delta y}(\rho v)^n_{i,j+\frac{1}{2}} \quad \text{if} \quad (\rho v)^n_{i,j+\frac{1}{2}} \geq 0, \tag{7.27}$$

subject to the constraint $0 \leq \tilde{\mu} \leq 1$. If ρ is a uniform pseudodensity, it can be divided out of (7.27), in which case the right side becomes the Courant number μ, whose integer and fractional parts are s and $\tilde{\mu}$, respectively. If ρ is variable, s may be determined through trial and error by systematically increasing the number of terms in the summation, beginning at zero.

Let $\tilde{v} = \tilde{\mu}\Delta y/\Delta t$. The scalar mass flux at $(x_i, y_{j+1/2})$ is computed as

$$F^y_{i,j+\frac{1}{2}}(\phi^n) = \frac{1}{\Delta t}\sum_{k=1}^{s}(\rho^n\phi^n\Delta y)_{i,j-k+1} + \tilde{F}^y_{i,j-s+\frac{1}{2}}(\phi^n), \tag{7.28}$$

where $\tilde{F}^y$ is the tracer flux arising from the noninteger part of the Courant number. $\tilde{F}^y$ is computed using the standard piecewise-parabolic-method flux (5.63) with μ replaced by $\tilde{\mu}$ and c by $\rho^n_{i,j-s}\tilde{v}$ to yield

$$\tilde{F}^y_{i,j-s+\frac{1}{2}}(\phi^n) = \rho^n_{i,j-s}\tilde{v}\left(a_0 + \frac{a_1}{2}|\tilde{\mu}| + \frac{a_2}{3}\tilde{\mu}^2\right). \tag{7.29}$$

Here a_0, a_1, and a_2 are defined as in Sect. 5.6.3 except that wherever "j" appears as an index of ϕ, it is replaced by "$i, j - s$". Note that $F^y_{i,j+1/2}(1) = (\rho v)^n_{i,j+1/2}$.

If $v_{i+1/2} < 0$, the preceding expressions are modified by choosing $\tilde{\mu}$ and s such that $-1 \leq \tilde{\mu} \leq 0$ and

$$\tilde{\mu}\rho^n_{i,j-s+1} - \sum_{k=s}^{-1} \rho^n_{i,j-k} = \frac{\Delta t}{\Delta y}(\rho v)^n_{i,j+\frac{1}{2}}. \tag{7.30}$$

Note that $s \leq 0$ in this case. Analogous to (7.28) and (7.29), the scalar mass flux at $(x_i, y_{j+1/2})$ is computed as

$$F^y_{i,j+\frac{1}{2}}(\phi^n) = \frac{1}{\Delta t}\sum_{k=s}^{-1}(\rho^n\phi^n\Delta y)_{i,j-k} + \tilde{F}^y_{i,j-s+\frac{1}{2}}(\phi^n), \tag{7.31}$$

where

$$\tilde{F}^y_{i,j-s+\frac{1}{2}}(\phi^n) = \rho^n_{i,j-s+1}\tilde{v}\left(a_0 + \frac{a_1}{2}|\tilde{\mu}| + \frac{a_2}{3}\tilde{\mu}^2\right), \tag{7.32}$$

with a_0, a_1, and a_2 again defined as in Sect. 5.6.3 except that wherever "$j+1$" appears as an index of ϕ, it is replaced by "$i, j-s+1$".

A mass-consistent splitting can then be evaluated as

$$\rho^s_{i,j} = \rho^n_{i,j} - \frac{\Delta t}{\Delta x}\left[(\rho u)^n_{i+\frac{1}{2},j} - (\rho u)^n_{i-\frac{1}{2},j}\right], \tag{7.33}$$

$$(\rho\phi)^s_{i,j} = (\rho\phi)^n_{i,j} - \frac{\Delta t}{\Delta x}\left[F^x_{i+\frac{1}{2},j}(\phi^n) - F^x_{i-\frac{1}{2},j}(\phi^n)\right], \tag{7.34}$$

$$\phi^s_{i,j} = (\rho\phi)^s_{i,j}/\rho^s_{i,j}, \tag{7.35}$$

$$\rho^{n+1}_{i,j} = \rho^s_{i,j} - \frac{\Delta t}{\Delta y}\left[(\rho v)^n_{i,j+\frac{1}{2}} - (\rho v)^n_{i,j-\frac{1}{2}}\right], \tag{7.36}$$

$$(\rho\phi)^{n+1}_{i,j} = (\rho\phi)^s_{i,j} - \frac{\Delta t}{\Delta y}\left[F^y_{i,j+\frac{1}{2}}(\phi^s) - F^y_{i,j-\frac{1}{2}}(\phi^s)\right], \tag{7.37}$$

$$\phi^{n+1}_{i,j} = (\rho\phi)^{n+1}_{i,j}/\rho^{n+1}_{i,j}. \tag{7.38}$$

This scheme is not positive definite, but the fluxes $\tilde{F}^x$ and $\tilde{F}^y$ can be limited to prevent the development of overshoots and undershoots as discussed in Sect. 5.6.3. Nevertheless, it is often better to selectively limit $\tilde{F}^x$ and $\tilde{F}^y$ as described in Sect. 5.8, since selective limiting will preserve the amplitude of smooth extrema much better than a global limiter.

As in the Eulerian case, the selectively limited result is not, however, strictly positive definite, although the undershoots it produces are small. The positivity-preserving flux-corrected-transport algorithm described in Sect. 5.10.1 can be used to prevent the development of negative concentrations. The monotone solution required for the flux-corrected-transport approach is computed using the upstream flux

$$F^{y_{\rm up}}_{i,j+\frac{1}{2}}(\phi^n) = \begin{cases} \dfrac{1}{\Delta t}\displaystyle\sum_{k=1}^{s}(\rho^n\phi^n\Delta y)_{i,j-k+1} + \rho^n_{i,j-s}\tilde{v}\phi^n_{i,j-s} & \text{if } \tilde{v} \geq 0\,, \\ \dfrac{1}{\Delta t}\displaystyle\sum_{k=s}^{-1}(\rho^n\phi^n\Delta y)_{i,j-k} + \rho^n_{i,j-s+1}\tilde{v}\phi^n_{i,j-s+1} & \text{if } \tilde{v} < 0\,. \end{cases} \tag{7.39}$$

In addition, during the second split step (which may be either along x or along y depending on the alternation in the Strang splitting), ρ^n should be replaced by ρ^s in (7.27)–(7.32) and (7.39).[2] The use of ρ^s is necessary to keep the upstream solution monotone in cases with Courant numbers greater than unity.

Figure 7.3 shows the performance of this scheme on the cosine-bell test problem of Sect. 5.9.5 in which the initial distribution of a passive tracer is deformed in a sheared swirling flow that reverses symmetrically in time. The solution whose maximum Courant number is unity, shown in Fig. 7.3b, is identical to that in Fig. 5.26b. When the time step is increased by a factor of 4, the solution (plotted in Fig. 5.26a) is slightly improved. This improvement is largely due to the reduction in numerical

[2] Do not replace the ρ in $(\rho v)^n_{i,j+1/2}$ by a ρ^s.

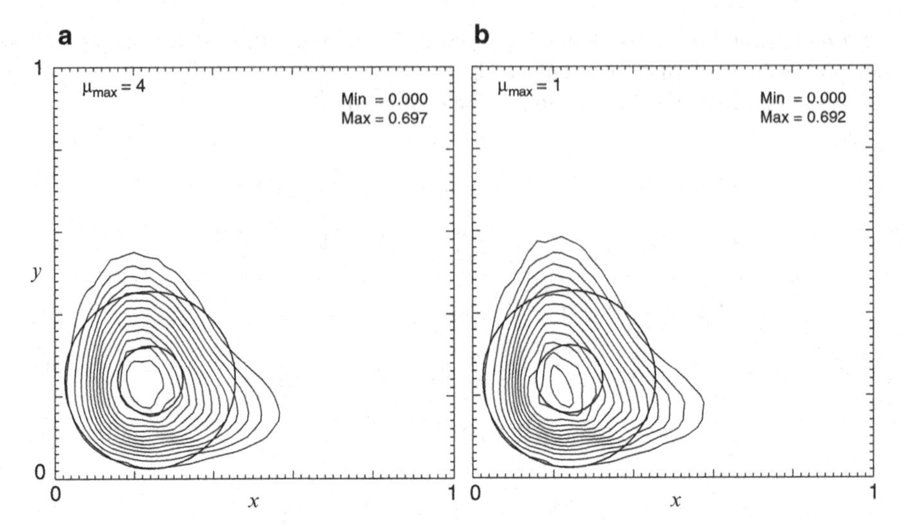

Fig. 7.3 Comparison of the exact solution at $t = 5$ with numerical solutions obtained using a mass-consistent Strang-split piecewise-parabolic-method scheme with selective limiting and positivity preservation when Δt is chosen such that the maximum one-dimensional Courant number μ is **a** 1 or **b** 4. The *heavy circles* are the 0.05 and 0.75 contour lines of the exact solution. The *thin solid lines* are contours of the numerical solution at intervals of 0.05, beginning with the 0.05 contour. The maximum and minimum values of ϕ at $t = 5$ are noted in the *upper right* of **a** and **b**

diffusion associated with the factor-of-4 reduction in the number of times the subcell distribution of the tracer is reconstructed from the cell-averaged ϕ^n. If the grid cell spacing is halved to 0.01, there is less numerical diffusion and the $\mu_{\max} = 1$ and $\mu_{\max} = 4$ solutions are almost identical. The main goal in using longer time steps is not better accuracy but rather improved efficiency, and that is certainly achieved when the time step is increased by a factor of 4. Some extra time is required to evaluate the multiply upstream Courant numbers and fluxes, so in this problem the net speed up gained by switching from a purely Eulerian code with $\mu_{\max} = 1$ to the multiply upstream variant with $\mu_{\max} = 4$ is roughly a factor of 3.

7.3 Forcing in the Lagrangian Frame

The forced scalar advection equation (7.1) provides a simple example in which to study the treatment of forcing terms in semi-Lagrangian schemes. Defining $\check{x}_j^{n-1}$ to be an estimate of the departure point of the fluid parcel at time t^{n-1} that arrives at (x_j, t^{n+1}), one may obtain second-order approximations to the forcing using the trapezoidal method (7.3), the leapfrog scheme

$$\frac{\phi\left(x_j, t^{n+1}\right) - \phi\left(\check{x}_j^{n-1}, t^{n-1}\right)}{2\Delta t} = S\left(\tilde{x}_j^n, t^n\right), \tag{7.40}$$

or the second-order Adams–Bashforth method

$$\frac{\phi\left(x_j, t^{n+1}\right) - \phi\left(\tilde{x}_j^n, t^n\right)}{\Delta t} = \frac{1}{2}\left[3S\left(\tilde{x}_j^n, t^n\right) - S\left(\check{x}_j^{n-1}, t^{n-1}\right)\right]. \qquad (7.41)$$

The most stable and accurate of these schemes is the trapezoidal method, but it may also require more work per time step because it is implicit.

The fundamental stability properties of each scheme can be analyzed by applying them to the prototype problem

$$\frac{d\psi}{dt} = \mathrm{i}\omega\psi + \lambda\psi, \qquad (7.42)$$

where ω and λ are real, ψ is complex, and the advecting velocity is constant, so

$$\frac{d}{dt} = \frac{\partial}{\partial t} + U\frac{\partial}{\partial x}.$$

If $\psi(x,0) = f(x)$, the solution to (7.42) is

$$\psi(x,t) = f(x - Ut)\mathrm{e}^{(\mathrm{i}\omega+\lambda)t},$$

which is nonamplifying for $\lambda \leq 0$. This prototype problem is similar to that used to assess regions of absolute stability in Sect. 2.2.1 except that (2.16) is an ordinary differential equation, whereas (7.42) is a partial differential equation. To simplify the stability analysis, the errors generated during the interpolation of ϕ to $\tilde{x}_j^n$ and $\check{x}_j^{n-1}$ will be neglected, in which case our results describe the limiting behavior of a family of semi-Lagrangian schemes that use increasingly accurate spatial interpolation.

First consider the case where the forcing is approximated with the trapezoidal scheme. Following the standard von Neumann stability analysis, the Fourier mode $A_k^n \mathrm{e}^{\mathrm{i}kj\Delta x}$ is substituted for $\phi(x_j, t^n)$ in the trapezoidal approximation to (7.42), which gives

$$A_k \mathrm{e}^{\mathrm{i}kj\Delta x} - \mathrm{e}^{\mathrm{i}k(j\Delta x - s)} = \frac{(\tilde{\lambda} + \mathrm{i}\tilde{\omega})}{2}\left(A_k \mathrm{e}^{\mathrm{i}kj\Delta x} + \mathrm{e}^{\mathrm{i}k(j\Delta x - s)}\right),$$

where $\tilde{\lambda} = \lambda\Delta t$, $\tilde{\omega} = \omega\Delta t$, and, as before, $s = x_j - \tilde{x}_j^n$. Solving for the magnitude of the amplification factor and noting that $|\mathrm{e}^{\mathrm{i}ks}| = 1$,

$$|A_k|^2 = \left|A_k \mathrm{e}^{\mathrm{i}ks}\right|^2 = \frac{\left(1 + \tilde{\lambda}/2\right)^2 + \tilde{\omega}^2/4}{\left(1 - \tilde{\lambda}/2\right)^2 + \tilde{\omega}^2/4}.$$

The scheme generates bounded solutions whenever the true solution is bounded, i.e., whenever $\tilde{\lambda} \leq 0$. This stability condition is independent of the Courant number $U\Delta t/\Delta x$, and the magnitude of the amplification factor is identical to that obtained if the ordinary differential equation (2.16) is approximated with a trapezoidal time difference.

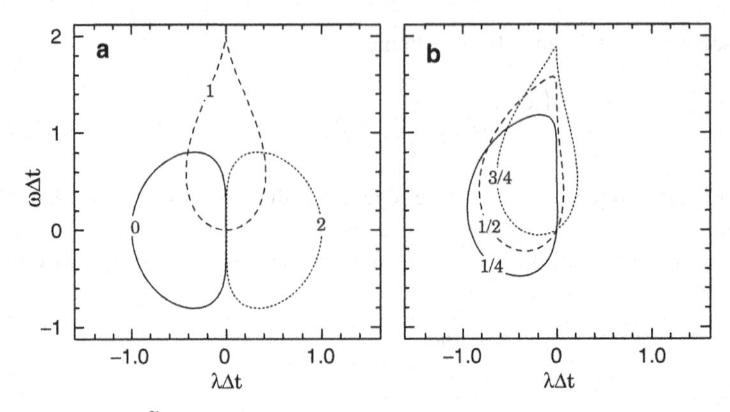

Fig. 7.4 Region of the $\tilde{\lambda}$–$\tilde{\omega}$ plane in which $2\Delta x$-wavelength solutions to (7.44) are nongrowing when **a** $U\Delta t/\Delta x$ is 0, 1, or 2, and **b** $U\Delta t/\Delta x$ is $1/4$, $1/2$, or $3/4$. The stable region lies inside each curve. The stable region for (7.41) is independent of $U\Delta t/\Delta x$ and is defined by the curve labeled *0* in **a**

A similar von Neumann analysis of the second-order Adams–Bashforth method yields

$$A_k^2 - A_k \mathrm{e}^{-\mathrm{i}ks} = \frac{(\tilde{\lambda} + \mathrm{i}\tilde{\omega})}{2}\left(3A_k \mathrm{e}^{-\mathrm{i}ks} - \mathrm{e}^{-2\mathrm{i}ks}\right).$$

Defining $\hat{A} = A_k \mathrm{e}^{\mathrm{i}ks}$ and $\gamma = \tilde{\lambda} + \mathrm{i}\tilde{\omega}$,

$$\hat{A}^2 - \hat{A}\left(\frac{3\gamma}{2} + 1\right) + \frac{\gamma}{2} = 0. \tag{7.43}$$

This quadratic equation is identical to that obtained when the ordinary differential equation (2.16) is approximated using the second-order Adams–Bashforth method. Those values of $\tilde{\lambda}$ and $\tilde{\omega}$ for which the second-order Adams–Bashforth method generates nongrowing solutions lie within the solid curve in Fig. 7.4a. Since $|A_k| = |\hat{A}|$, the amplification factor is independent of the Courant number.

If the leapfrog scheme (7.40) is used to approximate (7.42), the stability condition becomes $\tilde{\lambda} = 0$ and $|\tilde{\omega}| < 1$, which is once again independent of the Courant number and is identical to that for a leapfrog approximation to the ordinary differential equation (2.16). All three of the preceding methods, (7.40), (7.3), and (7.41), yield amplification factors for this prototype problem that are independent of the Courant number because the advecting velocity is a constant, errors in the polynomial interpolation are ignored, and the integration is performed using data lying along the backward trajectory. If the integration does not use data lying along a backward trajectory, the maximum stable time step will depend on the Courant number, and the stability criteria can become far more restrictive. In the case of Adams–Bashforth-type approximations, this consideration has not always been recognized. Alternatives to (7.41) of the form

$$\frac{\phi\left(x_j, t^{n+1}\right) - \phi\left(\tilde{x}_j^n, t^n\right)}{\Delta t} = \frac{3S\left([x_j + \tilde{x}_j^n]/2, t^n\right) - S\left([x_j + \tilde{x}_j^n]/2, t^{n-1}\right)}{2} \tag{7.44}$$

and

$$\frac{\phi\left(x_j,t^{n+1}\right)-\phi\left(\tilde{x}_j^n,t^n\right)}{\Delta t}$$
$$=\frac{1}{4}\Big[3S\left(x_j,t^n\right)+3S\left(\tilde{x}_j^n,t^n\right)-S\left(x_j,t^{n-1}\right)-S\left(\tilde{x}_j^n,t^{n-1}\right)\Big] \tag{7.45}$$

have been used in atmospheric models. These are both second-order accurate, and (7.45) is potentially more efficient because it requires one spatial interpolation fewer than either (7.41) or (7.44). Both schemes can, however, be substantially less stable than (7.41) when integrations are performed using Courant numbers larger than order unity.

Applying (7.44) to the prototype problem (7.42) and performing a von Neumann stability analysis yields the following quadratic equation for the amplification factor:

$$A_k^2-A_k\left(\frac{3\gamma}{2}\mathrm{e}^{-iks/2}+\mathrm{e}^{-iks}\right)+\frac{\gamma}{2}\mathrm{e}^{-iks/2}=0. \tag{7.46}$$

In contrast to the results obtained previously for schemes that use data lying along a back trajectory, the amplification factor for this scheme does depend on the Courant number. The region of the $\tilde{\omega}$–$\tilde{\lambda}$ plane in which $|A_k|\leq 1$ is plotted in Fig. 7.4 for several values of ks between 0 and 2π. The most severe stability constraints are typically imposed by the $2\Delta x$ wave, for which these values of ks correspond to Courant numbers $U\Delta t/\Delta x$ between 0 and 2. As the Courant number increases, the region of absolute stability rotates clockwise around the origin in the $\tilde{\lambda}$–$\tilde{\omega}$ plane. When $U\Delta t/\Delta x=2$, all $2\Delta x$ waves that should properly damp are amplified, and the only $2\Delta x$ waves that damp are those that should amplify. The instability that develops in the solutions to (7.44) as the Courant number increases may be qualitatively understood to result from a failure to match the numerical domain of dependence for the data used to compute $S(\psi)$ with the numerical domain of dependence for the true solution (see Sect. 3.2.3).

Polynomial interpolation damps the interpolated field. It is easy to show that this damping further stabilizes the trapezoidal approximation (see Problem 5), but the influence of this damping on the stability of the Adams–Bashforth-type schemes is harder to determine. Numerical simulations are shown in Fig. 7.5 for a series of tests in which (7.41), (7.44), and (7.45) were used to obtain approximate solutions to (7.42) on a periodic spatial domain $0\leq x\leq 1$. In each test the numerical approximation to ψ at the departure point is obtained by cubic Lagrange interpolation; $\Delta x=0.01$, $\Delta t=0.01$, $\lambda=0$, and the initial condition is

$$\psi(x,0)=\begin{cases}\left\{\left[(x-c)^2-w^2\right]/w^2\right\}^2+0i, & \text{if } |x-c|\leq w,\\ 0+0i, & \text{otherwise,}\end{cases}$$

where $w=0.1$ and $c=0.5$. Thus, at $t=0$ the real part of ψ is a smooth unit-amplitude pulse 20 grid points wide and the imaginary part of ψ is zero.

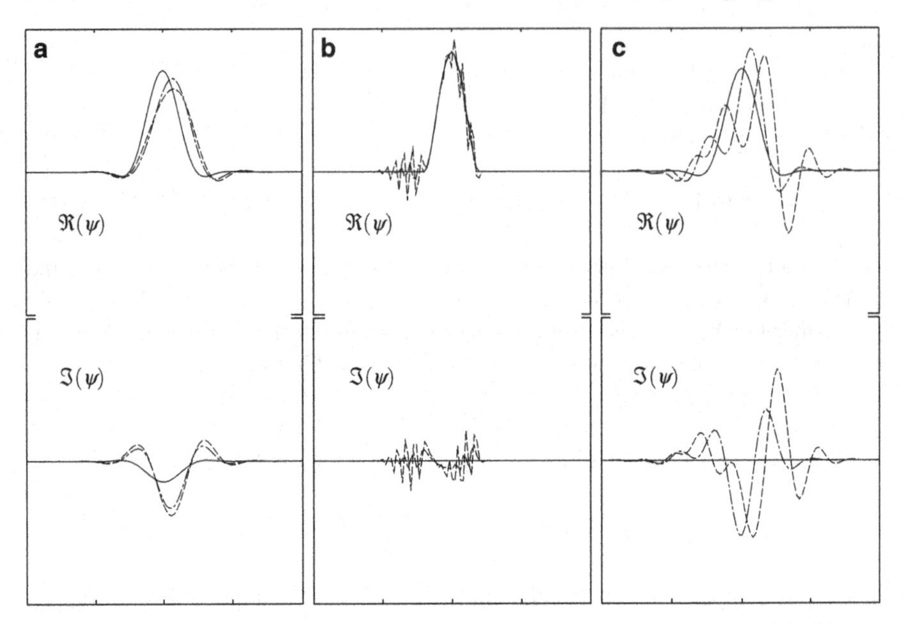

Fig. 7.5 Real and imaginary parts of the numerical solution to (7.42) at $t = 20$ for **a** $U\Delta t/\Delta x = 0.5$, $\omega\Delta t \approx -0.06$, **b** $U\Delta t/\Delta x = 1.0$, $\omega\Delta t \approx -0.003$, and **c** $U\Delta t/\Delta x = 2.5$, $\omega\Delta t \approx -0.006$. The *solid curves*, *dashed curves*, and *dot-dashed curves* show the solutions computed using (7.41), (7.44), and (7.45), respectively. Data are plotted for the interval [0,1] along the horizontal axis; the vertical axis spans the interval $[-1.2, 1.2]$

In the first case, shown in Fig. 7.5a, $U\Delta t/\Delta x = 0.5$, $\omega\Delta t = -\pi/50$, and the solution is plotted at $t = 20$, at which time the energy in the initial pulse has circled the periodic domain ten times and oscillated back and forth between the real and imaginary parts of ψ 20 times. The correct solution is identical to the initial condition: $\Re(\psi)$ is a unit-amplitude pulse centered in the domain and $\Im(\psi)$ is zero everywhere. Although second-order Adams–Bashforth time differencing generates growing solutions to ordinary differential equations describing purely oscillatory motion, all three numerical solutions shown in Fig. 7.5a have been damped by the diffusion in the cubic interpolation. The effect of the accelerative phase-speed error in the second-order Adams–Bashforth time difference is apparent in the plot of $\Im(\psi)$, which shows that all three solutions develop a negative pulse when the correct solution should be exactly zero. The phase error generated by (7.41) is, however, significantly smaller than that produced by (7.44) and (7.45) and is confined to the time coordinate, whereas the phase errors in the solutions to (7.44) and (7.45) appear in both time and space.

The Courant number is increased to unity and $\omega\Delta t$ is decreased to $-\pi/1{,}000$ in the second test case, shown in Fig. 7.5b. At $t = 20$ the correct solution is again identical to the initial condition. Since the wind speed is constant and the Courant number is unity, the semi-Lagrangian advection is exact, and the only source of error is in the integration of the forcing. The solution obtained using (7.41) is stable and almost perfect, whereas the solutions produced by (7.44) and (7.45) are corrupted by growing $2\Delta x$ disturbances. These unstable $2\Delta x$ waves completely dominate the solution computed using (7.44) by $t = 27$ and that obtained using (7.45) by $t = 34$.

In the third case, shown in Fig. 7.5c, the Courant number is 2.5, $\omega\Delta t = -\pi/500$, and the exact solution at $t = 20$ is once again identical to the initial condition. Although it has been somewhat diffused by the cubic interpolation in the semi-Lagrangian advection, the solution produced by (7.41) is free of instability and noticeable phase-speed error. In contrast, the solutions obtained using (7.44) and (7.45) are both contaminated by unstable long-wavelength disturbances.

Instabilities develop in the second and third cases even though the magnitude of the forcing is very small ($|\omega\Delta t| < 0.01$). These results, together with the stability analysis presented in Fig. 7.4, demonstrate the need to compute forcing terms that depend on the solution, i.e., forcing of the form $S(\psi)$, using data along the backward trajectory.

7.4 Systems of Equations

One of the most important applications of semi-Lagrangian methods in atmospheric science is in global weather prediction. This application was pioneered by Robert (1981, 1982), who showed that the equations describing large-scale atmospheric motion could be efficiently integrated using semi-Lagrangian methods in conjunction with a semi-implicit approximation of those terms in the governing equations representing the pressure gradient and velocity divergence. The essential elements of the semi-Lagrangian semi-implicit method will be explored in this section by examining numerical approximations to simple shallow-water equations. The shallow-water system also provides a convenient example in which to illustrate the difference between the semi-Lagrangian approach and the classical method of characteristics.

7.4.1 Comparison with the Method of Characteristics

Semi-Lagrangian approximations to the equations describing the advection and reaction of chemical tracers, such as (7.40), may be regarded as an algorithm for numerically implementing the classical method of characteristics (Courant et al. 1952; see also Gustafsson et al. 1995). The advection equation is, however, a special case because the characteristic curves are identical to the fluid parcel trajectories. The semi-Lagrangian method retains its simplicity and practical utility in more complicated applications precisely because the evolution of the flow continues to be computed following fluid parcel trajectories. The classical method of characteristics, on the other hand, becomes unwieldy or impossible in more general problems where the evolution of the flow along the characteristic curves may be more complicated or characteristic curves may not even be defined. As a simple example, consider the nonlinear one-dimensional shallow-water system

$$\frac{\partial}{\partial t}\begin{pmatrix} u \\ h \end{pmatrix} + \begin{pmatrix} u & g \\ h & u \end{pmatrix}\frac{\partial}{\partial x}\begin{pmatrix} u \\ h \end{pmatrix} = 0.$$

A three-time-level semi-Lagrangian approximation to the preceding equation can be written in the form

$$\frac{u^+ - u^-}{2\Delta t} = -g\left(\frac{\partial h}{\partial x}\right)^0, \tag{7.47}$$

$$\frac{h^+ - h^-}{2\Delta t} = -h^0\left(\frac{\partial u}{\partial x}\right)^0, \tag{7.48}$$

where the superscripts "+," "0," and "−" denote evaluation of the function at the points (x_j, t^{n+1}), $(\tilde{x}_j^n, t^n)$, and $(\check{x}_j^{n-1}, t^{n-1})$, respectively. As before, $\tilde{x}_j^n$ is determined by numerically integrating (7.2) backward over a time interval Δt subject to the initial condition $x(t^{n+1}) = x_j$, and $\check{x}_j^{n-1}$ is determined by a similar backward integration over the period $2\Delta t$. The spatial derivatives $\partial u/\partial x$ and $\partial h/\partial x$ are evaluated by centered differences on the regular mesh and then interpolated to $\tilde{x}_j^n$. As long as the solution remains smooth, the numerical evaluation of this system is no more difficult than the integration of a pair of forced advection equations of the form (7.40).

Considerably more computational effort is required to solve this problem using the classical method of characteristics. To implement the method of characteristics, the nonlinear shallow-water equations are transformed as described in connection with (1.8) to yield the system

$$\frac{\partial}{\partial t}\begin{pmatrix} d \\ e \end{pmatrix} + \begin{pmatrix} d & 0 \\ 0 & e \end{pmatrix}\frac{\partial}{\partial x}\begin{pmatrix} d \\ e \end{pmatrix} = \mathbf{B}\begin{pmatrix} d \\ e \end{pmatrix}; \tag{7.49}$$

here $d = u - \sqrt{gh}$, $e = u + \sqrt{gh}$, and

$$\mathbf{B} = -\mathbf{T}^{-1}\left[\frac{\partial \mathbf{T}}{\partial t} + \begin{pmatrix} u & g \\ h & u \end{pmatrix}\frac{\partial \mathbf{T}}{\partial x}\right],$$

$$\mathbf{T}^{-1} = \begin{pmatrix} 1 & -\sqrt{g/h} \\ 1 & \sqrt{g/h} \end{pmatrix},$$

$$\mathbf{T} = \frac{1}{2}\begin{pmatrix} 1 & 1 \\ -\sqrt{h/g} & \sqrt{h/g} \end{pmatrix}.$$

The numerical integration of this system requires more computation than that for the semi-Lagrangian scheme because the right side of (7.49) is much more complicated than the right sides of (7.47) and (7.48). Additional effort must also be expended in the trajectory calculations because the evaluation of (7.49) requires the computation of two backward trajectories per grid point (one along each characteristic curve), whereas the semi-Lagrangian method requires only one backward trajectory per grid point.

7.4.2 Semi-implicit Semi-Lagrangian Schemes

If latitudinally varying Coriolis forces are included in the shallow-water equations, the resulting system can support both Rossby and gravity waves. Most physically significant large-scale atmospheric circulations have timescales similar to those of the Rossby waves and much longer than those of the gravity waves. As a consequence, the maximum stable time step dictated by the CFL condition for gravity waves is often much smaller than that required to accurately simulate the physically significant phenomena. A considerable increase in efficiency can be realized by using semi-implicit time differencing to remove the stability constraint imposed by rapid gravity-wave propagation. Semi-implicit time differencing is discussed in detail in Sect. 8.2. In the following we will focus on just one aspect of the semi-implicit method, namely, how it improves the stability of semi-Lagrangian solutions to the one-dimensional shallow-water equations.

First, consider the stability properties of the linearized equivalent to (7.47) and (7.48). If the mean wind and fluid depth are constants denoted by U and H, respectively, a finite-difference approximation to the linearized system may be written as

$$\frac{u^+ - u^-}{2\Delta t} = -g\left(\frac{\partial h}{\partial x}\right)^0, \tag{7.50}$$

$$\frac{h^+ - h^-}{2\Delta t} = -H\left(\frac{\partial u}{\partial x}\right)^0, \tag{7.51}$$

where u and h now denote the amplitudes of the perturbation velocity and free surface displacement. Defining auxiliary variables a and b such that $a^+ = u^0$ and $b^+ = h^0$ and substituting wave solutions of the form

$$\begin{pmatrix} u \\ h \\ a \\ b \end{pmatrix}^n e^{ikx}$$

into the preceding system yields the linear system

$$\begin{pmatrix} u \\ h \\ a \\ b \end{pmatrix}^{n+1} = e^{-iks}\begin{pmatrix} 0 & -2i\tilde{g} & 1 & 0 \\ -2i\tilde{H} & 0 & 0 & 1 \\ 1 & 0 & 0 & 0 \\ 0 & 1 & 0 & 0 \end{pmatrix}\begin{pmatrix} u \\ h \\ a \\ b \end{pmatrix}^n,$$

where $s = U\Delta t$, $\tilde{g} = gk\Delta t$, and $\tilde{H} = Hk\Delta t$. Let λ be an eigenvalue of the amplification matrix for this scheme, and define $\tilde{\lambda} = \lambda e^{iks}$ and $\tilde{c} = \sqrt{gH}k\Delta t$. Then

$$\tilde{\lambda}^4 + (4\tilde{c}^2 - 2)\tilde{\lambda}^2 + 1 = 0$$

and

$$\tilde{\lambda}^2 = 1 - 2\tilde{c}^2 \pm 2i\tilde{c}\left(1 - \tilde{c}^2\right)^{1/2}.$$

Two of the $\tilde{\lambda}$ are associated with gravity waves and two are associated with computational modes. The magnitude of $\tilde{\lambda}^2$ is unity whenever $|\tilde{c}| \leq 1$, and since $|\lambda| = |\tilde{\lambda}|$, it follows that the eigenvalues of the amplification matrix are bounded by unity whenever $|\tilde{c}| \leq 1$. If the spatial derivatives are evaluated using the centered-difference operator δ_{2x}, this stability condition becomes $\sqrt{gH}\Delta t/\Delta x < 1$, where strict inequality is required to ensure that the norm of the amplification matrix is power-bounded (see Sect. 4.1.1.1). In contrast to the result obtained for an Eulerian scheme (4.13), the stability of the leapfrog semi-Lagrangian approximation (7.50) and (7.51) depends only on a Courant number defined with respect to the intrinsic gravity-wave phase speed and does not depend on the speed of the mean flow.

An unconditionally stable method can be obtained if the forcing terms in the linearized system are approximated by trapezoidal time differencing to yield the approximation

$$\frac{u^+ - u^0}{\Delta t} = -\frac{g}{2}\left[\left(\frac{\partial h}{\partial x}\right)^+ + \left(\frac{\partial h}{\partial x}\right)^0\right], \tag{7.52}$$

$$\frac{h^+ - h^0}{\Delta t} = -\frac{H}{2}\left[\left(\frac{\partial u}{\partial x}\right)^+ + \left(\frac{\partial u}{\partial x}\right)^0\right]. \tag{7.53}$$

The eigenvalues for the amplification matrix associated with this scheme are obtained by substituting solutions of the form

$$\begin{pmatrix} u \\ h \end{pmatrix}^n e^{ikx}$$

into the (7.52) and (7.53) to yield

$$\lambda = e^{-iks}\left(\frac{4 - \tilde{c}^2 \pm 4i\tilde{c}}{4 + \tilde{c}^2}\right). \tag{7.54}$$

This scheme is unconditionally stable, since $|\lambda| = 1$ for all Δt, and the norm of the amplification matrix is power-bounded because its eigenvectors are linearly independent (and it can therefore be transformed into a diagonal matrix).

The right side of (7.53) becomes nonlinear if the preceding trapezoidal scheme is generalized to approximate the nonlinear shallow-water equations. To avoid solving a coupled system of nonlinear algebraic equations at every time step, the total fluid depth is often split into a constant reference value and a perturbation, and the velocity divergence multiplying the perturbation is evaluated using leapfrog time differencing. Letting $h(x,t) = H + \eta(x,t)$, this approach leads to the following three-time-level scheme:

$$\frac{u^+ - u^-}{2\Delta t} = -\frac{g}{2}\left[\left(\frac{\partial \eta}{\partial x}\right)^+ + \left(\frac{\partial \eta}{\partial x}\right)^-\right], \tag{7.55}$$

$$\frac{\eta^+ - \eta^-}{2\Delta t} = -\frac{H}{2}\left[\left(\frac{\partial u}{\partial x}\right)^+ + \left(\frac{\partial u}{\partial x}\right)^-\right] - \eta^0\left(\frac{\partial u}{\partial x}\right)^0. \tag{7.56}$$

A complete stability analysis of the full nonlinear system is very difficult, but the stability of a linearized system in which

$$u(x,t) = U + u'(x,t), \qquad \eta(x,t) = \overline{\eta} + \eta'(x,t)$$

can be performed following essentially the same steps as those detailed in Sect. 8.2.3. This analysis shows that the linearized system is stable for all Δt, provided that $|\overline{\eta}| \le H$.

One disadvantage of the preceding scheme is that it is potentially half as efficient as a two-time-level method. Both the trajectory calculation and the trapezoidal difference in the preceding scheme are computed over a time interval of $2\Delta t$. To evaluate these terms with the same accuracy obtained in a two-level scheme such as (7.52) and (7.53), the time step used in (7.55) and (7.56) must be half that used in the two-level scheme. One way to obtain an $O\left[(\Delta t)^2\right]$ approximation to the nonlinear shallow-water equations that preserves the efficiency of the linearized system (7.52) and (7.53) is to use the second-order Adams–Bashforth method to evaluate the portion of the velocity divergence multiplying η, in which case the finite-difference equations become

$$\frac{u^+ - u^0}{\Delta t} = -\frac{g}{2}\left[\left(\frac{\partial \eta}{\partial x}\right)^+ + \left(\frac{\partial \eta}{\partial x}\right)^0\right], \tag{7.57}$$

$$\frac{\eta^+ - \eta^0}{\Delta t} = -\frac{H}{2}\left[\left(\frac{\partial u}{\partial x}\right)^+ + \left(\frac{\partial u}{\partial x}\right)^0\right] - \frac{3}{2}\eta^0\left(\frac{\partial u}{\partial x}\right)^0 + \frac{1}{2}\eta^-\left(\frac{\partial u}{\partial x}\right)^-. \tag{7.58}$$

A stability analysis similar to that for the linearized version of (7.55) and (7.56) can be performed by linearizing the preceding equations about the same basic state:

$$u(x,t) = U + u'(x,t), \qquad \eta(x,t) = \overline{\eta} + \eta'(x,t).$$

Dropping the primes on the perturbation variables, letting $\overline{\eta} = \alpha H$, $s = U\Delta t$, and defining $a^+ = u^0$, one can write the linearized system in the form

$$\begin{pmatrix} 1 & \mathrm{i}\tilde{g}/2 & 0 \\ \mathrm{i}\tilde{H}/2 & 1 & 3\mathrm{i}\alpha\tilde{H}/2 \\ 0 & 0 & 1 \end{pmatrix} \begin{pmatrix} u \\ \eta \\ a \end{pmatrix}^{n+1}$$

$$= \mathrm{e}^{-\mathrm{i}ks} \begin{pmatrix} 1 & -\mathrm{i}\tilde{g}/2 & 0 \\ -\mathrm{i}\tilde{H}/2 & 1 & \mathrm{i}\alpha\tilde{H}/2 \\ 1 & 0 & 0 \end{pmatrix} \begin{pmatrix} u \\ \eta \\ a \end{pmatrix}^{n},$$

where $\tilde{g} = gk\Delta t$ and $\tilde{H} = Hk\Delta t$. Let λ be an eigenvalue of the amplification matrix for this scheme and define $\tilde{\lambda} = \lambda \mathrm{e}^{\mathrm{i}ks}$, $\beta^2 = \tilde{g}\tilde{H}/4$, and $\mu = \alpha\beta^2/(1+\beta^2)$; then $\tilde{\lambda}$ satisfies the cubic equation

$$\tilde{\lambda}^3 + \left(3\mu - \frac{2(1-\beta^2)}{1+\beta^2}\right)\tilde{\lambda}^2 + (2\mu+1)\tilde{\lambda} - \mu = 0. \tag{7.59}$$

Simmons and Temperton (1997) obtained this cubic equation as part of a more extensive analysis of the stability of similar semi-Lagrangian approximations to the equations governing three-dimensional stratified flow. As noted by Simmons and Temperton, one root of (7.59) is a real number associated with a computational mode, and the other two roots are complex conjugates associated with rightward- and leftward-propagating gravity waves.

If $\alpha = 0$, then $\mu = 0$, so the linearized equations reduce to (7.52) and (7.53); the computational mode vanishes, and the remaining eigenvalues are given by (7.54). Let $\tilde{\lambda}_0$ denote the value of $\tilde{\lambda}$ when $\mu = 0$:

$$\tilde{\lambda}_0 = \frac{1-\beta^2 \pm 2\mathrm{i}\beta}{1+\beta^2} = \frac{1 \pm \mathrm{i}\beta}{1 \mp \mathrm{i}\beta}.$$

For small values of μ the stability of this scheme can be determined by expanding the $\tilde{\lambda}$ in powers of μ. The eigenvalue for the computational mode is $\tilde{\lambda} = \mu + O(\mu^2)$. Since $|\mu|$ is small by assumption, this mode is rapidly damped. Expanding the $\tilde{\lambda}$ for the gravity-wave modes in powers of μ,

$$\tilde{\lambda} = \tilde{\lambda}_0 + \tilde{\lambda}_1\mu + O(\mu^2), \tag{7.60}$$

and substituting the preceding into (7.59) yields

$$\tilde{\lambda}_1 = \frac{1-3\tilde{\lambda}_0}{\tilde{\lambda}_0 - 1}. \tag{7.61}$$

Since $|\lambda| = |\tilde{\lambda}|$, the square of the magnitude of the eigenvalues of the amplification matrix is

$$\tilde{\lambda}\tilde{\lambda}^* = \tilde{\lambda}_0\tilde{\lambda}_0^* + 2\mu\Re(\tilde{\lambda}_0\tilde{\lambda}_1^*) + O(\mu^2).$$

Substituting (7.60) and (7.61) into the preceding equation yields

$$|\lambda|^2 = |\tilde{\lambda}|^2 = 1 + \frac{4\beta^2}{1+\beta^2}\mu + O(\mu^2),$$

which implies that at least for small values of $|\mu|$, the scheme will be stable if $\mu < 0$ and unstable if $\mu > 0$. Note that μ has the same sign as α and for all Δt, $|\mu| < |\alpha|$. Thus, for sufficiently small values of α, the scheme is unconditionally stable when $\alpha < 0$ and unconditionally unstable when $\alpha > 0$. Numerical evaluation of the roots of (7.59) verifies that the scheme is damping independent of Δt for $-1 < \alpha < 0$. In the context of the original nonlinear problem, this analysis shows that a necessary condition for the scheme to be stable independent of Δt is that the reference fluid depth H exceed the maximum height of the actual free-surface displacement.

An alternative to the decomposition of the total fluid depth into $H + \eta$ is to remove the nonlinearity in the velocity divergence by linearizing h about its value at the preceding time step, in which case (7.58) is replaced by

$$\frac{h^+ - h^0}{\Delta t} = -\frac{h^0}{2}\left[\left(\frac{\partial u}{\partial x}\right)^+ + \left(\frac{\partial u}{\partial x}\right)^0\right]. \tag{7.62}$$

This approach requires the solution of a linear algebraic system with a more complicated coefficient structure than that generated by (7.58), and particularly in two- or three-dimensional problems, the increase in the complexity of the coefficient matrix can be an impediment to numerical efficiency. Promising results have, nevertheless, been obtained using preconditioned conjugate residual solvers (Skamarock et al. 1997), suggesting that this approach can be a viable alternative in those applications where a suitable preconditioning operator can be determined. Yet another possibility was pursued by Bates et al. (1995), who used a nonlinear multigrid method to solve the nonlinear finite-difference equations generated by a true trapezoidal approximation to the full shallow-water continuity equation.

7.5 Alternative Trajectories

As noted by Smolarkiewicz and Pudykiewicz (1992), the numerical solution can be integrated forward in time along trajectories other than those associated with the standard Eulerian and Lagrangian coordinate frames. Any function $\psi(\mathbf{x}, t)$ with continuous derivatives satisfies the relation

$$\psi(\mathbf{x}_j, t^{n+1}) - \psi(\tilde{\mathbf{x}}, t^n) = \int_C \left(\nabla\psi, \frac{\partial\psi}{\partial t}\right) \cdot (\mathrm{d}\mathbf{x}, \mathrm{d}t), \tag{7.63}$$

where $\nabla\psi$ is the gradient of ψ with respect to the spatial coordinates, and C is an arbitrary contour connecting the points $(\tilde{\mathbf{x}}, t^n)$ and $(\mathbf{x}_j, t^{n+1})$. If the time evolution of ψ is governed by the partial differential equation

$$\frac{\partial \psi}{\partial t} + \mathbf{v} \cdot \nabla \psi = S,$$

(7.63) may be expressed in the form

$$\psi(\mathbf{x}_j, t^{n+1}) = \psi(\tilde{\mathbf{x}}, t^n) + \int_C \nabla \psi \cdot (\mathrm{d}\mathbf{x} - \mathbf{v}\mathrm{d}t) + \int_C S \, \mathrm{d}t. \tag{7.64}$$

Lagrangian and semi-Lagrangian schemes approximate this equation by choosing the integration path to be a fluid parcel trajectory, in which case the first integral in (7.64) is zero, and $\tilde{\mathbf{x}}$ is the departure point of the trajectory arriving at $(\mathbf{x}_j, t^{n+1})$. Eulerian schemes approximate this equation by choosing C to be independent of $\mathbf{x}$, in which case (7.64) becomes

$$\psi(\mathbf{x}_j, t^{n+1}) = \psi(\mathbf{x}_j, t^n) + \int_C (S - \mathbf{v} \cdot \nabla \psi) \, \mathrm{d}t.$$

As an alternative to the pure Lagrangian and Eulerian approaches, one may choose $\tilde{\mathbf{x}}$ to coincide with the grid point that is closest to the departure point of the fluid parcel trajectory arriving at $(\mathbf{x}_j, t^{n+1})$. Two such methods will be considered in the following section. In the first method, C is a straight line in $\mathbf{x}$–t space; in the second approach C is deformed into the union of the true fluid parcel trajectory and a series of straight lines in the hyperplane $t = t^n$. In both of these alternatives the interpolation of ψ to the departure point is accomplished by solving an advection equation rather than by conventional interpolation.

7.5.1 A Noninterpolating Leapfrog Scheme

Some damping is produced in all the previously described semi-Lagrangian schemes when the prognostic fields are interpolated to the departure point. Numerical solutions to the forced one-dimensional advection equation (7.1) can be obtained without interpolation using a modified semi-Lagrangian algorithm due to Ritchie (1986). Let $\check{x}_j^{n-1}$ be the estimated x coordinate of the departure point of a trajectory originating at time t^{n-1} and arriving at (x_j, t^{n+1}); $\check{x}_j^{n-1}$ is calculated by integrating (7.2) backward over a time interval of $2\Delta t$ using the initial condition $x(t^{n+1}) = x_j$. Define p as the integer for which x_{j-p} is the grid point closest to $\check{x}_j^{n-1}$, and let u' be a residual velocity such that

$$u = \frac{p\Delta x}{2\Delta t} + u'.$$

Then (7.1) can be expressed as

$$\frac{\partial \psi}{\partial t} + \frac{p\Delta x}{2\Delta t}\frac{\partial \psi}{\partial x} = -u'\frac{\partial \psi}{\partial x} + S(\psi). \tag{7.65}$$

The velocities on the left side of the preceding equation carry a fluid parcel an integral number of grid points over a time interval $2\Delta t$, so the left side of (7.65) can be evaluated as a Lagrangian derivative without numerical error. The right side of (7.65) can be integrated using a leapfrog time difference with the forcing evaluated at $\check{x}_j^{n-1}/2$, which is the midpoint of a back trajectory computed with respect to the velocity $p\Delta x/(2\Delta t)$. Depending on whether p is even or odd, $\check{x}_j^{n-1}/2$ will either coincide with a grid point or lie halfway between two grid points. A second-order finite-difference approximation to (7.65) can therefore be written in the form

$$\frac{\phi_j^{n+1} - \phi_{j-p}^{n-1}}{2\Delta t} = \begin{cases} -u'\delta_{2x}\phi_{j-p/2}^n + S(\phi_{j-p/2}^n) & \text{if } p \text{ is even;} \\ -u'\delta_x\phi_{j-p/2}^n + \left\langle S(\phi_{j-p/2}^n)\right\rangle^x & \text{if } p \text{ is odd.} \end{cases} \tag{7.66}$$

The stability of the constant-coefficient equivalent of the preceding scheme can be easily investigated. Suppose that the source term is zero and the wind speed is U; substituting the discrete Fourier mode

$$\phi_j^n = \mathrm{e}^{\mathrm{i}(kj\Delta x - \omega n\Delta t)}$$

into (7.66) and invoking the assumption that $S = 0$ yields the discrete-dispersion relation

$$\sin\left[\omega\Delta t - k(U-u')\Delta t\right] = \begin{cases} u'\Delta t\,\dfrac{\sin(k\Delta x)}{\Delta x} & \text{if } p \text{ is even;} \\[2ex] 2u'\Delta t\,\dfrac{\sin(k\Delta x/2)}{\Delta x} & \text{if } p \text{ is odd.} \end{cases} \tag{7.67}$$

By the choice of p,

$$\frac{\Delta x}{2} > |x_j - \check{x}_j^{n-1} - p\Delta x| = |2U\Delta t - p\Delta x|,$$

so

$$\left|\frac{u'\Delta t}{\Delta x}\right| = \left|U - \frac{p\Delta x}{2\Delta t}\right|\frac{\Delta t}{\Delta x} < \frac{1}{4},$$

which implies that the right side of (7.67) is bounded by one half, ω is real, and the scheme is stable independent of the value of Δt.

If the velocity is a function of space and time, the residual Courant number $|u'\Delta t/\Delta x|$ will vary as a function of x and t. Let the subscript "*" indicate that the function is evaluated at the midpoint of the trajectory between x_j and x_{j-p}, in which case

$$u_* = \begin{cases} u_{j-p/2}^n & \text{if } p \text{ is even;} \\ \left\langle u_{j-p/2}^n\right\rangle^x & \text{if } p \text{ is odd.} \end{cases} \tag{7.68}$$

According to (7.67), a necessary condition for the stability of the variable-velocity algorithm is $|u'_*\Delta t/\Delta x| < 1/2$. This condition is not automatically satisfied unless the deviation of u_* from the average velocity along the back trajectory is sufficiently small. Let $\overline{u} = (x_j - \check{x}_j^{n-1})/(2\Delta t)$ be the average velocity required to arrive at the true departure point and define the deviation of the velocity at the midpoint as $\nu = u_* - \overline{u}$. Then

$$u'_* = u_* - \frac{p\Delta x}{2\Delta t} = \overline{u} - \frac{p\Delta x}{2\Delta t} + \nu,$$

and by the choice for p,

$$\left|\frac{u'_*\Delta t}{\Delta x}\right| < \frac{1}{4} + \left|\frac{\nu\Delta t}{\Delta x}\right|.$$

Thus, for the variable-velocity algorithm to be stable, the deviation of the velocity about the mean along the backward trajectory must be small enough that $|\nu\Delta t/\Delta x| \leq 1/4$.

This stability constraint can be removed if the departure point is computed as suggested by Ritchie (1986) by choosing

$$p = \text{nearest integer to } \frac{2\Delta t}{\Delta x}u_*. \tag{7.69}$$

The preceding expression is an implicit equation for p because u_* is the function of p defined by (7.68). Although this approach stabilizes the scheme in the sense that it keeps the solution bounded, it is still subject to problems if the deviation of u_* from the average velocity along the back trajectory is too large. In particular, the solution to (7.69) need not be unique if

$$\left|\frac{\partial u}{\partial x}\right|\Delta t > \frac{1}{2}.$$

The nonuniqueness of the solution to (7.69) is particularly apparent if the velocity field is defined by the relation $u[(j+n)\Delta x] = -n\Delta x/\Delta t$ at all grid points in a neighborhood surrounding x_j, since (7.69) is then satisfied by any integer p.

7.5.2 Interpolation via Parameterized Advection

Semi-Lagrangian methods will not preserve the nonnegativity of an initially nonnegative tracer concentration field if conventional quadratic or higher-order polynomial interpolation is used to determine the value of ψ at the departure point. Positive-definite semi-Lagrangian schemes can be obtained if the interpolating functions are required to satisfy appropriate monotonicity and convex–concave shape-preserving constraints (Williamson and Rasch 1989). As noted by Smolarkiewicz and Rasch (1991), positive-definite results can also be obtained if the interpolation step in the standard semi-Lagrangian algorithm is recast as a parameterized advection problem that is approximated using one of the positive-definite advection schemes discussed in Sect. 5.10. If a strictly positive-definite result is not required, overshoots and undershoots in the vicinity of poorly resolved gradients can still be minimized by approximating the solution to the parameterized advection equation using any of the various flux-limited or flux-corrected advection schemes discussed in Chap. 5.

If $f(x)$ is a continuously differentiable function, the value of f at some arbitrary point $\tilde{x}$ can be estimated from its value on a regularly spaced mesh by computing a numerical solution to the constant-coefficient advection problem

$$\frac{\partial \Upsilon}{\partial \tau} + \frac{\partial \Upsilon}{\partial x} = 0 \tag{7.70}$$

subject to the initial condition $\Upsilon(x, 0) = f(x)$. The solution to this advection problem is $\Upsilon(x, \tau) = f(x - \tau)$. Let x_p be the x coordinate of the grid point nearest to $\tilde{x}$ and define $\alpha = x_p - \tilde{x}$; then

$$\Upsilon(x_p, \alpha) = f(x_p - \alpha) = f(\tilde{x}).$$

Figure 7.6 illustrates how the initial distribution of Υ is shifted along the x coordinate so that desired value of $f(\tilde{x})$ becomes the value of Υ at grid point x_p when $\tau = \alpha$.

If a single time step is used to integrate forward or backward over the interval $\Delta\tau = \alpha$, the magnitude of the Courant number associated with this integration will be $|\alpha/\Delta x|$. Since $|\alpha/\Delta x| < 1/2$ by the definition of α, stable estimates of $f(\tilde{x})$ can be obtained in a single time step using most of the wide variety of schemes available for the numerical approximation of (7.70). Of course, there is no advantage in using this approach if (7.70) is solved using an elementary scheme such as the Lax–Wendroff method, which will yield exactly the same result as would be obtained if $f(\tilde{x})$ were interpolated from the quadratic polynomial passing through the points $f(x_{p+1})$, $f(x_p)$, and $f(x_{p-1})$. As discussed previously, the advantage of this approach lies in the possibility of using positive-definite or flux-limited advection schemes to eliminate spurious negative concentrations or minimize undershoots and overshoots in the semi-Lagrangian solution.

The use of parameterized advection equations to replace the interpolation step in conventional semi-Lagrangian methods can be interpreted as a method for advancing the solution to the new time level by integrating (7.64) along a specially

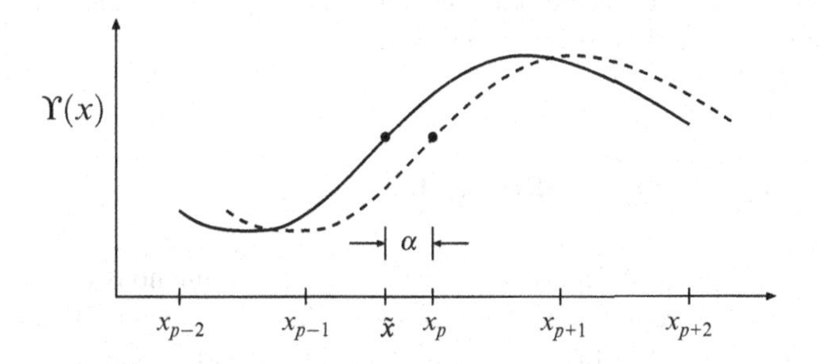

Fig. 7.6 Interpolation via the solution of a constant-wind-speed advection problem. The initial condition is indicated by the *solid line*; the solution after translation a distance α is indicated by the *dashed line*

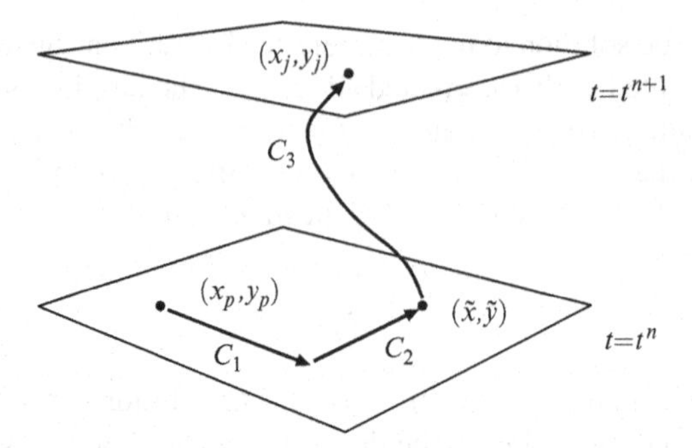

Fig. 7.7 Contour paths for the integration of (7.64); (x_j, y_j, t^{n+1}) is the arrival point, $(\tilde{x}, \tilde{y}, t^n)$ is the departure point, and (x_p, y_p, t^n) is the nearest grid point to the departure point

deformed contour between the arrival point and the nearest grid point to the departure point (Smolarkiewicz and Pudykiewicz 1992). To easily visualize the geometric structure of this contour, suppose that the spatial domain is two-dimensional and the spatial coordinates are x and y. Let $(\tilde{x}, \tilde{y})$ be the departure point of the trajectory originating at time t^n and arriving at (x_j, y_j, t^{n+1}), and let (x_p, y_p) be the coordinates of the node on the spatial grid that is nearest to $(\tilde{x}, \tilde{y})$. Since the contour integrals in (7.64) are path independent, $\psi(x_j, y_j, t^{n+1})$ can be evaluated by integrating along the path defined by the union of the three contours

$$
\begin{aligned}
C_1 &= (x_p + (\tilde{x} - x_p)\tau,\ y_p,\ t^n), \quad \tau \in [0, 1], \\
C_2 &= (\tilde{x},\ y_p + (\tilde{y} - y_p)\tau,\ t^n), \quad \tau \in [0, 1], \\
C_3 &= \left(\tilde{x} + \int_{t^n}^{t} u[x(s), y(s), s]\, ds,\ \tilde{y} + \int_{t^n}^{t} v[x(s), y(s), s]\, ds,\ t\right), \\
&\qquad t \in \left[t^n, t^{n+1}\right].
\end{aligned}
$$

A schematic diagram of this integration path is shown in Fig. 7.7. The integral of (7.64) over the contours C_1 and C_2 is independent of the true time variable t and yields the value of ψ at the departure point of the backward trajectory. The final integral along contour C_3 is the standard semi-Lagrangian evaluation of the integral of the forcing along a fluid parcel trajectory.

7.6 Eulerian or Semi-Lagrangian?

The relative efficiency of Eulerian and semi-Lagrangian methods can vary considerably between different physical applications. Semi-Lagrangian methods require more work per time step than their Eulerian counterparts because additional effort is required to compute the backward trajectories. Thus, to be more efficient, semi-Lagrangian methods must produce stable and accurate solutions using larger time steps than comparable Eulerian schemes. The feasibility of taking a large semi-Lagrangian time step is primarily determined by two factors: the ease with which an accurate trajectory can be computed that extends several grid intervals upstream and the extent to which the frequency of the forcing in the Lagrangian reference frame is reduced relative to that in the Eulerian frame.

One application where the semi-Lagrangian approach can have a distinct advantage is in the simulation of tracer transport in a smooth, slowly varying flow field. If the tracer is conservative and the flow is inviscid, there is no forcing in the Lagrangian reference frame, and the only factor determining the time step is the need to compute accurate backward trajectories. On the other hand, the highest-frequency forcing in the Eulerian reference frame, ω_E, is determined by the product of the velocity and the largest wave number resolved on the spatial mesh. Stability constraints (for explicit methods) and accuracy considerations require the time step of the Eulerian scheme to be small enough that $|\omega_E \Delta t| < O(1)$. It follows that if the spatial mesh required to adequately resolve the tracer field is much finer than that required to define the flow field, the maximum time step suitable for use with a semi-Lagrangian scheme can be much greater than that suitable for an Eulerian method.

Semi-Lagrangian methods also have an advantage in solving problems in spherical geometry. The most natural coordinate system for such problems is a latitude–longitude grid, but the convergence of the meridians near the poles greatly decreases the east–west distance between grid points in the polar regions. In applications such as global atmospheric modeling, the spatial scale of the disturbances near the poles is similar to that in middle latitudes, and the extra resolution in the polar regions is not required to accurately capture the meteorologically significant phenomena. The maximum stable time step of an Eulerian method must, nevertheless, be small enough to ensure that the CFL condition defined with respect to the wind speed is less than order unity in the polar regions. Semi-Lagrangian methods are free from this time-step restriction, although some care is required to accurately compute backward trajectories near the poles (Ritchie 1987; Williamson and Rasch 1989; McDonald and Bates 1989).

In those problems where the frequency of the forcing in the Lagrangian reference frame is similar to that in the Eulerian frame, semi-Lagrangian schemes tend to be at a disadvantage because accuracy considerations often require that both methods use similar time steps. In some cases, such as flow over a topographic barrier, forcing that is stationary in the Eulerian coordinate system is Doppler shifted to a higher frequency in the Lagrangian coordinate frame (Pinty et al. 1995; Héreil and Laprise 1996). One situation in which the frequency of the forcing is similar in both the Lagrangian and the Eulerian reference frames occurs in those shallow-water systems where the fluid velocities are much slower than the phase speeds of the gravity waves. In such systems both semi-Lagrangian and Eulerian methods must use essentially the same time step to accurately simulate the most rapidly moving waves.

In some applications the fastest-moving waves are not physically significant, and in these applications the semi-implicit approximation can be used to increase the time step in the numerical integration. When used in conjunction with the semi-implicit method, semi-Lagrangian schemes can be considerably more efficient than Eulerian methods. Semi-implicit semi-Lagrangian schemes have proved particularly useful in global atmospheric modeling. Ritchie et al. (1995) compared Eulerian and semi-Lagrangian versions of the semi-implicit global forecast model developed at the European Center for Medium Range Weather Forecasting and reported that "the semi-Lagrangian version with a 15-min time step gave an accuracy equivalent to that of an Eulerian version with a 3-min time step, giving an efficiency improvement of about a factor of four after allowing for the 20% ... [overhead for] the semi-Lagrangian computations."

It should be emphasized that the fastest-moving waves in the shallow-water system are artificially decelerated whenever semi-implicit integrations are performed using time steps significantly greater than those permitted by the CFL condition for gravity waves. This loss of accuracy occurs in both Eulerian semi-implicit and semi-Lagrangian semi-implicit models. In contrast, the increase in the time step permitted in semi-Lagrangian approximations to the pure advection equation is achieved without any inherent loss of accuracy because advective forcing generates a zero frequency response in the Lagrangian reference frame.

Problems

1. Show that the phase-speed error associated with the first-order semi-Lagrangian approximation (7.5) is

$$\frac{\tilde{\omega}}{\omega} = \frac{1}{(p+\alpha)k\Delta x}\left(pk\Delta x + \arctan\left[\frac{\alpha \sin k\Delta x}{1-\alpha(1-\cos k\Delta x)}\right]\right),$$

where $\tilde{\omega}$ and ω are the frequencies of the true and numerically approximated waves of wave number k. How does this error depend on the spatial resolution ($k\Delta x$) and the Courant number ($U\Delta t/\Delta x$)?

2. Show that the Lagrange interpolating polynomial in (7.12) is equivalent to the following Newton polynomial:

$$c_0 + (2-\alpha)\left[c_1 + (1-\alpha)\left(\frac{c_2}{2} - \alpha\frac{c_3}{6}\right)\right],$$

where

$$\begin{aligned}
c_0 &= \phi^n_{j-p-2},\\
c_1 &= \phi^n_{j-p-1} - c_0,\\
c_2 &= \phi^n_{j-p} - \phi^n_{j-p-1} - c_1,\\
c_3 &= \phi^n_{j-p+1} - 2\phi^n_{j-p} + \phi^n_{j-p-1} - c_2.
\end{aligned}$$

Compare the number of multiplications and additions required to evaluate the preceding equations with those required to evaluate (7.12).

3. Suppose that a semi-Lagrangian approximation to the constant-wind-speed advection equation uses quadratic polynomial interpolation as specified in (7.11).

(a) Derive the leading-order truncation error for this scheme.

(b) Determine the range of $U\Delta t/\Delta x$ for which the resulting scheme is identical to the Lax–Wendroff method (3.72). Also determine the values of $U\Delta t/\Delta x$ for which the scheme is identical to the method of Warming and Beam (3.79).

4. Determine the values of α for which the semi-Lagrangian approximation to the constant-wind-speed advection equation is stable when quadratic interpolation is used to evaluate $\phi(\tilde{x}_j^n, t^n)$ as in (7.11). Why is this scheme implemented by choosing p such that $|\alpha| \leq 1/2$?

5. Show that in comparison with a hypothetical scheme that exactly determines the value of ϕ at the departure point, the damping associated with the polynomial interpolation of $\phi(\tilde{x}_j^n, t^n)$ increases the stability of numerical approximations to (7.42) computed using the trapezoidal scheme (7.3).

6. Suppose that a noninterpolating three-time-level semi-Lagrangian scheme is used to compute approximate solutions to the variable-wind-speed advection equation

$$\frac{\partial \psi}{\partial t} + u(x)\frac{\partial \psi}{\partial x} = 0.$$

If the approximate solution is defined at the mesh points x_j and the velocity is available at both x_j and $x_{j+1/2}$, determine the strategy for choosing p that minimizes the truncation error in the resulting scheme. Should p be even, odd, or simply the integer such that $p\Delta x$ is closest to the departure point? Does this strategy yield stable solutions? How well does it generalize to two-dimensional problems?

7. Suppose that (7.44) is used to obtain approximate solutions to the prototype equation for forced scalar advection (7.42). How do the stability properties of the $2\Delta x$ waves compare with those of the $4\Delta x$ waves as a function of the Courant number $U\Delta t/\Delta x$?

8. Suppose that the two-level forward extrapolation (7.25) is replaced by the three-level scheme

$$u(t^{n+\frac{1}{2}}) = \frac{1}{8}\left[15u(t^n) - 10u(t^{n-1}) + 3u(t^{n-2})\right].$$

Determine the order of accuracy to which this method estimates $u(t^{n+1/2})$.

9. Show that backward trajectories computed with the Euler method in a spatially varying wind field can cross if $|\partial \mathbf{v}/\partial x|\Delta t$ exceeds unity.

Chapter 8
Physically Insignificant Fast Waves

One reason why explicit time differencing is widely used in the simulation of wave-like flows is that accuracy considerations and stability constraints often yield similar criteria for the maximum time step in numerical integrations of systems that support a single type of wave motion. Many fluid systems, however, support more than one type of wave motion, and in such circumstances accuracy considerations and stability constraints can yield very different criteria for the maximum time step. If explicit time differencing is used to construct a straightforward numerical approximation to the equations governing a system that supports several types of waves, the maximum stable time step will be limited by the Courant number associated with the most rapidly propagating wave, yet that rapidly propagating wave may be of little physical significance.

As an example, consider the Earth's atmosphere, which supports sound waves, gravity waves, and Rossby waves. Rossby waves propagate more slowly than gravity waves, which in turn move more slowly than sound waves. The maximum stable time step with which an explicit numerical method can integrate the full equations governing atmospheric motions will be limited by the Courant number associated with sound-wave propagation. If finite differences are used in the vertical, and the vertical grid spacing is 300 m, the maximum stable time step will be on the order of 1 s. Since sound waves have no direct meteorological significance, they need not be accurately simulated to obtain a good weather forecast. The quality of the weather forecast depends solely on the ability of the model to accurately simulate atmospheric disturbances that evolve on much slower timescales. Gravity waves can be accurately simulated with time steps on the order of 10–100 s; Rossby waves require a time step on the order of 500–5,000 s. To obtain a reasonably efficient numerical model for the simulation of atmospheric circulations, it is necessary to circumvent the stability constraint associated with sound-wave propagation and bring the maximum stable time step into closer agreement with the time-step limitations arising from accuracy considerations.

There are two basic approaches for circumventing the time-step constraint imposed by a rapidly moving, physically insignificant wave. The first approach is to approximate the full governing equations with a set of "filtered" equations that does

D.R. Durran, *Numerical Methods for Fluid Dynamics: With Applications to Geophysics*, Texts in Applied Mathematics 32, DOI 10.1007/978-1-4419-6412-0_8,

not support the rapidly moving wave. As an example, the full equations for stratified compressible flow might be approximated by the Boussinesq equations for incompressible flow. In this approach fundamental approximations to the original continuous equations are introduced prior to any numerical approximations that may be subsequently employed to discretize the filtered equations. The use of a filtered equation set may be motivated entirely by numerical considerations, or it may arise naturally from the standard approximations used in the study of a given physical phenomenon. Gravity waves, for example, are often studied in the context of Boussinesq incompressible flow to simplify and streamline the mathematical description of the problem.

The second approach for circumventing the time-step constraint imposed by a rapidly moving, physically insignificant wave leaves the continuous governing equations unmodified and relies on numerical techniques to stabilize the fast-moving wave. These numerical techniques achieve efficiency by sacrificing the accuracy with which the fast-moving wave is represented. Note that this approach is not appropriate in situations where the fast-moving wave needs to be accurately simulated, since small time steps are required to adequately resolve a fast-moving wave.

This chapter begins by examining techniques for the numerical solution of the Boussinesq equations, which constitute one of the most fundamental systems of filtered equations. Methods for the solution of a second system of filtered equations, the primitive equations, are presented in Sects. 8.5 and 8.6. Numerical methods for stabilizing the solution to problems that simultaneously support fast- and slow-moving waves are considered in Sects. 8.2–8.4. One of these techniques, the semi-implicit method, is frequently used to integrate the primitive equations in applications where the phenomena of primary interest are slow-moving Rossby waves. In such applications the numerical integration is stabilized with respect to two different types of physically insignificant, rapidly moving waves. Sound waves are filtered by the primitive-equation approximation, and the most rapidly moving gravity waves (and the Lamb wave) are stabilized by the semi-implicit time integration.

8.1 The Projection Method

The unapproximated mass-conservation equation (1.32) is a prognostic equation for the density that can be combined with the equation of state to form a prognostic equation for pressure. The equation of state can also be used to eliminate density from the momentum equations and thereby express the Euler equations as a closed system of five prognostic equations in five unknowns. When sound waves are filtered from the governing equations using the Boussinesq, anelastic, or pseudo-incompressible approximations, the approximate mass-conservation equations for these filtered systems do not depend on the local time derivative of the true density, and as a consequence, they cannot be used to obtain a prognostic equation for pressure. Since a prognostic equation is not available for the calculation of pressure, the filtered systems are not strictly hyperbolic, and their numerical solution cannot be obtained entirely through the use of explicit finite-difference schemes.

As an example, consider the Boussinesq equations (1.51), (1.56), and (1.60), which may be written in the form

$$\nabla \cdot \mathbf{v} = 0, \tag{8.1}$$

$$\frac{d\rho'}{dt} + w\frac{d\overline{\rho}}{dz} = 0, \tag{8.2}$$

$$\frac{\partial \mathbf{v}}{\partial t} + \frac{1}{\rho_0}\nabla p' = \mathbf{F}(\mathbf{v}, \rho'), \tag{8.3}$$

where

$$\mathbf{F}(\mathbf{v}, \rho') = -\mathbf{v} \cdot \nabla \mathbf{v} - g\frac{\rho'}{\rho_0}\mathbf{k}$$

and p' and ρ' are the deviations of the pressure and density from their values in a hydrostatically balanced reference state, $\overline{p}(z)$ and $\overline{\rho}(z)$. The unknown variables are the three velocity components, the perturbation density, and the perturbation pressure. There is no prognostic equation available to determine the time tendency of p'. The perturbation pressure field at a given instant can, however, be diagnosed from the instantaneous velocity and perturbation density fields by solving the Poisson equation

$$\nabla^2 p' = \rho_0 \nabla \cdot \mathbf{F}, \tag{8.4}$$

which can be derived by taking the divergence of (8.3) and then using (8.1). The perturbation pressure satisfying (8.4) is the instantaneous pressure distribution that will keep the evolving velocity field nondivergent.

8.1.1 Forward-in-Time Implementation

The *projection method* (Chorin 1968; Témam 1969) is a classical technique that may be used to obtain numerical solutions to the Boussinesq system. Suppose the momentum equation is integrated over a time interval Δt to yield

$$\int_{t^n}^{t^{n+1}} \frac{\partial \mathbf{v}}{\partial t}\, \mathrm{d}t = -\int_{t^n}^{t^{n+1}} \frac{1}{\rho_0}\nabla p'\, \mathrm{d}t + \int_{t^n}^{t^{n+1}} \mathbf{F}(\mathbf{v}, \rho')\, \mathrm{d}t, \tag{8.5}$$

where $t^n = n\Delta t$. Define the quantity $\tilde{p}^{n+1}$ such that

$$\Delta t\, \nabla \tilde{p}^{n+1} = \int_{t^n}^{t^{n+1}} \frac{1}{\rho_0}\nabla p'\, \mathrm{d}t.$$

Note that $\tilde{p}^{n+1}$ is not necessarily equal to the actual perturbation pressure at any particular time. Using the definition of $\tilde{p}^{n+1}$, one may write (8.5) as

$$\mathbf{v}^{n+1} - \mathbf{v}^n = -\Delta t\, \nabla \tilde{p}^{n+1} + \int_{t^n}^{t^{n+1}} \mathbf{F}(\mathbf{v}, \rho')\, \mathrm{d}t. \tag{8.6}$$

Define $\tilde{\mathbf{v}}$ such that

$$\tilde{\mathbf{v}} = \mathbf{v}^n + \int_{t^n}^{t^{n+1}} \mathbf{F}(\mathbf{v}, \rho')\, dt. \tag{8.7}$$

As noted by Orszag et al. (1986), the preceding integral can be conveniently evaluated using an explicit finite-difference scheme such as the third-order Adams–Bashforth method (2.80). Equations (8.6) and (8.7) imply that

$$\mathbf{v}^{n+1} = \tilde{\mathbf{v}} - \Delta t\, \nabla \tilde{p}^{n+1}, \tag{8.8}$$

which provides a formula for updating $\tilde{\mathbf{v}}$ to obtain the new velocity field $\mathbf{v}^{n+1}$ once $\tilde{p}^{n+1}$ has been determined.

A Poisson equation for $\tilde{p}^{n+1}$ that is analogous to (8.4) is obtained by taking the divergence of (8.8) and noting that $\nabla \cdot \mathbf{v}^{n+1} = 0$, in which case

$$\nabla^2 \tilde{p}^{n+1} = \frac{\nabla \cdot \tilde{\mathbf{v}}}{\Delta t}. \tag{8.9}$$

Boundary conditions for this equation are obtained by computing the dot product of the unit vector normal to the boundary ($\mathbf{n}$) with each term of (8.8) to yield

$$\frac{\partial \tilde{p}^{n+1}}{\partial n} = -\frac{1}{\Delta t}\, \mathbf{n} \cdot (\mathbf{v}^{n+1} - \tilde{\mathbf{v}}). \tag{8.10}$$

If there is no flow normal to the boundary, the preceding equation reduces to

$$\frac{\partial \tilde{p}^{n+1}}{\partial n} = \frac{\mathbf{n} \cdot \tilde{\mathbf{v}}}{\Delta t}, \tag{8.11}$$

which eliminates the implicit coupling between $\tilde{p}^{n+1}$ and $\mathbf{v}^{n+1}$ that is present in the general boundary condition (8.10). In this particularly simple case in which an inviscid fluid is bounded by rigid walls, the projection method is implemented by first updating (8.7), which accounts for the time tendencies produced by advection and buoyancy forces, and then solving (8.9) subject to the boundary conditions (8.11). As the final step of the algorithm, $\mathbf{v}^{n+1}$ is obtained by projecting $\tilde{\mathbf{v}}$ onto the subspace of nondivergent vectors using (8.8).

The preceding algorithm loses some of its simplicity when the computation of $\mathbf{v}^{n+1}$ is coupled with that of $\tilde{p}^{n+1}$, as would be the case if a wave-permeable boundary condition replaced the rigid-wall condition that $\mathbf{n} \cdot \mathbf{v}^{n+1} = 0$. In practice, the coupling between $\mathbf{v}^{n+1}$ and $\tilde{p}^{n+1}$ is eliminated by imposing some approximation to the full, implicitly coupled boundary condition. Coupling between $\mathbf{v}^{n+1}$ and $\tilde{p}^{n+1}$ may also occur when the projection method is applied to viscous flows with a no-slip condition at the boundary. The no-slip condition that $\mathbf{v} = 0$ at the boundary reduces (8.10) to

$$\frac{\partial \tilde{p}^{n+1}}{\partial n} = \frac{1}{\Delta t}\, \mathbf{n} \cdot \int_{t^n}^{t^{n+1}} -g\frac{\rho'}{\rho_0}\mathbf{k} + \nu \nabla^2 \mathbf{v}\, dt, \tag{8.12}$$

where viscous forcing is now included in the momentum equations and ν is the kinematic viscosity. High spatial resolution is often required to resolve the boundary layer in no-slip viscous flow. To maintain numerical stability in the high-resolution boundary layer without imposing an excessively strict limitation on the time step, the viscous terms are often integrated using implicit differencing[1] (Karniadakis et al. 1991). When the time integral of $\mathbf{F}(\mathbf{v}, \rho')$ includes viscous terms that are approximated using implicit finite differences, (8.12) is an implicit relation between $\tilde{p}^{n+1}$ and $\mathbf{v}^{n+1}$ whose solution is often computed via a fractional-step method. As noted by Orszag et al. (1986), the accuracy with which this boundary condition is approximated can significantly influence the accuracy of the overall solution. The design of optimal approximations to (8.12) has been the subject of considerable research. However, the emphasis in this book is not on viscous flow, and especially not on highly viscous flow in which the diffusion terms need to be treated implicitly for computational efficiency. The reader is referred to Boyd (1989) for further discussion of the use of the projection method in viscous no-slip flow.

8.1.2 Leapfrog Implementation

In atmospheric science the projection method is often implemented using leapfrog time differences, in which case (8.5) becomes

$$\mathbf{v}^{n+1} = \mathbf{v}^{n-1} - \frac{2\Delta t}{\rho_0}\nabla p'^n + 2\Delta t \mathbf{F}\left(\mathbf{v}^n, \rho'^n\right).$$

The solution procedure is very similar to the algorithm described in the preceding section. The velocity field generated by advection and buoyancy forces acting over the time period $2\Delta t$ is defined as

$$\tilde{\mathbf{v}} = \mathbf{v}^{n-1} + 2\Delta t \mathbf{F}\left(\mathbf{v}^n, \rho'^n\right);$$

then the Poisson equation for p'^n is

$$\nabla^2 p'^n = \frac{\rho_0}{2\Delta t}\nabla\cdot\tilde{\mathbf{v}},$$

and the velocity field is updated using the relation

$$\mathbf{v}^{n+1} = \tilde{\mathbf{v}} - 2\Delta t \nabla p'^n.$$

Some technique, such as Asselin time filtering (2.65), must also be used to prevent time-splitting instability in the leapfrog solution to nonlinear problems.

One difference between this approach and the standard projection method is that by virtue of the leapfrog time difference, the pressure field that ensures the

[1] Explicit time differencing can still be used for the advection terms because the wind speed normal to the boundary decreases as the fluid approaches the boundary.

nondivergence of $\mathbf{v}^{n+1}$ is the actual pressure at time $n\Delta t$. The pressure must, nevertheless, be updated at the same point in the integration cycle at which $\tilde{p}^{n+1}$ is obtained in the standard projection method, i.e., partway through the calculation of $\mathbf{v}^{n+1}$. Thus, the same problems with implicit coupling between the pressure and $\mathbf{v}^{n+1}$ arise in both the standard and the leapfrog projection methods. If viscosity is included in the momentum equations, stability considerations require that the contribution of viscosity to $\mathbf{F}(\mathbf{v}, \rho')$ be evaluated at time level $n-1$, so that the viscous terms are treated using forward differencing over a time interval of $2\Delta t$. This is not a particularly accurate way to represent the viscous terms and is not suitable for highly viscous flow, in which the viscous terms are more efficiently integrated using implicit time differences.

8.1.3 Solving the Poisson Equation for Pressure

Suppose that the Boussinesq equations are to be solved in a two-dimensional x–z domain and that the velocity and pressure variables are staggered as indicated in Fig. 4.6. Approximating the diagnostic pressure equation (8.9) using the standard five-point finite-difference stencil for the two-dimensional Laplacian, one obtains

$$\delta_x^2 \tilde{p}^{n+1} + \delta_z^2 \tilde{p}^{n+1} = \frac{1}{\Delta t}\left(\delta_x \tilde{u} + \delta_z \tilde{w}\right). \tag{8.13}$$

This is an implicit algebraic relation for the $\tilde{p}_{i,j}^{n+1}$. If pressure is defined at M grid points in x and N points in z, an $M \times N$ system of linear algebraic equations must be solved to determine the pressure. Let the unknown grid-point values of the pressure be ordered such that

$$\mathbf{p} = (\tilde{p}_{1,1}^{n+1}, \tilde{p}_{1,2}^{n+1}, \ldots, \tilde{p}_{1,N}^{n+1}, \tilde{p}_{2,1}^{n+1}, \ldots, \tilde{p}_{M,N}^{n+1}).$$

Then the system may be written as the matrix equation

$$\mathbf{A}\mathbf{p} = \mathbf{f}, \tag{8.14}$$

in which $\mathbf{f}$ is an identically ordered vector containing the numerically evaluated divergence of $\tilde{\mathbf{v}}$. The matrix $\mathbf{A}$ is very sparse, with only five nonzero diagonals. In practical applications the number of unknown pressures can easily exceed one million, and to solve (8.14) efficiently, it is important to take advantage of the sparseness of $\mathbf{A}$. Direct methods based on some variant of Gaussian elimination are therefore not appropriate. Direct methods for band matrices are also not suitable because the bandwidth of $\mathbf{A}$ is not 5, but $2N+1$, and direct methods for band matrices do not preserve sparseness within the band.

Direct solutions to (8.14) can, nevertheless, be efficiently obtained by exploiting the block structure of $\mathbf{A}$. For simplicity, suppose that (8.13) is to be solved subject to Dirichlet boundary conditions. Then the diagonal of $\mathbf{A}$ contains M copies of the $N \times N$ tridiagonal submatrix

$$
\begin{pmatrix}
-4 & 1 & & & & \\
1 & -4 & 1 & & & \\
 & \ddots & \ddots & \ddots & & \\
 & & \ddots & \ddots & \ddots & \\
 & & & 1 & -4 & 1 \\
 & & & & 1 & -4
\end{pmatrix},
$$

and the superdiagonal and subdiagonal are made up of $M - 1$ copies of the $N \times N$ identity matrix. This system can be efficiently solved using block-cyclic reduction (Golub and van Loan 1996, p. 177). Numerical codes for the solution of two- and three-dimensional Poisson equations subject to the most common types of boundary conditions may be accessed through the Internet at several cites, including the National Institute of Standards and Technology's Guide to Available Mathematical Software (NIST/GAMS, http://gams.nist.gov), the National Center for Atmospheric Research's Mathematical and Statistical Libraries (NCAR, http://www.cisl.ucar.edu/softlib/mathlib.html), and the Netlib Repository at the Oak Ridge National Laboratory (ORNL, http://www.netlib.org).

Numerical solutions to the anelastic equations (1.33), (1.52), and (1.64) can be obtained using the projection method in essentially the same manner as that for the Boussinesq system, except that the elliptic equation for the pressure in the anelastic system is slightly more complicated than a Poisson equation. After multiplication by $\overline{\rho}(z)$, the anelastic momentum equation (1.64) may be written

$$
\frac{\partial \overline{\rho} \mathbf{v}}{\partial t} + c_p \overline{\rho} \nabla (\overline{\theta} \pi') = \overline{\rho} \mathbf{F}, \tag{8.15}
$$

where

$$
\mathbf{F}(\mathbf{v}, \theta') = -\mathbf{v} \cdot \nabla \mathbf{v} + g \frac{\theta'}{\overline{\theta}} \mathbf{k}. \tag{8.16}
$$

An elliptic equation for pressure is obtained by taking the divergence of (8.15) and using the anelastic continuity equation

$$
\nabla \cdot (\overline{\rho} \mathbf{v}) = 0,
$$

to yield

$$
\nabla \cdot \left[\overline{\rho} \nabla (\overline{\theta} \pi') \right] = \nabla \cdot \left(\frac{\overline{\rho}}{c_p} \mathbf{F} \right).
$$

A linear system of algebraic equations for $\pi'_{i,j}$ is obtained after approximating the derivatives in the preceding expression by finite differences, but since $\overline{\rho}$ and $\overline{\theta}$ are functions of z, the structure of the coefficient matrix for this system is less uniform than that for the Boussinesq system. Nevertheless, the resulting linear system can still be efficiently solved by generalizations of the block-cyclic-reduction algorithm, and numerical codes for the solution of this problem appear in the previously noted software libraries.

When the projection method is used to solve the pseudo-incompressible equations (1.33), (1.37), and (1.54), the elliptic pressure equation becomes

$$\nabla \cdot \left[\bar{\rho}\bar{\theta}\theta\nabla\pi'\right] = \nabla \cdot \left(\frac{\bar{\rho}\bar{\theta}}{c_p}\mathbf{F}\right), \tag{8.17}$$

where **F** is once again defined by (8.16). The finite-difference approximation to this equation still produces a very sparse linear algebraic system, with only five nonzero diagonals, but since the coefficient of each second derivative includes the factor θ, which is an arbitrary function of x, y, and z, it is not possible to solve the system by block-cyclic reduction – iterative methods must be used. Iterative methods may also need to be employed to determine the pressure in the Boussinesq and anelastic systems when those equations are solved on a curvilinear grid (such as a terrain-following coordinate system) because the coefficient structure in the elliptic pressure equation is usually complicated by the coordinate transformation.

The two most commonly used techniques for the iterative solution of the sparse linear-algebraic systems that arise in computational fluid dynamics are the preconditioned conjugate gradient method and the multigrid method. The mathematical basis for both of these methods is very nicely reviewed in Ferziger and Perić (1997) and LeVeque (2007) and will not be covered in this text. Additional information about multigrid methods may be found in Briggs (1987), Hackbusch (1985), and, in the context of geophysical fluid dynamics, Adams et al. (1992). Conjugate-residual solvers are discussed in more detail in Golub and van Loan (1996) and in the context of atmospheric science in Smolarkiewicz and Margolin (1994) and Skamarock et al. (1997). Both multigrid and preconditioned conjugate residual solvers are available in the previously mentioned software libraries.

8.2 The Semi-implicit Method

As an alternative to filtering the governing equations to eliminate insignificant fast waves, one can retain the unapproximated governing equations and use numerical techniques to stabilize the simulation of the fast-moving waves. One common way to improve numerical stability is through the use of implicit time differences such as the backward and the trapezoidal methods.[2] Implicit methods can, however, produce rather inaccurate solutions when the time step is too large. It is therefore useful to analyze the effect of the time step on the accuracy of fully implicit solutions to wave-propagation problems before discussing the true semi-implicit method.

[2] Higher-order implicit schemes are, however, not necessarily more stable than related explicit methods. Backward and trapezoidal differencing are the first- and second-order members of the Adams–Moulton family of implicit time-integration schemes. The third- and fourth-order Adams–Moulton schemes generate amplifying solutions to the oscillation equation (2.19) for any choice of time step, whereas their explicit cousins, the third- and fourth-order Adams–Bashforth schemes, produce stable nonamplifying solutions whenever the time step is sufficiently small.

8.2.1 Large Time Steps and Poor Accuracy

Suppose that a differential–difference approximation to the one-dimensional advection equation

$$\frac{\partial \psi}{\partial t} + c\frac{\partial \psi}{\partial x} = 0 \tag{8.18}$$

is constructed in which finite differences are used to represent the time derivative, and the spatial derivative is not discretized. If the time derivative is approximated using leapfrog differencing such that

$$\frac{\phi^{n+1} - \phi^{n-1}}{2\Delta t} + c\left(\frac{d\phi}{dx}\right)^n = 0,$$

then wave solutions of the form

$$\phi^n(x) = \mathrm{e}^{\mathrm{i}(kx-\omega n\Delta t)} \tag{8.19}$$

must satisfy the semidiscrete dispersion relation

$$\omega = \frac{1}{\Delta t}\arcsin(ck\Delta t). \tag{8.20}$$

The phase speed of the leapfrog-differenced solution is

$$c_{\mathrm{lf}} = \frac{\omega}{k} = \frac{\arcsin(ck\Delta t)}{k\Delta t}. \tag{8.21}$$

The stability constraint $|ck\Delta t| < 1$ associated with the preceding leapfrog scheme can be avoided by switching to trapezoidal differencing. Many semi-implicit formulations use a combination of leapfrog and trapezoidal differencing, and in those formulations the trapezoidal time difference is computed over an interval of $2\Delta t$. To facilitate the application of this analysis to these semi-implicit formulations, and to compare the trapezoidal and leapfrog schemes more directly, (8.18) will be approximated using trapezoidal differencing over a $2\Delta t$-wide stencil such that

$$\frac{\phi^{n+1} - \phi^{n-1}}{2\Delta t} + \frac{c}{2}\left[\left(\frac{d\phi}{dx}\right)^{n+1} + \left(\frac{d\phi}{dx}\right)^{n-1}\right] = 0.$$

Wave solutions to this scheme must satisfy the dispersion relation

$$\omega = \frac{1}{\Delta t}\arctan(ck\Delta t). \tag{8.22}$$

The phase speed of the trapezoidally differenced solution is

$$c_{\mathrm{t}} = \frac{\arctan(ck\Delta t)}{k\Delta t}.$$

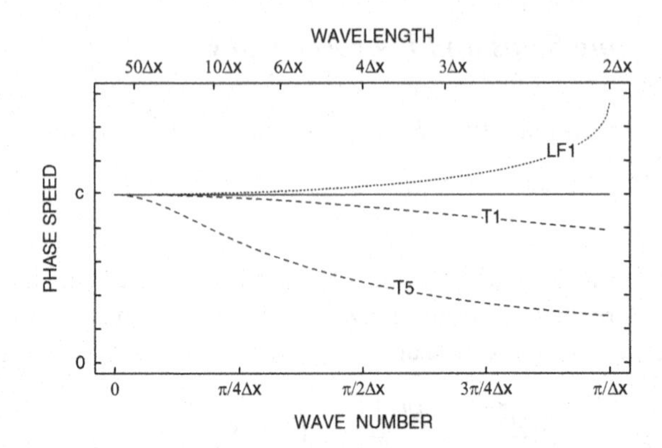

Fig. 8.1 Phase speed of leapfrog (*dotted line*) and $2\Delta t$-trapezoidal (*dashed line*) approximations to the advection equation when $c\Delta t/\Delta x = 1/\pi$ (*LF1* and *T1*), and for the trapezoidal solution when $c\Delta t/\Delta x = 5/\pi$ (*T5*)

The phase-speed errors generated by the leapfrog and $2\Delta t$-trapezoidal methods are compared in Fig. 8.1. The phase speed at a fixed Courant number is plotted as a function of spatial wave number, with the wave number axis scaled by $1/\Delta x$. These curves may therefore be interpreted as giving the phase speed that would be obtained if the spatial dependence of the numerical solution were represented by a Fourier spectral method with a cutoff wavelength of $2\Delta x$. When $c\Delta t/\Delta x = 1/\pi$, the errors generated by the leapfrog and the $2\Delta t$-trapezoidal methods are similar in magnitude and opposite in sign. The leapfrog scheme is unstable for Courant numbers greater than $1/\pi$, but solutions can still be obtained using the trapezoidal scheme. The phase-speed errors in the $2\Delta t$-trapezoidal solution computed with $c\Delta t/\Delta x = 5/\pi$ are, however, rather large. Even modes with relatively good spatial resolution, such as a $10\Delta x$ wave, are in significant error.

The deceleration generated by $2\Delta t$-trapezoidal differencing may be alternatively expressed in terms of the reduced phase speed

$$\hat{c} = c\cos(\omega\Delta t).$$

Then the $2\Delta t$-trapezoidal dispersion relation (8.22) assumes the form

$$\omega = \frac{1}{\Delta t}\arcsin(\hat{c}k\Delta t),$$

and the phase speed of the $2\Delta t$-trapezoidal solution becomes

$$c_{\mathrm{t}} = \frac{\omega}{k} = \frac{\arcsin(\hat{c}k\Delta t)}{k\Delta t}.$$

The preceding expressions differ from the corresponding expressions for the leapfrog scheme (8.20) and (8.21) in that the true propagation speed, c, has been replaced by the reduced speed $\hat{c}$. As the time step increases, $\hat{c}$ decreases, so $|\hat{c}k\Delta t|$

remains less than 1 and the numerical solution remains stable, but the relative error in $\hat{c}$ can become arbitrarily large. As a consequence, it is not possible to take advantage of the unconditional stability of the trapezoidal method by using very large time steps to solve wave-propagation problems unless one is willing to tolerate a considerable decrease in the accuracy of the solution.

8.2.2 A Prototype Problem

The loss of accuracy associated with poor temporal resolution that can occur using implicit numerical methods is not a problem if the poorly resolved waves are not physically significant. If the fastest-moving waves are insignificant, the accuracy constraints imposed on the time step by these waves can be ignored, and provided the scheme is unconditionally stable, a good solution can be obtained using any time step that adequately resolves the slower-moving features of primary physical interest. A simple but computationally inefficient way to ensure the unconditional stability of a numerical scheme is to use trapezoidal time differencing throughout the approximate equations. It is, however, more efficient to implicitly evaluate only those terms in the governing equations that are crucial to the propagation of the fast wave and to approximate the remaining terms with some explicit time-integration scheme. This is the fundamental strategy in the "semi-implicit" approach, which gains efficiency relative to a "fully implicit" method by reducing the complexity of the implicit algebraic equations that must be solved during each integration step. Semi-implicit differencing is particularly attractive when all the terms that are evaluated implicitly are linear functions of the unknown variables.

To investigate the stability of semi-implicit time-differencing schemes, consider a prototype ordinary differential equation of the form

$$\frac{d\psi}{dt} + \mathrm{i}\omega_{\mathrm{H}}\psi + \mathrm{i}\omega_{\mathrm{L}}\psi = 0. \tag{8.23}$$

This is simply a version of the oscillation equation (2.19) in which the oscillatory forcing is divided into high-frequency (ω_{H}) and low-frequency (ω_{L}) components. The division of the forcing into two terms may appear to be rather artificial, but the dispersion relation associated with wavelike solutions to more complex systems of governing equations (such as the shallow-water system discussed in the next section) often has individual roots of the form

$$\omega = \omega_{\mathrm{H}} + \omega_{\mathrm{L}},$$

and (8.23) serves as the simplest differential equation describing the time dependence of such waves.

The simplest semi-implicit approximation to (8.23) is

$$\frac{\phi^{n+1} - \phi^n}{\Delta t} + \mathrm{i}\omega_{\mathrm{H}}\phi^{n+1} + \mathrm{i}\omega_{\mathrm{L}}\phi^n = 0.$$

The stability and the accuracy of this scheme have already been analyzed in connection with (2.21); it is first-order accurate and is stable whenever $|\omega_{\mathrm{L}}| < |\omega_{\mathrm{H}}|$. Since $|\omega_{\mathrm{L}}| < |\omega_{\mathrm{H}}|$ by assumption, the method is stable for all Δt. The weakness of this scheme is its low accuracy. A more accurate second-order method can be obtained using the centered-in-time formula

$$\frac{\phi^{n+1} - \phi^{n-1}}{2\Delta t} + \mathrm{i}\omega_{\mathrm{H}}\left(\frac{\phi^{n+1} + \phi^{n-1}}{2}\right) + \mathrm{i}\omega_{\mathrm{L}}\phi^{n} = 0. \tag{8.24}$$

The stability of this method may be investigated by considering the behavior of oscillatory solutions of the form $\exp(-\mathrm{i}\omega n\Delta t)$, which satisfy (8.24) when

$$\sin\tilde{\omega} = \tilde{\omega}_{\mathrm{H}}\cos\tilde{\omega} + \tilde{\omega}_{\mathrm{L}}, \tag{8.25}$$

where

$$\tilde{\omega} = \omega\Delta t, \quad \tilde{\omega}_{\mathrm{H}} = \omega_{\mathrm{H}}\Delta t, \quad \text{and} \quad \tilde{\omega}_{\mathrm{L}} = \omega_{\mathrm{L}}\Delta t.$$

To solve for $\tilde{\omega}$, let $\tan\beta = \tilde{\omega}_H$, then (8.25) becomes

$$\sin\tilde{\omega} = \tan\beta\cos\tilde{\omega} + \tilde{\omega}_{\mathrm{L}},$$

or equivalently,

$$\sin\tilde{\omega}\cos\beta - \sin\beta\cos\tilde{\omega} = \tilde{\omega}_{\mathrm{L}}\cos\beta.$$

By the Pythagorean theorem, $\cos\beta = (1 + \tilde{\omega}_{\mathrm{H}}^2)^{-1/2}$, and the preceding equation reduces to

$$\sin(\tilde{\omega} - \beta) = \tilde{\omega}_{\mathrm{L}}(1 + \tilde{\omega}_{\mathrm{H}}^2)^{-1/2},$$

or equivalently,

$$\tilde{\omega} = \arctan(\tilde{\omega}_{\mathrm{H}}) + \arcsin\left(\tilde{\omega}_{\mathrm{L}}(1 + \tilde{\omega}_{\mathrm{H}}^2)^{-1/2}\right).$$

The semi-implicit scheme (8.24) will be stable when the $\tilde{\omega}$ satisfying this equation are real and distinct, which is guaranteed when

$$\tilde{\omega}_{\mathrm{L}}^2 \leq 1 + \tilde{\omega}_{\mathrm{H}}^2. \tag{8.26}$$

Since by assumption $|\omega_{\mathrm{L}}| \leq |\omega_{\mathrm{H}}|$, (8.24) is stable for all Δt. Note that (8.26) will also be satisfied whenever $|\omega_{\mathrm{L}}\Delta t| \leq 1$, implying that semi-implicit differencing permits an increase in the maximum stable time step relative to that for a fully explicit approximation even in those cases where $|\omega_{\mathrm{L}}| > |\omega_{\mathrm{H}}|$, because the terms approximated with the trapezoidal difference do not restrict the maximum stable time step.

As discussed in Sect. 3.5.1, semi-implicit time differencing may also be used to stabilize the diffusion operator in some advection–diffusion problems. The gain in the maximum stable time step achieved using a trapezoidal time difference for the diffusion term in conjunction with a leapfrog approximation to the advection is not, however, particularly impressive. A much more stable approximation to the advection–diffusion problem is obtained using the third-order Adams–Bashforth method to integrate the advection terms and trapezoidal differencing to approximate the diffusion term, but the advantages of the semi-implicit Adams–Bashforth–trapezoidal formulation do not carry over to the fast-wave–slow-wave problem in a completely straightforward manner. If, for example, λ is replaced by $\mathrm{i}\omega_{\mathrm{H}}$ in (3.95) so that the trapezoidally differenced term represents a fast-moving wave, the stability of the resulting scheme is still quite limited (the scheme is unstable for all $\omega_{\mathrm{H}}\Delta t$ greater than approximately 1.8).

8.2.3 Semi-implicit Solution of the Shallow-Water Equations

The shallow-water equations (1.25)–(1.27) support rapidly moving gravity waves. If there are spatial variations in the potential vorticity of the undisturbed system f/H, the shallow-water equations can also support slowly propagating potential-vorticity (or Rossby) waves. In many large-scale atmospheric and oceanic models, the Rossby waves are of greater physical significance than the faster-moving gravity waves, and the Rossby waves can be efficiently simulated using semi-implicit time differencing to accommodate the Courant–Friedrichs–Lewy (CFL) stability condition associated with gravity-wave propagation. The simplest example in which to illustrate the influence of semi-implicit differencing on the CFL condition for gravity waves is provided by (4.1) and (4.2), which are the one-dimensional shallow-water equations linearized about a reference state with a constant fluid velocity U and fluid depth H. If the mean-flow velocity is less than the phase speed of a shallow-water gravity wave, the numerical integration can be stabilized by evaluating those terms responsible for gravity-wave propagation with trapezoidal differencing; leapfrog differencing can be used for the remaining terms (Kwizak and Robert 1971). The terms essential to gravity-wave propagation are the hydrostatic pressure gradient $(g\partial h/\partial x)$ in (4.1) and the velocity divergence in (4.2), so the semi-implicit approximation to the linearized shallow-water system is

$$\delta_{2t}u^n + U\frac{du^n}{dx} + g\left\langle\frac{dh^n}{dx}\right\rangle^{2t} = 0, \tag{8.27}$$

$$\delta_{2t}h^n + U\frac{dh^n}{dx} + H\left\langle\frac{du^n}{dx}\right\rangle^{2t} = 0, \tag{8.28}$$

where the finite-difference operator δ_t and the averaging operator $\langle\ \rangle^t$ are defined by (A.1) and (A.2) in the Appendix. Solutions to (8.27) and (8.28) exist of the form $\mathrm{e}^{\mathrm{i}(kx-\omega j\Delta t)}$, provided that k and ω satisfy the semidiscrete dispersion relation

$$\sin\omega\Delta t = Uk\Delta t \pm ck\Delta t\cos\omega\Delta t,$$

in which $c = \sqrt{gH}$. This dispersion relation has the same form as (8.25), and as demonstrated in the preceding section, the method will be stable, provided that $|U| \leq c$, or equivalently, whenever the phase speed of shallow-water gravity waves exceeds the speed of the mean flow.

The Coriolis force has been neglected in the preceding shallow-water system, and as a consequence, there are no Rossby-wave solutions to (8.27) and (8.28). In a more general system that does include the Coriolis force,[3] semi-implicit time differencing leads to a system that is stable whenever the CFL condition for the Rossby waves is satisfied. This general case is examined in more detail in Problems 1–3 at the end of this chapter.

Instead of considering the complications introduced by the presence of Rossby waves, consider the nonlinear equivalent of the preceding linearized system:

$$\frac{\partial u}{\partial t} + u\frac{\partial u}{\partial x} + g\frac{\partial h}{\partial x} = 0, \tag{8.29}$$

$$\frac{\partial h}{\partial t} + u\frac{\partial h}{\partial x} + h\frac{\partial u}{\partial x} = 0. \tag{8.30}$$

As before, a semi-implicit algorithm can be obtained using trapezoidal time differences to evaluate the pressure gradient in (8.29) and the velocity divergence in (8.30). The term involving the velocity divergence is, however, nonlinear, and an implicit system of nonlinear algebraic equations will be generated if the time integral of $h\,\partial u/\partial x$ is approximated using the trapezoidal method. To avoid solving a nonlinear algebraic equation at every time step, one splits the velocity divergence in (8.30) into two terms such that

$$\frac{\partial \eta}{\partial t} + u\frac{\partial \eta}{\partial x} + H\frac{\partial u}{\partial x} + \eta\frac{\partial u}{\partial x} = 0,$$

where the total fluid depth has been divided into a constant-mean component H and a perturbation $\eta(x,t)$. The standard semi-implicit approximation to the preceding expression takes the form

$$\delta_{2t}\eta^n + u^n\frac{d\eta^n}{dx} + H\left(\frac{du^n}{dx}\right)^{2t} + \eta^n\frac{du^n}{dx} = 0; \tag{8.31}$$

only the linear term involving the constant depth H is treated implicitly. The time differencing of the nonlinear momentum equation is identical to that for the linearized system

$$\delta_{2t}u^n + u^n\frac{du^n}{dx} + g\left(\frac{d\eta^n}{dx}\right)^{2t} = 0. \tag{8.32}$$

Leaving aside possible problems with nonlinear instability, one would intuitively expect that solutions to (8.31) and (8.32) would be unconditionally stable, provided that $\max|u| < \sqrt{gH}$ and $\eta \ll H$, which is to say that stability would require the gravity-wave phase speed determined by the mean fluid depth to greatly exceed any local augmentation of the phase speed induced by a local increase in depth.

[3] The inclusion of the Coriolis force also requires the inclusion of an additional prognostic equation for the other component of the horizontal velocity.

The impact of a perturbation in the fluid depth on the stability of the semi-implicit scheme is most easily evaluated if (8.31) and (8.32) are linearized about a reference state with no mean flow and a horizontally uniform perturbation in the depth $\overline{\eta}$. The semi-implicit approximation to this linearized system is

$$\delta_{2t} u^n + g \left\langle \frac{d\eta^n}{dx} \right\rangle^{2t} = 0,$$

$$\delta_{2t} \eta^n + H \left\langle \frac{du^n}{dx} \right\rangle^{2t} + \overline{\eta} \frac{du^n}{dx} = 0.$$

Letting $\nu = \sqrt{gH}\, k\Delta t$ and $r = \overline{\eta}/H$, solutions to this system satisfy the dispersion relation

$$\sin^2 \omega\Delta t = \nu^2 (\cos^2 \omega\Delta t + r \cos \omega\Delta t),$$

which is a quadratic equation in $\cos \omega\Delta t$,

$$(\nu^2 + 1)\cos^2 \omega\Delta t + r\nu^2 \cos \omega\Delta t - 1 = 0, \tag{8.33}$$

whose roots are

$$\cos \omega\Delta t = \frac{-r\nu^2 \pm (r^2\nu^4 + 4\nu^2 + 4)^{1/2}}{2(\nu^2 + 1)}.$$

The scheme will be stable when ω is real. Since the radicand is always positive, the right side of the preceding expression is always real, and real solutions for ω are obtained when the magnitudes of both roots of (8.33) are less than or equal to unity.

Let $s = \cos \omega\Delta t$ be one of the roots of (8.33). The identity

$$(x - s_1)(x - s_2) = x^2 - (s_1 + s_2)x + s_1 s_2$$

implies that the sum and product of the roots of the quadratic equation (8.33) satisfy

$$s_1 + s_2 = \frac{-r\nu^2}{\nu^2 + 1} \quad \text{and} \quad s_1 s_2 = \frac{-1}{\nu^2 + 1}. \tag{8.34}$$

When $r = 0$,

$$|s_1| = |s_2| = \frac{1}{(\nu^2 + 1)^{1/2}} \leq 1,$$

and the scheme is stable. As $|r|$ increases, the magnitude of one of the roots eventually exceeds unity, and the scheme becomes unstable. The critical values of r beyond which the scheme becomes unstable may be determined by substituting $s_1 = 1$ and then $s_1 = -1$ into (8.34) to obtain the stability criterion $|r| \leq 1$ or $|\overline{\eta}| \leq H$. Instability will not occur unless the perturbation fluid depth exceeds the mean depth (in the $U = 0$ case). This may appear to be a very generous criterion; however, the local phase speed of a shallow-water gravity wave is

$$c = \sqrt{g(H + \overline{\eta})},$$

so numerical stability requires that the local wave speed be no faster than a factor of $\sqrt{2}$ times the mean wave speed. As will be discussed in the next section, the horizontal phase speed of quasi-hydrostatic internal gravity waves is proportional to the Brunt–Väisälä frequency, and the ratio of the local mean Brunt–Väisälä frequency in the polar regions of the Earth's atmosphere to that in the middle latitudes can easily exceed $\sqrt{2}$. Thus, it is generally necessary to specify the horizontally uniform reference state using numerical values from that region in the domain where gravity waves propagate at maximum speed. As a consequence, the reference-state stratification in most global atmospheric models is chosen to be isothermal (Simmons et al. 1978). The application of the semi-implicit method to global atmospheric models is discussed further in Sect. 8.6.5.

8.2.4 Semi-implicit Solution of the Euler Equations

Now consider how semi-implicit differencing can be used to eliminate the stability constraint imposed by sound waves in the numerical solution of the Euler equations for stratified flow. To present the numerical approach with a minimum of extraneous detail, it is useful to consider a simplified set of compressible equations that can be obtained from the linearized system (1.41)–(1.44) by the transformation of variables

$$u = \left(\frac{\overline{\rho}}{\rho_0}\right)^{1/2} u', \qquad P = \left(\frac{\overline{\rho}}{\rho_0}\right)^{1/2} c_p \overline{\theta} \pi' \tag{8.35}$$

$$w = \left(\frac{\overline{\rho}}{\rho_0}\right)^{1/2} w', \qquad b = \left(\frac{\overline{\rho}}{\rho_0}\right)^{1/2} \frac{g}{\overline{\theta}} \theta', \tag{8.36}$$

which removes the influence of the decrease in the mean density with height on the magnitudes of the dependent variables. This transformation does not symmetrize the system as nicely as (1.45) and (1.46), but P and b have more direct interpretations as normalized pressure and buoyancy than do the thermodynamic variables introduced in (1.45) and (1.46).

The transformed vertical momentum and pressure equations take the form

$$\left(\frac{\partial}{\partial t} + U\frac{\partial}{\partial x}\right) w + \left(\frac{\partial}{\partial z} + \Gamma\right) P = b,$$

and

$$\left(\frac{\partial}{\partial t} + U\frac{\partial}{\partial x}\right) P + c_s^2 \left[\frac{\partial u}{\partial x} + \left(\frac{\partial}{\partial z} - \Gamma\right) w\right] = 0,$$

where

$$\Gamma = \frac{1}{2\overline{\rho}}\frac{d\overline{\rho}}{dz} + \frac{g}{c_s^2} \qquad \text{and} \qquad c_s^2 = \frac{c_p}{c_v} R\overline{T}. \tag{8.37}$$

In an isothermal atmosphere at temperature T_0,

$$\Gamma = \frac{3}{14}\frac{g}{RT_0}.$$

If $T_0 = 0°\mathrm{C}$, $\Gamma = 2.7 \times 10^{-5}\ \mathrm{m}^{-1}$, implying that the vertical derivative in the operators

$$\left(\frac{\partial}{\partial z} \pm \Gamma\right)$$

will exceed the term involving Γ by an order of magnitude in all waves with vertical wavelengths shorter than 100 km. The terms involving Γ can therefore be neglected in most applications,[4] in which case the transformed Euler equations become

$$\left(\frac{\partial}{\partial t} + U\frac{\partial}{\partial x}\right)u + \frac{\partial P}{\partial x} = 0, \tag{8.38}$$

$$\left(\frac{\partial}{\partial t} + U\frac{\partial}{\partial x}\right)w + \frac{\partial P}{\partial z} = b, \tag{8.39}$$

$$\left(\frac{\partial}{\partial t} + U\frac{\partial}{\partial x}\right)b + N^2 w = 0, \tag{8.40}$$

$$\left(\frac{\partial}{\partial t} + U\frac{\partial}{\partial x}\right)P + c_s^2\left(\frac{\partial u}{\partial x} + \frac{\partial w}{\partial z}\right) = 0. \tag{8.41}$$

The preceding simplified compressible system (8.38)–(8.41) can also be regarded as the linearization of a "compressible Boussinesq" system in which

$$b = -g\frac{\rho - \overline{\rho}(z)}{\rho_0}, \quad P = \frac{p - \overline{p}(z)}{\rho_0}, \quad N^2 = -\frac{g}{\rho_0}\frac{d\overline{\rho}}{dz}. \tag{8.42}$$

As in the standard Boussinesq approximation, the compressible Boussinesq system neglects the influence of density variations on inertia while retaining the influence of density variations on buoyancy and assumes that buoyancy is conserved following a fluid parcel. In contrast to the standard Boussinesq system, the compressible Boussinesq system does retain the influence of density fluctuations on pressure and thereby allows the formation of the prognostic pressure equation (8.41).

Suppose that the simplified compressible system (8.38)–(8.41) is approximated using leapfrog time differencing and that the spatial derivatives are computed using a Fourier pseudospectral method. Waves of the form

$$(u, w, b, P) = (u_0, w_0, b_0, P_0)e^{i(kx+\ell z-\omega n\Delta t)}$$

[4] The isothermal atmosphere does support a free wave (known as the Lamb wave, see Sect. 8.5) that disappears in the limit $\Gamma \to 0$, but it is not necessary to account for the Lamb wave in this discussion of semi-implicit differencing.

are solutions to this system, provided that ω, k, and ℓ satisfy the dispersion relation

$$\hat{\omega}^4 - c_s^2\left(k^2 + \ell^2 + N^2/c_s^2\right)\hat{\omega}^2 + N^2k^2c_s^2 = 0,$$

where

$$\hat{\omega} = \frac{\sin\omega\Delta t}{\Delta t} - Uk.$$

This dispersion relation is quadratic in $\hat{\omega}^2$ and has solutions

$$\hat{\omega}^2 = \frac{c_s^2}{2}\left(k^2 + \ell^2 + \frac{N^2}{c_s^2} \pm \left[\left(k^2 + \ell^2 + \frac{N^2}{c_s^2}\right)^2 - \frac{4N^2k^2}{c_s^2}\right]^{1/2}\right). \qquad (8.43)$$

The positive root yields the dispersion relation for sound waves; the negative root yields the dispersion relation for gravity waves. The individual dispersion relations for sound and gravity waves may be greatly simplified whenever the last term inside the square root in (8.43) is much smaller than the first term. One condition sufficient for this simplification, which is easily satisfied in most atmospheric applications, is that $N^2/c_s^2 \ll \ell^2$. If $N^2/c_s^2 \ll \ell^2$, then

$$\frac{4N^2k^2}{c_s^2} \ll \frac{2N^2k^2}{c_s^2} + 2k^2\ell^2 \leq \left(k^2 + \ell^2 + \frac{N^2}{c_s^2}\right)^2, \qquad (8.44)$$

and therefore the sound-wave dispersion relation is well approximated by

$$\hat{\omega}^2 = c_s^2\left(k^2 + \ell^2 + N^2/c_s^2\right). \qquad (8.45)$$

Dividing the terms inside the square root in (8.43) by $\left(k^2 + \ell^2 + N^2/c_s^2\right)^2$ and again using (8.44), one may well approximate the gravity wave-dispersion relation as

$$\hat{\omega}^2 = \frac{N^2k^2}{k^2 + \ell^2 + N^2/c_s^2}. \qquad (8.46)$$

Consider the time-step limitation imposed by sound-wave propagation. Using the definition of $\hat{\omega}$, one may express (8.45) as

$$\sin\omega\Delta t = \Delta t\left(Uk \pm c_s\left(k^2 + \ell^2 + N^2/c_s^2\right)^{1/2}\right).$$

Stable leapfrog solutions are obtained when the right side of this expression is a real number whose absolute value is less than unity. A necessary condition for stability is that

$$\left(|U|k_{\text{max}} + c_s(k_{\text{max}}^2 + \ell_{\text{max}}^2)^{1/2}\right)\Delta t < 1, \qquad (8.47)$$

where k_{max} and ℓ_{max} are the magnitudes of the largest horizontal and vertical wave numbers retained in the truncation. Since the term involving N^2/c_s^2 is typically insignificant for the highest-frequency waves, (8.47) is also a good approximation to the sufficient condition for stability. In many applications the vertical resolution is much higher than the horizontal resolution, and the most severe restriction on the time step is associated with vertically propagating sound waves.

The dispersion relation for gravity waves (8.46) may be written as

$$\sin \omega \Delta t = \Delta t \left(Uk \pm \frac{Nk}{\left(k^2 + \ell^2 + N^2/c_s^2\right)^{1/2}} \right). \tag{8.48}$$

Because

$$\frac{N|k|}{\left(k^2 + \ell^2 + N^2/c_s^2\right)^{1/2}} \le c_s|k|,$$

the necessary condition for sound-wave stability (8.47) is sufficient to ensure the stability of the gravity waves. Although (8.47) guarantees the stability of the gravity-wave modes, it is far too restrictive. Since

$$\frac{N|k|}{\left(k^2 + \ell^2 + N^2/c_s^2\right)^{1/2}} \le N,$$

(8.48) implies the gravity waves will be stable, provided that

$$(|U|k_{\max} + N)\Delta t < 1.$$

This is also a good approximation to the necessary condition for stability, because the term involving N^2/c_s^2 is usually dominated by $k_{\max}^2$.

In most geophysical applications

$$c_s(k_{\max}^2 + \ell_{\max}^2)^{1/2} \gg |U|k_{\max} + N,$$

and the maximum stable time step with which the gravity waves can be integrated is therefore far larger than the time step required to maintain stability in the sound-wave modes. In such circumstances, the sound waves can be stabilized using a semi-implicit approximation in which the pressure-gradient and velocity-divergence terms are evaluated using trapezoidal differencing (Tapp and White 1976). The resulting semi-implicit system is

$$\delta_{2t} u^n + U \frac{\partial u^n}{\partial x} + \left\langle \frac{\partial P^n}{\partial x} \right\rangle^{2t} = 0, \tag{8.49}$$

$$\delta_{2t} w^n + U \frac{\partial w^n}{\partial x} + \left\langle \frac{\partial P^n}{\partial z} \right\rangle^{2t} = b^n, \tag{8.50}$$

$$\delta_{2t} b^n + U \frac{\partial b^n}{\partial x} + N^2 w^n = 0, \tag{8.51}$$

$$\delta_{2t} P^n + U \frac{\partial P^n}{\partial x} + c_s^2 \left(\left\langle \frac{\partial u^n}{\partial x} \right\rangle^{2t} + \left\langle \frac{\partial w^n}{\partial z} \right\rangle^{2t} \right) = 0. \tag{8.52}$$

Let $\hat{c}_s = c_s \cos(\omega \Delta t)$. Then the dispersion relation for the semi-implicit system is identical to that obtained for leapfrog differencing, except that c_s is replaced by $\hat{c}_s$ throughout (8.43). The dispersion relation for the sound-wave modes is

$$\hat{\omega}^2 = \hat{c}_s^2 \left(k^2 + \ell^2 + N^2/\hat{c}_s^2\right),$$

or

$$\sin \omega \Delta t = \Delta t \left(Uk \pm \hat{c}_s \left(k^2 + \ell^2 + N^2/\hat{c}_s^2\right)^{1/2}\right). \tag{8.53}$$

The most severe stability constraints are imposed by the shortest waves for which the term $N^2/\hat{c}_s^2$ can be neglected in comparison with $k^2 + \ell^2$. Neglecting $N^2/\hat{c}_s^2$, (8.53) becomes

$$\sin \omega \Delta t = Uk\Delta t \pm c_s \Delta t \left(k^2 + \ell^2\right)^{1/2} \cos \omega \Delta t,$$

which has the same form as (8.25), implying that the sound-wave modes are stable whenever

$$|Uk| \leq c_s \left(k^2 + \ell^2\right)^{1/2}.$$

A sufficient condition for the stability of the sound waves is simply that the flow be subsonic ($|U| \leq c_s$), or equivalently, that the Mach number be less than unity.

Provided that the flow is subsonic, the only constraint on the time step required to keep the semi-implicit scheme stable is that associated with gravity-wave propagation. The dispersion relation for the gravity waves in the semi-implicit system is

$$\hat{\omega}^2 = \frac{N^2 k^2}{k^2 + \ell^2 + N^2/\hat{c}_s^2}, \tag{8.54}$$

which differs from the result for leapfrog differencing only in the small term $N^2/\hat{c}_s^2$. Stable gravity-wave solutions to the semi-implicit system are obtained whenever

$$(|U|k_{\max} + N)\Delta t < 1,$$

which is the same condition obtained for the stability of the gravity waves using leapfrog differencing. Thus, as suggested previously, the semi-implicit scheme allows the compressible equations governing low-Mach-number flow to be integrated with a much larger time step than that allowed by fully explicit schemes. This increase in efficiency comes at a price; whenever the time step is much larger than that allowed by the CFL condition for sound waves, the sound waves are artificially decelerated by a factor of $\cos(\omega \Delta t)$. This error is directly analogous to that considered in Sect. 8.2.1, in which spurious decelerations were produced by fully implicit schemes using very large time steps. Nevertheless, in many practical applications the errors in the sound waves are of no consequence, and the quality of the solution is entirely determined by the accuracy with which the slower-moving waves are approximated.

How does semi-implicit differencing influence the accuracy of the gravity-wave modes? The only influence is exerted through the reduction in the speed of sound in the third term in the denominator of (8.54). This term is generally small and has little effect on the gravity waves unless $\omega\Delta t$ is far from zero and the waves are sufficiently long that $|k| + |\ell| \leq N/c_s$. It is actually rather hard to satisfy both of these requirements simultaneously. First, since $\omega\Delta t \leq 1$ for the stability of the mode in question, $\hat{c}_s$ can never drop below $0.54c_s$. Second, the maximum value of Δt is limited by the frequency of the most rapidly oscillating wave $\omega_{\max}$. In most applications the frequencies of the long waves are much lower than $\omega_{\max}$, so for all the long waves, $\omega\Delta t \ll \omega_{\max}\Delta t \leq 1$, and thus $\hat{c}_s \approx c_s$. As an example where the deviation of $\hat{c}_s$ from c_s is maximized, consider a basic state with $N = 0.02\ \mathrm{s}^{-1}$, $c_s = 318\ \mathrm{ms}^{-1}$, and $U = 0$, together with the mode $(k, \ell) = (N/c_s, 0.1N/c_s)$ and time steps in the range $0 \leq \Delta t \leq 1/N$. The approximate solution obtained using leapfrog time differencing exhibits an accelerative phase error that reaches 11% when $\Delta t = 1/N$. This accelerative phase-speed error is reduced by the semi-implicit method to a -5.7% decelerative error when $N = \Delta t$. The wave in this example is a rather pathological mode with horizontal and vertical wavelengths of 100 and 1,000 km, respectively. The difference between the leapfrog and semi-implicit gravity-wave solutions is much smaller in most realistic examples.

The semi-implicit differencing scheme (8.49)–(8.52) provides a way to circumvent the CFL stability criterion for sound-wave propagation without losing accuracy in simulation of the gravity-wave modes. In global-scale atmospheric models the gravity waves may actually be of minor physical significance, and the features of primary interest may evolve on an even slower timescale.[5] If the fastest-moving gravity-wave modes do not need to be accurately represented, it is possible to generalize the preceding semi-implicit scheme to allow even larger time steps by replacing (8.50) and (8.51) with

$$\delta_{2t} w^n + U\frac{\partial w^n}{\partial x} + \left\langle \frac{\partial P^n}{\partial z} \right\rangle^{2t} = \langle b^n \rangle^{2t},$$

$$\delta_{2t} b^n + U\frac{\partial b^n}{\partial x} + N^2 \langle w^n \rangle^{2t} = 0$$

(Cullen 1990; Tanguay et al. 1990). Note that the buoyancy forcing in the vertical-momentum equation and the vertical advection of the mean-state buoyancy in the buoyancy equation are now treated by trapezoidal differences. The gravity-wave dispersion relation for this generalized semi-implicit system is

$$\hat{\omega}^2 = \frac{N^2k^2\cos^2(\omega\Delta t)}{k^2 + \ell^2 + N^2/c_s^2},$$

or

$$\sin\omega\Delta t = \Delta t \left(Uk \pm \frac{Nk\cos(\omega\Delta t)}{\left(k^2 + \ell^2 + N^2/c_s^2\right)^{1/2}} \right).$$

[5] In particular, the most important features may consist of slow-moving Rossby waves, which appear as additional solutions to the Euler equations when latitudinal variations in the Coriolis force are included in the horizontal momentum equations.

This dispersion relation has the same form as that for the prototype semi-implicit scheme (8.25), and as discussed in connection with (8.26), stable solutions will be obtained, provided that $|Uk|\Delta t \leq 1$.

8.2.5 Numerical Implementation

The semi-implicit approximation to the compressible Boussinesq system discussed in the preceding section generates a system of implicit algebraic equations that must be solved at every time step. The solution procedure will be illustrated in a relatively simple example using the nonlinear compressible Boussinesq equations

$$\frac{db}{dt} + N^2 w = 0, \tag{8.55}$$

$$\frac{d\mathbf{v}}{dt} + \nabla P = b\mathbf{k}, \tag{8.56}$$

$$\frac{dP}{dt} + c_s^2 \nabla \cdot \mathbf{v} = 0. \tag{8.57}$$

The definitions of b, P, and N given in (8.42) may be used to show that (8.55) and (8.56) are identical to the buoyancy and momentum equations in the standard Boussinesq system (8.2) and (8.3). The standard incompressible continuity equation has been replaced by (8.57) and is recovered in the limit $c_s \to \infty$.

First, consider the situation where only the sound waves are stabilized by semi-implicit differencing and suppose that the spatial derivatives are not discretized. Then the resulting semi-implicit system has the form

$$b^{n+1} = b^{n-1} - 2\Delta t \left(\mathbf{v}^n \cdot \nabla b^n + N^2 w^n\right), \tag{8.58}$$

$$\mathbf{v}^{n+1} + \Delta t \nabla P^{n+1} = \mathbf{G}, \tag{8.59}$$

$$P^{n+1} + c_s^2 \Delta t \nabla \cdot \mathbf{v}^{n+1} = h. \tag{8.60}$$

Here

$$\mathbf{G} = \mathbf{v}^{n-1} - \Delta t \left[\nabla P^{n-1} - 2b^n \mathbf{k} + 2\mathbf{v}^n \cdot \nabla \mathbf{v}^n\right]$$

and

$$h = P^{n-1} - \Delta t \left[c_s^2 \nabla \cdot \mathbf{v}^{n-1} + 2\mathbf{v}^n \cdot \nabla P^n\right].$$

A single Helmholtz equation for P^{n+1} can be obtained by substituting the divergence of (8.59) into (8.60) to yield

$$\nabla^2 P^{n+1} - \frac{P^{n+1}}{(c_s \Delta t)^2} = \frac{\nabla \cdot \mathbf{G}}{\Delta t} - \frac{h}{(c_s \Delta t)^2}. \tag{8.61}$$

The numerical solution of this Helmholtz equation is trivial if the Fourier spectral method is employed in a rectangular domain or if spherical harmonic expansion functions are used in a global spectral model. If the spatial derivatives are approximated by finite differences, (8.61) yields a sparse linear-algebraic system that can be solved using the techniques described in Sect. 8.1.3. After (8.61) has been solved for P^{n+1}, the momentum equations can be stepped forward, and the buoyancy equation (8.58), which is completely explicit, can be updated to complete the integration cycle.

This implementation of the semi-implicit method is closely related to the projection method for incompressible Boussinesq flow. Indeed, in the limit $c_s \to \infty$ the preceding approach will be identical to the leapfrog projection method (described in Sect. 8.1.2) if $(P^{n+1} + P^{n-1})/2$ is replaced by P^n in (8.61). Although the leapfrog projection method and the semi-implicit method yield algorithms involving very similar algebraic equations, these methods are derived via very different approximation strategies. The projection method is an efficient way to solve a set of continuous equations that is obtained by filtering the exact Euler equations to eliminate sound waves. In contrast, the semi-implicit scheme is obtained by directly approximating the full compressible equations and using implicit time differencing to stabilize the sound waves. Neither approach allows one to correctly simulate sound waves, but both approaches allow the accurate and efficient simulation of the slower-moving gravity waves.

Now consider the version of the semi-implicit approximation in which those terms responsible for gravity-wave propagation are also approximated by trapezoidal differences; in this case (8.58) and (8.59) become

$$
\begin{aligned}
b^{n+1} + \Delta t N^2 w^{n+1} &= f_b, \\
\mathbf{v}^{n+1} + \Delta t \left(\nabla P^{n+1} - \mathbf{k} b^{n+1}\right) &= \tilde{\mathbf{G}},
\end{aligned}
\tag{8.62}
$$

where

$$
\begin{aligned}
f_b &= b^{n-1} - \Delta t \left[N^2 w^{n-1} + 2\mathbf{v}^n \cdot \nabla b^n\right], \\
\tilde{\mathbf{G}} &= \mathbf{v}^{n-1} - \Delta t \left[\nabla P^{n-1} - \mathbf{k} b^{n-1} + 2\mathbf{v}^n \cdot \nabla \mathbf{v}^n\right].
\end{aligned}
$$

The implicit coupling in the resulting semi-implicit system can be reduced to a single Helmholtz equation for P^{n+1} as follows. Let $\tilde{\mathbf{G}} = (g_u, g_v, g_w)$; then using (8.62) to substitute for b^{n+1} in the vertical-momentum equation, one obtains

$$
\left(1 + (N\Delta t)^2\right) w^{n+1} + \Delta t \frac{\partial P^{n+1}}{\partial z} = g_w + f_b \Delta t. \tag{8.63}
$$

Using the horizontal momentum equations to eliminate u and v from (8.60) yields

$$
\left(\frac{\partial^2}{\partial x^2} + \frac{\partial^2}{\partial y^2} - \frac{1}{(c_s \Delta t)^2}\right) P^{n+1} - \frac{1}{\Delta t}\frac{\partial w^{n+1}}{\partial z} = \frac{1}{\Delta t}\left(\frac{\partial g_u}{\partial x} + \frac{\partial g_v}{\partial y}\right) - \frac{h}{(c_s \Delta t)^2}.
$$

As the final step, w^{n+1} is eliminated between the two preceding equations to obtain

$$
\begin{aligned}
&\left[\left(1 + (N\Delta t)^2\right)\left(\frac{\partial^2}{\partial x^2} + \frac{\partial^2}{\partial y^2} - \frac{1}{(c_s \Delta t)^2}\right) + \frac{\partial^2}{\partial z^2}\right] P^{n+1} \\
&\quad = \left(1 + (N\Delta t)^2\right)\left[\frac{1}{\Delta t}\left(\frac{\partial g_u}{\partial x} + \frac{\partial g_v}{\partial y}\right) - \frac{h}{(c_s \Delta t)^2}\right] + \frac{\partial}{\partial z}\left(\frac{g_w}{\Delta t} + f_b\right).
\end{aligned}
$$

After this elliptic equation has been solved for P^{n+1}, u and v are updated using the horizontal momentum equations, w is updated using (8.63), and b is updated using (8.62).

Notice that the vertical advection of density in (8.62) is split between a term involving the mean vertical density gradient (N^2), which is treated implicitly, and a term involving the gradient of the perturbation density field ($\partial b/\partial z$), which is treated explicitly. As discussed in Sect. 8.2.3, when terms are split between a reference state that is treated implicitly and a perturbation that is treated explicitly, stability considerations demand that the term treated implicitly dominate the term treated explicitly. Thus, in most atmospheric applications the reference stability is chosen to be isothermal, thereby ensuring that $N^2 \geq \partial b/\partial z$. When semi-implicit differencing is used to integrate the complete Euler equations, the terms involving the pressure gradient and velocity divergence must also be partitioned into implicitly differenced terms involving a reference state and the remaining explicitly differenced perturbations. Since the speed of sound is relatively uniform throughout the atmosphere, it is easy to ensure that the terms evaluated implicitly dominate those computed explicitly and thereby guarantee that the scheme is stable.

8.3 Fractional-Step Methods

The semi-implicit method requires the solution of an elliptic equation for the pressure during each step of the integration. This can be avoided by splitting the complete problem into fractional steps and using a smaller time step to integrate the subproblem containing the terms responsible for the propagation of the fast-moving wave. Consider a general partial differential equation of the form

$$\frac{\partial \psi}{\partial t} = \mathscr{L}(\psi), \tag{8.64}$$

where $\mathscr{L}(\psi)$ contains the spatial derivatives and other forcing terms. Suppose that $\mathscr{L}(\psi)$ can be split into two parts

$$\mathscr{L}(\psi) = \mathscr{L}_1(\psi) + \mathscr{L}_2(\psi),$$

such that $\mathscr{L}_1$ and $\mathscr{L}_2$ contain those terms responsible for the propagation of slow- and fast-moving waves, respectively. As discussed in Sect. 4.3, if $\mathscr{L}$ does not depend on time, each of the individual operators can be formally integrated over an interval Δt to obtain

$$\psi(t+\Delta t) = \exp(\Delta t \mathscr{L}_1)\psi(t), \quad \psi(t+\Delta t) = \exp(\Delta t \mathscr{L}_2)\psi(t).$$

Following the notation in Sect. 4.3, let $\mathscr{F}_1(\Delta t)$ and $\mathscr{F}_2(\Delta t)$ be numerical approximations to the exact operators $\exp(\Delta t \mathscr{L}_1)$ and $\exp(\Delta t \mathscr{L}_2)$.

8.3.1 Completely Split Operators

In the standard fractional-step approach, the approximate solution is stepped forward over a time interval Δt using

$$\phi^s = \mathscr{F}_1(\Delta t)\phi^n, \tag{8.65}$$
$$\phi^{n+1} = \mathscr{F}_2(\Delta t)\phi^s, \tag{8.66}$$

but it is not necessary to use the same time step in each subproblem. If the maximum stable time step with which the approximate slow-wave operator (8.65) can be integrated is M times that with which the fast-wave operator (8.66) can be integrated, the numerical solution could be evaluated using the formula

$$\phi^{n+1} = [\mathscr{F}_2(\Delta t/M)]^M\, \mathscr{F}_1(\Delta t)\phi^n. \tag{8.67}$$

This approach can be applied to the linearized one-dimensional shallow-water system by writing (4.1) and (4.2) in the form

$$\frac{\partial \mathbf{r}}{\partial t} + \mathscr{L}_1(\mathbf{r}) + \mathscr{L}_2(\mathbf{r}) = \mathbf{0}, \tag{8.68}$$

where

$$\mathbf{r} = \begin{pmatrix} u \\ h \end{pmatrix}, \quad \mathscr{L}_1 = \begin{pmatrix} U\partial_x & 0 \\ 0 & U\partial_x \end{pmatrix}, \quad \mathscr{L}_2 = \begin{pmatrix} 0 & g\partial_x \\ H\partial_x & 0 \end{pmatrix},$$

and ∂_x denotes the partial derivative with respect to x. The first fractional step, which is an approximation to

$$\frac{\partial \mathbf{r}}{\partial t} + \mathscr{L}_1(\mathbf{r}) = \mathbf{0},$$

involves the solution of two decoupled advection equations. Since this is a fractional-step method, it is generally preferable to approximate the preceding equation with a two-time-level method. To avoid using implicit, unstable, or Lax–Wendroff methods, the first step can be integrated using the linearly third order version of the Runge–Kutta scheme (2.45):

$$\mathbf{r}^* = \mathbf{r}^n - \frac{\Delta t}{3}\mathscr{L}_1(\mathbf{r}^n), \tag{8.69}$$
$$\mathbf{r}^{**} = \mathbf{r}^n - \frac{\Delta t}{2}\mathscr{L}_1(\mathbf{r}^*), \tag{8.70}$$
$$\mathbf{r}^{n+1} = \mathbf{r}^n - \Delta t\, \mathscr{L}_1(\mathbf{r}^{**}). \tag{8.71}$$

This Runge–Kutta method is stable and damping for $|U|k_{\max}\Delta t < 1.73$, where $k_{\max}$ is the maximum retained wave number.

The second fractional step, which approximates

$$\frac{\partial \mathbf{r}}{\partial t} + \mathscr{L}_2(\mathbf{r}) = \mathbf{0},$$

can be efficiently integrated using forward–backward differencing. Defining $\Delta\tau = \Delta t/M$ as the length of a small time step, the forward–backward scheme is

$$\frac{u^{m+1} - u^m}{\Delta\tau} + g\frac{dh^m}{dx} = 0, \tag{8.72}$$

$$\frac{h^{m+1} - h^m}{\Delta\tau} + H\frac{du^{m+1}}{dx} = 0. \tag{8.73}$$

This scheme is stable for $ck_{\max}\Delta\tau < 2$ and is second-order accurate in time. Since the operators used in each fractional step commute, the complete method will be $O\left[(\Delta t)^2\right]$ accurate and stable whenever each of the individual steps is stable.[6]

Although the preceding fractional-step scheme works fine for the linearized one-dimensional shallow-water system, it does not generalize as nicely to problems in which the operators do not commute. As an example, consider the compressible two-dimensional Boussinesq equations, which could be split into the form (8.68) by defining

$$\mathbf{r} = \left(u\ w\ b\ P\right)^{\mathrm{T}},$$

$$\mathscr{L}_1 = \begin{pmatrix} \mathbf{v}\cdot\nabla & 0 & 0 & 0 \\ 0 & \mathbf{v}\cdot\nabla & 0 & 0 \\ 0 & 0 & \mathbf{v}\cdot\nabla & 0 \\ 0 & 0 & 0 & \mathbf{v}\cdot\nabla \end{pmatrix}, \quad \mathscr{L}_2 = \begin{pmatrix} 0 & 0 & 0 & \partial_x \\ 0 & 0 & -1 & \partial_z \\ 0 & N^2 & 0 & 0 \\ c_s^2\partial_x & c_s^2\partial_z & 0 & 0 \end{pmatrix},$$

where $\mathbf{v}$ is the two-dimensional velocity vector and $\nabla = (\partial/\partial x, \partial/\partial z)$. Suppose that N and c_s are constant and that the full nonlinear system is linearized about a reference state with a mean horizontal wind $U(z)$. The operators associated with this linearized system will not commute unless $\mathrm{d}U/\mathrm{d}z$ is zero.

As in the one-dimensional shallow-water system, the advection operator $\mathscr{L}_1$ can be approximated using the third-order Runge–Kutta method (8.69)–(8.71). The second fractional step may be integrated using trapezoidal differencing for the terms governing the vertical propagation of sound waves and forward–backward differencing for the terms governing horizontal sound-wave propagation and buoyancy oscillations. The resulting scheme is

$$\frac{u^{m+1} - u^m}{\Delta\tau} + \frac{\partial P^m}{\partial x} = 0, \tag{8.74}$$

$$\frac{w^{m+1} - w^m}{\Delta\tau} + \frac{\partial}{\partial z}\left(\frac{P^{m+1} + P^m}{2}\right) - b^m = 0, \tag{8.75}$$

$$\frac{b^{m+1} - b^m}{\Delta\tau} + N^2 w^{m+1} = 0, \tag{8.76}$$

$$\frac{P^{m+1} - P^m}{\Delta\tau} + c_s^2\frac{\partial u^{m+1}}{\partial x} + c_s^2\frac{\partial}{\partial z}\left(\frac{w^{m+1} + w^m}{2}\right) = 0. \tag{8.77}$$

[6] See Sect. 4.3 for a discussion of the impact of operator commutativity on the performance of fractional-step schemes.

Again let k_{max} be the magnitude of the maximum horizontal wave number retained in the truncation; then the preceding approximation to $\exp(\Delta\tau \mathscr{L}_2)$ is stable and nondamping provided $\max(c_s k_{max}, N)\Delta\tau < 2$. The trapezoidal approximation of the terms involving vertical derivatives does not significantly increase the computations required on each small time step because it generates a simple tridiagonal system of algebraic equations for the w^{m+1} throughout each vertical column within the domain. If the horizontal resolution is very coarse, so that $k_{max} \ll N/c_s$, further efficiency can also be obtained by treating the terms involving buoyancy oscillations with trapezoidal differencing. Since these terms do not involve derivatives, the resulting implicit algebraic system remains tridiagonal.

As an alternative to the trapezoidal method, the terms involving the vertical pressure gradient and the divergence of the vertical velocity could be integrated using forward–backward differencing, in which case the stability criterion for the small time step would include an additional term proportional to $c_s \ell_{max} \Delta\tau$, where ℓ_{max} is the maximum resolvable vertical wave number. It may be appropriate to use forward–backward differencing instead of the trapezoidal scheme in applications with identical vertical and horizontal grid spacing, but if the vertical resolution is much finer than the horizontal resolution, the additional stability constraint imposed by vertical sound-wave propagation will reduce the efficiency by requiring an excessive number of small time steps.

The performance of the preceding scheme is evaluated in a problem involving flow past a compact gravity-wave generator. The wave generator is modeled by including forcing terms in the momentum equations such that the nondiscretized versions of (8.74) and (8.75) take the form

$$\frac{du}{dt} + \frac{\partial P}{\partial x} = -\frac{\partial \Psi}{\partial z}, \tag{8.78}$$

$$\frac{dw}{dt} + \frac{\partial P}{\partial z} - b = \frac{\partial \Psi}{\partial x}, \tag{8.79}$$

where

$$\Psi(x,z,t) = E(x,z)\sin\omega t \sin k_1 x \cos \ell_1 z$$

and

$$E(x,z) = \begin{cases} \alpha\,(1+\cos k_2 x)\,(1+\cos \ell_2 z) & \text{if } |x| \le \pi/k_2 \text{ and } |z| \le \pi/\ell_2, \\ 0 & \text{otherwise.} \end{cases}$$

This forcing has no influence on the time tendency of the divergence, and as a consequence, it does not excite sound waves. The spatial domain is periodic at $x = \pm 50$ km and bounded by rigid horizontal walls at $z = \pm 5$ km. In the following tests $\Delta x = 250$ m, $\Delta z = 50$ m, $N = 0.01\,\mathrm{s}^{-1}$, $c_s = 350\,\mathrm{ms}^{-1}$, and the parameters defining the wave generator are $\alpha = 0.2$, $2\pi/k_1 = 10$ km, $2\pi/\ell_1 = 2.5$ km, $2\pi/k_2 = 11$ km, $2\pi/\ell_2 = 1.5$ km, and $\omega = 0.002\,\mathrm{s}^{-1}$. The forcing is evaluated every $\Delta\tau$ and applied to the solution on the small time step, $\Delta x = 250$ m, $\Delta z = 50$ m, $N = 0.01\,\mathrm{s}^{-1}$, and $c_s = 350\,\mathrm{ms}^{-1}$.

The spatial derivatives are approximated using centered differencing on a staggered grid identical to that shown in Fig. 4.6, except that b is colocated with the w points rather than the P points. As a consequence of the mesh staggering, the horizontal wave number obtained from the finite-difference approximations to the pressure gradient and velocity divergence is $(2/\Delta x)(\sin k\Delta x/2)$, and the small-step stability criterion is max $(2c_s/\Delta x + N)\Delta\tau < 2$. The horizontal wave number generated by the finite-difference approximation to the advection operator is $(\sin k\Delta x)/\Delta x$, so the large time step is stable when $|U|\Delta t/\Delta x < 1.73$. Strang splitting,

$$\phi^{n+1} = [\mathscr{F}_2(2\Delta t/M)]^{(M/2)}\, \mathscr{F}_1(\Delta t)\, [\mathscr{F}_2(2\Delta t/M)]^{(M/2)}\, \phi^n,$$

is used in preference to (8.67) to preserve $O\left[(\Delta t)^2\right]$ accuracy in those cases where $\mathscr{F}_1$ and $\mathscr{F}_2$ do not commute.

In the first simulation $\Delta t = 12.5$ s, there are 20 small time steps per large time step, and $U = 10\ \mathrm{ms}^{-1}$ throughout the domain. In this case $(2c_s\Delta x + N)\Delta\tau = 1.76$, so the small time step is being integrated using time steps near the stability limit. The horizontal velocity field and the pressure field obtained from this simulation are plotted in Fig. 8.2. The velocity field is essentially identical to that obtained using the full compressible equations. Very small errors are detectable in the pressure field, but the overall accuracy of the solution is excellent.

Now consider a second simulation that is identical to the first in every respect except that the mean wind U increases linearly from 5 to 15 ms^{-1} between the bottom and the top of the domain. The pressure perturbations that develop in this simulation are shown in Fig. 8.3a, along with streamlines for the forcing function Ψ. Spurious pressure perturbations appear throughout the domain. The correct pressure field is shown in Fig. 8.3d, which was computed using a scheme that will be described in the

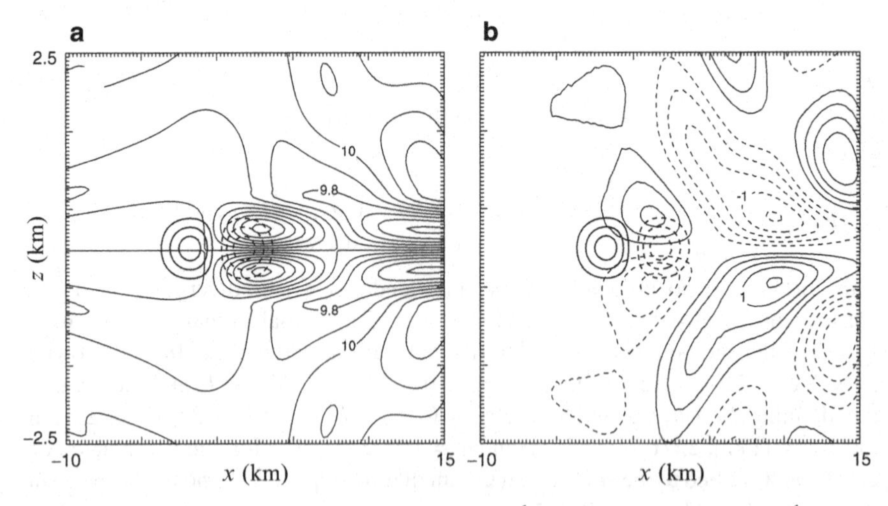

Fig. 8.2 **a** Contours of $U + u$ at intervals of $0.1\ \mathrm{ms}^{-1}$ and Ψ at intervals of $0.1\ \mathrm{s}^{-1}$ at $t = 8{,}000$ s. **b** As in **a** except that P is contoured at intervals of $0.25\ \mathrm{m}^2\,\mathrm{s}^{-2}$. No zero contour is shown for the P and Ψ fields. *Minor tick marks* indicate the location of the P points on the numerical grid. Only the central portion of the total computational domain is shown

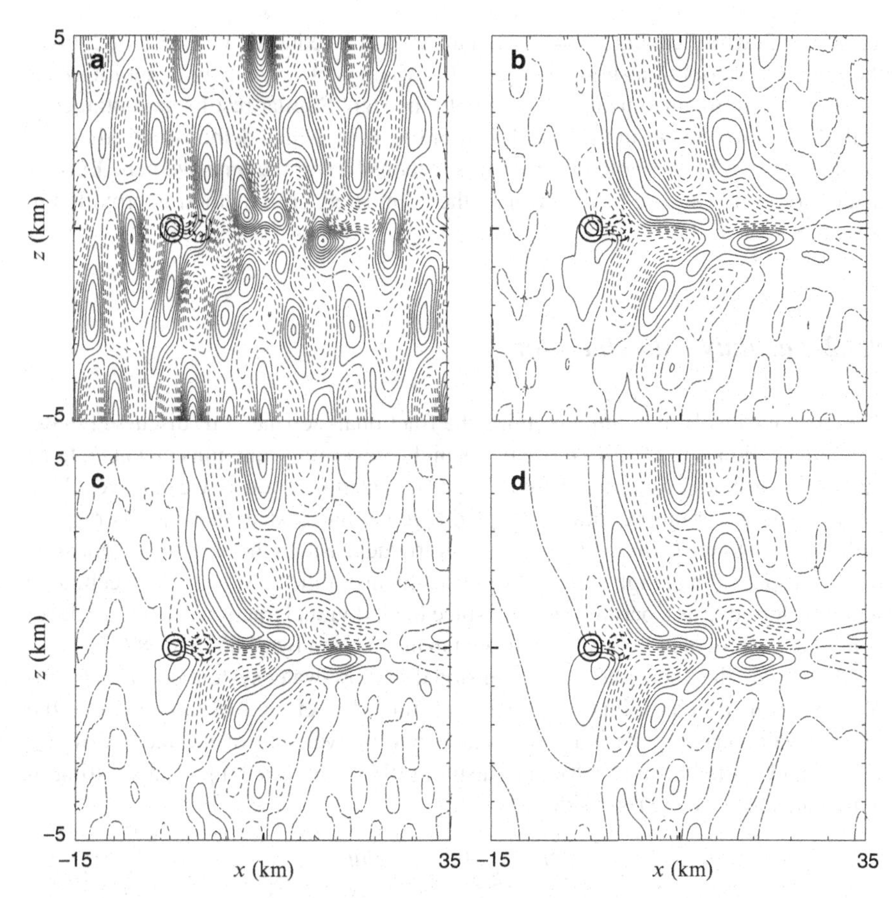

Fig. 8.3 Contours of P at intervals of $0.25\,\text{m}^2\,\text{s}^{-2}$ (the zero contour is *dot-dashed*) and Ψ at intervals of $0.15\,\text{s}^{-1}$ at $t = 3{,}000\,\text{s}$ for the case with vertical shear in the mean wind and **a** $\Delta t = 12.5\,\text{s}$, $M = 20$, **b** $\Delta t = 6.25\,\text{s}$, $M = 20$, and **c** $\Delta t = 6.25\,\text{s}$, $M = 10$. **d** The solution computed using the partially split method described in Sect. 8.3.2 with $\Delta t = 12.5\,\text{s}$, $M = 20$. *Tick marks* appear every 20 grid intervals

next subsection. Although the pressure field in Fig. 8.3a is clearly in error, most of the spurious signal in the pressure field relates to sound waves whose velocity perturbations are very weak. The velocity fields associated with all the solutions shown in Fig. 8.3 are essentially identical. The extrema in the pressure perturbations shown in Fig. 8.3a are approximately twice those in Fig. 8.3b–d and are growing very slowly, suggesting that the solution is subject to a weak instability. Since the operators for each fractional step do not commute, the stability of each individual operator no longer guarantees the stability of the overall scheme.

As will be discussed in Sect. 8.3.2, this scheme can be stabilized by damping the velocity divergence on the small time step. Divergence damping yields only a modest improvement in the solution, however, because the completely split method also has accuracy problems owing to inadequate temporal resolution. Cutting Δt by a factor of 2, while leaving $M = 20$ so that $\Delta\tau$ is also reduced by a factor of 2, gives the pressure distribution shown in Fig. 8.3b, which is clearly a significant improvement over that obtained using the original time step. Similar results are obtained if both Δt and M are cut in half, as shown in Fig. 8.3c, which demonstrates that it is the decrease in Δt, rather than $\Delta\tau$, that is responsible for the improvement. Further discussion of the source of the error in the completely split method is provided in Sect. 8.4.

8.3.2 Partially Split Operators

The first task involved in implementing the fractional-step methods discussed in the previous section is to identify those terms in the governing equations that need to be updated on a shorter time step. Having made this identification, one can leave all the terms in the governing equations coupled together and update those terms responsible for the slowly evolving processes less frequently than those terms responsible for the propagation of high-frequency physically insignificant waves. This technique will be referred to as a partial splitting, since the individual fractional steps are never completely decoupled in the conventional manner given by (8.65) and (8.66).

Once again the linearized one-dimensional shallow-water system provides a simple context in which to illustrate partial splitting. As before, it is assumed that the gravity-wave phase speed is much greater than the velocity of the mean flow U. Klemp and Wilhelmson (1978) and Tatsumi (1983) suggested a partial splitting in which the terms on the right sides of

$$\frac{\partial u}{\partial t} + g\frac{\partial h}{\partial x} = -U\frac{\partial u}{\partial x}, \tag{8.80}$$

$$\frac{\partial h}{\partial t} + H\frac{\partial u}{\partial x} = -U\frac{\partial h}{\partial x} \tag{8.81}$$

are updated as if the time derivative were being approximated using a leapfrog difference, but rather than the solution being advanced from time level $t-\Delta t$ to $t+\Delta t$ in a single step of length $2\Delta t$, the solution is advanced through a series of $2M$ "small time steps." During each small time step the terms on the right sides of (8.80) and (8.81) are held constant at their values at time level t and the remaining terms are updated using forward–backward differencing. Let m and n be time indices for the small and large time steps, respectively, and define $\Delta\tau = \Delta t/M$ as the length of a small time step. The solution is advanced from time level $n-1$ to $n+1$ in $2M$ small time steps of the form

$$\frac{u^{m+1} - u^m}{\Delta\tau} + g\frac{dh^m}{dx} = -U\frac{du^n}{dx},$$

$$\frac{h^{m+1} - h^m}{\Delta\tau} + H\frac{du^{m+1}}{dx} = -U\frac{dh^n}{dx}.$$

Note that the left sides of the preceding equations are identical to those appearing in the completely split scheme (8.72) and (8.73).

The complete small-step–large-step integration cycle for this problem can be written as a four-dimensional linear system as follows. Define $\hat{u}^m = u^n$, $\hat{h}^m = h^n$, and let

$$\mathbf{r} = (u, h, \hat{u}, \hat{h})^{\mathrm{T}}.$$

Then an individual small time step can be expressed in the form

$$\mathbf{r}^{m+1} = \mathbf{A}\mathbf{r}^m,$$

where

$$\mathbf{A} = \begin{pmatrix} 1 & -\tilde{g}\partial_x & -\tilde{U}\partial_x & 0 \\ -\tilde{H}\partial_x & 1 + \tilde{c}^2\partial^2_{xx} & \tilde{U}\tilde{H}\partial^2_{xx} & -\tilde{U}\partial_x \\ 0 & 0 & 1 & 0 \\ 0 & 0 & 0 & 1 \end{pmatrix},$$

and the tilde denotes multiplication of the parameter by $\Delta\tau$ (e.g., $\tilde{c} = c\Delta\tau$). At the beginning of the first small time step in a complete small-step–large-step integration cycle

$$\mathbf{r}^{m=1} = (u^{n-1}, h^{n-1}, u^n, h^n)^{\mathrm{T}}.$$

At the end of the $(2M)$th small step

$$\mathbf{r}^{m=2M} = (u^{n+1}, h^{n+1}, u^n, h^n)^{\mathrm{T}}.$$

Thus, if $\mathbf{S}$ is a matrix interchanging the first pair and second pair of elements in $\mathbf{r}$,

$$\mathbf{S} = \begin{pmatrix} 0 & 0 & 1 & 0 \\ 0 & 0 & 0 & 1 \\ 1 & 0 & 0 & 0 \\ 0 & 1 & 0 & 0 \end{pmatrix},$$

the complete small-step–large-step integration cycle is given by

$$\mathbf{r}^{n+1} = \mathbf{S}\mathbf{A}^{2M}\mathbf{r}^n.$$

Since the individual operators commute, the completely split approximation to this problem is stable whenever both of the individual fractional steps are stable. One might hope that the stability of the partially spilt method could also be guaranteed whenever the large- and small-step subproblems are stable. Unfortunately, there are many combinations of Δt and $\Delta\tau$ for which the partially split method is unstable even though the subproblems obtained by setting either U or c to zero are both stable (Tatsumi 1983; Skamarock and Klemp 1992). Suppose that the partially split scheme is applied to an individual Fourier mode with horizontal wave number k. Then the amplification matrix for an individual small time step is given by a matrix in which the partial-derivative operators in $\mathbf{A}$ are replaced by ik; let this matrix be denoted by $\hat{\mathbf{A}}$.

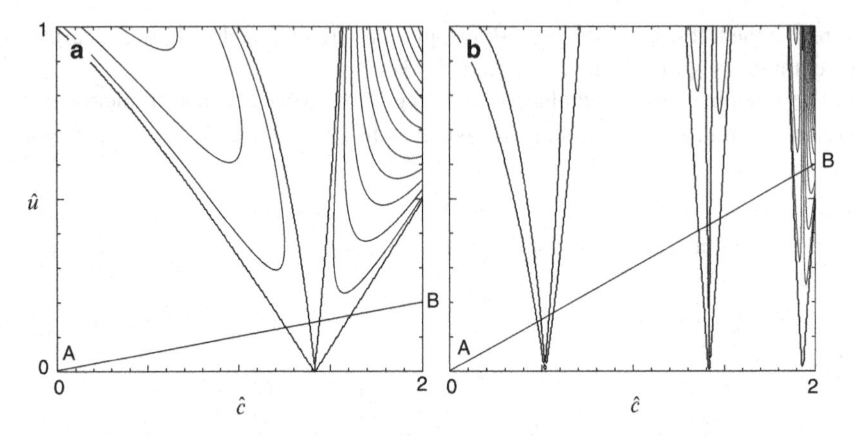

Fig. 8.4 Spectral radius of the amplification matrix for the partially split method contoured as a function of $\hat{c}$ and $\hat{u}$ for **a** $M = 1$ and **b** $M = 3$. Unstable regions are enclosed in the *wedged-shaped areas*. Contour intervals are 1.0 (*heavy line*), 1.2, 1.4, ... *Line AB* indicates the possible combinations of $\hat{c}$ and $\hat{u}$ that can be realized when $U/c = 1/10$ and M is specified as 1 or 3

Consider the case $M = 1$, for which the amplification matrix is $\mathbf{S}\hat{\mathbf{A}}^2$. The magnitude of the maximum eigenvalue, or spectral radius ρ_m, of $\mathbf{S}\hat{\mathbf{A}}^2$ is plotted in Fig. 8.4a as a function of $\hat{c} = ck\Delta\tau$ and $\hat{u} = Uk\Delta t$. The domain over which ρ_m is contoured, $0 \leq \hat{c} \leq 2$ and $0 \leq \hat{u} \leq 1$, is that for which the individual small- and large-step problems are stable. When $M = 1$, ρ_m exceeds unity and the partially split scheme is unstable throughout two regions of the $\hat{c}$–$\hat{u}$ plane whose boundaries intersect at $(\hat{c}, \hat{u}) = (\sqrt{2}, 0)$. If $U \ll c$, only a limited subset of the $\hat{c}$–$\hat{u}$ plane shown in Fig. 8.4a is actually relevant to the solution of the shallow-water problem. Once the number of small time steps per large time step have been fixed, the possible combinations of $\hat{u}$ and $\hat{c}$ will lie along a straight line of slope

$$\frac{\hat{u}}{\hat{c}} = \frac{U\Delta t}{c\Delta\tau} = M\frac{U}{c}.$$

Suppose that $U/c = 1/10$. Then if the partially split method is used with $M = 1$, the only possible combinations of $\hat{u}$ and $\hat{c}$ are those lying along line AB in Fig. 8.4a. The maximum stable value of $\Delta\tau$ is determined by the intersection of line AB and the left boundary of the leftmost region of instability. Thus, for $U/c = 1/10$ and $M = 1$, the stability requirement is that $\hat{c}$ be less than approximately 1.25.

As demonstrated in Fig. 8.4b, which shows contours of the spectral radius of $\mathbf{S}\hat{\mathbf{A}}^6$, the restriction on the maximum stable time step becomes more severe as M increases to 3. The regions of instability are narrower and the strength of the instability in each unstable region is reduced, but additional regions of instability appear, and the distance from the origin to the nearest region of instability decreases. When $M = 3$ and $U/c = 1/10$, the maximum stable value of $\hat{c}$ is roughly 0.48. Further reductions in the maximum stable value for $\hat{c}$ occur as M is increased, and as a consequence, the gain in computational efficiency that one might expect to achieve by increasing the number of small time steps per large time step is eliminated by a compensating decrease in the maximum stable value for $\Delta\tau$.

The partially split method has, nevertheless, been used extensively in many practical applications. The method has proved useful because in most applications it is very easy to remove these instabilities by using a filter. As noted by Tatsumi (1983) and Skamarock and Klemp (1992), the instability is efficiently removed by the Asselin time filter (2.65), which is often used in conjunction with leapfrog time differencing to prevent the divergence of the solution on the odd and even time steps. Other filtering techniques have also been suggested and will be discussed after considering a partially split approximation to the compressible Boussinesq system.

The equations evaluated at each small time step in a partially split approximation to the two-dimensional compressible Boussinesq equations linearized about a basic-state flow with Brunt–Väisälä frequency N and horizontal velocity U are

$$\frac{u^{m+1}-u^m}{\Delta\tau}+\frac{\partial P^m}{\partial x}=-U\frac{\partial u^n}{\partial x}-w^n\frac{\partial U}{\partial z}, \tag{8.82}$$

$$\frac{w^{m+1}-w^m}{\Delta\tau}+\frac{\partial}{\partial z}\left(\frac{P^{m+1}+P^m}{2}\right)-b^m=-U\frac{\partial w^n}{\partial x}, \tag{8.83}$$

$$\frac{b^{m+1}-b^m}{\Delta\tau}+N^2w^{m+1}=-U\frac{\partial b^n}{\partial x}, \tag{8.84}$$

$$\frac{P^{m+1}-P^m}{\Delta\tau}+c_s^2\frac{\partial u^{m+1}}{\partial x}+c_s^2\frac{\partial}{\partial z}\left(\frac{w^{m+1}+w^m}{2}\right)=-U\frac{\partial P^n}{\partial x}, \tag{8.85}$$

where, as before, m and n are the time indices associated with the small and large time steps. The left sides of these equations are identical to the small-time-step equations in the completely split method (8.74)–(8.77). The right sides are updated at every large time step.

This method is applied to the problem previously considered in connection with Fig. 8.2, in which fluid flows past a compact gravity-wave generator. The forcing from the wave generator appears in the horizontal and vertical momentum equations as in (8.78) and (8.79) and is updated on the small time step. In this test U is a constant 10 ms^{-1}, $\Delta t = 12.5$ s, and $\Delta\tau = 0.625$ s. The horizontal velocity field and the pressure field from this simulation are plotted in Fig. 8.5. The horizontal velocity field is very similar to, though slightly noisier than, that shown in Fig. 8.2a. The pressure field is, however, complete garbage. Indeed, it is surprising that errors of the magnitude shown in Fig. 8.5b can exist in the pressure field without seriously degrading the velocity field. These pressure perturbations are growing with time (the contour interval in Fig. 8.5b is twice that in Fig. 8.2b); the velocity field eventually becomes very noisy, and the solution eventually blows up.

This instability can be prevented by applying an Asselin time filter (2.65) at the end of each small-step–large-step integration cycle. Skamarock and Klemp (1992) have shown that filtering coefficients on the order of $\gamma = 0.1$ may be required to stabilize the partially split solution to the one-dimensional shallow-water system. A value of $\gamma = 0.1$ is sufficient to completely remove the noise in the pressure field and to eliminate the instability in the preceding test. Nevertheless, as discussed in Sect. 2.4.2, Asselin filtering reduces the accuracy of the leapfrog scheme to $O(\Delta t)$,

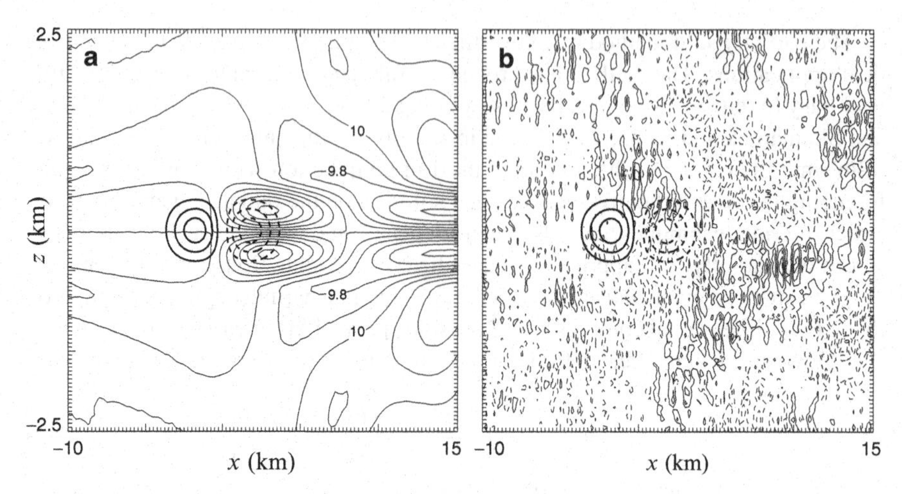

Fig. 8.5 **a** Contours of $U + u$ at intervals of $0.1\,\mathrm{m\,s^{-1}}$ and Ψ at intervals of $0.1\,\mathrm{s^{-1}}$ at $t = 8,000$ s. **b** As in **a** except that P is contoured at intervals of $0.5\,\mathrm{m^2\,s^{-2}}$

so it is best not to rely exclusively on the Asselin filter to stabilize the partially split approximation. Other techniques for stabilizing the preceding partially split approximation include divergence damping and forward biasing the trapezoidal integral of the vertical derivative terms (8.83) and (8.85). Forward biasing the trapezoidal integration is accomplished without additional computational effort by replacing those terms of the form $(\phi^{m+1} + \phi^m)/2$ with

$$\left(\frac{1+\epsilon}{2}\right)\phi^{m+1} + \left(\frac{1-\epsilon}{2}\right)\phi^m,$$

where $0 \le \epsilon \le 1$. A value of $\epsilon = 0.2$ provides an effective filter that does not noticeably modify the gravity waves (Durran and Klemp 1983).

Since trapezoidal time differencing is used only to approximate the vertical derivatives, forward biasing those derivatives will not damp horizontally propagating sound waves. Skamarock and Klemp (1992) recommended including a "divergence damper" in the momentum equations such that the system of equations that is integrated on the small time step becomes

$$\begin{aligned}
\frac{\partial u}{\partial t} + \frac{\partial P}{\partial x} - \alpha_x \frac{\partial \delta}{\partial x} &= F_u,\\
\frac{\partial w}{\partial t} + \frac{\partial P}{\partial z} - b - \alpha_z \frac{\partial \delta}{\partial z} &= F_w,\\
\frac{\partial b}{\partial t} + N^2 w &= F_b,\\
\frac{\partial P}{\partial t} + c_s^2 \delta &= F_p,
\end{aligned} \tag{8.86}$$

where

$$\delta = \frac{\partial u}{\partial x} + \frac{\partial w}{\partial z},$$

and F_u, F_w, F_b, and F_p represent the forcing terms that are updated every Δt. Damping coefficients of $\alpha_x = 0.001(\Delta x)^2/\Delta\tau$ and $\alpha_z = 0.001(\Delta z)^2/\Delta\tau$ removed all trace of noise and instability in the test problem shown in Fig. 8.5 without a supplemental Asselin filter.

The role played by divergence damping in stabilizing the small-time-step integration in the partially split method can be appreciated by noting that if a single damping coefficient α is used in all components of the momentum equation, the divergence satisfies

$$\frac{\partial \delta}{\partial t} + \nabla^2 P - \alpha \nabla^2 \delta = G, \tag{8.87}$$

where $G = -\nabla \cdot (\mathbf{v} \cdot \nabla \mathbf{v}) + \partial b/\partial z$. Eliminating the pressure between (8.86) and (8.87), one obtains

$$\frac{\partial^2 \delta}{\partial t^2} - \alpha \nabla^2 \frac{\partial \delta}{\partial t} - c_s^2 \nabla^2 \delta = \frac{\partial G}{\partial t} - \nabla^2 F_p.$$

The forcing on the right side of this equation will tend to produce divergence in an initially nondivergent flow. Substituting a single Fourier mode into the homogeneous part of this equation, one obtains the classic equation for a damped harmonic oscillator:

$$\frac{d^2 \tilde{\delta}}{dt^2} + \alpha \kappa^2 \frac{d\tilde{\delta}}{dt} + c_s^2 \kappa^2 \tilde{\delta} = 0, \tag{8.88}$$

where $\tilde{\delta}(t)$ is the amplitude and $\kappa = \sqrt{k^2 + \ell^2}$. The damping increases with wave number and is particularly effective in eliminating the high-wave-number modes at which the instability in the partially split method occurs. Gravity waves, on the other hand, are not significantly impacted by the divergence damper because the velocity field in internal gravity waves is almost nondivergent. Skamarock and Klemp (1992) showed that divergence damping only slightly reduces the amplitude of the gravity waves.

At this point it might appear that the partially split methods is inferior to the completely split method considered previously, since filters are required to stabilize the partially split approximation in situations where the completely split scheme performs quite nicely. Recall, however, that the completely split method does not generate usable solutions to the compressible Boussinesq equations when there is a vertical shear in the basic-state horizontal velocity impinging on the gravity-wave generator. The same filtering strategies that stabilize the partially split method in the no-shear problem remain effective in the presence of vertical wind shear. This is demonstrated in Fig. 8.3d, which shows the pressure perturbations in the test case with vertical shear as computed by the partially split method using a divergence damper with the values of α_x and α_z given previously. Even the best completely split solutions (Fig. 8.3b, c) contain widespread regions of spurious low-amplitude perturbations with horizontal wavelengths slightly less than 5 km ($20\Delta x$), whereas the partially split solution shows no evidence of such waves. Results similar to those in Fig. 8.3d may also be obtained using Asselin time filtering with $\alpha = 0.1$ in lieu of the divergence damper. The advantages of the partially split method are not connected with its performance in the simplest test cases, for which it can indeed be inferior to a completely split approximation, but in its adaptability to more complex problems.

One might inquire whether divergence damping can also be used to stabilize the completely split approximation to the test case with vertical shear in the horizontal wind. The norm of the amplification matrix for the large-time-step third-order Runge–Kutta integration (8.69)–(8.71) is strictly less than unity for all sufficiently small Δt. Divergence damping makes the norm of the amplification matrix for the small time step strictly less than unity for all sufficiently small $\Delta\tau$ and thereby stabilizes the completely split scheme by guaranteeing that the norm of the amplification matrix for the overall scheme will be less than unity. Nevertheless, divergence damping only modestly improves the solution obtained with the completely split scheme; the pressure field remains very noisy and completely unacceptable.[7] The fundamental problem with the completely split method appears to be one of inaccuracy, not instability. This will be discussed further in the next section.

The linearly third order Runge–Kutta scheme (8.69)–(8.71) can provide a simple, accurate alternative to leapfrog time differencing for use on the large time step in partially split integrations (Wicker and Skamarock 2002), and it has replaced the leapfrog scheme in several operational codes. To clarify how (8.69)–(8.71) are modified for use as the large-time-step integrator in a partially split problem, let the small time step again be defined such that $\Delta\tau = \Delta t/M$, where M must now be a multiple of 6. Let ${}_1\mathbf{r}^m$ be the vector of unknowns at the start of the mth small time step during the first Runge–Kutta iteration, which is initialized by setting ${}_1\mathbf{r}^1 = \mathbf{r}^n$. The mth small time step of the this iteration has the form

$$ {}_1\mathbf{r}^{m+1} = {}_1\mathbf{r}^m - \Delta\tau\left(\mathscr{L}_1(\mathbf{r}^n) + \mathscr{L}_2({}_1\mathbf{r}^m, {}_1\mathbf{r}^{m+1})\right). \tag{8.89} $$

As before, $\mathscr{L}_1$ and $\mathscr{L}_2$ contain the terms responsible for the low- and high-frequency forcing, respectively. After $M/3$ small time steps, the solution to (8.89) is projected forward to time $t^n + \Delta t/3$. The low-frequency forcing is then evaluated using this new estimated solution, and the second Runge–Kutta iteration is stepped forward from time t^n to $t^n + \Delta t/2$ in $M/2$ steps, beginning with ${}_2\mathbf{r}^1 = \mathbf{r}^n$. The mth small time step of this iteration is

$$ {}_2\mathbf{r}^{m+1} = {}_2\mathbf{r}^m - \Delta\tau\left(\mathscr{L}_1({}_1\mathbf{r}^{M/3+1}) + \mathscr{L}_2({}_2\mathbf{r}^m, {}_2\mathbf{r}^{m+1})\right). $$

[7] One way to appreciate the difference in the effectiveness of divergence damping in the completely and partially split schemes is to note the difference in wavelength at which spurious pressure perturbations appear in each solution. The partially split scheme generates errors at much shorter wavelengths than those produced by the completely split method (compare Figs. 8.3a, 8.5b), and the short-wavelength features are removed more rapidly by the divergence damper.

Following a similar update of the large-time-step forcing with the estimated solution at $t^n + \Delta t/2$, the mth small time step of the final Runge–Kutta iteration, which integrates from t^n to t^{n+1} in M steps, becomes

$$_3\mathbf{r}^{m+1} = {}_3\mathbf{r}^m - \Delta\tau\left(\mathscr{L}_1({}_2\mathbf{r}^{M/2+1}) + \mathscr{L}_2({}_3\mathbf{r}^m, {}_3\mathbf{r}^{m+1})\right),$$

where $_3\mathbf{r}^1 = \mathbf{r}^n$ and $\mathbf{r}^{n+1} = {}_3\mathbf{r}^{M+1}$. Several other alternatives to leapfrog-based partial splitting have also been proposed (Gassmann 2005; Park and Lee 2009; Wicker 2009).

8.4 Summary of Schemes for Nonhydrostatic Models

One way to compare the preceding methods for increasing the efficiency of numerical models for the simulation of fluids that support physically insignificant sound waves is to compare the way each approximation treats the velocity divergence. As before, the mathematics of this discussion will be streamlined by using the compressible Boussinesq equations (8.55)–(8.57) as a simple model for the Euler equations. The pressure and the divergence in the compressible Boussinesq system satisfy

$$\frac{\partial P}{\partial t} + c_s^2\delta = F_p, \tag{8.90}$$

$$\frac{\partial \delta}{\partial t} + \nabla^2 P = G, \tag{8.91}$$

where $\delta = \nabla\cdot\mathbf{v}$, $F_p = -\mathbf{v}\cdot\nabla P$, and $G = -\nabla\cdot(\mathbf{v}\cdot\nabla\mathbf{v}) + \partial b/\partial z$. The semi-implicit method approximates the left sides of the preceding equations with a stable trapezoidal time difference. Sound waves are artificially slowed when large time steps are used in this trapezoidal difference, but the gravity-wave modes are accurately approximated. The implicit coupling in the trapezoidal difference leads to a Helmholtz equation for the pressure that must be solved at every time step.

The prognostic pressure equation (8.90) is discarded in the incompressible Boussinesq approximation, and the local time derivative of the divergence is set to zero in (8.91). This leads to a Poisson equation for pressure that must be solved at every time step. The computational effort required to evaluate the pressure is similar to that required by the semi-implicit method. The Boussinesq system does, however, have the advantage of allowing a wider choice of methods for the integration of the remaining oscillatory forcing terms, which are approximated using leapfrog differencing in the conventional semi-implicit method.

The pressure fields generated by the Boussinesq projection method and the semi-implicit method for the test problem (8.78) and (8.79) are compared in Fig. 8.6. As in Fig. 8.3 the basic-state horizontal flow is vertically sheared from $U = 5\ \mathrm{m\ s^{-1}}$ at the bottom to $U = 15\ \mathrm{m\,s^{-1}}$ at the top of the domain. In the projection method,

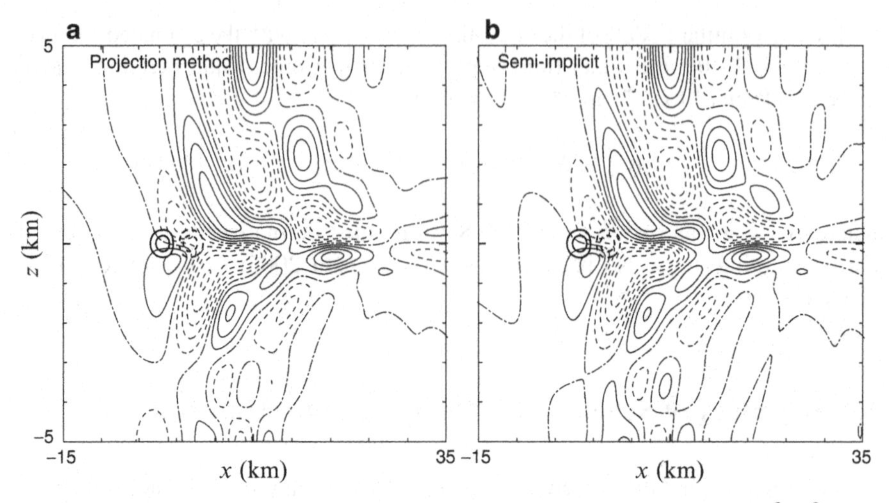

Fig. 8.6 As in Fig. 8.3. Contours at $t = 3{,}000$ s of **a** $\tilde{p}$ and **b** P at intervals of $0.25\ \mathrm{m^2\,s^{-2}}$. Also plotted are Ψ at intervals of $0.15\ \mathrm{s^{-1}}$. Solutions are obtained using **a** the Boussinesq projection method and **b** the semi-implicit method

the integral (8.7) is evaluated using the third-order Adams-Bashforth method with a time step of 10 s. The semi-implicit method is integrated using a 12.5-s time step. The pressure fields generated by both of these methods look very similar to that produced by the partially split method (Fig. 8.3d) and show no evidence of the noise produced using the completely split method (Figs. 8.3a–c).

The elliptic pressure equations that appear when the semi-implicit or projection methods are used are most efficiently solved by sophisticated algorithms such as block-cyclic reduction, conjugate gradient, or multigrid methods. One may think of the small-time-step procedure used in the fractional-step methods as a sort of specialized iterative solver for the Helmholtz equation obtained using the conventional semi-implicit method. The difference in the character of the solution obtained by the completely split and the partially split methods can be appreciated by considering the behavior of the divergence during the small-time-step integration.

During the small-time-step portion of the completely split method the divergence satisfies

$$\frac{\partial^2 \delta}{\partial t^2} - c_s^2 \nabla^2 \delta = \frac{\partial^2 b}{\partial t \partial z}.$$

The initial conditions for δ are those at the beginning of each small-time-step cycle, and since divergence is typically generated by the operators evaluated on the large time step, the initial δ is nonzero. This divergence is propagated without loss during the small-time-step integration (except for minor modification by the buoyancy forcing) and tends to accumulate over a series of large-step–small-step cycles. The test in which the completely split scheme performs well is the case in which the basic-state horizontal velocity is uniform throughout the fluid. When U is constant, the linearized advection operator merely produces a Galilean translation of the fluid that does not generate any divergence. (Recall that the forcing from the wave generator was computed on the small time step.) Nonlinear advection can, of course, generate divergence, as can the linearized advection operator when there is vertical shear in the basic-state wind, and these are the circumstances in which the completely split method produces spurious sound waves.

In contrast, the divergence is almost zero at the start of the first small time step of the partially split method, and only small changes in the divergence are forced during each individual small step. Moreover, the divergence forcing on each small time step closely approximates that which would appear in an explicit small-time-step integration of the full compressible equations, *provided* that the amplitude of all the sound waves is negligible in comparison to slower modes. The divergence damper ensures that the amplitude of the sound waves remains small and thereby preserves the stability and accuracy of the solution.

8.5 The Quasi-Hydrostatic Approximation

Large-scale atmospheric and oceanic motions are very nearly in hydrostatic balance, and as a consequence, they are well described by an approximate set of governing equations in which the full vertical-momentum equation is replaced by the hydrostatic relation

$$\frac{\partial p}{\partial z} = -\rho g. \tag{8.92}$$

The resulting quasi-hydrostatic system[8] is not a hyperbolic system of partial differential equations[9] because there is no prognostic equation for the vertical velocity. The vertical velocity and the pressure are obtained from diagnostic relations.

The quasi-hydrostatic approximation eliminates sound waves, although as will be discussed below, the quasi-hydrostatic approximation does not remove all horizontally propagating acoustic modes. In those large-scale geophysical applications where the numerical resolution along the vertical coordinate is much finer than the horizontal resolution, explicit finite-difference approximations to the quasi-hydrostatic system can be integrated much more efficiently than comparable approximations to either the nonhydrostatic Boussinesq equations or the

[8] The quasi-hydrostatic approximation is often simply referred to as the "hydrostatic approximation." Vertical accelerations would be exactly zero if the fluid were truly in hydrostatic balance, but the quasi-hydrostatic system permits vertical accelerations that are small in comparison with the individual terms retained in (8.92). The use of "quasi-hydrostatic" to describe this approximate equation set is completely analogous to the use of "quasi-geostrophic" to describe a flow in which the horizontal pressure gradients are roughly in geostrophic balance, but horizontal accelerations are not exactly zero.

[9] There has been some concern about the well-posedness of initial-boundary-value problems involving the quasi-hydrostatic equations (Oliger and Sundström 1978). It is not clear how to reconcile these concerns with the successful forecasts obtained twice daily at several operational centers for at least two decades using limited-area weather prediction models based on the quasi-hydrostatic governing equations.

nonhydrostatic compressible equations. The considerable improvement in model efficiency associated with the use of the quasi-hydrostatic governing equations does not, however, apply to semi-implicit models because these models can easily be modified to compute semi-implicit approximations to the full nonhydrostatic compressible equations without significantly increasing the computational overhead (Cullen 1990; Tanguay et al. 1990).

The influence of the quasi-hydrostatic approximation on wave propagation and the stability criteria for explicit finite-difference approximations may be determined by examining solutions to the two-dimensional quasi-hydrostatic equations linearized about a resting isothermally stratified basic state. For this problem, small-amplitude perturbations in the x–z plane satisfy

$$\frac{\partial u}{\partial t} + \frac{\partial P}{\partial x} = 0, \tag{8.93}$$

$$\left(\frac{\partial}{\partial z} + \Gamma\right) P = b, \tag{8.94}$$

$$\frac{\partial b}{\partial t} + N^2 w = 0, \tag{8.95}$$

$$\frac{\partial P}{\partial t} + c_s^2 \left[\frac{\partial u}{\partial x} + \left(\frac{\partial}{\partial z} - \Gamma\right) w\right] = 0, \tag{8.96}$$

where u, w, b, P, Γ, and c_s are defined by (8.35)–(8.37). Waves of the form

$$(u, w, b, P) = (u_0, w_0, b_0, P_0)e^{i(kx+mz-\omega t)}$$

are solutions to this system, provided that

$$\omega^2 = \frac{N^2 k^2}{m^2 + \Gamma^2 + N^2/c_s^2}.$$

This is the standard dispersion relation for two-dimensional gravity waves, except that a term k^2 is missing from the denominator. This term is insignificant when $k \ll m$ and the wave is almost hydrostatic, but the absence of this term can lead to a serious overestimate of the gravity-wave phase speed of modes for which $k \gg m$.

Although there are no conventional sound-wave solutions to the quasi-hydrostatic system, a horizontally propagating acoustic mode known as the Lamb wave is supported by both the quasi-hydrostatic and the nonhydrostatic equations. The vertical velocity and buoyancy perturbations in a Lamb wave in an isothermal atmosphere are zero, and the pressure and horizontal velocity perturbations have the form

$$(u, P) = (u_0, P_0)e^{ik(x \pm c_s t) - \Gamma z}.$$

This may be verified by noting that when $w = 0$, (8.93)–(8.96) reduce to

$$\frac{\partial^2 P}{\partial t^2} - c_s^2 \frac{\partial^2 P}{\partial x^2} = 0 \quad \text{and} \quad \frac{\partial}{\partial t}\left(\frac{\partial}{\partial z} + \Gamma\right) P = 0.$$

If leapfrog time differencing is used to create a differential–difference approximation to (8.93)–(8.96), a necessary and sufficient condition for the stability of the Lamb-wave mode is

$$c_s \Delta t k_{max} < 1, \tag{8.97}$$

where k_{max} is the magnitude of the maximum horizontal wave number resolved by the numerical model. This condition is also sufficient to guarantee the stability of the gravity-wave modes, since for these modes

$$\sin^2(\omega \Delta t) = \frac{(N \Delta t k)^2}{m^2 + \Gamma^2 + N^2/c_s^2} \leq (c_s \Delta t k_{max})^2.$$

In many geophysical applications the vertical resolution is much higher than the horizontal resolution, in which case (8.97) allows a much larger time step than that permitted by the stability condition for the leapfrog approximation to the full non-hydrostatic compressible equations (given by (8.47) with $U = 0$).

8.6 Primitive Equation Models

The exact equations governing global and large-scale atmospheric flows are often approximated by the so-called *primitive equations*. The primitive equations differ from the exact governing equations in that the quasi-hydrostatic assumption is invoked, small "curvature" and Coriolis terms involving the vertical velocity are neglected in the horizontal momentum equations, and the radial distance between any point within the atmosphere and the center of the Earth is approximated by the mean radius of the Earth. Taken together, these approximations yield a system that conserves both energy and angular momentum (Lorenz 1967, p. 16).

The primitive equations governing inviscid adiabatic atmospheric motion may be expressed using height as the vertical coordinate as follows. Let x, y, and z be spatial coordinates that increase eastward, northward, and upward, respectively. Let $\mathbf{u} = (dx/dt, dy/dt)$ be the horizontal velocity vector, f the Coriolis parameter, $\mathbf{k}$ an upward-directed unit vector parallel to the z-axis, and ∇_z the gradient with respect to x and y along surfaces of constant z. Then the rate of change of horizontal momentum in the primitive equation system is governed by

$$\frac{d\mathbf{u}}{dt} + f\mathbf{k} \times \mathbf{u} + \frac{1}{\rho}\nabla_z p = \mathbf{0},$$

where

$$\frac{d()}{dt} = \frac{\partial()}{\partial t} + \mathbf{u} \cdot \nabla_z() + w\frac{\partial()}{\partial z}.$$

The continuity equation is

$$\frac{\partial \rho}{\partial t} + \nabla_z \cdot (\rho \mathbf{u}) + \frac{\partial \rho w}{\partial z} = 0, \tag{8.98}$$

and the thermodynamic equation may be written

$$\frac{dT}{dt} - \frac{\omega}{c_p \rho} = 0, \tag{8.99}$$

where $\omega = \mathrm{d}p/\mathrm{d}t$ is the change in pressure following a fluid parcel. The preceding system of equations for the unknown variables $\mathbf{u}$, w, p, ω, ρ, and T may be closed using the hydrostatic relation (8.92) and the equation of state $p = \rho RT$.

8.6.1 Pressure and σ Coordinates

The primitive equations are often solved in a coordinate system in which geometric height is replaced by a new vertical coordinate $\zeta(x, y, z, t)$. Simple functions that have been used to define ζ include the hydrostatic pressure and the potential temperature. The most commonly used vertical coordinates in current operational models are generalized functions of the hydrostatic pressure.

The primitive equations may be expressed with respect to a different vertical coordinate as follows. Suppose that $\zeta(x, y, z, t)$ is the new vertical coordinate and that ζ is a monotone function of z for all fixed x, y, and t with a unique inverse $z(x, y, \zeta, t)$. Defining ∇_ζ as the gradient operator with respect to x and y along surfaces of constant ζ and applying the chain rule to the identity

$$p[x, y, z(x, y, \zeta, t), t] = p(x, y, \zeta, t)$$

yields

$$\nabla_z p + \frac{\partial p}{\partial z} \nabla_\zeta z = \nabla_\zeta p.$$

Using the hydrostatic relation (8.92) and defining the geopotential $\phi = gz$,

$$\nabla_z p = \nabla_\zeta p + \rho \nabla_\zeta \phi,$$

and the horizontal momentum equations in the transformed coordinates become

$$\frac{d\mathbf{u}}{dt} + f\mathbf{k} \times \mathbf{u} + \nabla_\zeta \phi + \frac{RT}{p} \nabla_\zeta p = \mathbf{0}, \tag{8.100}$$

where

$$\frac{d()}{dt} = \frac{\partial ()}{\partial t} + \mathbf{u} \cdot \nabla_\zeta () + \dot{\zeta} \frac{\partial ()}{\partial \zeta} \tag{8.101}$$

and $\dot{\zeta} = \mathrm{d}\zeta/\mathrm{d}t$. The thermodynamic equation in the transformed coordinates is identical to (8.99), except that the total time derivative is computed using (8.101). The hydrostatic equation may be written as

$$\frac{\partial \phi}{\partial \zeta} = -\frac{RT}{p}\frac{\partial p}{\partial \zeta}.$$

The continuity equation in the transformed coordinate system can be determined by transforming the partial derivatives in (8.98) (Kasahara 1974). It is perhaps simpler to derive the continuity equation directly from first principles. Let $\mathscr{V}$ be a fixed volume defined with respect to the time-independent spatial coordinates x, y, and z, and let $\mathbf{n}$ be the outward-directed unit vector normal to the surface $\mathscr{S}$ enclosing $\mathscr{V}$. Since the rate of change of mass in the volume $\mathscr{V}$ is equal to the net mass flux through $\mathscr{S}$,

$$\begin{aligned}\frac{\partial}{\partial t}\int_{\mathscr{V}} \rho \,\mathrm{d}V &= -\int_{\mathscr{S}} \rho \mathbf{v}\cdot\mathbf{n}\,\mathrm{d}A \\ &= -\int_{\mathscr{V}} \nabla\cdot(\rho\mathbf{v})\,\mathrm{d}V, \end{aligned} \tag{8.102}$$

where $\mathbf{v}$ is the three-dimensional velocity vector. Equation (5.111), which states the general relationship between the divergence in Cartesian coordinates and curvilinear coordinates, implies that

$$\nabla\cdot(\rho\mathbf{v}) = \frac{1}{J}\nabla_\zeta\cdot(J\rho\mathbf{u}) + \frac{1}{J}\frac{\partial}{\partial\zeta}\left(J\rho\dot{\zeta}\right),$$

where J is the Jacobian of the transformation between (x, y, z) and (x, y, ζ), which in this instance is simply $\partial z/\partial\zeta$. In the transformed coordinates

$$\mathrm{d}V = \frac{\partial z}{\partial \zeta}\,\mathrm{d}x\,\mathrm{d}y\,\mathrm{d}\zeta,$$

and since the boundaries of $\mathscr{V}$ do not depend on time, (8.102) may be expressed as

$$\int\!\!\int\!\!\int_{\mathscr{V}} \left[\frac{\partial}{\partial t}\left(\rho\frac{\partial z}{\partial\zeta}\right) + \nabla_\zeta\cdot\left(\rho\mathbf{u}\frac{\partial z}{\partial\zeta}\right) + \frac{\partial}{\partial\zeta}\left(\rho\dot{\zeta}\frac{\partial z}{\partial\zeta}\right)\right]\mathrm{d}x\,\mathrm{d}y\,\mathrm{d}\zeta = 0.$$

Using the hydrostatic equation (8.92) to eliminate ρ from the preceding equation, and noting that the integrand must be identically zero because the volume $\mathscr{V}$ is arbitrary, the continuity equation becomes

$$\frac{\partial}{\partial t}\left(\frac{\partial p}{\partial\zeta}\right) + \nabla_\zeta\cdot\left(\mathbf{u}\frac{\partial p}{\partial\zeta}\right) + \frac{\partial}{\partial\zeta}\left(\dot{\zeta}\frac{\partial p}{\partial\zeta}\right) = 0.$$

Now consider possible choices for ζ. In most respects, the simplest system is obtained by choosing $\zeta = p$; this eliminates one of the two terms that make up the pressure gradient in (8.100) and reduces the continuity equation to the simple diagnostic relation

$$\nabla_p\cdot\mathbf{u} + \frac{\partial\omega}{\partial p} = 0.$$

The difficulty with pressure coordinates arises at the lower boundary because the pressure at the surface of the Earth is a function of horizontal position and time. As a consequence, constant-pressure surfaces intersect the lower boundary of the domain in an irregular manner that changes as a function of time. To simplify the lower-boundary condition, Phillips (1957) suggested choosing $\zeta = \sigma = p/p_s$, where p_s is the surface pressure. The upper and lower boundaries in a σ-coordinate model coincide with the coordinate surfaces $\sigma = 0$ and $\sigma = 1$, and $\dot{\sigma} = 0$ at both the upper and lower boundaries.

The σ-coordinate equations include prognostic equations for $\mathbf{u}$, T, and p_s and diagnostic equations for $\dot{\sigma}$, ϕ, and ω. The prognostic equations for the horizontal velocity and the temperature are

$$\frac{d\mathbf{u}}{dt} + f\mathbf{k}\times\mathbf{u} + \nabla_\sigma\phi + \frac{RT}{p_s}\nabla_\sigma p_s = \mathbf{0} \tag{8.103}$$

and

$$\frac{dT}{dt} = \frac{\kappa T}{\sigma p_s}\omega, \tag{8.104}$$

where

$$\frac{d()}{dt} = \frac{\partial()}{\partial t} + \mathbf{u}\cdot\nabla_\sigma() + \dot{\sigma}\frac{\partial()}{\partial\sigma}.$$

The continuity equation in σ coordinates takes the form of a prognostic equation for the surface pressure:

$$\frac{\partial p_s}{\partial t} + \nabla_\sigma\cdot(p_s\mathbf{u}) + \frac{\partial}{\partial\sigma}(p_s\dot{\sigma}) = 0. \tag{8.105}$$

Recalling that $\dot{\sigma}$ is zero at $\sigma = 0$ and $\sigma = 1$, one can integrate (8.105) over the depth of the domain to obtain

$$\frac{\partial p_s}{\partial t} = -\int_0^1 \nabla_\sigma\cdot(p_s\mathbf{u})\,\mathrm{d}\sigma. \tag{8.106}$$

A diagnostic equation for the vertical velocity $\dot{\sigma}$ is obtained by integrating (8.105) from the top of the domain to level σ, which yields

$$\dot{\sigma}(\sigma) = -\frac{1}{p_s}\left[\sigma\frac{\partial p_s}{\partial t} + \int_0^\sigma \nabla_\sigma\cdot(p_s\mathbf{u})\,\mathrm{d}\tilde{\sigma}\right]. \tag{8.107}$$

A diagnostic equation for ω can be derived by noting that

$$\omega = \frac{d}{dt}(\sigma p_s) = \dot{\sigma}p_s + \sigma\frac{\partial p_s}{\partial t} + \sigma\mathbf{u}\cdot\nabla_\sigma p_s,$$

and thus

$$\omega(\sigma) = \sigma \mathbf{u} \cdot \nabla_\sigma p_s - \int_0^\sigma \nabla_\sigma \cdot (p_s \mathbf{u})\, \mathrm{d}\tilde{\sigma}. \tag{8.108}$$

The geopotential is determined by integrating the hydrostatic equation

$$\frac{\partial \phi}{\partial (\ln \sigma)} = -RT \tag{8.109}$$

from the surface to level σ, which gives

$$\phi(\sigma) = g z_s - R \int_1^\sigma T\, \mathrm{d}(\ln \tilde{\sigma}), \tag{8.110}$$

where $z_s(x, y)$ is the elevation of the topography.

The primary disadvantage of the σ-coordinate system is that it makes the accurate computation of horizontal pressure gradients difficult over steep topography. This problem arises because surfaces of constant σ tilt in regions where there are horizontal variations in surface pressure, and such variations are most pronounced over steep topography. When $\nabla_\sigma p_s \neq 0$, some portion of the vertical pressure gradient is projected onto each of the two terms $\nabla_\sigma \phi$ and $(RT/p_s)\nabla_\sigma p_s$. The vertical pressure gradient will not exactly cancel between these terms because of numerical errors, and over steep topography the noncanceling residual can be comparable to the true horizontal pressure gradient because the vertical gradient of atmospheric pressure is several orders of magnitude larger than the horizontal gradient. The pressure-gradient error in a σ-coordinate model is not confined to the lower levels near the topography, but it may be reduced at upper levels using a hybrid vertical coordinate that transitions from σ coordinates to p coordinates at some level (or throughout some layer) in the interior of the domain (Sangster 1960; Simmons and Burridge 1981). Although they are widely used in operational weather and climate models (Williamson and Olson 1994; Ritchie et al. 1995; Kiehl et al. 1996), these hybrid coordinates complicate the solution of the governing equations and will not be considered here.

Several other approaches have also been suggested to minimize the errors generated over topography in σ-coordinate models. Phillips (1973) and Gary (1973) suggested performing the computations using a perturbation pressure defined with respect to a hydrostatically balanced reference state. Finite-difference schemes have been proposed that guarantee exact cancelation of the vertical pressure gradient between the last two terms in (8.103) whenever the vertical profiles of temperature and pressure have a specified functional relation, such as $T = a \ln(p) + b$ (Corby et al. 1972; Nakamura 1978; Simmons and Burridge 1981). Mesinger (1984) suggested using "η coordinates," in which the mountain slopes are discretized as vertical steps at the grid interfaces with flat terrain between each step. More details and additional techniques for the treatment of pressure-gradient errors over mountains in quasi-hydrostatic atmospheric models are presented in the review by Mesinger and Janić (1985).

8.6.2 Spectral Representation of the Horizontal Structure

Global primitive-equation models often use spherical harmonics to represent the latitudinal and longitudinal variation of the forecast variables. In the following sections we present the basic numerical procedures for creating a spectral approximation to the σ-coordinate equations in a global atmospheric model. The approach is similar to that in Hoskins and Simmons (1975) and Bourke (1974), which may be consulted for additional details. The latitudinal and longitudinal variations in each field will be approximated using spherical harmonics, and the vertical variations will be represented using grid-point methods.

As was the case for the global shallow-water model described in Sect. 6.4.4, the spectral representation of the horizontal velocity field is facilitated by expressing the horizontal momentum equations in terms of the vertical vorticity ζ and the divergence δ. To integrate this system easily using semi-implicit time differencing, it is also helpful to divide the temperature into a horizontally uniform reference state and a perturbation such that $T = \overline{T}(\sigma) + T'$. Using the identity (6.78) and taking the divergence of (8.103) yields

$$\frac{\partial \delta}{\partial t} - \mathbf{k}\cdot\nabla\times(\zeta + f)\mathbf{u} + \nabla\cdot\left(\dot{\sigma}\frac{\partial \mathbf{u}}{\partial \sigma} + RT'\nabla(\ln p_s)\right) + \nabla^2\left(\phi + \frac{\mathbf{u}\cdot\mathbf{u}}{2} + R\overline{T}\ln p_s\right) = 0. \tag{8.111}$$

Again using (6.78) and taking the vertical component of the curl of (8.103), one obtains

$$\frac{\partial \zeta}{\partial t} + \nabla\cdot(\zeta + f)\mathbf{u} + \mathbf{k}\cdot\nabla\times\left(\dot{\sigma}\frac{\partial \mathbf{u}}{\partial \sigma} + RT'\nabla(\ln p_s)\right) = 0. \tag{8.112}$$

Following the notation used in Sect. 6.4.4, let χ be the velocity potential and ψ the stream function for the horizontal velocity. Let λ be the longitude, θ the latitude, and $\mu = \sin\theta$. Define the operator

$$\mathscr{H}(M,N) = \frac{1}{a}\left(\frac{1}{1-\mu^2}\frac{\partial M}{\partial \lambda} + \frac{\partial N}{\partial \mu}\right),$$

where a is the mean radius of the Earth. Then using the formula for the horizontal divergence in spherical coordinates,

$$\begin{aligned}\nabla M &= \frac{1}{a\cos\theta}\frac{\partial M}{\partial \lambda}\mathbf{i} + \frac{1}{a}\frac{\partial M}{\partial \theta}\mathbf{j}\\ &= \frac{1}{a(1-\mu^2)^{1/2}}\left(\frac{\partial M}{\partial \lambda}\mathbf{i} + (1-\mu^2)\frac{\partial M}{\partial \mu}\mathbf{j}\right),\end{aligned}$$

and the relations (6.82)–(6.85), one may express the prognostic equations for the σ-coordinate system in the form

$$\frac{\partial \nabla^2 \chi}{\partial t} = \mathscr{H}(B, -A) - 2\Omega \left(\frac{U}{a} - \mu \nabla^2 \psi \right)$$
$$- \nabla^2 \left(\phi + \frac{U^2 + V^2}{2(1 - \mu^2)} + R\overline{T} \ln p_s \right), \tag{8.113}$$

$$\frac{\partial \nabla^2 \psi}{\partial t} = -\mathscr{H}(A, B) - 2\Omega \left(\frac{V}{a} + \mu \nabla^2 \chi \right), \tag{8.114}$$

$$\frac{\partial T'}{\partial t} = -\mathscr{H}(UT', VT') + T' \nabla^2 \chi - \dot{\sigma} \frac{\partial T}{\partial \sigma} + \frac{\kappa T \omega}{\sigma p_s}, \tag{8.115}$$

$$\frac{\partial}{\partial t}(\ln p_s) = -\frac{U}{a(1 - \mu^2)} \frac{\partial}{\partial \lambda}(\ln p_s) - \frac{V}{a} \frac{\partial}{\partial \mu}(\ln p_s) - \nabla^2 \chi - \frac{\partial \dot{\sigma}}{\partial \sigma}, \tag{8.116}$$

where

$$U = u \cos \theta = (1 - \mu^2) \mathscr{H}(\chi, -\psi),$$
$$V = v \cos \theta = (1 - \mu^2) \mathscr{H}(\psi, \chi),$$
$$A = U \nabla^2 \psi + \dot{\sigma} \frac{\partial V}{\partial \sigma} + \frac{RT'}{a} (1 - \mu^2) \frac{\partial}{\partial \mu}(\ln p_s).$$
$$B = V \nabla^2 \psi - \dot{\sigma} \frac{\partial U}{\partial \sigma} - \frac{RT'}{a} \frac{\partial}{\partial \lambda}(\ln p_s).$$

The preceding system of equations is formulated using $\ln p_s$ instead of p_s as the prognostic variable to make the term $(RT/p_s)\nabla_\sigma p_s$ into a binary product of the prognostic variables and thereby facilitate the alias-free evaluation of the pressure-gradient force via the spectral transform method.

At each σ level, the unknown functions ψ, χ, T', and ϕ are approximated using a truncated series of spherical harmonics. The unknown function $\ln p_s$ is also approximated by a spherical harmonic expansion. Expressions for the time tendencies of the expansion coefficients for each spherical harmonic are obtained using the transform method in a manner analogous to that for the global shallow-water model described in Sect. 6.4.4. As an example, suppose that the stream function and velocity potential at a given σ level are expanded in spherical harmonics as in (6.90) and (6.91). Then, using the notation defined in Sect. 6.4.4, the equation for $\partial \psi_{m,n}/\partial t$ is once again given by (6.95) except that $\hat{A}_m$ and $\hat{B}_m$ now satisfy

$$U \nabla^2 \psi + \dot{\sigma} \frac{\partial V}{\partial \sigma} + \frac{RT'}{a} (1 - \mu^2) \frac{\partial}{\partial \mu}(\ln p_s) = \sum_{m=-M}^{M} \hat{A}_m e^{im\lambda} \tag{8.117}$$

and

$$V \nabla^2 \psi - \dot{\sigma} \frac{\partial U}{\partial \sigma} - \frac{RT'}{a} \frac{\partial}{\partial \lambda}(\ln p_s) = \sum_{m=-M}^{M} \hat{B}_m e^{im\lambda}. \tag{8.118}$$

The spectral form of the tendency equations for the velocity potential, the perturbation temperature, and the surface pressure may be found in Bourke (1974) and will not be given here. Note that the vertical advection terms in (8.117) and (8.118) involve the product of three spatially varying functions (since $\dot{\sigma}$ itself depends on the product of two spatially varying functions). The standard transform method cannot be used to transform these triple products between wave-number and physical space without incurring some numerical error. This "aliasing" error is nevertheless very small (Hoskins and Simmons 1975).

8.6.3 Vertical Differencing

The most significant modifications required to extend the shallow-water algorithm to a σ-coordinate model are those associated with the computation of the vertical derivatives. The vertical derivatives are computed using finite differences at that stage of the integration cycle when all the unknown variables are available on the physical mesh. As in (8.117) and (8.118), the results from these finite-difference computations are then combined with the other binary products computed on the physical mesh, and the net forcing is transformed back to wave-number space.

A convenient and widely used vertical discretization for the σ-coordinate equations is illustrated in Fig. 8.7 for a model with N vertical levels. The upper and lower boundaries are located at $\sigma = 0$ and

$$\sigma = 1 = \sum_{k=1}^{N} \Delta\sigma_k,$$

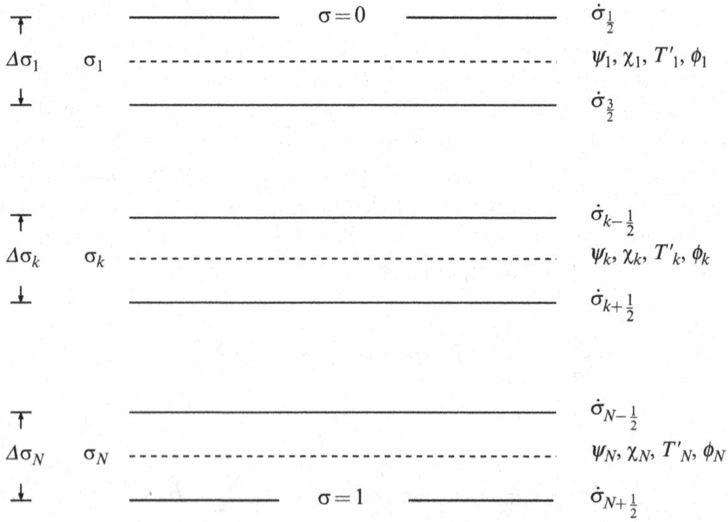

$\sigma=0$ $\dot{\sigma}_{\frac{1}{2}}$

$\Delta\sigma_1$ σ_1 $\psi_1, \chi_1, T'_1, \phi_1$

$\dot{\sigma}_{\frac{3}{2}}$

$\dot{\sigma}_{k-\frac{1}{2}}$

$\Delta\sigma_k$ σ_k $\psi_k, \chi_k, T'_k, \phi_k$

$\dot{\sigma}_{k+\frac{1}{2}}$

$\dot{\sigma}_{N-\frac{1}{2}}$

$\Delta\sigma_N$ σ_N $\psi_N, \chi_N, T'_N, \phi_N$

$\sigma=1$ $\dot{\sigma}_{N+\frac{1}{2}}$

Fig. 8.7 Vertical distribution of the unknown variables on a σ-coordinate grid. The thickness of the σ layers need not be uniform. The center of each layer is at level $\sigma = \sigma_k$ and is indicated by the *dashed lines*. Note that the vertical index k increases with σ and decreases with geometric height

where $\Delta\sigma_k$ is the width of the kth σ layer. The stream function, velocity potential, temperature, and geopotential are defined at the center of each σ layer, and the velocity $\dot{\sigma}$ is defined at the interface between each layer. The vertical derivatives appearing in (8.113)–(8.115) involve variables, such as the temperature, that are defined at the center of each σ layer. These derivatives are approximated such that

$$\begin{aligned}\left(\dot{\sigma}\frac{\partial T}{\partial \sigma}\right)_k &\approx \langle \dot{\sigma}_k \delta_\sigma T_k \rangle^\sigma \\ &= \dot{\sigma}_{k+\frac{1}{2}}\left(\frac{T_{k+1}-T_k}{\Delta\sigma_{k+1}+\Delta\sigma_k}\right)+\dot{\sigma}_{k-\frac{1}{2}}\left(\frac{T_k-T_{k-1}}{\Delta\sigma_k+\Delta\sigma_{k-1}}\right).\end{aligned} \tag{8.119}$$

The preceding differencing is the generalization of the "averaging scheme" discussed in Sect. 4.4 to the nonuniform staggered mesh shown in Fig. 8.7.

The vertical derivative of $\dot{\sigma}$ in (8.116) is approximated as

$$\left(\frac{\partial \dot{\sigma}}{\partial \sigma}\right)_k \approx \delta_\sigma \dot{\sigma}_k = \frac{\dot{\sigma}_{k++\frac{1}{2}} - \dot{\sigma}_{k--\frac{1}{2}}}{\Delta\sigma_k}. \tag{8.120}$$

Defining $G_k = \nabla_\sigma \cdot \mathbf{u}_k + \mathbf{u}_k \cdot \nabla_\sigma(\ln p_s)$, the preceding expression implies that the vertically discretized approximation to the surface-pressure-tendency equation (8.106) is

$$\frac{\partial}{\partial t}(\ln p_s) = -\sum_{k=1}^{N} G_k \Delta\sigma_k, \tag{8.121}$$

and that (8.107) is approximated as

$$\dot{\sigma}_{k+\frac{1}{2}} = \left(\sum_{j=1}^{k}\Delta\sigma_j\right)\sum_{j=1}^{N} G_j\Delta\sigma_j - \sum_{j=1}^{k} G_j \Delta\sigma_j. \tag{8.122}$$

The hydrostatic equation (8.109) is approximated as

$$\frac{\phi_{k+1}-\phi_k}{\ln\sigma_{k+1}-\ln\sigma_k} = -\frac{R}{2}(T_{k+1}+T_k),$$

except in the half-layer between the lowest σ level and the surface, where

$$\frac{\phi_N-\phi_s}{\ln\sigma_N} = -RT_N.$$

Defining $\alpha_N = -\ln\sigma_N$ and $\alpha_k = 1/2\ln(\sigma_{k+1}/\sigma_k)$ for $1 \le k < N$, the discrete analogue of (8.110) becomes

$$\phi_k = \phi_s + R\left(\sum_{j=k}^{N}\alpha_j T_j + \sum_{j=k+1}^{N}\alpha_{j-1}T_j\right). \tag{8.123}$$

Finally, as suggested by Corby et al. (1972), the vertical discretization for the ω equation (8.108) is chosen to preserve the energy-conservation properties of the vertically integrated continuous equations. Such conservation is achieved if

$$\frac{\omega_k}{\sigma_k p_s} = \mathbf{u}_k \cdot \nabla_\sigma(\ln p_s) - \frac{\alpha_k}{\Delta\sigma_k}\sum_{j=1}^{k} G_j \Delta\sigma_j - \frac{\alpha_{k-1}}{\Delta\sigma_k}\sum_{j=1}^{k-1} G_j \Delta\sigma_j. \tag{8.124}$$

8.6.4 Energy Conservation

Why does (8.124) give better energy-conservation properties than the simpler formula that would result if both α_k and α_{k-1} were replaced by $\Delta\sigma_k/(2\sigma_k)$? To answer this question it is necessary to review the energy-conservation properties of the continuous σ-coordinate primitive equations. Our focus is on the vertical discretization, so it is helpful to obtain a conservation law for the vertically integrated total energy per unit horizontal area. Using the hydrostatic equation (8.92) and the definition $\sigma p_s = p$, one may express the vertical integral of the sum of the kinetic[10] and internal energy per unit volume as

$$\begin{aligned}\int_{z_s}^{\infty} \rho\left(\frac{\mathbf{u}\cdot\mathbf{u}}{2} + c_v T\right) \mathrm{d}z &= -\frac{1}{g}\int_{p_s}^{0}\left(\frac{\mathbf{u}\cdot\mathbf{u}}{2} + c_v T\right) \mathrm{d}p \\ &= \frac{p_s}{g}\int_0^1\left(\frac{\mathbf{u}\cdot\mathbf{u}}{2} + c_v T\right) \mathrm{d}\sigma.\end{aligned}$$

Using the hydrostatic equation twice and integrating by parts, the vertical integral of the potential energy becomes

$$\int_{z_s}^{\infty} \rho g z \, \mathrm{d}z = \int_0^{p_s} \frac{\phi}{g}\mathrm{d}p = \frac{1}{g}\int_0^{\phi_s p_s} \mathrm{d}(\phi p) - \int_{\infty}^{\phi_s} \frac{p}{g}\mathrm{d}\phi = \frac{\phi_s p_s}{g} + \int_0^{p_s}\frac{p}{g\rho}\mathrm{d}p.$$

Recalling that $p = \rho R T$,

$$\int_{z_s}^{\infty} \rho g z \, \mathrm{d}z = \frac{\phi_s p_s}{g} + \frac{p_s}{g}\int_0^1 RT \, \mathrm{d}\sigma,$$

and the total vertically integrated energy per unit area is

$$\mathscr{E} = \frac{\phi_s p_s}{g} + \frac{p_s}{g}\int_0^1\left(\frac{\mathbf{u}\cdot\mathbf{u}}{2} + c_p T\right) \mathrm{d}\sigma.$$

The total time derivatives in the momentum and thermodynamic equations must be written in flux form to obtain a conservation law governing $\mathscr{E}$. For any scalar γ,

[10] Note that as a consequence of the primitive-equation approximation, the vertical velocity does not appear as part of the kinetic energy.

$$p_s \frac{d\gamma}{dt} = p_s \frac{\partial \gamma}{\partial t} + p_s \mathbf{u} \cdot \nabla_\sigma \gamma + p_s \dot{\sigma} \frac{\partial \gamma}{\partial \sigma} + \gamma \left[\frac{\partial p_s}{\partial t} + \nabla_\sigma \cdot (p_s \mathbf{u}) + \frac{\partial}{\partial \sigma}(p_s \dot{\sigma}) \right]$$
$$= \frac{\partial}{\partial t}(p_s \gamma) + \nabla_\sigma \cdot (p_s \gamma \mathbf{u}) + \frac{\partial}{\partial \sigma}(p_s \gamma \dot{\sigma}), \quad (8.125)$$

where the quantity in square brackets is zero by the pressure-tendency equation (8.105). Adding $p_s c_p$ times the thermodynamic equation (8.104) to the dot product of $p_s \mathbf{u}$ and the momentum equation (8.103) and using (8.125), one obtains

$$\frac{\partial E}{\partial t} + \nabla_\sigma \cdot (E\mathbf{u}) + \frac{\partial}{\partial \sigma}(E\dot{\sigma}) + \mathbf{u} \cdot p_s \nabla_\sigma \phi + \mathbf{u} \cdot RT \nabla_\sigma p_s - \frac{RT\omega}{\sigma} = 0, \quad (8.126)$$

where $E = p_s(\mathbf{u} \cdot \mathbf{u}/2 + c_p T)$. Defining

$$F = -\phi \nabla_\sigma \cdot (p_s \mathbf{u}) + \mathbf{u} \cdot RT \nabla_\sigma p_s - \frac{RT\omega}{\sigma}, \quad (8.127)$$

one may express (8.126) as

$$\frac{\partial E}{\partial t} + \nabla_\sigma \cdot [(E + p_s \phi)\mathbf{u}] + \frac{\partial}{\partial \sigma}(E\dot{\sigma}) + F = 0. \quad (8.128)$$

The forcing F may be written as the vertical divergence of a flux as follows. Substituting for ω using (8.108),

$$F = -\phi \nabla_\sigma \cdot (p_s \mathbf{u}) + \frac{RT}{\sigma} \int_0^\sigma \nabla_\sigma \cdot (p_s \mathbf{u}) \, d\tilde{\sigma},$$

and then substituting for RT/σ from the hydrostatic equation,

$$F = -\phi \nabla_\sigma \cdot (p_s \mathbf{u}) - \frac{\partial \phi}{\partial \sigma} \int_0^\sigma \nabla_\sigma \cdot (p_s \mathbf{u}) \, d\tilde{\sigma}$$
$$= -\frac{\partial}{\partial \sigma} \left[\phi \int_0^\sigma \nabla_\sigma \cdot (p_s \mathbf{u}) \, d\tilde{\sigma} \right].$$

Thus,

$$\int_0^1 F \, d\sigma = -\phi_s \int_0^1 \nabla_\sigma \cdot (p_s \mathbf{u}) \, d\sigma$$
$$= \phi_s \int_0^1 \left(\frac{\partial p_s}{\partial t} + \frac{\partial}{\partial \sigma}(\dot{\sigma} p_s) \right) d\sigma$$
$$= \frac{\partial}{\partial t}(\phi_s p_s).$$

The preceding expression may be used to derive a conservation law for $\mathscr{E}$ by integrating (8.128) over the depth of the domain and applying the boundary condition $\dot{\sigma} = 0$ at the upper and lower boundaries to obtain

$$\frac{\partial \mathscr{E}}{\partial t} + \frac{1}{g} \nabla_\sigma \cdot \int_0^1 (E + p_s \phi)\mathbf{u} \, d\sigma = 0. \quad (8.129)$$

Of course, (8.129) also implies that if the horizontal domain is periodic, or if there is no flow normal to the lateral boundaries, the σ-coordinate primitive equations conserve the domain-integrated total energy

$$\int\int\left[\frac{\phi_s p_s}{g}+\frac{p_s}{g}\int_0^1\left(\frac{\mathbf{u}\cdot\mathbf{u}}{2}+c_pT\right)\mathrm{d}\sigma\right]\mathrm{d}x\,\mathrm{d}y. \tag{8.130}$$

The domain-integrated total energy is not, however, exactly conserved by global spectral models. As discussed in Sect. 6.2.3, a Galerkin spectral approximation to a prognostic equation for an unknown function γ will generally conserve the domain integral of γ^2, provided that the domain integral of γ^2 is also conserved by the continuous equations and time-differencing errors are neglected. Unfortunately, the conservation of the squares of the prognostic variables in (8.113)–(8.116) does not imply exact conservation of the total energy. Practical experience has, nevertheless, shown that the deviations from exact energy conservation generated by the spectral approximation of the horizontal derivatives is very small. The nonconservation introduced by the semi-implicit time differencing used in most global primitive-equation models has also been shown to be very small (Hoskins and Simmons 1975). Nonconservative formulations of the vertical finite differencing can, however, have a significantly greater impact on the global energy conservation. This appears to be a particularly important issue if long-time integrations are conducted using global climate models with poor vertical resolution.

The energy-conservation properties of the vertical discretization given by (8.119)–(8.124) will therefore be isolated from the nonconservative effects of the spectral approximation and the time differencing by considering a system of differential–difference equations in which only those terms containing vertical derivatives are discretized. Except for the terms involving vertical derivatives, the total-energy equation for the semidiscrete system must be identical to (8.128) because the time and horizontal derivatives are exact.

The semidiscrete system will therefore conserve total energy, provided that it satisfies the discrete analogues of

$$\int_0^1\frac{\partial}{\partial\sigma}(E\dot\sigma)\,\mathrm{d}\sigma=0 \tag{8.131}$$

and

$$\int_0^1 F\,\mathrm{d}\sigma=\frac{\partial}{\partial t}(\phi_s p_s). \tag{8.132}$$

The integrand in (8.131) appears in the total-energy equation (8.128) as a mathematical simplification of a linear combination of the vertical derivative terms in the momentum, thermodynamic, and surface-pressure-tendency equations such that

$$\frac{\partial}{\partial\sigma}(E\dot\sigma)\,\mathrm{d}\sigma=\mathbf{u}\cdot\dot\sigma\frac{\partial\mathbf{u}}{\partial\sigma}+\frac{\mathbf{u}\cdot\mathbf{u}}{2}\frac{\partial\dot\sigma}{\partial\sigma}+c_p\dot\sigma\frac{\partial T}{\partial\sigma}+c_pT\frac{\partial\dot\sigma}{\partial\sigma}.$$

When the vertical derivatives on the right side of the preceding equation are approximated using (8.119) and (8.120), their summation over the depth of the domain is exactly zero. This may be demonstrated for the pair of terms involving T by noting that since $\dot{\sigma}_{1/2} = \dot{\sigma}_{N+1/2} = 0$,

$$\sum_{k=1}^{N} \left[\langle \dot{\sigma}_k \delta_\sigma T_k \rangle^\sigma + T_k \delta_\sigma \dot{\sigma}_k \right] \Delta\sigma_k = \sum_{k=1}^{N} \left[\dot{\sigma}_{k+\frac{1}{2}} \left(\frac{\Delta\sigma_k T_{k+1} + \Delta\sigma_{k+1} T_k}{\Delta\sigma_{k+1} + \Delta\sigma_k} \right) \right.$$
$$\left. - \dot{\sigma}_{k-\frac{1}{2}} \left(\frac{\Delta\sigma_{k-1} T_k + \Delta\sigma_k T_{k-1}}{\Delta\sigma_k + \Delta\sigma_{k-1}} \right) \right] = 0.$$

A similar relation holds for the two terms involving the horizontal velocity (see Problem 7).

Now consider the discrete analogue of (8.132), or equivalently,

$$\int_0^1 \frac{F}{p_s} \, d\sigma = \phi_s \frac{\partial}{\partial t} (\ln p_s).$$

Defining $G = \nabla_\sigma \cdot \mathbf{u} + \mathbf{u} \cdot \nabla_\sigma (\ln p_s)$ and substituting for F using (8.127) yields

$$\int_0^1 \left(-\phi G + \mathbf{u} \cdot RT \nabla_\sigma (\ln p_s) - \frac{RT\omega}{\sigma p_s} \right) d\sigma = \phi_s \frac{\partial}{\partial t} (\ln p_s).$$

The discrete form of this integral equation may be obtained using (8.121) and (8.124) and is algebraically equivalent to

$$\sum_{k=1}^{N} (\phi_k - \phi_s) G_k \Delta\sigma_k = \sum_{k=1}^{N} RT_k \left(\alpha_k \sum_{j=1}^{k} G_j \Delta\sigma_j + \alpha_{k-1} \sum_{j=1}^{k-1} G_j \Delta\sigma_j \right).$$

It may be verified that the preceding expression is indeed an algebraic identity by substituting for $\phi_k - \phi_s$ from the discrete form of the hydrostatic equation (8.123) and using the relation

$$\sum_{k=1}^{N} \sum_{j=1}^{k} a_k b_j = \sum_{k=1}^{N} \sum_{j=k}^{N} a_j b_k.$$

In addition to conserving total energy, the preceding vertical discretization also conserves total mass (see Problem 8). This scheme does not conserve the integrated angular momentum or the integrated potential temperature. Arakawa and Lamb (1977), Simmons and Burridge (1981), and Arakawa and Konor (1996) described alternative vertical discretizations that conserve angular momentum, potential temperature, or various other vertically integrated functions.

8.6.5 Semi-implicit Time Differencing

Computational efficiency can be enhanced by using semi-implicit time differencing to integrate the preceding primitive-equation model. The semi-implicit method can be implemented in σ-coordinate primitive-equation models as follows. Let $\mathbf{d}$ be a column vector whose kth element is the function $\nabla_\sigma^2 \chi$ at level σ_k. Similarly, define $\bar{\mathrm{t}}$, $\mathbf{t}$, and $\mathbf{h}$ to be column vectors containing the σ-level values of the functions $R\overline{T}$, T, and ϕ. Let $\mathbf{h}_\mathrm{s}$ be a column vector in which every element is ϕ_s. Then the vertically discretized equations for the divergence, temperature, surface-pressure tendency, and geopotential may be written in the form

$$\frac{\partial \mathbf{d}}{\partial t} = \mathbf{f}_\mathrm{d} - \nabla_\sigma^2 \left(\mathbf{h} + \bar{\mathrm{t}} \ln p_\mathrm{s} \right), \tag{8.133}$$

$$\frac{\partial \mathbf{t}}{\partial t} = \mathbf{f}_\mathrm{t} - \mathbf{H}\mathbf{d}, \tag{8.134}$$

$$\frac{\partial}{\partial t}(\ln p_s) = f_p - \mathbf{p}^\mathrm{T}\, \mathbf{d}, \tag{8.135}$$

$$\mathbf{h} = \mathbf{h}_\mathrm{s} + \mathbf{G}\mathbf{t}. \tag{8.136}$$

Here $\mathbf{G}$ and $\mathbf{H}$ are matrices and $\mathbf{p}$ is a column vector, none of which depend on λ, μ, or t. The thermodynamic equation (8.134) is partitioned such that all terms containing the product of $\overline{T}(\sigma)$ and the divergence are collected in $\mathbf{Hd}$.

Equation (8.121) implies that

$$\mathbf{p}^\mathrm{T} = (\Delta\sigma_1, \Delta\sigma_2, \ldots, \Delta\sigma_N),$$

and (8.123) requires

$$\frac{\mathbf{G}}{R} = \begin{pmatrix} \alpha_1 & \alpha_1 + \alpha_2 & \alpha_2 + \alpha_3 & \cdots \\ 0 & \alpha_2 & \alpha_2 + \alpha_3 & \cdots \\ 0 & 0 & \alpha_3 & \cdots \\ 0 & 0 & 0 & \cdots \\ \vdots & \vdots & \vdots & \end{pmatrix}.$$

Let $h_{r,s}$ denote the sth element in the rth row of $\mathbf{H}$. Then according to (8.115), $h_{r,s}$ is determined by the contribution of the divergence at level s to $\dot{\sigma}\partial\overline{T}/\partial\sigma - \kappa\overline{T}\omega/(\sigma p_\mathrm{s})$ at level r. Define a step function such that $\mathscr{S}(x) = 1$ if $x \geq 0$ and $\mathscr{S}(x) = 0$ otherwise. Then from (8.119), (8.122), and (8.124),

$$\begin{aligned} \frac{h_{r,s}}{\Delta\sigma_s} = {} & \frac{\kappa\overline{T}_r \mathscr{S}(r-s)}{\Delta\sigma_r}\left[\alpha_r + \mathscr{S}(r-s-1)\alpha_{r-1}\right] \\ & - \left(\frac{\overline{T}_{r+1} - \overline{T}_r}{\Delta\sigma_{r+1} + \Delta\sigma_r}\right)\left(\mathscr{S}(r-s) - \sum_{j=1}^{r} \Delta\sigma_j\right) \end{aligned}$$

$$-\left(\frac{\overline{T}_r - \overline{T}_{r-1}}{\Delta\sigma_r + \Delta\sigma_{r-1}}\right)\left(\mathscr{S}(r-s-1) - \sum_{j=1}^{r-1}\Delta\sigma_j\right).$$

The remaining terms in (8.121) and the vertically discretized versions of (8.113) and (8.115) are gathered into f_p, $\mathbf{f_d}$, and $\mathbf{f_t}$, respectively.

A single equation for the divergence may be obtained by eliminating **t**, **h**, and $\ln p_s$ from (8.133)–(8.136) to give

$$\left(\frac{\partial^2}{\partial t^2} - \mathbf{B}\nabla_\sigma^2\right)\mathbf{d} = \frac{\partial \mathbf{f_d}}{\partial t} - \nabla_\sigma^2\left(\mathbf{G}\mathbf{f_t} + f_p\bar{\mathfrak{t}}\right), \tag{8.137}$$

where $\mathbf{B} = \mathbf{G}\mathbf{H} + \bar{\mathfrak{t}}\mathbf{p}^{\mathrm{T}}$. The solutions to the homogeneous part of this equation comprise the set of gravity waves supported by the vertically discretized model. Hoskins and Simmons (1975) presented plots showing the vertical structure of each of the gravity-wave modes in a five-layer model. For typical atmospheric profiles of $\overline{T}(\sigma)$ the fastest mode propagates at a speed on the order of 300 ms^{-1} and thereby imposes a severe constraint on the maximum stable time step with which these equations can be integrated using explicit time differencing.

Since the fastest-moving gravity waves do not need to be accurately simulated to obtain an accurate global weather forecast, (8.133)–(8.135) can be efficiently integrated using a semi-implicit scheme in which those terms that combine to form the left side of (8.137) are integrated using the trapezoidal method over a time interval of $2\Delta t$. The formulae that result from this semi-implicit approximation are

$$\delta_{2t}\mathbf{d}^n = \mathbf{f_d}^n - \nabla_\sigma^2\left[\langle\mathbf{h}^n\rangle^{2t} + \bar{\mathfrak{t}}\,\langle(\ln p_s)^n\rangle^{2t}\right], \tag{8.138}$$

$$\delta_{2t}\mathbf{t}^n = \mathbf{f_t}^n - \mathbf{H}\,\langle\mathbf{d}^n\rangle^{2t}, \tag{8.139}$$

$$\delta_{2t}(\ln p_s)^n = f_p^n - \mathbf{p}^T\,\langle\mathbf{d}^n\rangle^{2t}. \tag{8.140}$$

Using the relation $\delta_{2t}\gamma^n = (\langle\gamma^n\rangle^{2t} - \gamma^{n-1})/\Delta t$ together with (8.136), (8.139), and (8.140) to eliminate $\langle\mathbf{h}^n\rangle^{2t}$ and $\langle(\ln p_s)^n\rangle^{2t}$ from (8.138) gives

$$\begin{aligned}&\left[\mathbf{I} - (\Delta t)^2\mathbf{B}\nabla_\sigma^2\right]\langle\mathbf{d}^n\rangle^{2t}\\ &\quad = \mathbf{d}^{n-1} + \Delta t\mathbf{f_d}^n - \nabla_\sigma^2\left\{\Delta t\left[\mathbf{h}^{n-1} + \bar{\mathfrak{t}}(\ln p_s)^{n-1}\right] + (\Delta t)^2\left[\mathbf{G}\mathbf{f_t}^n + \bar{\mathfrak{t}}f_p^n\right]\right\}.\end{aligned} \tag{8.141}$$

Let $\boldsymbol{\chi}_{r,s}$ be a column vector whose kth element is the coefficient of $Y_{r,s}$ in the series expansion for the velocity potential at level k. Since the spherical harmonics are eigenfunctions of the horizontal Laplacian operator on the sphere, (8.141) is equivalent to a linear-algebraic system for $\boldsymbol{\chi}_{r,s}^{n+1}$ of the form

$$\left[\mathbf{I} + (\Delta t)^2\frac{s(s+1)}{a^2}\mathbf{B}\right]\boldsymbol{\chi}_{r,s}^{n+1} = \mathbf{f},$$

where $\mathbf{f}$ does not involve the values of any unknown functions at time $(n+1)\Delta t$. The N unknown variables in this relatively small linear system can be determined by Gaussian elimination. Additional efficiency can be achieved by exploiting the fact that the coefficient matrix is constant in time, so its "LU" decomposition into upper and lower triangular matrices need only be computed once.

Some of the forcing terms that are responsible for gravity-wave propagation in the σ-coordinate equations are nonlinear. To obtain the preceding linear-algebraic equation for $\chi_{r,s}^{n+1}$ these terms have been decomposed into a linear part and a nonlinear perturbation by splitting the total temperature into a constant horizontally uniform reference temperature $\overline{T}(\sigma)$ and a perturbation. As discussed in Sect. 8.2.3, this decomposition imposes a constraint on the stability of the semi-implicit solution that, roughly speaking, requires the speed of the fastest-moving gravity wave supported by the actual atmospheric structure to be only modestly faster than the speed of the fasting-moving gravity wave in the reference state. This stability constraint is usually satisfied by choosing an isothermal profile for the reference state, i.e., $\overline{T}(\sigma) = T_0$ (Simmons et al. 1978). A typical value for T_0 is 300 K.

Problems

1. Consider small-amplitude shallow-water motions on a "mid-latitude β-plane." In the following, x and y are horizontal coordinates oriented east–west and north–south, respectively; the Coriolis parameter is approximated as $f_0 + \beta y$ where f_0 and β are constant; g is the gravitational acceleration; $U > 0$ is a constant mean flow from west to east; u' and v' are the perturbation west-to-east and south-to-north velocities; h is the perturbation displacement of the free surface. Define the vorticity ζ and the divergence δ as

$$\zeta = \frac{\partial v'}{\partial x} - \frac{\partial u'}{\partial y}, \qquad \delta = \frac{\partial u'}{\partial x} + \frac{\partial v'}{\partial y}.$$

Assume that the mean flow is in geostrophic balance,

$$U = -\frac{g}{f_0}\frac{\partial \overline{h}}{\partial y},$$

and that there is a mean north–south gradient in the bottom topography equal to the mean gradient in the height of the free surface, $\partial\overline{h}/\partial y$, so that the mean fluid depth is a constant H. The linearized shallow-water equations for this system are

$$\left(\frac{\partial}{\partial t}+U\frac{\partial}{\partial x}\right)\zeta+f\delta+\beta v=0,$$

$$\left(\frac{\partial}{\partial t}+U\frac{\partial}{\partial x}\right)\delta-f\zeta+\beta(U+u)+g\left(\frac{\partial^2 h}{\partial x^2}+\frac{\partial^2 h}{\partial y^2}\right)=0,$$

$$\left(\frac{\partial}{\partial t}+U\frac{\partial}{\partial x}\right)h+H\delta=0. \quad (8.142)$$

The terms involving β in the preceding vorticity and divergence equations can be approximated[11] as

$$\left(\frac{\partial}{\partial t}+U\frac{\partial}{\partial x}\right)\zeta+f_0\delta+\frac{\beta g}{f_0}\frac{\partial h}{\partial x}=0, \quad (8.143)$$

$$\left(\frac{\partial}{\partial t}+U\frac{\partial}{\partial x}\right)\delta-f_0\zeta+g\left(\frac{\partial^2 h}{\partial x^2}+\frac{\partial^2 h}{\partial y^2}\right)=0. \quad (8.144)$$

(a) Show that waves of the form

$$(\zeta,\delta,h)=(\zeta_0,\delta_0,h_0)e^{i(kx+\ell y-\omega t)}$$

are solutions to the preceding system if they satisfy the dispersion relation

$$(\omega-Uk)^2=c^2(k^2+\ell^2)+f_0^2+\frac{k\beta c^2}{\omega-Uk},$$

where $c^2=gH$.

(b) Show that if $|\beta/c|\ll k^2$, the individual solutions to this dispersion relation are well approximated by the solutions to either the inertial-gravity-wave dispersion relation

$$(\omega-Uk)^2=c^2(k^2+\ell^2)+f_0^2$$

or the Rossby-wave dispersion relation

$$\omega=Uk-\frac{\beta k}{k^2+\ell^2+f_0^2/c^2}.$$

2. Suppose that the time derivatives in (8.142)–(8.144) are approximated by leapfrog differencing and the spatial dependence is represented by a Fourier spectral or pseudospectral method. Recall that we have assumed $U>0$, as would be the case in the middle latitudes of the Earth's atmosphere.

(a) Determine the constraints on Δt required to keep the gravity waves stable and show that $(U+c)k\Delta t\le 1$ is a necessary condition for stability.

[11] The approximations used to obtain (8.143) and (8.144) are motivated by the desire to obtain a clean dispersion relation rather than a straightforward scale analysis.

(b) Determine the constraints on Δt required to keep the Rossby waves stable. Let K be the magnitude of the maximum vector wave number retained in the truncation, i.e.,

$$K = \max_{k,\ell} \sqrt{k^2 + \ell^2}.$$

Show that $UK\Delta t \leq 1$ is a sufficient condition for the stability of the Rossby waves unless

$$K^2 \leq \frac{\beta}{2U} - \frac{f_0^2}{c^2}.$$

3. Suppose that (8.142)–(8.144) are integrated using the semi-implicit scheme

$$\delta_{2t}\zeta^n + U\frac{\partial \zeta^n}{\partial x} + f_0\delta^n + \frac{\beta g}{f_0}\frac{\partial h^n}{\partial x} = 0,$$

$$\delta_{2t}\delta^n + U\frac{\partial \delta^n}{\partial x} - f_0\zeta^n + g\left\langle \frac{\partial^2 h^n}{\partial x^2} + \frac{\partial^2 h^n}{\partial y^2}\right\rangle^{2t} = 0,$$

$$\delta_{2t}h^n + U\frac{\partial h^n}{\partial x} + H\langle \delta^n \rangle^{2t} = 0.$$

(a) Determine the conditions under which the gravity waves are stable.

(b) Determine the conditions under which the Rossby waves are stable.

(c) Discuss the impact of semi-implicit differencing on the accuracy of the Rossby-wave and gravity-wave modes.

4. Two-dimensional sound waves in a neutrally stratified atmosphere satisfy the linearized equations

$$\frac{\partial u}{\partial t} + c_s\frac{\partial P}{\partial x} = 0.$$

$$\frac{\partial w}{\partial t} + c_s\frac{\partial P}{\partial z} = 0.$$

$$\frac{\partial P}{\partial t} + c_s\left(\frac{\partial u}{\partial x} + \frac{\partial w}{\partial z}\right) = 0,$$

where $P = p'/(\rho_0 c_s)$. Let this system be approximated using forward–backward differencing for the horizontal gradients and trapezoidal differencing for the vertical gradients such that

$$u^{m+1} = u^m - c_s\Delta\tau\frac{\partial P^m}{\partial x},$$

$$w^{m+1} = w^m - c_s\frac{\Delta\tau}{2}\frac{\partial}{\partial z}\left(P^{m+1} + P^m\right),$$

$$P^{m+1} = P^m - c_s\Delta\tau\left(\frac{\partial u^{m+1}}{\partial x} - \frac{1}{2}\frac{\partial}{\partial z}\left(w^{m+1} + w^m\right)\right).$$

Consider an individual Fourier mode with spatial structure $\exp i(kx + \ell z)$ and show that the eigenvalues of the amplification matrix for this scheme are unity and

$$\frac{4 - \tilde{\ell}^2 - 2\tilde{k}^2 \pm 2\sqrt{(\tilde{k} - 2)(\tilde{k} + 2)(\tilde{k}^2 + \tilde{\ell}^2)}}{\tilde{\ell}^2 + 4},$$

where $\tilde{k} = c_s k \Delta\tau$ and $\tilde{\ell} = c_s \ell \Delta\tau$. What is the stability condition that ensures that this method will not have any eigenvalues with absolute values exceeding unity?

5. Compare the errors generated in gravity waves using semi-implicit differencing with those produced in a compressible Boussinesq system in which the true speed of sound c_s is artificially reduced to $\tilde{c}_s$ in an effort to increase efficiency by increasing the maximum stable value for $\Delta\tau$. *Hint:* Consider waves with wave numbers on the order of $N/\tilde{c}_s$ but larger than N/c_s.

6. The oscillations of the damped-harmonic oscillator (8.88) are "overdamped" when $\alpha^2\kappa^2 > c_s^2$. Suppose that the mesh is isotropic with grid interval Δ, the Courant number for sound-wave propagation on the small time step is $1/2$, and $\alpha = \gamma\Delta^2/\Delta\tau$. Estimate the minimum value of α required make the divergence damper overdamp a mode resolved on the numerical mesh. Do the values of α_x and α_z used in the test problem shown in Fig. 8.3d overdamp any of the resolved modes in that test problem?

7. Show that discrete integral of the finite-difference approximation to the vertical divergence of the vertical advective flux of kinetic energy,

$$\sum_{k=1}^{N} \left[\left\langle \langle \mathbf{u}_k \rangle^\sigma \cdot \dot{\sigma}_k \, \delta_\sigma \mathbf{u}_k \right\rangle^\sigma + \frac{\mathbf{u}_k \cdot \mathbf{u}_k}{2} \, \delta_\sigma \dot{\sigma}_k \right] \Delta\sigma_k,$$

is zero.

8. Examine the mass-conservation properties of σ-coordinate primitive-equation models.

(a) Show that the vertically integrated mass per unit area in a hydrostatically balanced atmosphere is p_s/g.

(b) Show that the vertical finite-difference scheme for the σ-coordinate primitive-equation model described in Sect. 8.6.3 will conserve total mass if nonconservative effects due to time differencing and the horizontal spectral representation are neglected.

(c) Suppose that instead of predicting $\ln p_s$ as in (8.116), the actual surface pressure were predicted using (8.105). Show that except for nonconservative effects due to time differencing, total mass will be exactly conserved in a numerical model in which the Galerkin spectral method is used to evaluate the horizontal derivatives, and the vertical derivative is approximated by (8.120). This approach is not used in practice because it generates a noisier solution than that obtained using $\ln p_s$ as the prognostic variable (Kiehl et al. 1996, p. 15).

Chapter 9
Nonreflecting Boundary Conditions

If the boundary of a computational domain coincides with a true physical boundary, an appropriate boundary condition can generally be derived from physical principles and can be implemented in a numerical model with relative ease. It is, for example, easy to derive the condition that the fluid velocity normal to a rigid boundary must vanish at that boundary, and if the shape of the boundary is simple, it is easy to impose this condition on the numerical solution. More serious difficulties may be encountered if the computational domain terminates at some arbitrary location within the fluid. When possible, it is a good idea to avoid artificial boundaries by extending the computational domain throughout the entire fluid. Nevertheless, in many problems the phenomena of interest occur in a localized region, and it is impractical to include all of the surrounding fluid in the numerical domain. As a case in point, one would not simulate an isolated thunderstorm with a global atmospheric model just to avoid possible problems at the lateral boundaries of a limited domain. Moreover, in a fluid such as the atmosphere there is no distinct upper boundary, and any numerical representation of the atmosphere's vertical structure will necessarily terminate at some arbitrary level.

When the computational domain is terminated at an arbitrary location within a larger body of fluid, the conditions imposed at the edge of the domain are intended to mimic the presence of the surrounding fluid. The boundary conditions should therefore allow outward-traveling disturbances to pass through the boundary without generating spurious reflections that propagate back toward the interior. Boundary conditions designed to minimize spurious backward reflection are known as *nonreflecting*, *open*, *wave-permeable*, or *radiation* boundary conditions. The terminology "radiation boundary condition" is due to Sommerfeld (1949, p. 189), who defined it as the condition that "the sources must be *sources*, not *sinks*, of energy. The energy which is radiated from the sources must scatter to infinity; *no energy may be radiated from infinity into . . . the field*." As formulated by Sommerfeld, the radiation condition applies at infinity; however, in all practical computations a boundary condition must be imposed at some finite distance from the energy source, and this creates two problems. The first problem is that the radiation condition itself may not properly describe the physical behavior occurring at an arbitrarily designated

D.R. Durran, *Numerical Methods for Fluid Dynamics: With Applications to Geophysics*, Texts in Applied Mathematics 32, DOI 10.1007/978-1-4419-6412-0_9,

location within the fluid when that location is only a finite distance from the energy source. The second problem is that it is typically more difficult to express the radiation condition mathematically at a boundary that is only a finite distance from the energy source.

The radiation condition is obviously not appropriate in situations where energy must be transmitted inward through the boundary; yet inward radiation may even be required in problems where the surrounding fluid is initially quiescent and all the initial disturbances are contained within the computational domain. Two nonlinear waves that propagate past an artificial boundary within a fluid may interact outside the artificial boundary and generate an inward-propagating disturbance that should reenter the domain. This point was emphasized by Hedstrom (1979), who provided a simple example from compressible gas dynamics demonstrating that a shock overtaking a contact discontinuity must generate an echo that propagates back toward the wave generator. Numerical simulations of thunderstorms provide another example where the documented sensitivity of numerical simulations to the lateral boundary conditions (Clark 1979; Hedley and Yau 1988) may be not only a consequence of poorly approximating the radiation condition, but also the result of inadequately representing important feedbacks on the convection arising through interactions with the storm's environment that occur outside the numerical domain.

Errors resulting from a failure to incorporate inward-propagating signals generated by real physical processes occurring outside the boundaries of the computational domain cannot be avoided without enlarging the domain. Nevertheless, nonreflecting boundary conditions often become reasonable approximations to the true physical boundary condition as the size of the computational domain increases, provided that the local energy density of a disturbance arriving at the boundary is reduced as the result of wave dispersion or absorption in the large domain. When the disturbances arriving at the boundary are sufficiently weak, the governing equations in the region near the boundary can be approximated by their linearized equivalents, and it can be relatively easy to ensure that the radiation condition correctly describes the boundary conditions for the linearized system. The situation is particularly simple in the case of constant-coefficient linear hyperbolic systems, for which radiation boundary conditions are clearly appropriate because the characteristic curves for such systems are straight lines that cannot exit and subsequently reenter the domain.

Even when it is clear that the radiation condition is appropriate, it is not always easy to translate Sommerfeld's physical description into a mathematical formula. As noted in the monograph by Givoli (1992), it is generally easier to express the radiation condition at infinity mathematically than to formulate the same condition for a boundary that is only a finite distance from the wave source. In fact, in most practical problems it is impossible to express the radiation condition exactly as an algebraic or differential equation involving the prognostic variables in the neighborhood of the boundary. Instead of the exact radiation condition, one must use an approximation, and this approximation introduces an error in the mathematical model of the physical system that is distinct from the errors subsequently incurred when constructing a discrete approximation to the mathematical model. As a consequence, the numerical solution will not converge to the correct solution to the underlying physical problem as the spatial and temporal mesh is refined, but rather to the correct solution to the approximate mathematical problem determined by the approximate radiation condition.

The first topic that will be considered in this chapter is therefore the mathematical formulation of well-posed radiation boundary conditions. We begin with examples where this can be done exactly and then consider problems where approximations are required. After formulating exact or approximate radiation boundary conditions for the continuous problem, we will consider their numerical implementation.

9.1 One-Dimensional Flow

Exact open boundary conditions can be obtained for certain simple one-dimensional systems. Two such problems will be considered in this section, the linear advection equation and the linearized shallow-water system. The shallow-water system provides an instructive example illustrating the construction of exact radiation boundary conditions for the continuous equations. In contrast, there is no need to explicitly determine a radiation boundary condition for the nondiscretized linear advection equation because a well-posed mathematical formulation of the advection problem does not require any outflow boundary condition. The linear advection equation is, however, useful for investigating the influence of various numerical approximations to the exact outflow boundary condition on the accuracy and stability of the discretized solution.

9.1.1 Well-Posed Initial-Boundary-Value Problems

Consider the linearized one-dimensional advection equation

$$\frac{\partial \psi}{\partial t} + U\frac{\partial \psi}{\partial x} = 0, \tag{9.1}$$

and suppose ψ is to be determined throughout some limited domain $0 \leq x \leq L$. For the sake of illustration, assume that $U > 0$. If initial data are given for the domain $0 \leq x \leq L$, how should boundary conditions at $x = 0$ and $x = L$ be specified to yield a well-posed problem?[1]

Since ψ is the solution to a homogeneous hyperbolic equation with constant coefficients, ψ will be constant along the characteristic curves $x - Ut = x_0$. The initial data will therefore uniquely determine ψ in the shaded triangular region in Fig. 9.1. The solution in the remainder of the strip $t > 0, 0 \leq x \leq L$ is determined by the values of $\psi(0,t)$ at the inflow boundary. Thus, to uniquely determine the solution (one

[1] As discussed in Sect. 1.3.2, a well-posed problem is one in which a unique solution to a given partial differential equation exists and depends continuously on the initial- and boundary-value data.

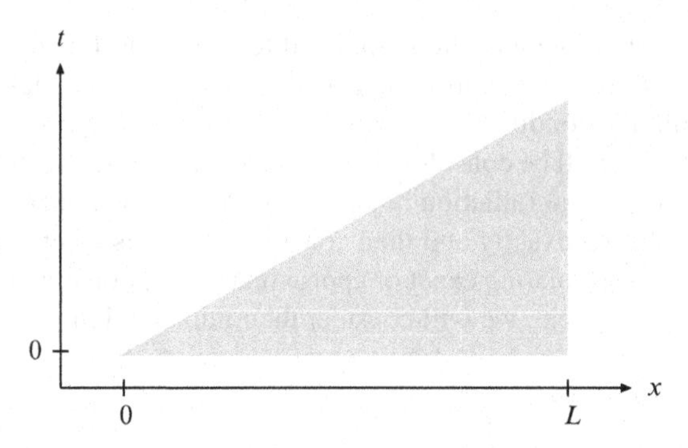

Fig. 9.1 Portion of the x–t plane (*shaded region*) in which the solution to (9.1) is determined by the initial data $\psi(x, 0)$, $0 \leq x \leq L$

prerequisite for well-posedness), a boundary condition must be imposed at $x = 0$. On the other hand, no differentiable function will be able to satisfy an arbitrary boundary condition imposed at $x = L$, because $\psi(L, t)$ is already determined by the governing equation (9.1), the initial condition, and the boundary condition at $x = 0$. Since no solution to the overspecified problem exists, it is not well posed. A consistent solution might be obtained if the correct value of $\psi(x, t)$ is imposed at $x = L$, but even in this case the solution will not depend continuously on the boundary conditions, since it will cease to exist if the downstream boundary values are perturbed, and the problem remains ill posed.

The preceding example may seem obvious: A boundary condition is required at inflow; no condition should be imposed at outflow. Physical intuition is less likely to yield an obvious answer in the case of the one-dimensional shallow-water equations. If η is gravity times the displacement of the free surface about its equilibrium height H, and u and U are, respectively, the perturbation and constant basic-state fluid velocities, the linearized shallow-water equations are

$$\frac{\partial}{\partial t}\begin{pmatrix} u \\ \eta \end{pmatrix} + \begin{pmatrix} U & 1 \\ c^2 & U \end{pmatrix} \frac{\partial}{\partial x}\begin{pmatrix} u \\ \eta \end{pmatrix} = 0, \tag{9.2}$$

where $c = \sqrt{gH}$. The preceding equation is a homogeneous hyperbolic system of partial differential equations of the form

$$\frac{\partial \mathbf{q}}{\partial t} + \mathbf{A}\frac{\partial \mathbf{q}}{\partial x} = \mathbf{0}.$$

The matrices

$$\mathbf{T}^{-1} = \begin{pmatrix} 1 & -1/c \\ 1 & 1/c \end{pmatrix}, \qquad \mathbf{T} = \begin{pmatrix} 1/2 & 1/2 \\ -c/2 & c/2 \end{pmatrix}$$

may be used to transform (9.2) to

$$\frac{\partial \mathbf{T}^{-1}\mathbf{q}}{\partial t} + \mathbf{T}^{-1}\mathbf{A}\mathbf{T}\frac{\partial \mathbf{T}^{-1}\mathbf{q}}{\partial x} = \mathbf{0},$$

which is the system of two decoupled scalar equations

$$\frac{\partial}{\partial t}\begin{pmatrix} d \\ e \end{pmatrix} + \begin{pmatrix} U-c & 0 \\ 0 & U+c \end{pmatrix} \frac{\partial}{\partial x}\begin{pmatrix} d \\ e \end{pmatrix} = \mathbf{0} \tag{9.3}$$

for the Riemann invariants $d = u - \eta/c$ and $e = u + \eta/c$. Each of these scalar equations is a partial differential equation of the form (9.1) and will be well posed if a boundary value is specified at inflow and no value is specified at outflow. A well-posed shallow-water problem is therefore obtained by specifying d at the boundary through which $U-c$ is directed inward and e at the boundary where $U+c$ is directed inward. Suppose that $U > 0$ and solutions are sought on the interval $0 \leq x \leq L$. In the "supercritical" case $c < U$, both d and e should be specified at $x = 0$ in a manner directly analogous to the scalar advection problem.

In many geophysical applications $c > |U|$; the flow is "subcritical," and well-posed boundary conditions have the general form

$$e(0,t) = \alpha_1 d(0,t) + f_1(t), \qquad d(L,t) = \alpha_2 e(L,t) + f_2(t). \tag{9.4}$$

The terms $f_i(t)$ represent external forcing, whereas the terms involving α_i allow information carried along the outward-directed characteristic to be incorporated in the boundary condition. The boundary condition at $x = 0$ can be rewritten as

$$(1-\alpha_1)u(0,t) + (1+\alpha_1)\eta(0,t)/c = f_1(t),$$

thereby demonstrating that the value of α_i determines how the forcing is apportioned between u and η. If $\alpha_i = -1$, the boundary conditions on d and e reduce to conditions on u. If $\alpha_i = 1$, the forcing determines η. When $\alpha_i = 0$, $f(t)$ specifies values for the Riemann invariants d and e.

Well-posed boundary conditions for one-dimensional hyperbolic systems with more unknowns may be determined by the same transformation procedure. Since the system is hyperbolic, the coefficient matrix of the spatial derivative term can be diagonalized by a suitable change of variables. Each component of the diagonalized system will have the form (9.1) and will require a boundary value at inflow and no value at outflow. Each positive eigenvalue of the coefficient matrix will therefore be associated with a Riemann invariant requiring a boundary value at $x = 0$, and each negative eigenvalue will necessitate the specification of a boundary value at $x = L$.

9.1.2 The Radiation Condition

Radiation boundary conditions can be easily imposed in the one-dimensional shallow-water system by transforming the equations to the diagonal form (9.3) and setting the incoming Riemann invariant to zero at each boundary. Unfortunately, this approach does not easily generalize to two-dimensional shallow-water flow. In many practical problems it is simpler to retain the velocities and the height (or pressure) field as the unknown prognostic variables and to develop open boundary conditions involving these variables.

Consider, therefore, the problem of expressing the radiation boundary condition at $x = L$ in terms of u and η instead of the Riemann invariants in a case where $c > |U|$. Since the incoming characteristic is zero at $x = L$ for all $t > 0$, $d(x,t)$ must be zero throughout the wedge-shaped region of the x–t plane defined by the inequalities $(U - c)t + L < x < L$. Thus, for all $t > 0$,

$$u(x,t) = \frac{\eta(x,t)}{c} \tag{9.5}$$

in a small neighborhood of the $x = L$ boundary. In the same neighborhood of $x = L$, the equation for the outward-directed characteristic is

$$\frac{\partial}{\partial t}\left(u + \frac{\eta}{c}\right) + (U + c)\frac{\partial}{\partial x}\left(u + \frac{\eta}{c}\right) = 0. \tag{9.6}$$

Using (9.5) to eliminate η/c from the preceding equation yields the radiation boundary condition for u at $x = L$,

$$\frac{\partial u}{\partial t} + (U + c)\frac{\partial u}{\partial x} = 0. \tag{9.7}$$

An identical radiation boundary condition for η,

$$\frac{\partial \eta}{\partial t} + (U + c)\frac{\partial \eta}{\partial x} = 0, \tag{9.8}$$

can be derived using (9.5) to eliminate u from (9.6). Similar conditions may also be obtained $x = 0$; they are

$$\frac{\partial u}{\partial t} + (U - c)\frac{\partial u}{\partial x} = 0 \tag{9.9}$$

and

$$\frac{\partial \eta}{\partial t} + (U - c)\frac{\partial \eta}{\partial x} = 0. \tag{9.10}$$

The radiation boundary conditions (9.7) and (9.8) have the following alternative derivation and interpretation as one-way wave equations. The general solution for the perturbation velocity in the linearized shallow-water system may be expressed as

$$u = F_{\mathrm{r}}[x - (U + c)t] + F_{\mathrm{l}}[x - (U - c)t].$$

The first component of the general solution, F_{r}, represents a wave traveling to the right, and the second component represents a wave traveling to the left. The partial differential equation (9.7) imposed at the boundary is satisfied by solutions of the form F_{r} but does not admit solutions of the form F_{l}. All reflection at the right boundary may be eliminated by ensuring that no leftward-propagating waves are present at the boundary, and this may be achieved by imposing (9.7) and (9.8) at the right boundary. Because all the solutions to (9.7) propagate in the same direction, that equation is sometimes known as a *one-way* wave equation. One-way wave equations are particularly useful in problems involving several spatial dimensions.

9.1.3 Time-Dependent Boundary Data

In some applications it is important to allow changes in the surrounding fluid to influence the interior solution while simultaneously radiating outward-propagating waves through the boundary without reflection. Suppose that $u_e(x,t)$ and $\eta_e(x,t)$ are the velocity and the scaled free-surface displacement in a large domain containing the subdomain $0 \le x \le L$, that $c > |U|$, and that information about the solution in the large domain is to be used to generate boundary conditions for a simulation of the linearized shallow-water equations inside the subdomain. It is not generally possible simply to set $u(L,t) = u_e(L,t)$ and $\eta(L,t) = \eta_e(L,t)$ without generating spurious reflections in those waves attempting to pass outward through the boundary at $x = L$. Well-posed boundary conditions are obtained by specifying the value of the incoming Riemann invariant. At the right boundary $d(L,t)$ is set to $d_e(L,t)$, where

$$d_e(x,t) = u_e(x,t) - \eta_e(x,t)/c,$$

and at the left boundary $e(0,t)$ is set to $e_e(0,t)$, where

$$e_e(x,t) = u_e(x,t) + \eta_e(x,t)/c.$$

As an alternative to the direct specification of the incoming Riemann invariants, the data from the large domain can be incorporated in the one-way wave equations for u and η as follows. For all $t > 0$, the value of the incoming Riemann invariant in a small neighborhood of $x = L$ will equal $d_e(x,t)$. Thus, in this neighborhood

$$u - \frac{\eta}{c} = u_e - \frac{\eta_e}{c}. \tag{9.11}$$

Solving the preceding equation for η and substituting the result in the equation for the outward-directed Riemann invariant (9.6) yields

$$\left(\frac{\partial}{\partial t} + (U+c)\frac{\partial}{\partial x}\right)(u - u_e) = -\frac{1}{2}\left(\frac{\partial}{\partial t} + (U+c)\frac{\partial}{\partial x}\right)\left(u_e + \frac{\eta_e}{c}\right).$$

The right side of the preceding equation is zero by (9.6), since $u_e + \eta_e/c$ is the positive-phase-speed Riemann invariant for the large-scale flow. Thus, the boundary condition at $x = L$ reduces to

$$\left(\frac{\partial}{\partial t} + (U+c)\frac{\partial}{\partial x}\right)(u - u_e) = 0. \tag{9.12}$$

A similar derivation, in which (9.11) is used to eliminate u from (9.6) gives the boundary condition on η as

$$\left(\frac{\partial}{\partial t} + (U + c)\frac{\partial}{\partial x}\right)(\eta - \eta_e) = 0. \tag{9.13}$$

In summary, the correct way to impose large-scale information at the boundaries of the limited domain is not to simply set $u(L,t) = u_e(L,t)$ and $\eta(L,t) = \eta_e(L,t)$, but rather to apply the radiation conditions (9.12) and (9.13) to the perturbation of u and η about the values in the larger-scale flow (Carpenter 1982).

9.1.4 Reflections at an Artificial Boundary: The Continuous Case

Although exact nonreflecting boundary conditions were derived in Sect. 9.1.2 for the one-dimensional shallow-water problem, exact formulae for nonreflecting boundary conditions are not generally available in more complicated problems. Even in the relatively simple case of two-dimensional shallow-water flow some approximation to the exact radiation boundary condition is required to obtain a useful relationship involving the prognostic variables and their derivatives at the lateral boundary (see Sect. 9.2). As a consequence of these approximations, errors are introduced in the mathematical model of the physical system before the governing equations are discretized. The errors generated by such approximations are considered in this section.

Consider the one-dimensional shallow-water equations on the domain $0 \le x \le L$ and suppose that a zero-gradient condition is used to approximate the correct radiation boundary condition at the lateral boundaries. The strength of the reflections generated at the $x = 0$ boundary may be analyzed by examining the behavior of a unit-amplitude incident wave as it reflects off the boundary. Suppose that the perturbation velocity has the form

$$u(x,t) = \sin\left[k_i(x - (U - c)t)\right] + r\sin\left[k_r(x - (U + c)t) + \phi\right].$$

The first term in the preceding equation represents the unit-amplitude wave propagating toward the boundary at $x = 0$, and the second term represents an arbitrary reflected wave propagating away from that boundary. The amplitude, wave number, and phase of this reflected wave are determined by the boundary condition at $x = 0$. If the zero-gradient condition $\partial u/\partial x = 0$ is enforced at $x = 0$, then

$$k_i\cos\left[k_i(U - c)t\right] + rk_r\cos\left[k_r(U + c)t + \phi\right] = 0. \tag{9.14}$$

Since this equation must hold for all t, the two terms must be linearly dependent functions of time, which implies that the frequencies of the incident and reflected waves must be identical and that $\phi = 0$ (or $\phi = \pi$, which will only flip the sign of r). Equating the frequencies yields a formula for the wave number of the reflected wave

$$k_{\mathrm{r}} = k_{\mathrm{i}} \left(\frac{U - c}{U + c} \right).$$

This, together with (9.14), implies that the amplitude of the reflected wave is

$$r = -\frac{k_{\mathrm{i}}}{k_{\mathrm{r}}} = -\frac{U + c}{U - c}.$$

If $U > 0$, the wavelength and the amplitude of the reflected wave exceed those of the incident wave. After the initial reflection at $x = 0$, the reflected wave will travel across the domain and experience a second reflection at $x = L$. If a zero gradient condition is also specified at the right boundary, a similar analysis will show that the reflection coefficient at $x = L$ is

$$\hat{r} = -\frac{U - c}{U + c}.$$

After reflecting off both the left and the right boundaries, the amplitude of the spurious wave will be $r\hat{r} = 1$. No energy has escaped through the lateral boundaries! Indeed, the net reflection generated by specifying $\partial u/\partial x = 0$ at both boundaries is just as strong as that obtained from a pair of rigid lateral boundaries at which $u = 0$. The same type of reflection is also obtained by specifying $\partial \eta/\partial x = 0$. Evidently, $\partial u/\partial x = 0$ and $\partial \eta/\partial x = 0$ are not acceptable substitutes for the correct nonreflecting boundary conditions (9.9) and (9.10).

9.1.5 Reflections at an Artificial Boundary: The Discretized Case

Once physically appropriate well-posed boundary conditions have been formulated, the problem is ready for numerical solution. Unfortunately, additional difficulties develop when the spatial derivatives in the continuous problem are replaced with finite differences. The first difficulty is that finite-difference formulae often require boundary data where none are specified in the well-posed continuous problem. The second difficulty is that most finite-difference formulae support nonphysical modes that propagate in a direction opposite to the correct physical solution and may thereby blur the distinction between inflow and outflow boundaries.

To examine the effects of spatial discretization, suppose that the advection equation (9.1) is approximated by the differential–difference equation

$$\frac{d\phi_j}{dt} + U \left(\frac{\phi_{j+1} - \phi_{j-1}}{2\Delta x} \right) = 0 \tag{9.15}$$

at $j = 0, \ldots, N$ discrete grid points in the finite domain $0 \le x \le L$. The preceding differential–difference equation cannot be used to calculate ϕ_0 and ϕ_N because the centered difference cannot be evaluated at a boundary. Suppose that $U < 0$, so $x = L$ is an inflow boundary. Then a boundary condition $\psi(L, t) = f(t)$ must be imposed to render the continuous problem complete and well posed. This same inflow boundary condition may be used to specify ϕ_N, but the numerical calculation of the solution at the other boundary will be a problem. There is no boundary condition in the formulation of the nondiscretized initial-boundary-value problem (IBVP) that can be used to specify ϕ_0. Since $\psi(0,t)$ is determined by the interior solution, one might try to estimate $\psi(0,t)$ by extrapolating the interior solution outward to the boundary. The simplest extrapolation is

$$\phi_0 = \phi_1. \tag{9.16}$$

This, of course, is a zero-gradient condition, and on the basis of the earlier discussion of zero-gradient conditions in the continuous shallow-water problem, one might expect it to produce reflection.

The numerical boundary condition (9.16) does indeed produce some reflection, but the situation is fundamentally different from that in the shallow-water system. When $c > |U|$, the shallow-water equations support physical modes that propagate both to the right and to the left. Specifying $\partial u/\partial x = 0$ or $\partial \eta/\partial x = 0$ generates reflection from the outgoing physical mode into the incoming physical mode because these conditions are inadequate approximations to the correct open boundary conditions in the shallow-water system. In contrast to the shallow-water system, all physical solutions to the scalar advection equation (9.1) travel in the same direction. Any reflections that occur when (9.16) is used as a boundary condition for the differential–difference equation (9.15) involve nonphysical numerical modes.

In the remainder of this section we will examine the interactions that may occur between physical and nonphysical modes as a result of the numerical approximations that are made at an open boundary. This investigation will focus on the simplest dynamical system in which such interactions occur – the problem of scalar advection. A similar, though more tedious, analysis may be performed for shallow-water flow after allowing for the fact that centered second-order finite-difference approximations to the spatial derivatives in the one-dimensional shallow-water system support two physical and two nonphysical modes. Under such circumstances, a shallow-water wave arriving at a boundary may simultaneously reflect into two inward-moving waves. If $c > |u|$, one reflected wave will be a physical mode and the other will be a nonphysical mode.

As discussed in Sect. 3.3.1, solutions of the form

$$\phi(x,t) = \mathrm{e}^{\mathrm{i}(kj\Delta x - \omega t)}$$

satisfy the differential–difference equation (9.15), provided that

$$\omega = U\frac{\sin k\Delta x}{\Delta x}. \tag{9.17}$$

The group velocities of these waves are given by

$$\frac{\partial \omega}{\partial k} = U\cos k\Delta x.$$

For each ω there are two different wave numbers that satisfy the dispersion relation (9.17): k_1 and $k_2 = \pi/\Delta x - k_1$. Since the group velocities of these two waves are equal in magnitude but opposite in sign, it is useful to divide the waves resolvable on the discrete grid into two categories: physical modes, for which $0 \leq k\Delta x < \pi/2$, and nonphysical modes, for which $\pi/2 \leq k\Delta x \leq \pi$. According to this division of the modes, a group (or wave packet) of physical modes propagates in the same direction as the true physical solution, whereas a group of nonphysical modes propagates backward ($4\Delta x$ waves have zero group velocity). For each ω except $U/\Delta x$ one of the roots of (9.17) is a physical mode and one is a nonphysical mode.

To determine the strength of the reflection introduced by the extrapolation boundary condition (9.16), suppose that a disturbance oscillating at the frequency ω is present at the left boundary. This disturbance, being a solution of the interior differential–difference equation, must have the form

$$\left(\alpha e^{ikj\Delta x} + \beta e^{i(\pi - k\Delta x)j}\right) e^{-i\omega t}.$$

Let the first term in the preceding equation represent the incident wave (with $\alpha = 1$) and the second term the reflected wave (with $\beta = r$). Then the disturbance has the form

$$\left(e^{ikj\Delta x} + r(-1)^j e^{-ikj\Delta x}\right) e^{-i\omega t}. \tag{9.18}$$

Although the interior differential–difference equation will support a single-mode solution with $r = 0$, the boundary condition (9.16) requires nonzero amplitude in both modes. The amplitude of the reflected wave may be evaluated by substituting (9.18) into (9.16), which yields

$$1 + r = e^{ik\Delta x} - re^{-ik\Delta x}.$$

Solving for r, one obtains

$$r = ie^{ik\Delta x} \tan \frac{k\Delta x}{2},$$

in which case

$$|r| = \left| \tan \frac{k\Delta x}{2} \right|.$$

If the incident wave is well resolved, then $k\Delta x \ll 1$, and there is very little reflection. The magnitude of the reflection coefficient rises to unity for the $4\Delta x$ wave, and it approaches infinity for a $2\Delta x$ wave. It may appear that the large reflection coefficients associated with very short waves make the zero-gradient condition unusable, but if (9.16) is applied at the outflow boundary, no waves shorter than $4\Delta x$ will ever reach that boundary because the group velocities of these short waves are directed upstream. The zero-gradient condition is therefore a possibly useful outflow boundary condition for the advection equation. On the other hand, the large reflection coefficients associated with the shortest waves do render (9.16) unsuitable as an inflow boundary condition because waves with wavelengths between $2\Delta x$ and $4\Delta x$ propagate upstream and rapidly amplify when they encounter the inflow boundary.

Nitta (1962) and Matsuno (1966a) used a similar procedure to analyze the reflections generated by a wide variety of boundary conditions. The magnitude of the reflection coefficient generated by a fixed boundary value, such as $\phi_0 = 0$, is unity. Although fixed boundary values are not appropriate at outflow, they are useful at the inflow boundary, where the only waves that encounter the boundary will be low-amplitude nonphysical modes. Second-order extrapolation,

$$\phi_0 = 2\phi_1 - \phi_2, \tag{9.19}$$

has reflection $|r| = |\tan(k\Delta x/2)|^2$. The general nth-order extrapolation

$$\sum_{m=0}^{n}(-1)^m\left(\frac{n!}{m!(n-m)!}\right)\phi_m = 0$$

reflects with amplitude $|r| = |\tan(k\Delta x/2)|^n$. Higher-order extrapolation reduces the reflection of well-resolved physical waves, but it increases the reflection of non-physical modes.

As an alternative to extrapolation, one might employ a one-sided finite difference at an outflow boundary, replacing (9.15) by

$$\frac{d\phi_0}{dt} + U\left(\frac{\phi_1 - \phi_0}{\Delta x}\right) = 0. \tag{9.20}$$

The reflections generated by this scheme may once again be analyzed by substituting the general solution (9.18) into (9.20). The result is

$$-\mathrm{i}\omega(1+r) + \frac{U}{\Delta x}(\mathrm{e}^{\mathrm{i}k\Delta x} - r\mathrm{e}^{-\mathrm{i}k\Delta x}) - \frac{U}{\Delta x}(1+r) = 0.$$

After substituting for ω from the dispersion relation (9.17), the preceding equation simplifies to

$$|r| = \left|\frac{1-\cos k\Delta x}{1+\cos k\Delta x}\right| = \tan^2\left(\frac{k\Delta x}{2}\right).$$

Thus, the magnitude of the reflection coefficient generated by one-sided differencing is identical to that produced by second-order extrapolation.

The interaction of physical and nonphysical modes with the lateral boundaries is illustrated in Fig. 9.2, which is patterned after an example in Trefethen (1985). The differential–difference advection equation was solved on the domain $0 \le x \le 1$, with $U = 1$, $\Delta x = 0.01$, and the initial distribution of ϕ_j was given by the rectified Gaussian

$$\phi_j = 0.5\left[1 + (-1)^j\right]\mathrm{e}^{-400(x-0.5)^2}.$$

The inflow condition was $\phi_0 = 0$, and the outflow condition was $\phi_N = \phi_{N-1}$. The time derivative was approximated using trapezoidal time differencing with a small Courant number. The initial condition is shown in Fig. 9.2a. Figure 9.2b shows the solution shortly before the disturbances encounter the lateral boundaries; the direction in which each disturbance is propagating is indicated by an arrow.

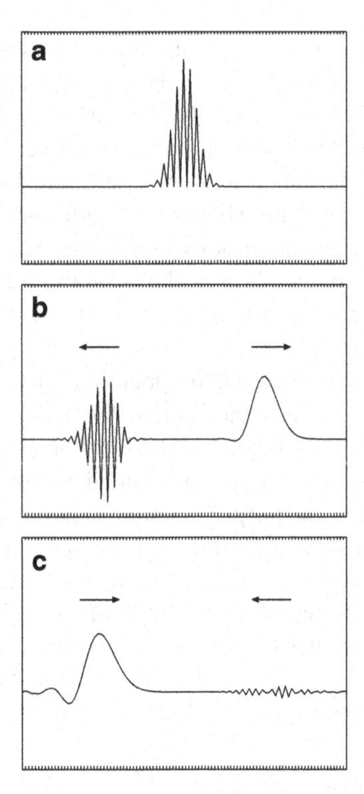

Fig. 9.2 Approximate solution to the differential–difference equation (9.15) at time **a** 0, **b** 0.25, and **c** 0.75

The long-wave component of the solution travels toward the outflow boundary in a physically correct manner, but the $2\Delta x$ wave packet propagates upwind at speed $-U$. After the disturbances have reflected off the lateral boundaries, the solution appears as shown in Fig. 9.2c. Physical modes are reflected into nonphysical modes by the outflow boundary condition, but since the initial disturbance is well resolved, the reflection is relatively weak. The left-moving nonphysical mode is reflected, without loss of amplitude, into a right-moving physical mode at the inflow boundary. The strong reflection at the left boundary is not necessarily indicative of a deficiency in the inflow boundary condition. Only nonphysical modes lead to difficulties at the inflow boundary, and even those modes are not amplified. One can eliminate the problem at the upstream boundary by filtering the nonphysical modes out of the interior solution.

In addition to the evaluation of the reflection coefficient, one can also examine the formal accuracy of the numerical boundary condition by substituting Taylor series expansions into the discretized formula in the usual manner (see Sect. 2.1). Gustafsson (1975) showed that there is generally no degradation in the overall order of convergence if the boundary conditions are formulated with one order less accuracy than that used for the interior finite differences. In many practical problems the order of accuracy of the discretized approximation to the radiation condition is actually a relatively minor concern. The numerical errors generated at the boundary often involve reflections into poorly resolved short waves, and the accuracy achieved in the representation of such poorly resolved features is not reliably determined by the leading-order term in the truncation error. Moreover, in complex multidimensional problems the most serious errors are often introduced in the mathematical formulation of approximate radiation boundary conditions, and it is not necessary to implement these approximate conditions with highly accurate numerical formulae.

In fact, the optimal nonreflecting boundary condition for a discretized problem is not generally a high-order approximation to the exact one-way wave equation for the continuous problem, but rather the one-way wave equation whose discrete dispersion relation best replicates the propagation characteristics of the interior finite-difference scheme (Engquist and Majda 1979). As an example, suppose that solutions are sought to the one-dimensional advection equation using the semidiscrete approximation (9.15). The optimal boundary condition would be one that masks the apparent change in the propagation medium generated by the difference between the finite-difference formulae at the boundary and the interior of the domain. The dispersion relation for the one-way wave equation that will pass solutions to this semidiscrete system through the boundary with zero reflection would be $\omega = U \sin(k\Delta x)/\Delta x$ rather than the exact condition $\omega = Uk$. Unfortunately, techniques are not currently available to create a practical boundary condition with either of these dispersion relations.

9.1.6 Stability in the Presence of Boundaries

Pure initial-value, or *Cauchy*, problems are posed on infinite or periodic domains. Methods for analyzing the stability of finite-difference approximations to Cauchy problems were presented in Sect. 3.2. Unfortunately, the imposition of a boundary condition at the edge of a limited domain can destabilize numerical methods that are stable approximations to the Cauchy problem. In this section we will examine the additional properties that must be satisfied to ensure the stability of numerical approximations to IBVPs.

As in Sect. 3.2, the present analysis will be restricted to linear problems with constant coefficients, for which the theory is simplest and most complete. The easiest way to determine the stability of a linear constant-coefficient initial-value problem on the unbounded domain $-\infty < x < \infty$ is to use von Neumann's method, which examines the behavior of individual waves of the form

$$e^{i(kj\Delta x - \omega n\Delta t)}. \tag{9.21}$$

According to the von Neumann criterion, a numerical method is stable if it does not amplify any of the modes resolved on the numerical mesh. Defining $\kappa = e^{ik\Delta x}$ and $z = e^{-i\omega\Delta t}$, one may write the grid-point values associated with each wave as $\kappa^j z^n$. The von Neumann stability of the Cauchy problem requires $|z| \leq 1$ for all $|\kappa| = 1$. Spatial distributions other than pure sinusoidal waves may also be represented in the form $\kappa^j z^n$. If $|\kappa| < 1$, the expression (9.21) represents an oscillation whose amplitude grows exponentially as $x \to -\infty$. There is no need to worry about the stability of this type of mode in the Cauchy problem, because it is not an admissible solution on a periodic or unbounded domain. If, however, the same equations are to be solved as an IBVP on the half-infinite domain $x \geq 0$, then the mode is admissible (since $|\kappa|^j \to 0$ as $j \to \infty$). The stability requirement for IBVPs on the half-bounded domain $x \geq 0$ must therefore be extended to the *Godunov–Ryabenki* condition that $|z| \leq 1$ for all $|\kappa| \leq 1$. (If the domain is $-\infty < x \leq 0$, the modes to be examined are those for which $|\kappa| \geq 1$.)

Although the Godunov–Ryabenki condition is necessary for stability, it is not sufficient to guarantee that numerical solutions to IBVPs are stable. General conditions for stability were given by Gustafsson et al. (1972), who noted that in practice, the most important stability question involves the behavior of modes for which $|z| = 1$ and $|\kappa| = 1$. These modes are undamped waves; they can lead to instability if the interior finite-difference scheme together with the numerical boundary condition permits unforced waves to propagate inward through the boundary. Trefethen (1983) showed that the Gustafsson–Kreiss–Sundström (GKS) stability condition is essentially equivalent to the requirement that (1) the interior difference formula is stable for the Cauchy problem, (2) the model (including the boundary conditions) admits no eigensolutions that amplify with each time step by a constant factor z with $|z| > 1$ (i.e., the Godunov–Ryabenki stability condition is satisfied), and (3) the model (including the boundary conditions) admits no unforced waves with group velocities directed inward through the boundaries of the domain.

To see how a numerical method might allow waves to spontaneously propagate inward through a lateral boundary, suppose that the advection equation (9.1) is to be solved on the semi-infinite domain $0 \leq x \leq \infty$, and that $U < 0$, so $x = 0$ is an outflow boundary. The interior solution will be approximated using leapfrog-time centered-space differencing

$$\phi_j^{n+1} - \phi_j^{n-1} + \mu(\phi_{j+1}^n - \phi_{j-1}^n) = 0, \tag{9.22}$$

where $\mu = U\Delta t/\Delta x$. The outflow boundary condition will be determined by extrapolation, $\phi_0^n = \phi_1^n$. The numerical dispersion relation for the interior finite-difference scheme (9.22) is

$$\sin \omega \Delta t = \mu \sin k \Delta x. \tag{9.23}$$

Substituting a wave of the form (9.21) into the extrapolation boundary condition yields

$$1 = e^{ik\Delta x}. \tag{9.24}$$

The free modes of this finite-difference IBVP consist of those waves for which ω and k simultaneously satisfy (9.23) and (9.24), and they include the mode $(\omega, k) = (\pi/\Delta t, 0)$, whose group velocity,

$$\frac{\partial \omega}{\partial k} = U \frac{\cos k \Delta x}{\cos \omega \Delta t},$$

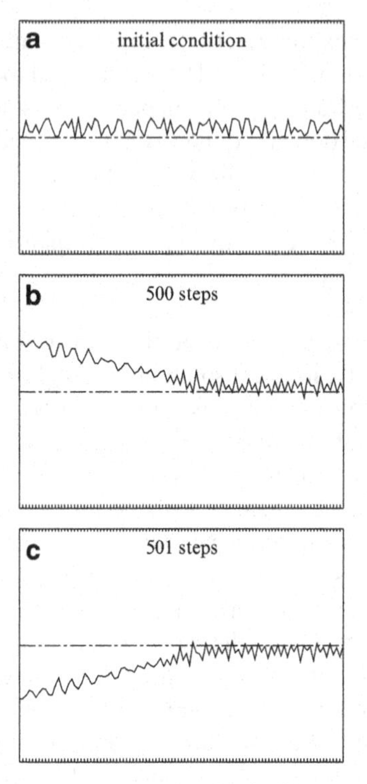

Fig. 9.3 Solution to (9.1) at **a** the initial time, and after **b** 500 and **c** 501 time steps

is $-U$. The mode $(\pi/\Delta t, 0)$ is therefore a free mode of the discretized problem that can propagate inward through the downstream boundary, and according to Trefethen's interpretation of the GKS stability criteria, the scheme should be unstable. This instability is demonstrated in Fig. 9.3, which shows a numerical integration of (9.22) on the interval $0 \leq x \leq 1$. The initial data, plotted in Fig. 9.3a, are random numbers with amplitudes between 0 and 0.2. The wind speed is $U = -1$, so $x = 1$ is an inflow boundary at which ϕ^n_{100} is fixed at zero. The extrapolation condition $\phi^n_0 = \phi^n_1$ is applied at outflow; $\Delta x = 0.01$, and the time step is fixed such that the Courant number is 0.9. Figure 9.3b and c shows the solution at two adjacent time steps after the instability has developed at the downstream boundary. As suggested by the preceding analysis, the growing mode has a period of $2\Delta t$ and a long horizontal wavelength.

To stabilize the numerical solution, it is necessary to change either the interior finite-difference equation or the boundary condition. Suppose the boundary condition is replaced by $\phi^n_0 = \phi^{n-1}_1$, which is a backward extrapolation in both space and time. Substitution of the wave solution (9.21) into this boundary condition yields

$$1 = e^{ik\Delta x} e^{i\omega\Delta t},$$

or

$$k\Delta x + \omega\Delta t = 2n\pi, \qquad n = 0, \pm 1, \pm 2, \ldots. \tag{9.25}$$

This condition is not satisfied by the previously troublesome mode $(\omega, k) = (\pi/\Delta t, 0)$. The only resolvable modes that simultaneously satisfy (9.25) and the dispersion relation (9.23) are $(\omega, k) = (0, 0)$ and $(\pi/\Delta t, \pi/\Delta x)$, both of which have outward-directed group velocities. The numerical IBVP is therefore stable.

The numerical solution may be alternatively stabilized without modifying the original boundary condition $\phi_0^n = \phi_1^n$ if the leapfrog time difference is replaced by the trapezoidal scheme:

$$\phi_j^{n+1} - \phi_j^n + \frac{\mu}{4}\left(\phi_{j+1}^{n+1} - \phi_{j-1}^{n+1} + \phi_{j+1}^n - \phi_{j-1}^n\right) = 0.$$

The dispersion relation and group velocity for the preceding trapezoidal scheme are

$$\tan\left(\frac{\omega\Delta t}{2}\right) = \frac{\mu}{2}\sin k\Delta x \quad \text{and} \quad \frac{\partial\omega}{\partial k} = U\cos(k\Delta x)\cos^2\left(\frac{\omega\Delta t}{2}\right).$$

The only resolvable modes satisfying the boundary condition (9.24) are those for which $k\Delta x = 0$. The group velocities of all such modes have the same sign as U and are directed outward through the downstream boundary. Since the numerical scheme does not support free modes with inward-directed group velocities, it is stable.

Observe that the extrapolation condition $\phi_0^n = \phi_1^n$ would never be stable for an inflow boundary, because the group velocity of the zero-wave-number physical mode $(\omega, k) = (0, 0)$ is directed inward through the boundary, and this mode satisfies both the boundary condition and the interior finite-difference scheme. Such instability might have been anticipated from the reflection-coefficient analysis in the preceding section, which indicated that the reflection coefficient associated with zero-order extrapolation at an inflow boundary becomes infinite as $k\Delta x \to \pi$. The connection between the reflection coefficient and stability is, however, somewhat complex. Boundary conditions associated with infinite coefficients are always unstable, but GKS stability does not require $|r| \leq 1$. Further discussion of the relation between reflection coefficients and instability appears in Trefethen (1985). The stability, or lack thereof, of several basic methods for the numerical solution of the advection equation is presented in a series of examples by Goldberg and Tadmor (1985).

In most situations, two boundaries are present, and one needs stable methods for closed spatial domains such as $0 \leq x \leq L$. Gustafsson et al. (1972) showed that the stability of the two-boundary problem can be determined by analyzing each boundary separately. If the boundary condition at $x = 0$ and the interior finite-difference formulae are GKS-stable approximations to an IBVP on the domain $0 \leq x < \infty$, and if the boundary condition at $x = L$ is also GKS stable for problems on the domain $-\infty < x \leq 0$, then the two-boundary problem is GKS stable. The basic reason why each boundary can be considered separately is that the simultaneous presence of two boundaries does not admit new types of eigensolutions beyond those whose stability was already tested in the pair of single-boundary problems. Nevertheless, GKS stability does not guarantee that the solution to the two-boundary problem will be completely satisfactory. In particular, it is possible for a method to be GKS stable but to generate reflections with $|r| > 1$. If the reflection coefficient at both boundaries exceeds unity, the solution may grow with time. Such growth is not a true instability, in the sense that it need not prevent the convergence of integrations performed over some *fixed time interval*, since, in theory, the error can be made arbitrarily small in the limit $\Delta x \to 0$, $\Delta t \to 0$. Nevertheless, in most practical situations it is advisable to choose a method with $|r| \leq 1$ at both boundaries.

9.2 Two-Dimensional Shallow-Water Flow

Exact and practically useful mathematical formulae for the specification of nonreflecting boundary conditions are seldom available in multidimensional problems. The difficulties are readily apparent in two relatively simple examples: two-dimensional shallow-water flow and two-dimensional vertically stratified flow. Boundary conditions for the two-dimensional shallow-water system will be discussed in this section. Stratified flow will be considered in Sect. 9.3.

The two-dimensional shallow-water equations, linearized about a basic state with constant mean flow (U, V) and depth H, may be written

$$\begin{pmatrix} u \\ v \\ \eta \end{pmatrix}_t + \begin{pmatrix} U & 0 & 1 \\ 0 & U & 0 \\ c^2 & 0 & U \end{pmatrix} \begin{pmatrix} u \\ v \\ \eta \end{pmatrix}_x + \begin{pmatrix} V & 0 & 0 \\ 0 & V & 1 \\ 0 & c^2 & V \end{pmatrix} \begin{pmatrix} u \\ v \\ \eta \end{pmatrix}_y = 0, \tag{9.26}$$

where the notation follows that used in Sect. 9.1.1. Suppose that a solution is sought in the limited domain $0 \leq x \leq L$, $0 \leq y \leq L$. In contrast to the one-dimensional case, the two-dimensional system cannot be reduced to a set of three scalar equations because no transformation of variables will simultaneously diagonalize both coefficient matrices. Nevertheless, the coefficient matrix multiplying the x-derivative can be diagonalized through the same change of variables used in the one-dimensional problem. Defining $d \equiv u - \eta/c$ and $e \equiv u + \eta/c$ as before, the two-dimensional system becomes

$$\begin{pmatrix} d \\ v \\ e \end{pmatrix}_t + \begin{pmatrix} U - c & 0 & 0 \\ 0 & U & 0 \\ 0 & 0 & U + c \end{pmatrix} \begin{pmatrix} d \\ v \\ e \end{pmatrix}_x + \begin{pmatrix} V & -c & 0 \\ -c/2 & V & c/2 \\ 0 & c & V \end{pmatrix} \begin{pmatrix} d \\ v \\ e \end{pmatrix}_y = 0.$$

This equation is useful for determining the number of boundary conditions that should be specified at the x-boundaries. In the case $c > U > 0$, the signal in v and e is propagating inward through the boundary at $x = 0$, and the signal in d is propagating inward through the boundary at $x = L$. To obtain a well-posed problem, one might therefore attempt to specify two conditions at $x = 0$ of the form

$$e(0, t) = \alpha_1 d(0, t) + f_1(t), \qquad v(0, t) = \alpha_2 d(0, t) + f_2(t),$$

and one condition at $x = L$ of the form

$$d(L,t) = \alpha_3 e(L,t) + \alpha_4 v(L,t) + f_3(t). \tag{9.27}$$

This approach follows the guideline that the number of conditions specified at each boundary should be equal to the number of eigenvalues associated with outward propagation through each boundary. Even so, not all choices of $\alpha_1, \ldots, \alpha_4$ yield a well-posed problem. Oliger and Sundström (1978) and Sundström and Elvius (1979) provided details about various allowable values for $\alpha_1, \ldots, \alpha_4$. One way to obtain a well-posed problem is by choosing $\alpha_1 = \alpha_2 = \alpha_3 = \alpha_4 = 0$, which specifies e and v at inflow and d at outflow.

Unlike the one-dimensional case, it is not clear what values of e, v, and d should be prescribed to prevent waves impinging on the boundary from partially reflecting into an inward-propagating mode. Suppose that the incoming variable d is set to zero at $x = L$, and suppose that a wave of the form $(u, v, \eta) = (\hat{u}, \hat{v}, \hat{\eta}) \exp[\mathrm{i}(kx + \ell y - \omega t)]$ is approaching this boundary as shown in Fig. 9.4. The amplitudes $\hat{u}$ and $\hat{\eta}$ are related by the linearized x-momentum equation (first row of (9.26)), which for the simplest case with no mean flow yields $\hat{u} = (k/\omega)\hat{\eta}$. Using this expression for $\hat{u}$,

$$d(L,y,t) = \left(\frac{k}{\omega} - \frac{1}{c}\right)\hat{\eta}e^{\mathrm{i}(kL+\ell y-\omega t)} = \frac{1}{c}\left(\frac{k}{(k^2+\ell^2)^{1/2}} - 1\right)\hat{\eta}e^{\mathrm{i}(kL+\ell y-\omega t)},$$

where the dispersion relation for shallow-water waves (9.29) has been used to replace ω to obtain the second equality. Thus, the condition $d(L,y,t) = 0$ cannot be satisfied without the simultaneous presence of a second reflected wave unless $\ell = 0$, or equivalently, unless the outward-propagating wave is traveling at right angles to the boundary.

9.2.1 One-Way Wave Equations

Engquist and Majda (1977) suggested that an effective approximation to the true radiation boundary condition in the two-dimensional shallow-water system could be obtained using a one-way wave equation whose solutions are waves with group velocities directed outward through the boundary. To minimize spurious reflection at the boundary, this one-way wave equation should be designed such that its solutions approximate the outward-directed waves in the shallow-water system as closely as possible.

Consider the linearized shallow-water equations for a basic state with no mean flow, which reduce to

$$\frac{\partial^2 \eta}{\partial t^2} - c^2\left(\frac{\partial^2 \eta}{\partial x^2} + \frac{\partial^2 \eta}{\partial y^2}\right) = 0. \tag{9.28}$$

Substituting solutions of the form

$$\eta(x, y, t) = \eta_0 e^{i(kx+\ell y-\omega t)}$$

(where k and ℓ may be positive or negative, but $\omega \geq 0$ to avoid redundancy) into (9.28) gives the shallow-water dispersion relation

$$\omega^2 = c^2 \left(k^2 + \ell^2\right), \tag{9.29}$$

or equivalently,

$$k = \pm\frac{\omega}{c}\left(1 - \frac{c^2\ell^2}{\omega^2}\right)^{1/2}. \tag{9.30}$$

The group velocity parallel to the x-axis is

$$c_{g_x} = \frac{\partial \omega}{\partial k} = kc\left(k^2 + \ell^2\right)^{-1/2}.$$

No plus or minus sign appears in the preceding equation because ω is nonnegative – the sign of the group velocity is determined by the sign of k. Spurious reflection can be eliminated at the right boundary by requiring that all waves present at $x = L$ propagate energy in the positive x direction or, equivalently, that their dispersion relation be given by the positive root of (9.30), so that

$$k = \frac{\omega}{c}\left(1 - \frac{c^2\ell^2}{\omega^2}\right)^{1/2}. \tag{9.31}$$

If ℓ were zero, (9.31) would reduce to $\omega = kc$, which is the dispersion relation associated with a one-way wave equation of the form

$$\frac{\partial \eta}{\partial t} + c\frac{\partial \eta}{\partial x} = 0. \tag{9.32}$$

This is the radiation boundary condition obtained in Sect. 9.1.2 for the one-dimensional shallow-water system, and it is also an exact radiation condition for two-dimensional waves propagating directly parallel to the x-axis, but it is only an approximation to the correct boundary condition for those waves that strike the boundary at nonnormal angles of incidence.

When ℓ is not zero, (9.31) ceases to be the dispersion relation for any differential equation because it contains a square root.[2] Engquist and Majda proposed approximating (9.31) with an algebraic expression that *is* the dispersion relation for some differential equation and using that differential equation as an approximate nonreflecting boundary condition. Let

---

[2] The relation (9.31) is sometimes described as a dispersion relation for a *pseudodifferential* operator.

$$\gamma^2 = \frac{c^2\ell^2}{\omega^2} = \frac{\ell^2}{k^2+\ell^2}.$$

Under the assumption that γ is small, the lowest-order approximation to the square root in (9.31) is simply

$$(1-\gamma^2)^{1/2} \approx 1,$$

which reduces (9.31) to the dispersion relation for one-dimensional shallow-water flow and yields the boundary condition (9.32).

Engquist and Majda's second-order approximation is

$$(1-\gamma^2)^{1/2} \approx 1 - \frac{\gamma^2}{2},$$

which yields the dispersion relation

$$\omega^2 - ck\omega - \frac{c^2\ell^2}{2} = 0.$$

The partial differential equation associated with this dispersion relation,

$$\frac{\partial^2\eta}{\partial t^2} + c\frac{\partial^2\eta}{\partial t\,\partial x} - \frac{c^2}{2}\frac{\partial^2\eta}{\partial y^2} = 0, \tag{9.33}$$

can be imposed as an approximate radiation boundary condition at $x = L$. The well-posedness of this boundary condition was demonstrated by Engquist and Majda (1977); see also Trefethen and Halpern (1986).

The benefits associated with the use of (9.33), instead of the first-order scheme (9.32), can be assessed by computing the magnitude of the spurious reflection generated by each scheme as a function of the angle at which outward-propagating waves strike the lateral boundary. Consider again the situation shown in Fig. 9.4, in which a wave is approaching the boundary at $x = L$. The lines of constant phase are parallel to the wave troughs and crests and satisfy $F(x, y) = 0$, where

$$F(x,y) = kx + \ell y - C,$$

C is a constant, and, in the case shown in Fig. 9.4, k and ℓ are positive. The group velocity of the wave is both perpendicular to the lines of constant phase and parallel to the wave-number vector (k, ℓ), since

$$\left(\frac{\partial\omega}{\partial k}, \frac{\partial\omega}{\partial \ell}\right) = \frac{c^2}{\omega}(k,\ell) = \frac{c^2}{\omega}\nabla F.$$

Let θ be the angle by which the propagation of the wave deviates from the direction normal to the boundary. Then $\tan\theta = \ell/k$, and from the dispersion relation (9.29),

$$k = \frac{\omega}{c}\cos\theta, \quad \ell = \frac{\omega}{c}\sin\theta. \tag{9.34}$$

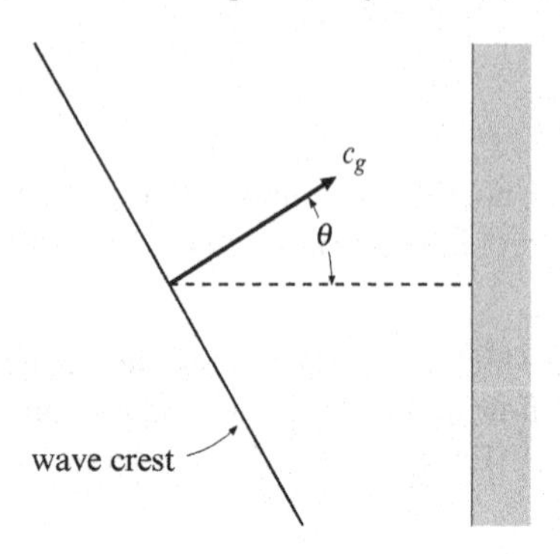

Fig. 9.4 Wave crest approaching the "east" boundary at $x = L$

Suppose that a wave of unit amplitude in η strikes the boundary and is reflected as a wave of amplitude r. Since both the first-order and the second-order one-way wave equations are linear functions of η, they cannot be satisfied at $x = L$ unless the incident and reflected waves are linearly dependent functions of y and t. The frequency and the wave number parallel to the y-axis must therefore be the same in the incident and reflected waves. Then it follows from the dispersion relation (9.29) that the wave numbers parallel to the x-axis have the same magnitude and opposite sign and that the sum of the incident and reflected waves may be expressed in the form

$$\eta(x, y, t) = \mathrm{e}^{\mathrm{i}(kx+\ell y-\omega t)} + r\mathrm{e}^{\mathrm{i}(-kx+\ell y-\omega t)}. \tag{9.35}$$

Here the first term represents the wave approaching the boundary, and the second term represents the reflected wave.

To evaluate the magnitude of the reflected wave, (9.35) may be substituted into the first-order boundary condition (9.32) to obtain

$$r = -\left(\frac{\omega - kc}{\omega + kc}\right)\mathrm{e}^{2\mathrm{i}kL}.$$

Using (9.34),

$$|r| = \left|\frac{\omega - kc}{\omega + kc}\right| = \left|\frac{1 - \cos\theta}{1 + \cos\theta}\right|,$$

showing that the first-order condition is perfectly nonreflecting when the waves approach the boundary along a line normal to the boundary.

The second-order Engquist and Majda boundary condition can be rewritten in the form

$$\left(\frac{\partial}{\partial t} + c\frac{\partial}{\partial x}\right)^2 \eta = 0 \tag{9.36}$$

by eliminating η_{yy} from (9.33) using the shallow-water wave equation (9.28). Substituting (9.35) into the preceding equation, one obtains the reflection coefficient for the second-order scheme,

$$|r| = \left| \frac{1 - \cos\theta}{1 + \cos\theta} \right|^2,$$

which is once again perfectly nonreflecting when the waves are propagating perpendicular to the boundary. Higdon (1986) observed that if the first-order Engquist–Majda boundary condition (9.32) is modified to require

$$\cos\alpha \frac{\partial \eta}{\partial t} + c \frac{\partial \eta}{\partial x} = 0,$$

the reflection coefficient becomes

$$|r| = \left| \frac{\cos\alpha - \cos\theta}{\cos\alpha + \cos\theta} \right|,$$

which is perfectly nonreflecting for waves striking the boundary at an angle α. Higdon also noted that higher-order nonreflecting boundary conditions may be written in the form

$$\left[\prod_{j=1}^{p} \left(\cos\alpha_j \frac{\partial}{\partial t} + c \frac{\partial}{\partial x} \right) \right] \eta = 0. \qquad (9.37)$$

The preceding family of schemes includes the second-order Engquist and Majda formulation (for which $p = 2$ and $\alpha_1 = \alpha_2 = 0$). The advantage of (9.37) arises from the fact that

$$|r| = \prod_{j=1}^{p} \left| \frac{\cos\alpha_j - \cos\theta}{\cos\alpha_j + \cos\theta} \right|,$$

and thus the generalized scheme is perfectly nonreflecting for waves arriving at each of the angles α_j, whereas the original Engquist and Majda formulations are only perfectly nonreflecting for waves propagating perpendicular to the boundary.

The reflectivity of several schemes is illustrated in Fig. 9.5, in which the magnitude of the reflection coefficient is plotted as a function of the angle at which the waves propagate into the boundary. The advantages of the various second-order methods over the first-order method (9.32) are clearly evident; however, the difference between the various second-order schemes is more subtle. The second-order method with perfect transmission at $\theta = 0°$ and 45° appears to offer the best overall performance for waves striking the boundary at angles between 0° and 50°.

9.2.2 Numerical Implementation

The second-order approximation of Engquist and Majda (9.33) requires the evaluation of derivatives parallel to the boundary, which can be a problem near the corners of a rectangular domain. Engquist and Majda (1979) suggested using a first-order approximation at the corner and at the two mesh points closest to the corner. They assumed that the waves propagated into the corner along a diagonal, in which case the boundary condition at the "northeast" corner of a rectangular domain would be

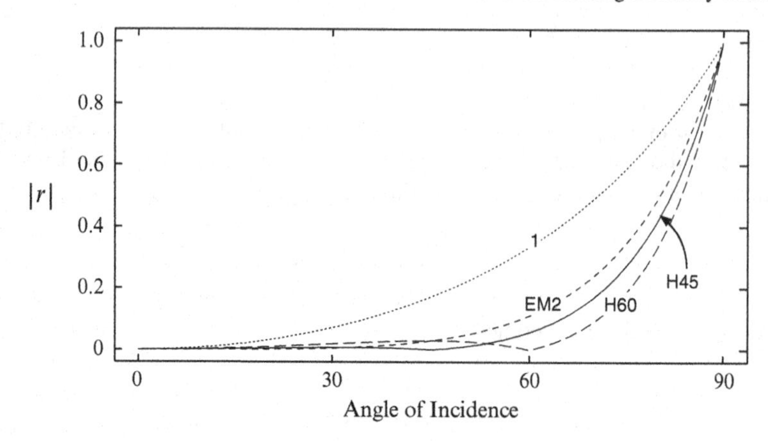

Fig. 9.5 Dependence of the reflection coefficient on the angle of incidence (θ) for shallow-water waves in which the lateral boundary condition is given by the first-order condition (*1*), the second-order Engquist and Majda condition (*EM2*), or Higdon's second-order formulation with perfect transmission at $\theta = 0°$ and $\theta = 45°$ (*H45*) or $\theta = 0°$ and $\theta = 60°$ (*H60*)

$$\frac{\partial}{\partial t} + \frac{c}{\sqrt{2}}\left(\frac{\partial}{\partial x} + \frac{\partial}{\partial y}\right)\eta = 0.$$

This equation can be approximated using upstream differencing.

Higdon's higher-order boundary condition (9.37) does not involve derivatives parallel to the boundary and thereby avoids problems in the corners. Care must nevertheless be taken to ensure the stability of the numerical approximation to the higher-order derivatives that appear in (9.37) when $p \geq 2$. Define $\phi^n_{r,s}$ as the numerical approximation to $\eta(r\Delta x, s\Delta y, n\Delta t)$ and S^-_j as a shift operator with respect to the jth coordinate such that $S^-_x(\phi^n_{r,s}) = \phi^n_{r-1,s}$ and $S^-_t(\phi^n_{r,s}) = \phi^{n-1}_{r,s}$. The pth-order radiation boundary condition at the "east" boundary may be approximated as

$$\left[\prod_{j=1}^{p}\left\{\left(\frac{I - S^-_t}{\Delta t}\right) + c_j\left(\frac{I - S^-_x}{\Delta x}\right)\right\}\right]\phi^n_{N,s} = 0, \qquad (9.38)$$

where N is the x index of the eastern boundary point and $c_j = c/\cos\alpha_j$. The preceding equation may be solved to yield a formula for $\phi^n_{N,s}$. For $p \geq 2$ there is an implicit coupling between the interior and boundary values at time $n\Delta t$. This coupling does not, however, lead to a loss of efficiency, provided that the solution on the interior points can be updated before the points on the boundary, as would be the case if an explicit finite-difference scheme were used to approximate the governing equation in the interior. Higdon (1987) showed that (9.38) is stable when used in conjunction with a centered second-order approximation to the interior finite-difference equation of the form

$$\delta_t^2\phi - c^2\left(\delta_x^2\phi + \delta_y^2\phi\right) = 0.$$

9.3 Two-Dimensional Stratified Flow

The incompressible Boussinesq equations linearized about a reference state with a uniform horizontal wind U can be written in the form

$$\left(\frac{\partial}{\partial t} + U\frac{\partial}{\partial x}\right)u + \frac{\partial P}{\partial x} = 0, \tag{9.39}$$

$$\left(\frac{\partial}{\partial t} + U\frac{\partial}{\partial x}\right)w + \frac{\partial P}{\partial z} = b, \tag{9.40}$$

$$\left(\frac{\partial}{\partial t} + U\frac{\partial}{\partial x}\right)b + N^2 w = 0, \tag{9.41}$$

$$\frac{\partial u}{\partial x} + \frac{\partial w}{\partial z} = 0, \tag{9.42}$$

where b, P, and N^2 are defined according to (8.42). Suppose that solutions to these equations are sought in the limited domain $-L \leq x \leq L$ and $-H \leq z \leq H$ and that open boundary conditions are to be imposed at the edges of the domain. As in the two-dimensional shallow-water system, difficulties are immediately encountered in trying to formulate exact open boundary conditions for the continuous equations. Our first goal will be, therefore, to derive approximate open boundary conditions for the nondiscretized problem, after which we will consider numerical methods for using these boundary conditions in conjunction with the fully discretized equations.

9.3.1 Lateral Boundary Conditions

The chief difficulty in formulating open lateral boundary conditions for two-dimensional stratified flow is usually attributed to the dispersive nature of internal gravity waves. The incompressible Boussinesq system (9.39)–(9.42) supports solutions of the form

$$\begin{pmatrix} u \\ w \\ b \\ P \end{pmatrix} = \begin{pmatrix} \hat{u} \\ \hat{w} \\ \hat{b} \\ \hat{P} \end{pmatrix} e^{i(kx+mz-\omega t)}, \tag{9.43}$$

provided that the wave numbers and frequencies satisfy the dispersion relation

$$\omega = Uk \pm \frac{Nk}{(k^2+m^2)^{1/2}}. \tag{9.44}$$

These waves are dispersive, since their phase speed $\omega(k^2+m^2)^{-1/2}$ is a function of the spatial wave numbers k and m. (See Whitham 1974 for further details on dispersive waves.) The difficulties that arise in formulating open lateral boundary conditions for these waves are, however, very similar to those encountered in formulating boundary conditions for nondispersive shallow-water waves in two dimensions. Those difficulties do not arise from wave dispersion per se, but rather from the fact that the x trace speed[3] for the outward-propagating wave is a function of k and m that cannot be manipulated to form the exact dispersion relation for a partial differential equation.

To determine the dispersion relation governing the outward-propagating wave, note that the horizontal group velocity for internal gravity waves is

$$\frac{\partial \omega}{\partial k} = U \pm \frac{Nm^2}{(m^2 + k^2)^{3/2}}. \tag{9.45}$$

Temporarily assume that $U = 0$; then the mode associated with the positive root in (9.44) will have positive group velocity, and a perfect open boundary condition at $x = L$ will require that all waves present at the boundary satisfy the dispersion relation

$$\omega = \frac{Nk}{(k^2 + m^2)^{1/2}}. \tag{9.46}$$

Since the preceding expression is not the dispersion relation for a partial differential equation, it cannot be used to express the open boundary condition at $x = L$ in terms of the physical variables on the computational mesh. To obtain an approximate open boundary condition, define $\tilde{c}$ as the x trace speed, ω/k. If the dependence of $\tilde{c}$ on m and k could be neglected, the dispersion relation for the correct open boundary condition would become $\omega = \tilde{c}k$, which is the dispersion relation for the familiar one-way wave equation

$$\left(\frac{\partial}{\partial t} + \tilde{c}\frac{\partial}{\partial x}\right)\psi = 0. \tag{9.47}$$

Of course, the dependence of $\tilde{c}$ on m and k cannot be properly ignored, and as a consequence, (9.47) is only an approximate open boundary condition that will induce spurious reflection in all waves for which

$$\tilde{c} \neq \frac{N}{(k^2 + m^2)^{1/2}}.$$

In most practical applications several different waves are simultaneously present at the boundary, and no single value of $\tilde{c}$ will correctly radiate all of the waves.

Considerable effort has, nevertheless, been devoted to devising estimates for $\tilde{c}$ that minimize the reflections generated by (9.47). In the hydrostatic limit, which often applies in geophysical problems, $k^2 \ll m^2$ and $\tilde{c} \to N/m$, so the task of estimating $\tilde{c}$ reduces to that of estimating the vertical wave number of those modes striking the boundary. Pearson (1974) suggested using a fixed value of $\tilde{c}$ equal to the x trace speed of the dominant vertical mode. As an alternative, Orlanski (1976) suggested calculating $\tilde{c}$ at a point just inside the boundary using the relation

[3] The x trace speed, ω/k, is the apparent phase speed of the wave parallel to the x-axis. Unless the wave is propagating parallel to the x-axis, the x trace speed exceeds the true phase speed $\omega(k^2 + m^2)^{-1/2}$.

$$\tilde{c} = -\frac{\partial \psi/\partial t}{\partial \psi/\partial x}. \tag{9.48}$$

The results of this calculation must be limited to values in the interval $0 \leq \tilde{c} \leq \Delta x/\Delta t$ to preserve the stability of the commonly used upstream approximation to (9.47). Orlanski's approach has the virtue of avoiding the specification of arbitrary parameters, but it amounts to little more than an extrapolation procedure, since the same equation is applied at slightly different locations on the space–time grid to determine both $\tilde{c}$ and $\partial\psi/\partial t$. Moreover, (9.47) has no a priori validity in the continuously stratified problem, since there is generally no correct value of $\tilde{c}$ to diagnose via (9.48).

Durran et al. (1993) investigated the effectiveness with which (9.48) diagnosed the phase velocity of numerically simulated shallow-water waves, for which $\tilde{c}$ has the well-defined value of $\sqrt{gH}$. Except during the passage of a trough or crest, a simple finite-difference approximation to (9.48) proved capable of diagnosing a reasonable approximation to $\sqrt{gH}$ in a control simulation on a very large periodic domain in which all waves propagating past the point of the calculation were really traveling in the positive x direction. Attempts to perform the same diagnostic calculation for $\tilde{c}$ in conjunction with the imposition of a radiation boundary condition in a second simulation were, however, a complete failure. Small errors in the initial diagnosis of $\tilde{c}$ generated weak reflected waves. These reflected waves generated increasing errors in the calculation of $\tilde{c}$ because (9.47) does not apply at locations where both rightward- and leftward-propagating waves are present. The additional error in $\tilde{c}$ increased the amplitude of the spurious reflected waves and induced a positive feedback that rapidly destroyed the reliability of the $\tilde{c}$ calculation. The majority of the values computed from (9.48) were outside the stability limits for the upstream method and had to be reset to either zero or $\Delta t/\Delta x$. Durran et al. (1993) also considered tests in which the physical system supported several different modes moving at different trace speeds and concluded that it is best to use a fixed $\tilde{c}$.

When the mean horizontal wind is not zero, $\tilde{c}$ is replaced by the Doppler-shifted trace speeds $U + \tilde{c}$ and $U - \tilde{c}$ at the upstream and downstream boundaries, respectively. One should also ensure that these Doppler-shifted trace speeds are actually directed out of the domain by choosing $|\tilde{c}| > |U|$ at the inflow boundary. In some applications the dominant upstream-propagating mode may have a different intrinsic trace speed from the dominant downstream-propagating mode, and in such circumstances it can be useful to specify the Doppler-shifted trace speeds as $U + \tilde{c}_1$ and $U - \tilde{c}_2$ without requiring $\tilde{c}_1 = \tilde{c}_2$.

Higdon (1994) suggested that an improved approximation to the radiation boundary condition for dispersive waves can be obtained by replacing the basic one-way wave equation (9.47) with the product of a series of one-way operators of the form

$$\left[\prod_{j=1}^{p}\left(\frac{\partial}{\partial t} + \tilde{c}_j\frac{\partial}{\partial x}\right)\right]\psi = 0, \tag{9.49}$$

where the set of $\tilde{c}_j$ is chosen to span the range of x trace speeds associated with the waves appearing at $x = L$. The preceding expression is a generalization of Higdon's radiation boundary condition for shallow-water waves (9.37). Although this scheme does not seem to have been used as a boundary condition for limited-area models of stably stratified flow, Higdon successfully used to it simulate dispersive shallow-water waves on an f-plane.

The spurious reflection generated by the Higdon boundary condition can be determined by considering a unit-amplitude wave striking the boundary at $x = L$. To satisfy (9.49) exactly, a reflected wave of amplitude r must also be present, in which case the total solution may be expressed as

$$\psi(x,z,t) = \mathrm{e}^{\mathrm{i}(kx+mz-\omega t)} + r\mathrm{e}^{\mathrm{i}(\tilde{k}x+\tilde{m}z-\tilde{\omega}t)}.$$

The frequencies and vertical wave numbers of the incident and reflected waves must be identical, or the preceding expression will not satisfy (9.49) for arbitrary values of z and t. Since $\tilde{m} = m$ and $\tilde{\omega} = \omega$, the dispersion relation implies that $\tilde{k} = -k$. Using these results to substitute

$$\psi(x,z,t) = \mathrm{e}^{\mathrm{i}(kx+mz-\omega t)} + r\mathrm{e}^{\mathrm{i}(-kx+mz-\omega t)}$$

into (9.49) and recalling the definition $\tilde{c} = \omega/k$, one obtains

$$|r| = \prod_{j=1}^{p} \left| \frac{\tilde{c} - \tilde{c}_j}{\tilde{c} + \tilde{c}_j} \right|.$$

If the waves are quasi-hydrostatic, $\tilde{c} = N/m$, and the reflection coefficient may be alternatively expressed as

$$|r| = \prod_{j=1}^{p} \left| \frac{m - m_j}{m + m_j} \right|, \tag{9.50}$$

where m_j is the vertical wave number of a quasi-hydrostatic wave moving parallel to the x-axis at speed $\tilde{c}_j$. The reflections generated by the Higdon boundary condition are plotted as a function of the vertical wave number of the incident wave in Fig. 9.6. The waves are assumed to be quasi-hydrostatic, so the reflection coefficient is determined by (9.50). Perfect transmission is achieved whenever the vertical wave number of the incident wave matches one of the m_j. When the perfectly transmitted waves in the two-operator scheme are chosen such that $m_2 = 3m_1$, the range of vertical wave numbers that are transmitted with minimal reflection is much larger than that obtained using the standard one-way wave equation. The range of wave numbers that undergo minimal spurious reflection can be further increased by using the three-operator scheme, which was configured such that $m_2 = 3m_1$ and $m_3 = 9m_1$ in the example plotted in Fig. 9.6.

The Higdon boundary condition can be implemented in the numerical approximations to the momentum and buoyancy equations (9.39)–(9.41) using the finite-difference formula (9.38). The numerical formulation of the boundary condition for pressure is less obvious, because (9.38) is implicit whenever $p \geq 2$. This implicitness need not reduce the efficiency of the numerical integration when the solution in the interior can be updated prior to the evaluation of the boundary condition, as will

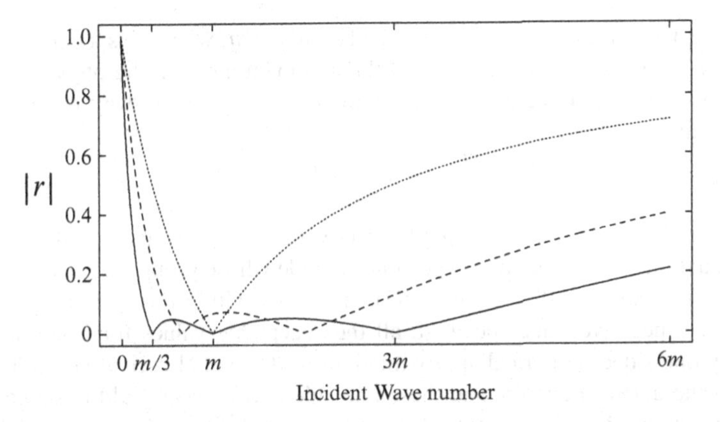

Fig. 9.6 Reflection coefficients generated by the Higdon lateral boundary condition (9.49) as a function of the incident vertical wave number for p equal to 1 (*dotted curve*), 2 (*dashed curve*), and 3 (*solid curve*). The waves are assumed to be quasi-hydrostatic

be the case if (9.39)–(9.41) are approximated using explicit time differencing. The pressure, however, is evaluated by solving a Poisson equation, and the derivation of a stable pressure boundary condition for the Poisson equation from (9.38) may be nontrivial. An alternative approach for formulating the pressure boundary condition is investigated in Problems 8 and 9.

9.3.2 Upper Boundary Conditions

Open upper or lower boundary conditions for the incompressible Boussinesq equations cannot be formulated in the same manner as the lateral boundary conditions because the vertical trace speed of an internal gravity wave is frequently directed opposite to its vertical group velocity. The difference in the direction of the vertical trace speed and group velocity is particularly evident when the waves are quasi-hydrostatic and $U = 0$. In such circumstances the dispersion relation simplifies to $\omega = \pm Nk/m$; the vertical trace speed is

$$\frac{\omega}{m} = \pm\frac{Nk}{m^2}, \tag{9.51}$$

and the vertical group velocity is

$$\frac{\partial\omega}{\partial m} = \mp\frac{Nk}{m^2}, \tag{9.52}$$

which is equal in magnitude to the trace speed but opposite in sign.

Let us attempt to follow the procedure used in the preceding section to obtain an approximate open boundary condition at the upper boundary. As a first step, the correct dispersion relation is approximated as $\omega = \tilde{c}m$, where $\tilde{c}$ is a constant equal to the estimated vertical trace speed of the dominant mode in the physical system. The relation $\omega = \tilde{c}m$ is the exact dispersion relation for a one-way wave equation of the form

$$\left(\frac{\partial}{\partial t} + \tilde{c}\frac{\partial}{\partial z}\right)\psi = 0. \tag{9.53}$$

To avoid reflections from the upper boundary, the value of $\tilde{c}$ used in this one-way wave equation must be the trace speed for a mode whose group velocity is directed upward. According to (9.51) and (9.52), any mode with a positive group velocity will have a negative trace speed, so all the acceptable values for $\tilde{c}$ are negative. Yet only one-sided numerical approximations to the spatial derivative can be evaluated at the upper boundary, and these one-sided differences yield unstable finite-difference approximations to (9.53) whenever $\tilde{c} < 0$, because the numerical domain of dependence does not include the domain of dependence of the true solution.

9.3.2.1 Boundary Conditions That Are Nonlocal in Space, but Local in Time

The problems with the one-way wave equation (9.53) can be avoided by following the strategy of Klemp and Durran (1983) and Bougeault (1983), in which a radiation upper boundary condition is imposed through a diagnostic relationship between the pressure and the vertical velocity at the top boundary of the domain. This relationship is derived by substituting a wave of the form (9.43) into the horizontal momentum equation (9.39) to yield

$$-\,\mathrm{i}(\omega - Uk)\hat{u} + \mathrm{i}k\hat{P} = 0. \tag{9.54}$$

Substituting the same wave into the continuity equation (9.42), one obtains

$$\mathrm{i}k\hat{u} + \mathrm{i}m\hat{w} = 0.$$

Eliminating $\hat{u}$ between the two preceding equations gives

$$\hat{P} = -\left(\frac{\omega - Uk}{k}\right)\left(\frac{m}{k}\right)\hat{w}. \tag{9.55}$$

After making the quasi-hydrostatic approximation, one can write the dispersion relation (9.44) as

$$\frac{\omega - Uk}{k} = \pm\frac{N}{m}.$$

The vertical group velocity for quasi-hydrostatic waves, which is given by (9.52), will depend on both the sign of k and the choice of the positive or negative root. The choice of sign required to limit the dispersion relation to those waves with upward group velocity is

$$\frac{\omega - Uk}{k} = -\mathrm{sgn}(k)\frac{N}{m}. \tag{9.56}$$

Substituting the preceding equation into (9.55) yields

$$\hat{P} = \frac{N}{|k|}\hat{w}. \tag{9.57}$$

This relationship between $\hat{P}$ and $\hat{w}$ is nonlocal, in the sense that it cannot be imposed as an algebraic or differential relation involving P and w on the physical mesh.

The one-way dispersion relations (9.46) and (9.56) are also nonlocal formulae that might be useful as radiation boundary conditions if it were possible to transform the values of $w(x,z,t)$ on the computational mesh to and from the space of dual variables $\hat{w}(k,m,\omega)$. There is, however, no way to determine the frequency dependence of the dual variables, because the grid-point data are never simultaneously available at more than one or two time levels, and as a consequence, (9.46) and (9.56) have no direct practical utility. In contrast, the nonlocal condition (9.57) can be easily used in practical computations, because neither ω nor m appears in that formula. All that is required to use this boundary condition in a numerical model with periodic lateral boundaries is to compute the Fourier transform of the w values along the top row of the computational mesh, evaluate $\hat{P}$ using (9.57), and then obtain the values of P along the top row of the computational mesh from an inverse Fourier transform. Further details about the numerical implementation of this boundary condition are provided in Sect. 9.3.3.

The boundary condition (9.57) perfectly transmits linear quasi-hydrostatic waves through the upper boundary, but nonhydrostatic waves will be partially reflected back into the domain. The strength of the partial reflection can be determined as follows. Since the vertical group velocity obtained without making the quasi-hydrostatic approximation is

$$\frac{\partial \omega}{\partial m} = \mp \frac{Nkm}{(k^2+m^2)^{3/2}},$$

the gravity-wave dispersion relation can be limited to those waves with upward group velocities by taking the negative root in (9.44) when $\operatorname{sgn}(k) = \operatorname{sgn}(m)$ and the positive root when $\operatorname{sgn}(k) = -\operatorname{sgn}(m)$. In both cases, using the dispersion relation for the wave with upward group velocity to substitute for $(\omega - Uk)/k$ in (9.55) yields the correct radiation condition for nonhydrostatic waves,

$$\hat{P} = \frac{N}{|k|}s\hat{w}, \tag{9.58}$$

where

$$s = \frac{|m|}{(k^2+m^2)^{1/2}}.$$

A similar derivation shows that the pressure and vertical velocity in downward-propagating waves are correlated such that

$$\hat{P} = -\frac{N}{|k|}s\hat{w}. \tag{9.59}$$

Now suppose that a nonhydrostatic wave of the form (9.43) encounters the top boundary. To satisfy the quasi-hydrostatic open boundary condition, a second downward-propagating wave with the same horizontal wave number must also be present at the upper boundary. Letting the subscripts 1 and 2 denote the incident and reflected waves, respectively, (9.57) becomes

$$(\hat{P}_1 + \hat{P}_2) = \frac{N}{|k|}(\hat{w}_1 + \hat{w}_2). \tag{9.60}$$

The horizontal momentum equation (9.54), which is also satisfied at the upper boundary, implies that both of these waves have the same frequency because they have the same horizontal wave number. Since both waves have the same ω and k, the dispersion relation requires their vertical wave numbers to differ by a factor of -1, and as a consequence, s is identical for both waves. Substituting for $\hat{P}_1$ and $\hat{P}_2$ from (9.58) and (9.59), respectively, (9.60) becomes

$$s(\hat{w}_1 - \hat{w}_2) = \hat{w}_1 + \hat{w}_2.$$

Let r be the ratio of the vertical-velocity amplitude in the reflected wave to that in the incident wave; then the preceding equation implies that

$$|r| = \left|\frac{s-1}{s+1}\right|.$$

In the hydrostatic limit, $s \to 1$ and there is no reflection. The reflection increases as the waves become less hydrostatic, but even when $k = m$, the reflection coefficient remains a relatively modest 0.17.

In fully three-dimensional problems (9.57) generalizes to

$$\hat{P} = \frac{N}{\sqrt{k^2 + \ell^2}}\hat{w}, \tag{9.61}$$

where ℓ is the wave number parallel to the y-axis. This boundary condition is evaluated in the same manner as that for the two-dimensional problem (9.57), except that two-dimensional Fourier transforms are computed with respect to the x and y coordinates along the top boundary of the domain. Extensions of the preceding approach to include the effects of a constant Coriolis force were suggested by Garner (1986). Rasch (1986) provided a further generalization suitable for both gravity and Rossby waves.

9.3.2.2 Boundary Conditions That Are Local in Both Space and Time

The relations (9.57) and (9.61) are best suited to problems in laterally periodic domains, for which the horizontal wave numbers present on the numerical mesh can be computed exactly using fast Fourier transforms. These boundary conditions can also be used in conjunction with open lateral boundaries, but some type of periodic completion must be assumed to allow the computation of the Fourier transforms. A simple assumption of false periodicity in the w and p fields at the topmost level usually gives adequate results when the main disturbance is located in the central portion of the domain. Nevertheless, this assumption introduces a modest erroneous coupling between the upstream and downstream boundaries.

The use of Fourier transforms and the assumption of false periodicity can be eliminated by approximating the factor of $|k|$ in (9.57) with an algebraic expression that converts (9.57) into a dispersion relation for a partial differential equation that can be solved on the physical mesh. Let $|k|$ be replaced by the rational function

$$\frac{a_1 + a_2 k^2}{1 + a_3 k^2}, \tag{9.62}$$

where a_1, a_2, and a_3 are constants chosen to ensure that (9.62) is a good approximation to $|k|$ over the entire range of wave numbers that need to be transmitted through the upper boundary. It is convenient to express the arbitrary constants a_1, a_2, and a_3 in terms of the three wave numbers, k_1, k_2, and k_3, for which (9.62) would be exactly equal to $|k|$. Then

$$a_1 = \frac{k_1 k_2 k_3}{D}, \qquad a_2 = \frac{k_1 + k_2 + k_3}{D}, \qquad a_3 = \frac{1}{D},$$

where

$$D = k_1 k_2 + k_1 k_3 + k_2 k_3.$$

Numerical tests have suggested that an effective strategy for choosing appropriate values for k_1, k_2, and k_3 (and thereby specifying a_1, a_2, and a_3) is to let k_3 be the largest horizontal wave number likely to appear with significant amplitude in the perturbations at the upper boundary and then choose $k_2 = k_3/3$ and $k_1 = k_3/9$. Replacing $|k|$ by (9.62) in (9.57) and taking an inverse Fourier transform yields the local differential equation

$$\left(a_1 - a_2 \frac{\partial^2}{\partial x^2}\right) P = N \left(1 - a_3 \frac{\partial^2}{\partial x^2}\right) w. \tag{9.63}$$

The reflection generated by (9.63) in a vertically propagating quasi-hydrostatic gravity wave of horizontal wave number k is

$$|r| = \left|\frac{k - k_1}{k + k_1}\right| \left|\frac{k - k_2}{k + k_2}\right| \left|\frac{k - k_3}{k + k_3}\right|.$$

This reflection coefficient has the same form as that for the three-operator Higdon scheme, except that the wave number parallel to the boundary is k in the preceding expression and is m in (9.50). The dependence of r on the horizontal wave number of an incident quasi-hydrostatic wave is given by the solid curve in Fig. 9.6, except that the m's appearing in the labels for the horizontal axis should be replaced with k's. Recall, however, that Higdon's one-way wave equation is not suitable for use at the upper boundary because the phase speed and group velocity of internal gravity waves are generally opposite in sign. Equation (9.63) provides an alternative that avoids instability while achieving the same degree of wave transmission through the upper boundary.

The preceding local boundary condition is generalized to three-dimensional problems by approximating the factor $\sqrt{k^2 + \ell^2}$ in (9.61) with

$$\frac{a_1 + a_2(k^2 + \ell^2)}{1 + a_3(k^2 + \ell^2)}.$$

The approximate radiation upper boundary condition that results is

$$\left[a_1 - a_2\left(\frac{\partial^2}{\partial x^2} + \frac{\partial^2}{\partial y^2}\right)\right] P = N\left[1 - a_3\left(\frac{\partial^2}{\partial x^2} + \frac{\partial^2}{\partial y^2}\right)\right] w.$$

9.3.3 Numerical Implementation of the Radiation Upper Boundary Condition

First, consider the nonlocal formulation (9.57) and suppose that the numerical solution of the Boussinesq system is to be obtained using the projection method on the staggered mesh shown in Fig. 4.6. An upper boundary condition can be obtained for the Poisson equation (8.9) as follows. Fourier transform w^n along the top computational level of the domain, compute the Fourier coefficients for pressure from the relation $\hat{p}^n = N\rho_0\hat{w}^n/|k|$, and inverse-transform to obtain p^n at a level $\Delta z/2$ above the uppermost row of p points.[4] These values of p^n are used as approximations to $\tilde{p}^{n+1}$ along the boundary and provide a Dirichlet boundary condition for (8.9). Ideally, one would formulate the Dirichlet boundary condition using an exact expression for the boundary values of $\tilde{p}^{n+1}$, but recall that $\tilde{p}^{n+1}$ does not represent the actual pressure at any given time. Approximating $\tilde{p}^{n+1}$ with p^n has yielded satisfactory results in tests conducted by this author. After the Poisson equation has been solved, the interior velocities are updated from (8.8). As a final step, the w^{n+1} along the upper boundary can be obtained from the nondivergence condition, or by linear extrapolation from below. In most applications, the mean vertical velocity is zero, and no other upper boundary conditions are required to obtain numerical solutions to the linearized equations.

As noted by Bougeault (1983), stability considerations require that (9.57) be implemented in a compressible model using implicit time differencing. As an example, suppose that solutions are to be obtained using the partially time-split approximation to the "compressible" Boussinesq equations (8.82)–(8.85) on the staggered grid shown in Fig. 4.6 and that N is the vertical index of the topmost row of pressure and buoyancy points. The boundary condition can be cast into an implicit expression for $P_{N-1/2}$, which represents the pressure at the same vertical level as $w_{N-1/2}$. A vertically discretized approximation to (8.83) at level $N - 1/2$ may be written as

[4] Here P is replaced by p/ρ_0 to match the terminology in Sect. 8.3.2.

$$\frac{w_{N-\frac{1}{2}}^{m+1} - w_{N-\frac{1}{2}}^{m}}{\Delta\tau} + \frac{1}{2}\left(\frac{P_{N-\frac{1}{2}}^{m+1} - P_{N-1}^{m+1}}{\Delta z/2} + \frac{P_{N}^{m} - P_{N-1}^{m}}{\Delta z}\right) = b_{N-\frac{1}{2}}^{m} - U\frac{\partial w_{N-\frac{1}{2}}^{n}}{\partial x}, \tag{9.64}$$

where m and n are the indices for the small and large time steps, respectively. The vertically discretized pressure equation for grid level $N-1$ is

$$\frac{P_{N-1}^{m+1} - P_{N-1}^{m}}{\Delta\tau} + \frac{c_s^2}{2}\left(\frac{w_{N-\frac{1}{2}}^{m+1} - w_{N-\frac{3}{2}}^{m+1}}{\Delta z} + \frac{w_{N-\frac{1}{2}}^{m} - w_{N-\frac{3}{2}}^{m}}{\Delta z}\right) = -c_s^2\frac{\partial u_{N-1}^{m+1}}{\partial x} - U\frac{\partial P_{N-1}^{n}}{\partial x},$$

which can be used to substitute for P_{N-1}^{m+1} in (9.64) to obtain

$$\left(1+\frac{\tilde{c}^2}{2}\right)w_{N-\frac{1}{2}}^{m} - \frac{\tilde{c}^2}{2}w_{N-\frac{3}{2}}^{m+1} = F_{N-\frac{1}{2}} - \frac{\Delta\tau}{\Delta z}P_{N-\frac{1}{2}}^{m+1}, \tag{9.65}$$

where $\tilde{c} = c_s\Delta\tau/\Delta z$ and $F_{N-1/2}$ is the sum of all those terms that can be explicitly evaluated at this stage of the integration cycle. Equation (9.65) closes the tridiagonal system for w^{m+1} generated by the implicit coupling between the discretized pressure and vertical momentum equations throughout each vertical column in the interior of the domain. After the forward elimination sweep of the tridiagonal solver described in Sect. A.2.1,

$$w_{N-\frac{1}{2}}^{m+1} = \gamma - \beta P_{N-\frac{1}{2}}^{m+1}, \tag{9.66}$$

where

$$\beta = p\Delta\tau/\Delta z, \qquad \gamma = p\left(F_{N-\frac{1}{2}} + \tilde{c}^2 f/2\right),$$

p is as defined at the last iteration in the loop ($j = jmx$), and f is the array element $f(jmx-1)$. Taking the Fourier transform of (9.66) and using (9.57), the radiation condition becomes

$$\hat{P}_{k,N-\frac{1}{2}}^{m+1} = \left(\frac{|k|}{N} + \beta\right)^{-1}\hat{\gamma}_k,$$

which allows the computation of $\hat{P}_{N-1/2}^{m+1}$ from the Fourier transform of γ. After $P_{N-1/2}^{m+1}$ has been obtained by inverse transforming, the computation of $w_{N-1/2}^{m+1}$ is completed using (9.66), and the remaining w^{m+1} are updated during the backward pass of the tridiagonal solver. The P^{m+1} in the interior of the domain are updated using these w^{m+1}, and finally, P_N^{m+1} is computed from $P_{N-1/2}^{m+1}$ and P_{N-1}^{m+1} by linear extrapolation.

The approximate local radiation condition (9.63) is implemented in essentially the same manner as the nonlocal condition (9.57), except that instead of Fourier transforming w, applying (9.57), and then inverse transforming to obtain the boundary values for the pressure, (9.63) is solved to compute the grid-point values of P directly from the grid-point values of w. A tridiagonal system for the grid-point values of P along the top boundary is obtained when the second derivatives in (9.63) are approximated by a standard three-point finite difference. Special conditions are required at those points adjacent to the lateral boundaries, where it can be advantageous to use the less accurate approximation

$$\left(b_1 - b_2\frac{\partial^2}{\partial x^2}\right) P = Nw.$$

This relation, which is derived from (9.57) using the approximation $|k| \approx b_1 + b_2 k^2$, does not require any assumption about the horizontal variation of w at the lateral boundaries. Some assumption is nevertheless required about the variation of P near the boundary, and satisfactory results have been obtained by setting $\partial P/\partial x$ to zero in the upper corners of the domain.

9.4 Wave-Absorbing Layers

One way to prevent outward-propagating disturbances from reflecting back into the domain when they encounter the boundary is to place a wave-absorbing layer at the edge of the domain. Wave-absorbing layers are conceptually simple and are particularly attractive in applications for which appropriate radiation boundary conditions have not been determined. Wave-absorbing layers also allow "large-scale" time tendencies to be easily imposed at the lateral boundaries of the domain. These large-scale tendencies might be generated by a previous or concurrent coarse-resolution simulation on a larger spatial domain. The chief disadvantage of the absorbing-layer approach is that the absorber often needs to be rather thick to be effective, and significant computational effort may be required to compute the solution on the mesh points within a thick absorbing layer. Considerable engineering may also be required to ensure that a wave-absorbing layer performs adequately in a given application.

Suppose that a radiation upper boundary condition for the two-dimensional linearized Boussinesq equations (9.39)–(9.42) is to be approximated using a wave-absorbing layer of thickness D. This absorbing layer can be created by defining a vertically varying viscosity $\alpha(z)$ and adding viscous terms of the form $\alpha(z)\partial^2 u/\partial x^2$, $\alpha(z)\partial^2 w/\partial x^2$, and $\alpha(z)\partial^2 b/\partial x^2$ to the right sides of (9.39), (9.40), and (9.41), respectively. The viscosity is zero in the region $z \le H$ within which the solution is to be accurately approximated and increases gradually with height throughout the layer $H < z \le H + D$. A simple rigid-lid condition, $w = 0$, can be imposed at the top of the absorbing layer. The performance of this absorbing layer is largely determined by the vertical profile of the artificial viscosity $\alpha(z)$ and the total absorbing-layer depth D. The total viscosity in the absorbing layer must be sufficient to dissipate a wave before it has time to propagate upward through the absorbing layer, reflect off the rigid upper boundary, and travel back down through the depth of the layer. It might, therefore, appear advantageous simply to set α to the maximum value permitted by the stability constraints of the finite-difference scheme.

Reflections will also occur, however, when a wave encounters a rapid change in the propagation characteristics of its medium, and as a consequence, reflection will be produced if the artificial viscosity increases too rapidly with height. The only way to make the total damping within the wave absorber large while keeping the gradient of $\alpha(z)$ small is to use a relatively thick wave-absorbing layer.

The reflectivity of a wave-absorbing layer also depends on the characteristics of the incident wave. Klemp and Lilly (1978) examined the reflections produced by a wave-absorbing layer at the upper boundary in a problem where quasi-hydrostatic vertically propagating gravity waves were generated by continuously stratified flow over topography. They found that although a very thin absorbing layer could be tuned to efficiently remove a single horizontal wave number, considerably deeper layers were required to uniformly minimize the reflection over a broad range of wave numbers. To ensure that the absorbing layer was sufficiently deep, and to guarantee that the numerical solution was adequately resolved within the wave-absorbing layer, Klemp and Lilly devoted the entire upper half of their computational domain to the absorber. The efficiency of their numerical model could have been increased by a factor of 2 if the wave-absorbing layer had been replaced with the radiation upper boundary condition described in the preceding section.

The finding that effective wave-absorbing layers must often be rather thick was also supported by Israeli and Orszag (1981), who examined both viscous and Rayleigh-damping absorbing layers for the linearized shallow-water system of the form

$$\frac{\partial u}{\partial t} + \frac{\partial \eta}{\partial x} = \nu(x)\frac{\partial^2 u}{\partial x^2} - R(x)u, \tag{9.67}$$

$$\frac{\partial \eta}{\partial t} + c^2\frac{\partial u}{\partial x} = 0 \tag{9.68}$$

on the domain $-L \leq x \leq L$. Boundary conditions were specified for $u(-L,t)$ and $u(L,t)$; no boundary conditions were specified for η. Since there is one inward-directed characteristic at each boundary, the specification of u at each boundary yields a well-posed problem for all nonnegative ν. Numerical solutions to the preceding system can be conveniently obtained without requiring numerical boundary conditions for η by using a staggered mesh where the outermost u points are located on the boundaries and the outermost η points are $\Delta x/2$ inside those boundaries (see Sect. 4.1.2). Israeli and Orszag demonstrated that better results could be obtained using Rayleigh damping ($R > 0$, $\nu = 0$) than by using viscous damping ($\nu > 0$, $R = 0$) because the erroneous backward reflection induced by the Rayleigh damping is less scale dependent than that generated by viscous damping. Israeli and Orszag also suggested that superior results could be obtained by using a wave-absorbing layer in combination with a one-way wave equation at the actual boundary. When using both techniques in combination, one must modify the one-way wave equation to account for the dissipation near the boundary. For example, an approximate one-way wave equation for the right-moving wave supported by (9.67) and (9.68) with $\nu = 0$ and R constant is

$$\frac{\partial u}{\partial t} + c\frac{\partial u}{\partial x} = -\frac{R}{2}u. \tag{9.69}$$

The problem considered by Israeli and Orszag is somewhat special in that it can be numerically integrated without specifying any boundary conditions for η. If a mean current were present, so that the unapproximated linear system is described by (9.2), then numerical boundary conditions would also be required for η to evaluate $U\partial\eta/\partial x$. The specification of η at the boundary where the mean wind is directed inward, together with specification of u at each boundary, will lead to an overdetermined problem. Nevertheless, Davies (1976, 1983) suggested that wave-absorbing layers can have considerable practical utility even when they require overspecification of the boundary conditions.

As a simple example of overspecification, consider the one-dimensional scalar advection equation with $U > 0$. Davies's absorbing boundary condition for the outflow boundary corresponds to the mathematical problem of solving

$$\frac{\partial\psi}{\partial t} + U\frac{\partial\psi}{\partial x} = -R(x)(\psi - \psi_b) \tag{9.70}$$

on the domain $-L \le x \le L$ subject to the boundary conditions

$$\psi(-L,t) = s(t), \qquad \psi(L,t) = \psi_b(t).$$

The Rayleigh damper is constructed such that $R(x)$ is zero except in a narrow region in the vicinity of $x = L$. The boundary condition at $x = -L$ is required to uniquely determine the solution, since the characteristic curves intersecting that boundary are directed into the domain. The boundary condition at $x = L$ is, however, redundant, since no characteristics are directed inward through that boundary, and as noted by Oliger and Sundström (1978), the imposition of a boundary condition at $x = L$ renders the problem ill posed.

The practical ramifications of this ill-posedness are, however, somewhat subtle. One certainly cannot expect to obtain a numerical solution that converges to the unique solution to an ill-posed problem. This behavior is illustrated by the test problem shown in Fig. 9.7, in which (9.70) was approximated as

$$\frac{\phi_j^{n+1} - \phi_j^{n-1}}{2\Delta t} + U\frac{\phi_{j+1}^n - \phi_{j-1}^n}{2\Delta x} = R_j(\phi_j^{n+1} - \phi_N)$$

on the interval $-1 \le x \le 1.125$. Here N is the spatial index of the grid point on the right boundary, and the Rayleigh damping coefficient is defined as

$$R(x) = \begin{cases} 0, & \text{if } x \le 1, \\ \alpha\big(1 - \cos[8\pi(x-1)]\big), & \text{otherwise,} \end{cases}$$

so that the region $1 \le x \le 1.125$ contains the wave absorber. The Courant number was 0.5, U was 1, and ϕ was fixed at a constant value of zero at the right and left boundaries. The initial condition was

$$\psi(x,0) = \begin{cases} 0, & \text{if } |x - \frac{1}{2}| \ge 1, \\ \cos^2[\pi(x - \frac{1}{2})], & \text{otherwise.} \end{cases}$$

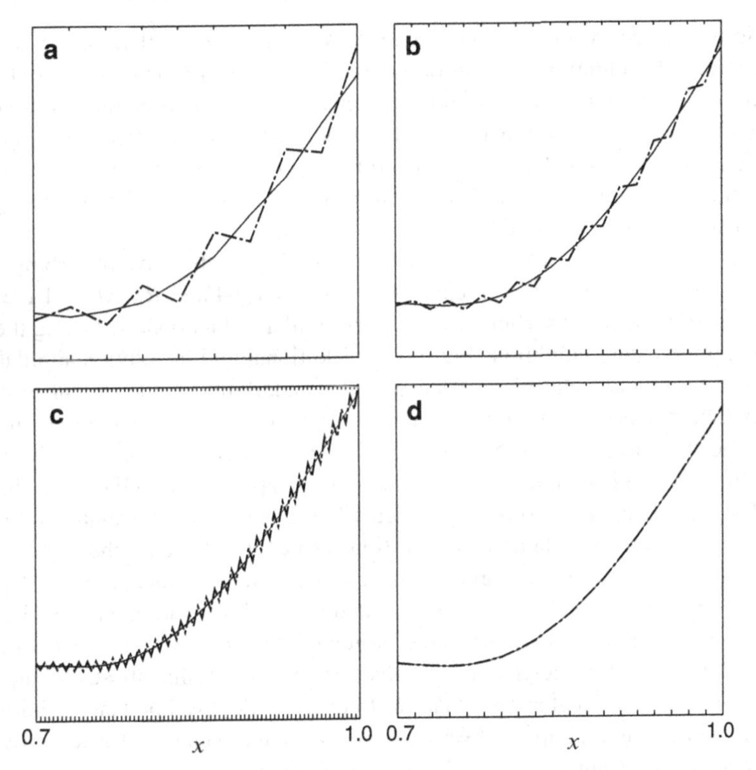

Fig. 9.7 Comparison of numerical solutions to the advection equation obtained using a Rayleigh damping wave absorber (*dot-dashed curve*) or linear extrapolation (*thin solid curve*) and Δx equal to **a** 1/32, **b** 1/64, or **c** 1/256. In **d** the magnitude of the Rayleigh damping coefficient is doubled and $\Delta x = 1/64$

A second solution, indicated by the thin solid line in Fig. 9.7, was obtained using the linear extrapolation boundary condition (9.19) at $x = L$ instead of using a wave-absorbing layer.

Figure 9.7 focuses on the trailing edge of the disturbance in the subdomain $0.7 \leq x \leq 1$. The time shown is $t = 3/4$, at which time three quarters of the initial pulse has passed into the wave-absorbing layer. The interface between the absorbing layer and the interior domain coincides with the right edge of each plot. The horizontal grid spacing for the simulation shown in Fig. 9.7a is $1/32$ and α is chosen such that $R_N \Delta t$ is unity. Considerable reflection is produced by the absorbing layer, which is only four grid points wide. Weaker reflection is also produced by the extrapolation boundary condition. Figure 9.7b shows the solutions to the same physical problem obtained after halving Δx, which increases the width of the absorbing layer to eight grid points. Since the Courant number is fixed, Δt is also halved, and the maximum value of $R_n \Delta t$ is reduced to 0.5. Both solutions are improved by this increase in resolution, but the sponge layer continues to produce significantly more reflection than that generated by the extrapolation boundary condition. The grid spacing is reduced by an additional factor of 4 in Fig. 9.7c, so that $\Delta x = 1/256$. This increase in numerical resolution continues to improve the solution obtained with the extrapolation boundary condition, but does not improve the solution obtained with the wave-absorbing layer. Further increases in the resolution do not make the solution computed with the wave-absorbing layer converge toward the correct solution. Such convergence is, however, exhibited by the solution obtained using the extrapolation boundary condition.

If the grid-spacing is sufficiently fine and α is doubled, the wave-absorbing layer can give much better results. For example, if $\Delta x = 1/64$ and $R_N \Delta t = 1$ the performance of the wave-absorbing layer is very similar to that obtained using the extrapolation boundary condition (Fig. 9.7d). Additional high-resolution simulations suggest that although the ill-posedness of the underlying mathematical problem prevents the numerical solution from uniformly converging to the correct solution within the absorbing layer as $\Delta x \rightarrow 0$ and $\Delta t \rightarrow 0$, the error *outside the absorbing layer* remains small and may be acceptable in some applications. Of course, the use of a Rayleigh-damping absorber is not actually recommended in situations such as this, where an exact open boundary condition can be formulated for the original partial differential equation. It is harder to make a clear-cut recommendation in the vast majority of cases, for which exact open boundary conditions are not available. It is certainly possible that over some range of numerical parameters, the errors generated by the ill-posed Rayleigh damping absorber can be less than those obtained using well-posed numerical approximations to overly reflective boundary conditions. Indeed, the Rayleigh-damping absorber appears to have been used successfully in a variety of studies. Caution is, nevertheless, advised.

9.5 Summary

When simulating the evolution of localized disturbances within a large body of fluid, one often has to limit the computational domain to an arbitrary subset of the total fluid. The conditions that are imposed at artificial boundaries within the fluid are often only approximate descriptions of the true physical behavior of the system and can be a significant source of error. Unless special information is available describing the solution outside the limited domain, the only practical boundary condition that can be specified is one that attempts to radiate disturbances through the boundary without spurious reflection. Nevertheless, the radiation condition is not always the correct physical boundary condition, because nonlinear interactions among the waves that have passed through an artificial boundary can generate a new wave that propagates backward and should reenter the domain.

Even when the radiation boundary condition correctly describes the physical processes occurring at an open boundary, it is frequently impossible to exactly express that boundary condition in a useful mathematical form. In many practical problems, the radiation boundary condition can be expressed only in a form that involves the frequencies and wave numbers of the incident disturbance, or equivalently, temporal and spatial integrals of the solution along the boundary. Such boundary conditions are nonlocal because the condition that must be imposed on the fields at a point $(\mathbf{x}_0, t_0)$ cannot be exactly evaluated from the data available in a limited neighborhood of $(\mathbf{x}_0, t_0)$. Boundary conditions that are nonlocal in time are particularly unsuitable because only data from a few time levels are routinely stored during the numerical solution of time-dependent problems.

Exact local radiation boundary conditions can be obtained for some one-dimensional problems, such as the linearized one-dimensional shallow-water system. In more complicated situations, approximate radiation boundary conditions can be obtained using approximate one-way wave equations. Exact radiation boundary conditions that are local in time but nonlocal in space are also available for a limited class of multidimensional problems. Examples include the upper boundary condition for the linearized quasi-hydrostatic waves discussed in connection with (9.57) and (9.61), and the radiation condition proposed by Grote and Keller (1996) for the solution of the three-dimensional wave equation

$$\frac{\partial^2 \psi}{\partial t^2} - \nabla \cdot (\nabla \psi) = 0$$

outside a spherical domain. Spherical boundaries are attractive in applications where the wave propagation is isotropic in the region external to the boundary. Spherical boundaries are less well suited to many geophysical problems in which gravity introduces a fundamental anisotropy in the waves that makes it advantageous to use different mathematical formulae at the lateral and the upper or lower boundaries. A comprehensive review of the wave-permeable boundary conditions employed in many different disciplines was given by Givoli (1991, 1992).

Once exact or approximate boundary conditions have been formulated for the continuous problem, they must be approximated for use in the numerical integration. Extra boundary conditions beyond those required to yield a well-posed problem may also be needed to evaluate finite differences near the boundary. Although these numerical boundary conditions can have a significant impact on the stability and accuracy of the solution, it can be easier to reduce these finite-differencing errors than the errors that originate from inadequate approximations to the correct open boundary condition.

Problems

1. Explain how to choose α_1 and f_1 in (9.4) to enforce a rigid-side-wall condition at the lateral boundary of the shallow-water system. (This is appropriate only when $U = 0$.)

2. Downstream differencing is obviously an unstable way to approximate an advection term at an inflow boundary. Suppose that one attempts to use backward time differencing to stabilize the approximation. As an indicator of the probable

stability of the result, use a von Neumann analysis to determine the stability of the backward approximation to the advection equation

$$\frac{\phi_j^{n+1} - \phi_j^n}{\Delta t} + U\frac{\phi_j^{n+1} - \phi_{j-1}^{n+1}}{\Delta x} = 0$$

on the unbounded domain $-\infty < x < \infty$. Consider the case $U < 0$.

3. Suppose that the one-dimensional constant-wind-speed advection equation is approximated as

$$(\delta_{2t} + U\delta_{2x})\,\phi_j^n = 0$$

in the interior of the domain and as

$$\left(\langle\delta_t\rangle^x + \tilde{U}\,\langle\delta_x\rangle^t\right)\phi_{N-\frac{1}{2}}^{n+\frac{1}{2}} = 0,$$

where N is the x index of the point on the downstream boundary. Show that setting

$$\tilde{U} = U\frac{\cos^2(k\Delta x/2)}{\cos^2(\omega\Delta t/2)}$$

will allow the discretized mode with wave number k and frequency ω to pass through the boundary without reflection. Explain why this technique cannot be used in practice to create a perfectly nonreflecting boundary.

4. Show that the second-order Higdon boundary condition (9.37) can be expressed in a form similar to the second-order Engquist and Majda condition (9.33) as

$$(1 + \cos\alpha_1\cos\alpha_2)\frac{\partial^2\eta}{\partial t^2} + c(\cos\alpha_1 + \cos\alpha_2)\frac{\partial^2\eta}{\partial t\,\partial x} - c^2\frac{\partial^2\eta}{\partial y^2} = 0.$$

5. Show that if c is the shallow-water gravity-wave phase speed and $|U| < c$, then

$$\frac{\partial^2\psi}{\partial t^2} + (U + c)\left(\frac{\partial^2\psi}{\partial t\,\partial x} - \frac{c}{2}\frac{\partial^2\psi}{\partial y^2}\right) = 0$$

is a one-way wave equation that admits only waves with x-component group velocities greater than zero. How do wave solutions to this equation compare with the solutions to the two-dimensional shallow-water equations linearized about a constant basic-state flow U parallel to the x-axis,

$$\left(\frac{\partial}{\partial t} + U\frac{\partial}{\partial x}\right)^2\psi - c^2\left(\frac{\partial^2}{\partial x^2} + \frac{\partial^2}{\partial y^2}\right)\psi = 0?$$

Consider the dispersion relations for each system. Also discuss the dependence of the sign of the x-component group velocity on U and the horizontal wave numbers.

6. Derive the approximate one-way wave equation (9.69) for the right-moving wave satisfying the Rayleigh-damped shallow-water equations (9.67) and (9.68) with $\nu = 0$ and R constant. Describe the conditions under which this is an accurate approximation.

7. Derive the conditions under which the vertical trace speed (ω/m) of internal gravity waves satisfying the dispersion relation (9.44) is in the same direction as the vertical group velocity.

8. Consider linear Boussinesq flow in the x–z plane. Show that a quasi-hydrostatic internal gravity wave propagating relative to the mean flow in the positive x-direction satisfies the relation $\hat{P} = N\hat{u}/|m|$, where the hat denotes a Fourier transform along the z coordinate and m is the vertical wave number. Use these results to derive an approximate partial differential equation relating P to u that could, in principle, be evaluated using the grid-point values of pressure and velocity.

9. Suppose that the linearized Boussinesq equations are to be solved in the domain $-L \leq x \leq L$. The relation derived in Problem 8 might serve as a radiation boundary condition along the boundary at $x = L$ when the basic-state horizontal wind speed is zero. Explain how this boundary condition might prove unsatisfactory if $U < 0$.

Appendix A
Numerical Miscellany

A.1 Finite-Difference Operator Notation

Complex finite-difference formulae are written in compact form using the following operator notation, which is similar to that used in Shuman and Hovermale (1968):

$$\delta_{nx} f(x) = \frac{f(x + n\Delta x/2) - f(x - n\Delta x/2)}{n\Delta x} \tag{A.1}$$

and

$$\langle f(x)\rangle^{nx} = \frac{f(x + n\Delta x/2) + f(x - n\Delta x/2)}{2}. \tag{A.2}$$

The grid-point approximation to the value of a continuous function $\psi(x,t)$ at the point $(j\Delta x, n\Delta t)$ is denoted by ϕ_j^n. Thus,

$$\delta_{2x}\phi_j^n = \frac{\phi_{j+1}^n - \phi_{j-1}^n}{2\Delta x}, \qquad \delta_x^2\phi_j^n = \frac{\phi_{j+1}^n - 2\phi_j^n + \phi_{j-1}^n}{(\Delta x)^2},$$

and

$$\langle\langle c_j\rangle^x \delta_x \phi_j^n\rangle^x = \frac{1}{4}\left[(c_{j+1} + c_j)\left(\frac{\phi_{j+1}^n - \phi_j^n}{\Delta x}\right) + (c_j + c_{j-1})\left(\frac{\phi_j^n - \phi_{j-1}^n}{\Delta x}\right)\right].$$

When it does not lead to ambiguity, the grid-point indices are omitted from finite-difference formulae expressed in operator notation. As an example,

$$\delta_{2t}\phi + \delta_{2x}\phi = 0$$

expands to

$$\frac{\phi_j^{n+1} - \phi_j^{n-1}}{2\Delta t} + \frac{\phi_{j+1}^n - \phi_{j-1}^n}{2\Delta x} = 0.$$

Simple finite-difference formulae are generally written out in expanded form.

D.R. Durran, *Numerical Methods for Fluid Dynamics: With Applications to Geophysics*, Texts in Applied Mathematics 32, DOI 10.1007/978-1-4419-6412-0_10,

A.2 Tridiagonal Solvers

Implicit finite-difference schemes for one-dimensional problems often lead to tridiagonal systems of linear algebraic equations. Tridiagonal systems can be solved very efficiently using the Thomas tridiagonal algorithm (Isaacson and Keller 1966). The algorithm can be extended to the periodic case as discussed in Strikwerda (1989). FORTRAN programs for both the standard and the periodic cases are given in the following subsections. A well-behaved solution will be obtained using the following algorithms whenever the tridiagonal systems are diagonally dominant, which will be the case if for all j, $|b_j| > |a_j| + |c_j|$, where a_j, b_j, and c_j are, respectively, the subdiagonal, diagonal, and superdiagonal entries in row j.

A.2.1 Code for a Tridiagonal Solver

```
      subroutine tridiag(jmx,a,b,c,f,q)

c     Solves a standard tridiagonal system
c
c     Definition of the variables:
c        jmx   = dimension of all the following arrays
c        a     = sub (lower) diagonal
c        b     = center diagonal
c        c     = super (upper) diagonal
c        f     = right hand side
c        q     = work array provided by calling program
c
c        a(1) and c(jmx) need not be initialized
c     The output is in f; a, b, and c are unchanged

      real a(*),b(*),c(*),f(*),q(*),p
      integer j,jmx

      c(jmx)=0.

c Forward elimination sweep

      q(1)=-c(1)/b(1)
      f(1)= f(1)/b(1)
```

```
      do j=2,jmx
        p= 1.0/( b(j)+a(j)*q(j-1) )
        q(j)= -c(j)*p
        f(j)=( f(j)-a(j)*f(j-1) )*p
      end do

c Backward pass

      do j=jmx-1,1,-1
        f(j)=f(j)+q(j)*f(j+1)
      end do

      return
      end
```

A.2.2 Code for a Periodic Tridiagonal Solver

```
      subroutine tridiag_per(jmx,a,b,c,f,q,s)
c
c     Solves a periodic tridiagonal system
c
c     Definition of the variables:
c        jmx  = dimension of all arrays
c        a    = sub (lower) diagonal
c        b    = center diagonal
c        c    = super (upper) diagonal
c        f    = right hand side
c        q    = work array provided by calling program
c        s    = work array provided by calling program
c
c     Output is in f; a, b, and c are unchanged

      real a(*),b(*),c(*),f(*),q(*),s(*),p,fmx
      integer j,jmx

      fmx=f(jmx)

c Forward elimination sweep

      q(1)=-c(1)/b(1)
      f(1)= f(1)/b(1)
      s(1)=-a(1)/b(1)
```

```
      do j=2,jmx
        p=1.0/(b(j)+a(j)*q(j-1))
        q(j)=-c(j)*p
        f(j)=(f(j)-a(j)*f(j-1))*p
        s(j)=-a(j)*s(j-1)*p
      end do

c Backward pass

      q(jmx)=0.0
      s(jmx)=1.0

      do j=jmx-1,1,-1
        s(j)=s(j)+q(j)*s(j+1)
        q(j)=f(j)+q(j)*q(j+1)
      end do

c Final pass

      f(jmx)=( fmx-c(jmx)*q(1)-a(jmx)*q(jmx-1) )/
     &       ( c(jmx)*s(1)+a(jmx)*s(jmx-1)+b(jmx) )

      do j=1,jmx-1
       f(j)=f(jmx)*s(j)+q(j)
      end do

      return
      end
```

References

Abarbanel S, Gottlieb D (1976) A note on the leap-frog scheme in two and three dimensions. J Comp Phys 21:351–355

Adams J, Garcia R, Gross B, Hack J, Haidvogel D, Pizzo V (1992) Applications of multigrid software in the atmospheric sciences. Mon Wea Rev 120:1447–1458

Arakawa A (1966) Computational design for long-term numerical integration of the equations of fluid motion: Two-dimensional incompressible flow. Part I. J Comp Phys 1:119–143

Arakawa A, Konor CS (1996) Vertical differencing of the primitive equations based on the Charney-Phillips grid in hybrid σ-p vertical coordinates. Mon Wea Rev 124:511–528

Arakawa A, Lamb VR (1977) Computational design of the basic dynamical processes of the UCLA general circulation model. Methods Comput Phys 17:173–265

Asselin RA (1972) Frequency filter for time integrations. Mon Wea Rev 100:487–490

Batchelor GK (1967) An Introduction to Fluid Dynamics. Cambridge University Press, Cambridge, 615 p

Bates JR, McDonald A (1982) Multiply-upstream, semi-Lagrangian advective schemes: analysis and application to a multi-level primitive equation model. Mon Wea Rev 110:1831–1842

Bates JR, Li Y, Brandt A, McCormick SF, Ruge J (1995) A global shallow-water numerical model based on the semi-Lagrangian advection of potential vorticity. Q J R Meteorol Soc 121:1981–2005

Bird RB, Stewart WE, Lightfoot EN (1960) Transport Phenomena. Wiley, New York, 780 p

Bjerknes V (1904) Das problem der Wettervorhersage, betrachtet vom Stanpunkte der Mechanik und der Physik Meteor Zeits 21:1–7

Blaisdell GA, Spyropoulos ET, Qin JH (1996) The effect of the formulation of nonlinear terms on aliasing error in spectral methods. Appl Numer Math 21:207–219

Blossey PN, Durran DR (2008) Selective monotonicity preservation in scalar advection. J Comp Phys 227:5160–5183

Blum EK (1962) A modification of the Runge-Kutta fourth-order method. Math Comp 16:176–187

Boris JP, Book DL (1973) Flux-corrected transport I: SHASTA, a fluid transport algorithm that works. J Comp Phys 11:38–69

Bougeault P (1983) A nonreflective upper boundary condition for limited-height hydrostatic models. Mon Wea Rev 111:420–429

Bourke W (1972) An efficient, one-level, primitive-equation spectral model. Mon Wea Rev 100:683–689

Bourke W (1974) A multi-level spectral model. I. Formulation and hemispheric integrations. Mon Wea Rev 102:687–701

Boussinesq J (1903) Théorie Analytique de la Chaleur, vol II. Gauthier-Villars, Paris, 625 p

Boyd JP (1989) Chebyshev and Fourier Spectral Methods. Springer, Berlin, 798 p

Briggs WL (1987) A Multigrid Tutorial. SIAM, Philadelphia, 88 p

Burks A, Burks A (1981) The ENIAC: first general purpose electronic computer. Ann Hist Comput 3:310–389

Carpenter KM (1982) Note on the paper 'Radiation conditions for the lateral boundaries of limited area models'. Q J R Meteorol Soc 108:717–719

Carrier GF, Pearson CE (1988) Partial Differential Equations: Theory and Technique, 2nd edn. Academic, Boston, 340 p

Charney JG, Fjörtoft R, von Neumann J (1950) Numerical integration of the barotropic vorticity equation. Tellus 2:237–254

Chorin AJ (1968) Numerical solution of the Navier-Stokes equations. Math Comp 22:745–762

Clappier A (1979) A correction method for use in multidimensional time-splitting advection algorithms: application to two- and three-dimensional transport. Mon Wea Rev 126:232–242

Clark TL (1979) Numerical simulations with a three-dimensional cloud model: lateral boundary conditions and multicellular severe storm simulations. J Atmos Sci 36:2191–2215

Cockburn B, Shu C (2001) Runge–Kutta discontinuous Galerkin methods for convection-dominated problems. J Sci Comp 16:173–261

Colella P (1990) Multidimensional upwind methods for hyperbolic conservation laws. J Comp Phys 87:171–200

Colella P, Sekora MD (2008) A limiter for PPM that preserves accuracy at smooth extrema. J Comp Phys 227:7069–7076

Colella P, Woodward P (1984) The piecewise-parabolic method (PPM) for gas-dynamical simulations. J Comp Phys 54:174–201

Cooley JW, Tukey JW (1965) An algorithm for the machine calculation of complex Fourier series. Math Comp 19:297–301

Corby GA, Gilchrist A, Newson RL (1972) A general circulation model of the atmosphere suitable for long period integration. Q J R Meteorol Soc 98:809–832

Courant R, Hilbert D (1953) Methods of Mathematical Physics, vol I. Interscience, New York, 562 p

Courant R, Isaacson E, Rees M (1952) On the solution of nonlinear hyperbolic differential equations by finite differences. Comm Pure Appl Math 5:243–255

Crandall MG, Majda A (1980a) The method of fractional steps for conservation laws. Math Comp 34:285–314

Crandall MG, Majda A (1980b) Monotone difference approximations for scalar conservation laws. Math Comp 34:1–21

Cullen MJP (1974) Integrations of the primitive equations on a sphere using the finite element method. Q J R Meteorol Soc 100:555–562

Cullen MJP (1982) The use of quadratic finite element methods and irregular grids in the solution of hyperbolic problems. J Comp Phys 45:221–245

Cullen MJP (1990) A test of a semi-implicit integration technique for a fully compressible non-hydrostatic model. Q J R Meteorol Soc 116:1253–1258

Cullen MJP, Hall CD (1979) Forecasting and general circulation results from finite element models. Q J R Meteorol Soc 105:571–592

Cullen MJP, Morton KW (1980) Analysis of evolutionary error in finite element and other methods. J Comp Phys 34:245–267

Dahlquist G, Björck A (1974) Numerical Methods. Prentice Hall, Englewood Cliffs, New Jersey, 573 p

Dalhquist G (1956) Numerical integration of ordinary differential equations. Math Scandinavica 4:33–50

Dalhquist G (1959) Stability and error bounds in the numerical integration of ordinary differential equations. Tech. rep., 85p

Dalhquist G (1963) A special stability problem for linear multistep methods. BIT 3:27–43

Davies HC (1976) A lateral boundary formulation for multi-level prediction models. Q J R Meteorol Soc 102:405–418

Davies HC (1983) Limitations of some common lateral boundary conditions used in regional NWP models. Mon Wea Rev 111:1002–1012

Donea J (1984) A Taylor-Galerkin method for convective transport problems. Int J Numer Methods Eng 20:101–119
Donea J, Quartapelle L, Selmin V (1987) An analysis of time discretization in the finite element solution of hyperbolic problems. J Comp Phys 70:463–499
Durran DR (1989) Improving the anelastic approximation. J Atmos Sci 46:1453–1461
Durran DR (1991) The third-order Adams-Bashforth method: an attractive alternative to leapfrog time differencing. Mon Wea Rev 119:702–720
Durran DR (2008) A physically motivated approach for filtering acoustic waves from the equations governing compressible stratified flow. J Fluid Mech 601:365–379
Durran DR, Klemp JB (1983) A compressible model for the simulation of moist mountain waves. Mon Wea Rev 111:2341–2361
Durran DR, Yang MJ, Slinn DN, Brown RG (1993) Toward more accurate wave-permeable boundary conditions. Mon Wea Rev 121:604–620
Easter RC (1993) Two modified versions of Bott's positive-definite numerical advection scheme. Mon Wea Rev 121:297–304
Eliasen E, Machenhauer B, Rasmussen E (1970) On a numerical method for integration of the hydrodynamical equations with a spectral representation of the horizontal fields. Report no. 2. Institut for Teoretisk mMteorologi, University of Copenhagen
Engquist B, Majda A (1977) Absorbing boundary conditions for the numerical simulation of waves. Math Comp 31:629–651
Engquist B, Majda A (1979) Radiation boundary conditions for acoustic and elastic wave calculations. Comm Pure Appl Math 32:313–357
Ferziger JH, Perić M (1997) Computational Methods for Fluid Dynamics. Springer, Berlin, 364 p
Fjørtoft R (1953) On the changes in the spectral distribution of kinetic energy for two-dimensional nondivergent flow. Tellus 5:225–230
Foreman MGG, Thomson RE (1997) Three-dimensional model simulations of tides and buoyancy currents along the west coast of Vancouver Island. J Phys Oceanogr 27:1300–1325
Fornberg B (1973) On the instability of leap-frog and Crank-Nicolson approximations to a nonlinear partial differential equation. Math Comp 27:45–57
Fornberg B (1996) A Practical Guide to Pseudospectral Methods. Cambridge University Press, Cambridge, 231 p
Gal-Chen T, Somerville RCJ (1975) On the use of a coordinate transformation for the solution of the Navier-Stokes equations. J Comp Phys 17:209–228
Garner ST (1986) A radiative upper boundary condition for f-plane models. Mon Wea Rev 114:1570–1577
Gary J (1979) Nonlinear instability. In: Numerical Methods Used in Atmospheric Models, GARP Publication Series No. 17, vol II, World Meteorological Organization, Geneva, pp 476–499
Gary JM (1973) Estimate of truncation error in transformed coordinate, primitive equation atmospheric models. J Atmos Sci 30:223–233
Gassmann A (2005) An improved two-time-level split-explicit integration scheme for non-hydrostatic compressible models. Meteorol Atmos Phys 88:23–38
Gill AE (1982) Atmosphere–Ocean Dynamics. Academic, Orlando, 662 p
Giraldo FX, Hesthaven JS, Warburton T (2002) Nodal high-order discontinous Galerkin methods for the spherical shallow water equations. J Comp Phys 181:499–525
Givoli D (1991) Non-reflecting boundary conditions. J Comp Phys 94:1–29
Givoli D (1992) Numerical Methods for Problems in Infinite Domains. Elsevier, Amsterdam, 299 p
Godlewski E, Raviart PA (1996) Numerical Approximation of Hyperbolic Systems of Conservation Laws. Springer, New York, 509 p
Godunov SK (1959) A difference scheme for numerical computation of discontinuous solutions of equations in fluid dynamics. Math Sb 47:271, also: Cornell Aero. Lab. translation
Goldberg M, Tadmor E (1985) New stability criteria for difference approximations of hyperbolic initial-boundary value problems. Lectures in Appl Math 22:177–192
Golub GH, van Loan CF (1996) Matrix Computations, 3rd edn. Johns Hopkins University Press, Baltimore, 694 p

Goodman JB, LeVeque RJ (1985) On the accuracy of stable schemes for 2D scalar conservation laws. Math Comp 45:15–21

Gottleib S (2005) On high order strong stability preserving Runge-Kutta and multi step time discretizations. J Sci Comput 25:105–128

Gottleib S, Shu CW (1998) Total variation diminishing Runge-Kutta schemes. Math Comp 67:73–85

Gottleib S, Shu CW, Tadmor E (2001) Strong stability-preserving high-order time discretization. SIAM Review 43:89–112

Gottlieb D, Orszag SA (1977) Numerical Analysis of Spectral Methods: Theory and Applications. SIAM, Philadelphia, 172 p

Gresho PM, Sani RL (1998) Incompressible Flow and the Finite Element Method. Wiley, New York, 1040 p

Gresho PM, Lee RL, Sani RL (1978) Advection-dominated flows with emphasis on the consequences of mass lumping. In: Finite Elements in Fluids, vol 3, Wiley, New York, pp 335–350

Grote MJ, Keller JB (1996) Nonreflecting boundary conditions for time-dependent scattering. J Comp Phys 127:52–65

Gustafsson B (1975) The convergence rate for difference approximations to mixed initial boundary value problems. Math Comp 29:396–406

Gustafsson B, Kreiss HO, Sundström A (1972) Stability theory of difference approximations for initial boundary value problems. II. Math Comp 26:649–686

Gustafsson B, Kreiss HO, Oliger J (1995) Time Dependent Problems and Difference Methods. Wiley, New York, 642 p

Hackbusch W (1985) Multi-grid Methods and Applications. Springer, Berlin, 377 p

Harten A (1983) High resolution schemes for hyperbolic conservation laws. J Comp Phys 49:357–393

Harten A, Hyman JM, Lax PD (1976) On finite-difference approximations and entropy conditions for shocks. Comm Pure Appl Math 29:297–322

Hedley M, Yau MK (1988) Radiation boundary conditions in numerical modelling. Mon Wea Rev 116:1721–1736

Hedstrom GW (1979) Nonreflecting boundary conditions for nonlinear hyperbolic systems. J Comp Phys 30:222–237

Héreil P, Laprise R (1996) Sensitivity of internal gravity wave solutions to the time step of a semi-implicit semi-Lagrangian nonhydrostatic model. Mon Wea Rev 124:972–999

Hesthaven JS, Warburton T (2008) Nodal Discontinuous Galerkin Methods. Springer, New York, 500 p

Higdon RL (1986) Absorbing boundary conditions for difference approximations to the multi-dimensional wave equation. Math Comp 47:437–459

Higdon RL (1987) Numerical absorbing boundary conditions for the wave equation. Math Comp 49:65–90

Higdon RL (1994) Radiation boundary conditions for dispersive waves. SIAM J Numer Anal 31:64–100

Hill DJ, Pullin DI (2004) Hybrid tuned center-difference-WENO method for large eddy simulations in the presence of strong shocks. J Comp Phys 194:435–450

Hirt CW, Amsdem AA, Cook JL (1974) An arbitrary Lagrangian–Eulerian computing method for all flow speeds. J Comp Phys 14:227–253

Holton JR (1992) An Introduction to Dynamic Meteorology, 3rd edn. Academic, San Diego, 507 p

Holton JR (2004) An Introduction to Dynamic Meteorology, 4th edn. Elsevier, Amsterdam, 535 p

Hong H, Liska R, Steinberg S (1997) Testing stability by quantifier elimination. J Symbolic Computation 24:161–187

Hoskins BJ, Bretherton FP (1972) Atmospheric frontogenesis models: Mathematical formulation and solution. J Atmos Sci 29:11–37

Hoskins BJ, Simmons AJ (1975) A multi-layer spectral model and the semi-implicit method. Q J R Meteorol Soc 101:637–655

Hundsdorfer W, Verwer J (2003) Numerical Solution of Time-Dependent Advection-Diffusion-Reaction Equations. Springer, Berlin, 471 p

Isaacson E, Keller HB (1966) Analysis of Numerical Methods. Wiley, New York, 541 p

Iserles A (1996) A First Course in the Numerical Analysis of Differential Equations. Cambridge University Press, Cambridge, 378 p

Iserles A, Strang G (1983) The optimal accuracy of difference schemes. Trans Am Math Soc 277:779–803

Israeli M, Orszag SA (1981) Approximation of radiation boundary conditions. J Comp Phys 41:115–135

Jannelli A, Fazio R (2006) Adaptive stiff solvers at low accuracy and complexity. J Comp Appl Math 191:246–258

Jiang GS, Shu CW (1996) Efficient implementation of weighted ENO schemes. J Comp Phys 126:202–228

Karniadakis GE, Sherwin SJ (2005) Spectral/hp Element Methods for Computational Fluid Dynamics, 2nd edn. Oxford University Press, Oxford, 655 p

Karniadakis GE, Israeli M, Orszag SA (1991) High-order splitting methods for the incompressible Navier-Stokes equations. J Comp Phys 97:414–443

Kasahara A (1974) Various vertical coordinate systems used for numerical weather prediction. Mon Wea Rev 102:509–522

Kevorkian J (1990) Partial Differential Equations: Analytical Solution Techniques. Wadsworth and Brooks Cole, Pacific Grove, California, 547 p

Kiehl JT, Hack JJ, Bonan GB, Boville BA, Briegleb BP, Williamson DL, Rasch PJ (1996) Description of the NCAR community climate model (CCM3). NCAR Technical Note NCAR/TN-420+STR, National Center for Atmospheric Research, 152 p

Klemp JB, Durran DR (1983) An upper boundary condition permitting internal gravity wave radiation in numerical mesoscale models. Mon Wea Rev 111:430–444

Klemp JB, Lilly DK (1978) Numerical simulation of hydrostatic mountain waves. J Atmos Sci 35:78–107

Klemp JB, Wilhelmson R (1978) The simulation of three-dimensional convective storm dynamics. J Atmos Sci 35:1070–1096

Kreiss H (1962) über die Stabilitätsdefinition für Differenzengleichungen die partielle Differentialgleichungen approximieren. Nordisk Tidskrift Informationsbehandling (BIT) 2:153–181

Krishnamurti TN, Kumar A, Yap KS, Dastoor AP, Davidson N, Shieng J (1990) Performance of a high-resolution mesoscale tropical prediction model. Adv Geophys 32:133–286

Kurganov A, Tadmor E (2000) New high-resolution central schemes for nonlinear conservation laws and convection-diffusion problems. J Comp Phys 160:241–282

Kurihara Y (1965) On the use of implicit and iterative methods for the time integration of the wave equation. Mon Wea Rev 93:33–46

Kuznecov NN, Vološin SA (1976) On monotone difference approximations for a first-order quasi-linear equation. Soviet Math Dokl 17:1203–1206

Kwizak M, Robert AJ (1971) A semi-implicit scheme for grid point atmospheric models of the primitive equations. Mon Wea Rev 99:32–36

Lander J, Hoskins BJ (1997) Believable scales and parameterizations in a spectral transform model. Mon Wea Rev 125:292–303

Lauritzen PH (2007) A stability analysis of finite-volume advection schemes permitting long time steps. Mon Wea Rev 135:2658–2673

Lax P (1971) Shock waves and entropy. In: Zarantonello EH (ed) Contributions to Nonlinear Analysis, Academic, New York, pp 603–634

Lax P, Richtmyer RD (1956) Survey of the stability of linear finite difference equations. Comm Pure Appl Math 9:267–293

Lax P, Wendroff B (1960) Systems of conservation laws. Comm Pure Appl Math 13:217–237

van Leer B (1974) Towards the ultimate conservative difference scheme II. Monotonicity and conservation combined in a second order scheme. J Comp Phys 14:361–370

van Leer B (1977) Towards the ultimate conservative difference scheme IV. A new approach to numerical convection. J Comp Phys 23:276–299

Lele SK (1992) Compact finite difference schemes with spectral-like resolution. J Comp Phys 103:16–42
Leonard BP, MacVean MK, Lock AP (1993) Positivity-preserving numerical schemes for multidimensional advection. Technical Memorandum 106055, ICOMP-93-05, NASA
Leonard BP, Lock AP, MacVean MK (1996) Conservative explicit unrestricted-time-step multidimensional constancy-preserving advection schemes. Mon Wea Rev 124:2588–2606
Leslie LM, Purser RJ (1995) Three-dimensional mass-conserving semi-Lagrangian scheme employing forward trajectories. Mon Wea Rev 123:2551–2566
LeVeque RJ (1992) Numerical Methods for Conservation Laws. Birkhäuser, Basel, 214 p
LeVeque RJ (1996) High-resolution conservative algorithms for advection in incompressible flow. SIAM J Numer Anal 33:627–665
LeVeque RJ (2002) Finite Volume Methods for Hyperbolic Problems. Cambridge University Press, Cambridge, 558 p
LeVeque RJ (2007) Finite Difference Methods for Ordinary and Partial Differential Equations. SIAM, Philadelphia, 341 p
Lin SJ, Rood RB (1996) Multidimensional flux-form semi-Lagrangian transport schemes. Mon Wea Rev 124:2046–2070
Lipps F (1990) On the anelastic approximation for deep convection. J Atmos Sci 47:1794–1798
Lipps F, Hemler R (1982) A scale analysis of deep moist convection and some related numerical calculations. J Atmos Sci 29:2192–2210
Lorenz EN (1967) The Nature and Theory of the General Circulation of the Atmosphere. World Meteorological Organization, Geneva, 161 p
Lynch DR, Gray WG (1979) A wave equation model for finite element tidal computations. Comput Fluids 7:207–228
MacCormack RW (1969) The effect of viscosity in hypervelocity impact cratering. AIAA Paper 69-352
Machenhauer B (1979) The spectral method. In: Numerical Methods Used in Atmospheric Models, GARP Publication Series No. 17, vol II, World Meteorological Organization, Geneva, pp 121–275
Magazenkov LN (1980) Trudy Glavnoi Geofizicheskoi Observatorii 410:120–129, (Transactions of the Main Geophysical Observatory)
Mahlman JD, Sinclair RW (1977) Tests of various numerical algorithms applied to a simple trace constituent air transport problem. In: Suffet IH (ed) Fate of Pollutants in Air and Water Environments, vol 8, Wiley, New York, part 1
Matsuno T (1966a) False reflection of waves at the boundary due to the use of finite differences. J Meteor Soc Jpn, Ser 2 44:145–157
Matsuno T (1966b) Numerical integrations of the primitive equations by a simulated backward difference method. J Meteor Soc Jpn, Ser 2 44:76–84
McDonald A (1984) Accuracy of multiply-upstream, semi-Lagrangian advective schemes. Mon Wea Rev 112:1267–1275
McDonald A, Bates JR (1989) Semi-Lagrangian integration of a gridpoint shallow-water model on the sphere. Mon Wea Rev 117:130–137
Merilees PE, Orszag SA (1979) The pseudospectral method. In: Numerical Methods Used in Atmospheric Models, GARP Publication Series No. 17, vol II, World Meteorological Organization, Geneva, pp 276–299
Mesinger F (1984) A blocking technique for representation of mountains in atmospheric models. Riv Meteor Aeronautica 44:195–202
Mesinger F, Janić ZI (1985) Problems and numerical methods of the incorporation of mountains in atmospheric models. Lectures in Appl Math 22:81–120
Miyakoda K (1962) Contribution to the numerical weather prediction. Computation with finite difference. Jpn J Geophys 3:75–190
Morinishi Y, Lund TS, Vasilyev OV, Moin P (1998) Fully conservative higher order finite difference schemes for incompressible flow. J Comp Phys 143:90–124

Morton KW, Parrott AK (1980) Generalised Galerkin methods for first-order hyperbolic equations. J Comp Phys 36:249–270
Nair RD, Coté J, Staniforth A (1999) Monotonic cascade interpolation for semi-Lagrangian advection. Q J R Meteorol Soc 125:197–212
Nair RD, Scroggs JS, Semazzi FH (2002) Efficient conservative global transport schemes for climate and atmospheric chemistry models. Mon Wea Rev 130:2059–2073
Nair RD, Thomas SJ, Loft RD (2005) A discontinuous Galekrin global shallow water model. Mon Wea Rev 133:876–888
Nakamura H (1978) Dynamical effects of mountains on the general circulation of the atmosphere. I. Development of finite-difference schemes suitable for incorporating mountains. J Meteor Soc Jpn 56:317–339
Nance LB (1997) On the inclusion of compressibility effects in the scorer parameter. J Atmos Sci 54:362–367
Nance LB, Durran DR (1994) A comparison of three anelastic systems and the pseudo-incompressible system. J Atmos Sci 51:3549–3565
Nessyahu H, Tadmor E (1990) Non-oscillatory central differencing for hyperbolic conservation laws. J Comp Phys 87:408–463
Nitta T (1962) The outflow boundary condition in numerical time integration of advective equations. J Meteor Soc Jpn 40:13–24
Norman MR, Nair RD (2008) Inherently conservative nonpolynomial-based remapping schemes: Application to semi-Lagrangian transport. Mon Wea Rev 136:5044–5061
Ogura Y, Phillips N (1962) Scale analysis for deep and shallow convection in the atmosphere. J Atmos Sci 19:173–179
Oleinik O (1957) Discontinuous solutions of nonlinear differential equations. Amer Math Soc Tranl, Ser 2 26:95–172
Oliger J, Sundström A (1978) Theoretical and practical aspects of some initial boundary value problems in fluid dynamics. SIAM J Appl Math 35:419–446
Oliveira A, Baptista AM (1995) A comparison of integration and interpolation Eulerian-Lagrangian methods. Int J Numer Methods Fluids 21:183–204
Orlanski I (1976) A simple boundary condition for unbounded hyperbolic flows. J Comp Phys 21:251–269
Orszag SA (1970) Transform method for calculation of vector-coupled sums: Application to the spectral form of the vorticity equation. J Atmos Sci 27:890–895
Orszag SA, Israeli M, Deville MO (1986) Boundary conditions for incompressible flows. J Scientific Comp 1:75–111
Park SH, Lee TY (2009) High-order time-integration schemes with explicit time-splitting methods. Mon Wea Rev 137:4047–4060
Pearson RA (1974) Consistent boundary conditions for numerical models of systems that admit dispersive waves. J Atmos Sci 31:1481–1489
Pedlosky J (1987) Geophysical Fluid Dynamics, 2nd edn. Springer, New York, 703 p
Phillips NA (1957) A coordinate system having some special advantages for numerical forecasting. J Meteor 14:184–185
Phillips NA (1959) An example of non-linear computational instability. In: The Atmosphere and Sea in Motion, Rossby Memorial Volume, Rockefeller Institute, New York, pp 501–504
Phillips NA (1973) Principles of large scale numerical weather prediction. In: Morel P (ed) Dynamic Meteorology, Riedel, Dordrecht, pp 1–96
Pinty JP, Benoit R, Richard E, Laprise R (1995) Simple tests of a semi-implicit semi-Lagrangian model on 2D mountain wave problems. Mon Wea Rev 123:3042–3058
Priestley A (1992) The Taylor-Galerkin method for the shallow-water equations on the sphere. Mon Wea Rev 120:3003–3015
Rančić M (1992) Semi-Lagrangian piecewise biparabolic scheme fro two-dimensional horizontal advection of a passive scalar. Mon Wea Rev 120:1394–1406
Rasch PJ (1986) Towards atmospheres without tops: absorbing upper boundary conditions for numerical models. Q J R Meteorol Soc 112:1195–1218

Reinecke PA, Durran DR (2009) The over-amplification of gravity waves in numerical solutions to flow over topography. Mon Wea Rev 137:1533–1549

Richardson LF (1922) Weather Prediction by Numerical Process. Cambridge University Press, Cambridge, 624 p, (Reprinted by Dover Publications, New York, 1965)

Ritchie H (1986) Eliminating the interpolation associated with the semi-Lagrangian scheme. Mon Wea Rev 114:135–146

Ritchie H (1987) Semi-Lagrangian advection on a Gaussian grid. Mon Wea Rev 115:608–619

Ritchie H, Temperton C, Simmons A, Hortal M, Davies T, Dent D, Hamrud M (1995) Implementation of the semi-Lagrangian method in a high-resolution version of the ECMWF forecast model. Mon Wea Rev 123:489–514

Robert AJ (1966) The integration of a low order spectral form of the primitive meteorological equations. J Meteor Soc Jpn 44:237–244

Robert AJ (1981) A stable numerical integration scheme for the primitive meteorological equations. Atmos Ocean 19:35–46

Robert AJ (1982) A semi-Lagrangian and semi-implicit numerical integration scheme for the primitive meteorological equations. J Meteor Soc Jpn 60:319–324

Roe PL (1985) Some contributions to the modeling of discontinuous flows. In: Lecture Notes in Applied Mathematics, vol 22, Springer, New York, pp 163–193

Rood RB (1987) Numerical advection algorithms and their role in atmospheric transport and chemistry models. Rev Geophys 25:71–100

Rosenbrock HH (1963) Some general implicit processes for the numerical solution of differential equations. Comput J 5:329–330

Sadourny R, Arakawa A, Mintz Y (1968) Integration of the nondivergent barotropic vorticity equation with an iscosahedral-hexagonal grid for the sphere. Mon Wea Rev 96:351–356

Saltzman J (1994) An unsplit 3D upwind method for hyperbolic conservation laws. J Comp Phys 115:153–168

Sandu A, Verwer JG, Blom JG, Spee EJ, Carmichael GR, Potra FA (1997) Benchmarking stiff ode solvers for atmospheric chemistry problems. 2. rosenbrock solvers. Atmos Environ 31:3459–3472

Sangster WE (1960) A method of representing the horizontal pressure force without reduction of station pressures to sea level. J Meteor 17:166–176

Schär C, Smolarkiewicz PK (1996) A synchronous and iterative flux-correction formalism for coupled transport equations. J Comp Phys 128:101–120

Schlesinger RE, Uccellini LW, Johnson DR (1983) The effects of the Asselin time filter on numerical solutions to the linearized shallow-water wave equations. Mon Wea Rev 111:455–467

Shu CW, Osher S (1988) Efficient implementation of essentially nonoscillatory shock-capturing schemes. J Comp Phys 77:439–471

Shuman FG, Hovermale JB (1968) An operational six-layer primitive equation model. J Appl Meteor 7:525–547

Simmons AJ, Burridge DM (1981) An energy and angular-momentum conserving vertical finite-difference scheme and hybrid vertical coordinates. Mon Wea Rev 109:758–766

Simmons AJ, Temperton C (1997) Stability of a two-time-level semi-implicit integration scheme for gravity wave motion. Mon Wea Rev 125:600–615

Simmons AJ, Hoskins BJ, Burridge DM (1978) Stability of the semi-implicit method of time integration. Mon Wea Rev 106:405–412

Skamarock WC (2006) Positive-definite and monotonic limiters for unrestricted-time-step transport schemes. Mon Wea Rev 134:2241–2250

Skamarock WC, Klemp JB (1992) The stability of time-split numerical methods for the hydrostatic and nonhydrostatic elastic equations. Mon Wea Rev 120:2109–2127

Skamarock WC, Smolarkiewicz PK, Klemp JB (1997) Precondition conjugate-residual solvers for Helmholtz equations in nonhydrostatic models. Mon Wea Rev 125:587–599

Smolarkiewicz PK (1983) A simple positive definite advection scheme with small implicit diffusion. Mon Wea Rev 111:479–486

Smolarkiewicz PK (1984) A fully multidimensional positive definite advection transport algorithm with small implicit diffusion. J Comp Phys 54:325–362
Smolarkiewicz PK (1989) Comment on "A positive definite advection scheme obtained by nonlinear renormalization of the advective fluxes". Mon Wea Rev 117:2626–2632
Smolarkiewicz PK, Grabowski WW (1990) The multidimensional positive definite advection transport algorithm: nonoscillatory option. J Comp Phys 86:355–375
Smolarkiewicz PK, Margolin LG (1993) On forward in time differencing for fluids: Extension to a curvilinear framework. Mon Wea Rev 121:1847–1859
Smolarkiewicz PK, Margolin LG (1994) Variational solver for elliptic problems in atmospheric flows. Appl Math and Comp Sci 4:527–551
Smolarkiewicz PK, Pudykiewicz JA (1992) A class of semi-Lagrangian algorithms for fluids. J Atmos Sci 49:2082–2096
Smolarkiewicz PK, Rasch RJ (1991) Monotone advection on the sphere: an Eulerian versus semi-Lagrangian approach. J Atmos Sci 48:793–810
Sommerfeld A (1949) Partial Differential Equations in Physics. Academic, New York, 335 p
Spiteri RJ, Ruuth SJ (2002) A new class of optimal high-order strong-stability-preserving time discretization methods. SIAM J Numer Anal 40:469–491
Staniforth AN (1987) Review: formulating efficient finite-element codes for flows in regular domains. Int J Numer Methods Fluids 7:1–16
Staniforth AN, Côté J (1991) Semi-Lagrangian integration schemes for atmospheric models—A review. Mon Wea Rev 119:2206–2223
Staniforth AN, Daley RW (1977) A finite-element formulation for vertical discretization of sigma-coordinate primitive equation models. Mon Wea Rev 105:1108–1118
Steppler J (1987) Quadratic Galerkin finite element schemes for the vertical discretization of numerical forecast models. Mon Wea Rev 115:1575–1588
Stevens DE, Bretherton CS (1996) A forward-in-time advection scheme and adaptive multilevel flow solver for nearly incompressible atmospheric flow. J Comp Phys 129:284– 295
Stoker JJ (1957) Water Waves; the Mathematical Theory with Applications. Interscience, New York, 567 p
Strang G (1964) Accurate partial difference methods. II. Nonlinear problems. Numer Mathematik 6:37–46
Strang G (1968) On the construction and comparison of difference schemes. SIAM J Numer Anal 5:506–517
Strang G, Fix GJ (1973) An Analysis of the Finite Element Method. Prentice Hall, New York, 306 p
Strikwerda JC (1989) Finite Difference Schemes and Partial Differential Equations. Wadsworth and Brooks/Cole, Pacific Grove, California, 386 p
Sun WY (1982) A comparison of two explicit time integration schemes applied to the transient heat equation. Mon Wea Rev 110:1645–1652
Sun WY, Yeh KS, Sun RY (1996) A simple semi-Lagrangian scheme for advection equations. Q J R Meteorol Soc 122:1211–1226
Sundström A, Elvius T (1979) Computational problems related to limited area modelling. In: Numerical Methods Used in Atmospheric Models, GARP Publication Series No. 17, vol II, World Meteorological Organization, Geneva, pp 379–416
Sweby PK (1984) High resolution schemes using flux limiters for hyperbolic conservation laws. SIAM J Numer Anal 21:995–1011
Tanguay M, Robert A, Laprise R (1990) A semi-implicit semi-Lagrangian fully compressible regional forecast model. Mon Wea Rev 118:1970–1980
Tapp MC, White PW (1976) A non-hydrostatic mesoscale model. Q J R Meteorol Soc 102:277–296
Tatsumi Y (1983) An economical explicit time integration scheme for a primitive model. J Meteor Soc Jpn 61:269–287
Témam R (1969) Sur l'approximation de la solution des équations Navier-Stokes par la méthode des pas fractionnaires. Archiv Ration Mech Anal 33:377–385

Temperton C, Staniforth A (1987) An efficient two-time-level semi-Lagrangian semi-implicit integration scheme. Q J R Meteorol Soc 113:1025–1039

Thomée V, Wendroff B (1974) Convergence estimates for Galerkin methods for variable coefficient initial value problems. SIAM J Numer Anal 11:1059–1068

Thompson PD (1983) A history of numerical weather prediction in the United States. Bull Amer Meteor Soc 64:755–769

Thuburn J (1996) Multidimensional flux-limited advection schemes. J Comp Phys 123:74–83

Trefethen LN (1983) Group velocity interpretation of the stability theory of Gustafsson, Kreiss and Sundström. J Comp Phys 49:199–217

Trefethen LN (1985) Stability of hyperbolic finite-difference models with one or two boundaries. Lect Appl Math 22:311–326

Trefethen LN, Halpern L (1986) Well-posedness of one-way wave equations and absorbing boundary conditions. Math Comp 47:421–435

Turkel E (1977) Symmetric hyperbolic difference schemes and matrix problems. Linear Algebra Appl 16:109–129

Vallis GK (2006) Atmospheric and Oceanic Fluid Dynamics. Cambridge University Press, Cambridge, 745 p

Verwer JG, Spee EJ, Blom JG, Hundsdorfer W (1999) A second-order Rosenbrock method applied to photochemical dispersion problems. SIAM J Sci Comput 20:1456–1480

Warming RF, Beam RM (1976) Upwind second-order difference schemes and applications in aerodynamic flows. AIAA J 14:1241–1249

Warming RF, Hyett BJ (1974) The modified equation approach to the stability and accuracy analysis of finite-difference methods. J Comp Phys 14:159–179

Whitham GB (1974) Linear and Nonlinear Waves. Wiley, New York, 636 p

Wicker LJ (2009) A two-step Adams–Bashforth–Moulton split-explicit integrator for compressible atmospheric models. Mon Wea Rev 137:3588–3595

Wicker LJ, Skamarock WC (2002) Time-splitting methods for elastic models using forward time schemes. Mon Wea Rev 130:2088–2097

Wiin-Nielsen A (1959) On the application of trajectory methods in numerical forecasting. Tellus 11:180–196

Williams PD (2009) A proposed modification to the Robert–Asselin time filter. Mon Wea Rev 137:2538–2546

Williamson DL (1968) Integration of the barotropic vorticity equation on a spherical geodesic grid. Tellus 20:642–653

Williamson DL (1979) Difference approximations for fluid flow on a sphere. In: Numerical Methods Used in Atmospheric Models, GARP Publication Series No. 17, vol II, World Meteorological Organization, Geneva, pp 53–120

Williamson DL, Olson JG (1994) Climate simulations with a semi-Lagrangian version of the NCAR community climate model. Mon Wea Rev 122:1594–1610

Williamson DL, Rasch P (1989) Two-dimensional semi-Lagrangian transport with shape-preserving interpolation. Mon Wea Rev 117:102–129

Williamson JH (1980) Low-storage Runge-Kutta schemes. J Comp Phys 35:48–56

Yih CS (1977) Fluid Mechanics: A Concise Introduction to the Theory. West River, Ann Arbor, 622 p

Zalesak ST (1979) Fully multidimensional flux-corrected transport algorithms for fluids. J Comp Phys 31:335–362

Zerroukat M, Wood N, Staniforth A (2002) SLICE: A semi-Lagrangian inherently conserving and efficient scheme for transport problems. Q J R Meteorol Soc 128:2801–2820

Zerroukat M, Wood N, Staniforth A (2005) A monotonic and positive-definite filter for a semi-Lagrangian inherently conserving and efficient (SLICE) scheme. Q J R Meteorol Soc 131:2923–2936

Zerroukat M, Wood N, Staniforth A (2006) The parabolic spline method (PSM) for conservative transport problems. Int J Numer Meth Fluids 51:1297–1318

Index